PROCESS IMAGING FOR AUTOMATIC CONTROL

ELECTRICAL AND COMPUTER ENGINEERING

A Series of Reference Books and Textbooks

1. Rational Fault Analysis, *edited by Richard Saeks and S. R. Liberty*
2. Nonparametric Methods in Communications, *edited by P. Papantoni-Kazakos and Dimitri Kazakos*
3. Interactive Pattern Recognition, *Yi-tzuu Chien*
4. Solid-State Electronics, *Lawrence E. Murr*
5. Electronic, Magnetic, and Thermal Properties of Solid Materials, *Klaus Schröder*
6. Magnetic-Bubble Memory Technology, *Hsu Chang*
7. Transformer and Inductor Design Handbook, *Colonel Wm. T. McLyman*
8. Electromagnetics: Classical and Modern Theory and Applications, *Samuel Seely and Alexander D. Poularikas*
9. One-Dimensional Digital Signal Processing, *Chi-Tsong Chen*
10. Interconnected Dynamical Systems, *Raymond A. DeCarlo and Richard Saeks*
11. Modern Digital Control Systems, *Raymond G. Jacquot*
12. Hybrid Circuit Design and Manufacture, *Roydn D. Jones*
13. Magnetic Core Selection for Transformers and Inductors: A User's Guide to Practice and Specification, *Colonel Wm. T. McLyman*
14. Static and Rotating Electromagnetic Devices, *Richard H. Engelmann*
15. Energy-Efficient Electric Motors: Selection and Application, *John C. Andreas*
16. Electromagnetic Compossibility, *Heinz M. Schlicke*
17. Electronics: Models, Analysis, and Systems, *James G. Gottling*
18. Digital Filter Design Handbook, *Fred J. Taylor*
19. Multivariable Control: An Introduction, *P. K. Sinha*
20. Flexible Circuits: Design and Applications, *Steve Gurley, with contributions by Carl A. Edstrom, Jr., Ray D. Greenway, and William P. Kelly*

21. Circuit Interruption: Theory and Techniques, *Thomas E. Browne, Jr.*
22. Switch Mode Power Conversion: Basic Theory and Design, *K. Kit Sum*
23. Pattern Recognition: Applications to Large Data-Set Problems, *Sing-Tze Bow*
24. Custom-Specific Integrated Circuits: Design and Fabrication, *Stanley L. Hurst*
25. Digital Circuits: Logic and Design, *Ronald C. Emery*
26. Large-Scale Control Systems: Theories and Techniques, *Magdi S. Mahmoud, Mohamed F. Hassan, and Mohamed G. Darwish*
27. Microprocessor Software Project Management, *Eli T. Fathi and Cedric V. W. Armstrong (Sponsored by Ontario Centre for Microelectronics)*
28. Low Frequency Electromagnetic Design, *Michael P. Perry*
29. Multidimensional Systems: Techniques and Applications, *edited by Spyros G. Tzafestas*
30. AC Motors for High-Performance Applications: Analysis and Control, *Sakae Yamamura*
31. Ceramic Motors for Electronics: Processing, Properties, and Applications, *edited by Relva C. Buchanan*
32. Microcomputer Bus Structures and Bus Interface Design, *Arthur L. Dexter*
33. End User's Guide to Innovative Flexible Circuit Packaging, *Jay J. Miniet*
34. Reliability Engineering for Electronic Design, *Norman B. Fuqua*
35. Design Fundamentals for Low-Voltage Distribution and Control, *Frank W. Kussy and Jack L. Warren*
36. Encapsulation of Electronic Devices and Components, *Edward R. Salmon*
37. Protective Relaying: Principles and Applications, *J. Lewis Blackburn*
38. Testing Active and Passive Electronic Components, *Richard F. Powell*
39. Adaptive Control Systems: Techniques and Applications, *V. V. Chalam*
40. Computer-Aided Analysis of Power Electronic Systems, *Venkatachari Rajagopalan*
41. Integrated Circuit Quality and Reliability, *Eugene R. Hnatek*
42. Systolic Signal Processing Systems, *edited by Earl E. Swartzlander, Jr.*
43. Adaptive Digital Filters and Signal Analysis, *Maurice G. Bellanger*
44. Electronic Ceramics: Properties, Configuration, and Applications, *edited by Lionel M. Levinson*

45. Computer Systems Engineering Management, *Robert S. Alford*
46. Systems Modeling and Computer Simulation, *edited by Naim A. Kheir*
47. Rigid-Flex Printed Wiring Design for Production Readiness, *Walter S. Rigling*
48. Analog Methods for Computer-Aided Circuit Analysis and Diagnosis, *edited by Takao Ozawa*
49. Transformer and Inductor Design Handbook: Second Edition, Revised and Expanded, *Colonel Wm. T. McLyman*
50. Power System Grounding and Transients: An Introduction, *A. P. Sakis Meliopoulos*
51. Signal Processing Handbook, *edited by C. H. Chen*
52. Electronic Product Design for Automated Manufacturing, *H. Richard Stillwell*
53. Dynamic Models and Discrete Event Simulation, *William Delaney and Erminia Vaccari*
54. FET Technology and Application: An Introduction, *Edwin S. Oxner*
55. Digital Speech Processing, Synthesis, and Recognition, *Sadaoki Furui*
56. VLSI RISC Architecture and Organization, *Stephen B. Furber*
57. Surface Mount and Related Technologies, *Gerald Ginsberg*
58. Uninterruptible Power Supplies: Power Conditioners for Critical Equipment, *David C. Griffith*
59. Polyphase Induction Motors: Analysis, Design, and Application, *Paul L. Cochran*
60. Battery Technology Handbook, *edited by H. A. Kiehne*
61. Network Modeling, Simulation, and Analysis, *edited by Ricardo F. Garzia and Mario R. Garzia*
62. Linear Circuits, Systems, and Signal Processing: Advanced Theory and Applications, *edited by Nobuo Nagai*
63. High-Voltage Engineering: Theory and Practice, *edited by M. Khalifa*
64. Large-Scale Systems Control and Decision Making, *edited by Hiroyuki Tamura and Tsuneo Yoshikawa*
65. Industrial Power Distribution and Illuminating Systems, *Kao Chen*
66. Distributed Computer Control for Industrial Automation, *Dobrivoje Popovic and Vijay P. Bhatkar*
67. Computer-Aided Analysis of Active Circuits, *Adrian Ioinovici*
68. Designing with Analog Switches, *Steve Moore*
69. Contamination Effects on Electronic Products, *Carl J. Tautscher*
70. Computer-Operated Systems Control, *Magdi S. Mahmoud*
71. Integrated Microwave Circuits, *edited by Yoshihiro Konishi*

72. Ceramic Materials for Electronics: Processing, Properties, and Applications, Second Edition, Revised and Expanded, *edited by Relva C. Buchanan*
73. Electromagnetic Compatibility: Principles and Applications, *David A. Weston*
74. Intelligent Robotic Systems, *edited by Spyros G. Tzafestas*
75. Switching Phenomena in High-Voltage Circuit Breakers, *edited by Kunio Nakanishi*
76. Advances in Speech Signal Processing, *edited by Sadaoki Furui and M. Mohan Sondhi*
77. Pattern Recognition and Image Preprocessing, *Sing-Tze Bow*
78. Energy-Efficient Electric Motors: Selection and Application, Second Edition, *John C. Andreas*
79. Stochastic Large-Scale Engineering Systems, *edited by Spyros G. Tzafestas and Keigo Watanabe*
80. Two-Dimensional Digital Filters, *Wu-Sheng Lu and Andreas Antoniou*
81. Computer-Aided Analysis and Design of Switch-Mode Power Supplies, *Yim-Shu Lee*
82. Placement and Routing of Electronic Modules, *edited by Michael Pecht*
83. Applied Control: Current Trends and Modern Methodologies, *edited by Spyros G. Tzafestas*
84. Algorithms for Computer-Aided Design of Multivariable Control Systems, *Stanoje Bingulac and Hugh F. VanLandingham*
85. Symmetrical Components for Power Systems Engineering, *J. Lewis Blackburn*
86. Advanced Digital Signal Processing: Theory and Applications, *Glenn Zelniker and Fred J. Taylor*
87. Neural Networks and Simulation Methods, *Jian-Kang Wu*
88. Power Distribution Engineering: Fundamentals and Applications, *James J. Burke*
89. Modern Digital Control Systems: Second Edition, *Raymond G. Jacquot*
90. Adaptive IIR Filtering in Signal Processing and Control, *Phillip A. Regalia*
91. Integrated Circuit Quality and Reliability: Second Edition, Revised and Expanded, *Eugene R. Hnatek*
92. Handbook of Electric Motors, *edited by Richard H. Engelmann and William H. Middendorf*
93. Power-Switching Converters, *Simon S. Ang*
94. Systems Modeling and Computer Simulation: Second Edition, *Naim A. Kheir*
95. EMI Filter Design, *Richard Lee Ozenbaugh*
96. Power Hybrid Circuit Design and Manufacture, *Haim Taraseiskey*

97. Robust Control System Design: Advanced State Space Techniques, *Chia-Chi Tsui*
98. Spatial Electric Load Forecasting, *H. Lee Willis*
99. Permanent Magnet Motor Technology: Design and Applications, *Jacek F. Gieras and Mitchell Wing*
100. High Voltage Circuit Breakers: Design and Applications, *Ruben D. Garzon*
101. Integrating Electrical Heating Elements in Appliance Design, *Thor Hegbom*
102. Magnetic Core Selection for Transformers and Inductors: A User's Guide to Practice and Specification, Second Edition, *Colonel Wm. T. McLyman*
103. Statistical Methods in Control and Signal Processing, *edited by Tohru Katayama and Sueo Sugimoto*
104. Radio Receiver Design, *Robert C. Dixon*
105. Electrical Contacts: Principles and Applications, *edited by Paul G. Slade*
106. Handbook of Electrical Engineering Calculations, *edited by Arun G. Phadke*
107. Reliability Control for Electronic Systems, *Donald J. LaCombe*
108. Embedded Systems Design with 8051 Microcontrollers: Hardware and Software, *Zdravko Karakehayov, Knud Smed Christensen, and Ole Winther*
109. Pilot Protective Relaying, *edited by Walter A. Elmore*
110. High-Voltage Engineering: Theory and Practice, Second Edition, Revised and Expanded, *Mazen Abdel-Salam, Hussein Anis, Ahdab El-Morshedy, and Roshdy Radwan*
111. EMI Filter Design: Second Edition, Revised and Expanded, *Richard Lee Ozenbaugh*
112. Electromagnetic Compatibility: Principles and Applications, Second Edition, Revised and Expanded, *David Weston*
113. Permanent Magnet Motor Technology: Design and Applications, Second Edition, Revised and Expanded, *Jacek F. Gieras and Mitchell Wing*
114. High Voltage Circuit Breakers: Design and Applications, Second Edition, Revised and Expanded, *Ruben D. Garzon*
115. High Reliability Magnetic Devices: Design and Fabrication, *Colonel Wm. T. McLyman*
116. Practical Reliability of Electronic Equipment and Products, *Eugene R. Hnatek*
117. Electromagnetic Modeling by Finite Element Methods, *João Pedro A. Bastos and Nelson Sadowski*
118. Battery Technology Handbook, Second Edition, *edited by H. A. Kiehne*
119. Power Converter Circuits, *William Shepherd and Li Zhang*

120. Handbook of Electric Motors: Second Edition, Revised and Expanded, *edited by Hamid A. Toliyat and Gerald B. Kliman*
121. Transformer and Inductor Design Handbook, *Colonel Wm T. McLyman*
122. Energy Efficient Electric Motors: Selection and Application, Third Edition, Revised and Expanded, *Ali Emadi*
123. Power-Switching Converters, Second Edition, *Simon Ang and Alejandro Oliva*
124. Process Imaging For Automatic Control, *edited by David M. Scott and Hugh McCann*
125. Handbook of Automotive Power Electronics and Motor Drives, *Ali Emadi*

PROCESS IMAGING FOR AUTOMATIC CONTROL

DAVID M. SCOTT
DuPont Company
Wilmington, Delaware, U.S.A.

HUGH MCCANN
University of Manchester
Manchester, UK

Boca Raton London New York Singapore

A CRC title, part of the Taylor & Francis imprint, a member of the Taylor & Francis Group, the academic division of T&F Informa plc.

Published in 2005 by
CRC Press
Taylor & Francis Group
6000 Broken Sound Parkway NW, Suite 300
Boca Raton, FL 33487-2742

Printed in the United States of America on acid-free paper
10 9 8 7 6 5 4 3 2 1

International Standard Book Number-10: 0-8247-5920-6 (Hardcover)
International Standard Book Number-13: 978-0-8247-5920-9 (Hardcover)
Library of Congress Card Number 2004061911

Library of Congress Cataloging-in-Publication Data

Process imaging for automatic control / edited by David M. Scott and Hugh McCann.
p. cm.
Includes bibliographical references and index.
ISBN 0-8247-5920-6 (alk. paper)
1. Tomography--Industrial applications. 2. Image processing--Industrial applications. I. McCann, Hugh. II. Scott, David M. III. Title.

TA417.25.P7497 2005
670.42'7--dc22 2004061911

Taylor & Francis Group
is the Academic Division of T&F Informa plc.

Visit the Taylor & Francis Web site at
http://www.taylorandfrancis.com

and the CRC Press Web site at
http://www.crcpress.com

Preface

Industry has traditionally relied on point sensors such as thermocouples and pressure gauges, allied to relatively superficial process models, to control its operations. However, as manufacturing processes and associated numerical models become increasingly complex, additional types of information are required. For example, typical process measurement needs now include contamination detection, particulate size and shape, concentration and density profile (in pipes and tanks), and thermal profile. In research and development of processes and products, detailed numerical models need to be validated by experimental determination of parameters such as the distributions of different flow phases and chemical concentration. In many cases, imaging systems are the only sensors that can provide the required information.

Process imaging is used to visualize events inside industrial processes. These events could be the mixing between two component materials, for example, or the completion of a chemical reaction. The image capture process can be conventional (e.g., directly acquired with a CCD camera), reconstructed (e.g., tomographic imaging), or abstract (sensor data represented as an image). New cameras, versatile tomographic technology, and increasingly powerful computer technology have made it feasible to apply imaging and image processing techniques to a wide range of process measurements. Process images contain a wealth of information about the structure and state of the process stream. For control applications, data can be extracted from such images and fed back to the process control system to optimize and maintain production.

This book, written by a collaboration of international experts in their respective fields, offers a broad perspective on the interdisciplinary topic of process imaging and its use in controlling industrial processes. Its aim is to provide an overview of recent progress in this rapidly developing area. Both academic and industrial points of view are included, and particular emphasis has been placed on the practical applications of this technology. This book will be of interest to process engineers, electrical engineers, and instrumentation developers, as well as plant designers and operators from the chemical, mineral, food, and nuclear industries. The discussion of tomographic technology will also be of particular interest to workers in the clinical sector.

We hope that, through reading this book, researchers in both academia and industry with an interest in this area will be encouraged and facilitated to pursue it further. They will be joining a large band of devotees who have already come a long way in this endeavor. By disseminating the state of the art of process imaging for automatic control, it is our deepest wish that the process engineering community will find many more useful applications for this exciting new technology.

David M. Scott
DuPont Central Research & Development
Wilmington, Delaware, U.S.A.

Hugh McCann
University of Manchester
Manchester, U.K.

Editors

David Scott is a physicist at the DuPont Company's main research facility in Wilmington, Delaware, where he has been developing industrial imaging applications for two decades. He joined DuPont in 1986 after completing his PhD in atomic and molecular physics at the College of William & Mary; he also holds the BA (Earlham College, 1981) and MS (William & Mary, 1984) degrees in physics. He initially worked on nondestructive evaluation of advanced composite materials through radioscopy (real-time x-ray imaging), x-ray computed tomography, and ultrasonic imaging. He also developed several new optical and ultrasonic sensors for gauging multilayer films and other industrial process applications. He started working on process applications of tomography in the early 1990s and was the sole non-EU participant at the early ECAPT process tomography conferences in Europe. He co-chaired the first two worldwide conferences on this topic in San Luis Obispo, California (1995) and Delft (1997).

In 1996 Dr. Scott was invited to establish a research group in the area of particle characterization and was appointed its group leader. His primary research interest is on-line characterization of particulate systems, and his research activities have included process tomography and in-line ultrasonic measurement of particle size. He collaborates internationally with several academic groups, and these collaborations have demonstrated the application of tomography in polymerization reactions and paste extrusion processes. The scope of his group at DuPont has expanded to include interfacial engineering and characterization of nanoparticle systems. Dr. Scott has published over 30 technical papers in peer-reviewed journals, presented keynote and plenary lectures at many international conferences, authored more than 15 company research reports, and edited several journal special issues. He holds several patents.

Hugh McCann has been deeply involved in measurement technique development, with heavy emphasis on multidimensional techniques, throughout a research career spanning more than 25 years. As professor of industrial tomography at the University of Manchester (formerly UMIST) since 1996, he now leads one of the world's foremost imaging research groups. He graduated from the University of Glasgow (BSc Physics, 1976, and PhD 1980) and was awarded the university's Michael Faraday Medal in 1976. For ten years, he worked in high energy particle physics at Glasgow, Manchester, CERN (Geneva) and DESY (Hamburg), to test and establish the so-called Standard Model of physics. During this time, he developed techniques to image particle interactions, based on bubble chambers and drift chambers. The JADE collaboration in which he worked at DESY was awarded a special prize of the European Physical Society in 1995 for discovery of the gluon in the early 1980s, and elucidation of its properties. In 1986, Dr. McCann embarked on ten years of research and development at the Royal Dutch/Shell group's Thornton Research Centre, and was the founding group leader of Shell's specialist engine measurements group. His research on in-situ engine measurement technology was recognized by the SAE Arch T. Colwell Merit Award in 1996.

At the University of Manchester, Dr. McCann has extended industrial tomography into the domain of specific chemical contrast, incorporating infrared absorption, and optical fluorescence. He has explored microwave tomography and has investigated electrical impedance tomography for medical applications. His current research is dominated by IR chemical species tomography and brain function imaging by electrical impedance tomography, and he collaborates intensively with a wide range of scientists and engineers in both academia and industry. Dr. McCann teaches undergraduate and postgraduate classes in measurement theory and instrumentation electronics. He was head of the department of electrical engineering and electronics (1999–2002), and chairman of U.K. professors and heads of electrical engineering (2003–2005). He has published more than 80 papers in peer-reviewed journals and many conference papers.

Contributors

James A. Coveney
G.K. Williams Research Centre for Extractive Metallurgy
Department of Chemical and Biomolecular Engineering
The University of Melbourne
Melbourne, Australia

Stephen Duncan
Department of Engineering Science
University of Oxford
Oxford, U.K.

Tomasz Dyakowski
School of Chemical Engineering and Analytical Science
University of Manchester
Manchester, U.K.

Neil B. Gray
G.K. Williams Research Centre for Extractive Metallurgy
Department of Chemical and Biomolecular Engineering
The University of Melbourne
Melbourne, Australia

Brian S. Hoyle
School of Electronic and Electrical Engineering
University of Leeds
Leeds, U.K.

Artur J. Jaworski
School of Mechanical, Aerospace, and Civil Engineering
University of Manchester
Oxford Road
Manchester, U.K.

Jari P. Kaipio
Department of Applied Physics
University of Kuopio
Kuopio, Finland

Antonis Kokossis
Centre for Process and Information Systems Engineering
University of Surrey
Guildford, Surrey, U.K.

Andrew K. Kyllo
G.K. Williams Research Centre for Extractive Metallurgy
Department of Chemical and Biomolecular Engineering
The University of Melbourne
Melbourne, Australia

Patrick Linke
Centre for Process and Information Systems Engineering
University of Surrey
Guildford, Surrey, U.K.

Matti Malinen
Department of Applied Physics
University of Kuopio
Kuopio, Finland

Hugh McCann
Department of Electrical Engineering and Electronics
University of Manchester
Manchester, U.K.

Jens-Uwe Repke
Institute of Process Dynamics and Operation
Technical University Berlin
Berlin, Germany

Anna R. Ruuskanen
Department of Applied Physics
University of Kuopio
Kuopio, Finland

David M. Scott
Central Research and Development
DuPont Company
Experimental Station
Wilmington, Delaware, U.S.A.

Aku Seppänen
Department of Applied Physics
University of Kuopio
Kuopio, Finland

Volker Sick
Department of Mechanical Engineering
University of Michigan–Ann Arbor
Ann Arbor, Michigan, U.S.A.

Erkki Somersalo
Institute of Mathematics
Helsinki University of Technology
Helsinki, Finland

Satoshi Someya
National Institute of Advanced Industrial Science and Technology (AIST)
Tsukuba, Ibaraki, Japan

Masahiro Takei
Department of Mechanical Engineering
Nihon University
Tokyo, Japan

Arto Voutilainen
Department of Applied Physics
University of Kuopio
Kuopio, Finland

Richard A. Williams
Institute of Particle Science & Engineering
School of Process, Environmental and Materials Engineering
University of Leeds
Leeds, West Yorkshire, U.K.

Günter Wozny
Institute of Process Dynamics and Operation
Technical University Berlin
Berlin, Germany

Dongming Zhao
Electrical and Computer Engineering
University of Michigan–Dearborn
Dearborn, Michigan, U.S.A.

Contents

1 The Challenge

David M. Scott and Hugh McCann

CONTENTS

1.1 MOTIVATION

The technology behind manufacturing and energy extraction processes is one of the principal keys to the prosperity of humankind. This technology provides an enormous range of products and facilities to enable us to live in a manner that could not have been imagined only a century ago. Besides the indisputable benefits that many of us enjoy, many challenges arise as well. Some are inherently political, such as the fair distribution of the benefits that accrue, and access to fossil fuel sources in countries that are themselves less able than others to enjoy the benefits. Others are fundamentally technological in nature, such as improving the efficiency of usage of fossil fuels, reducing the environmental impact of processes, improving the economic performance of manufacturing operations, and developing new processes that enable the manufacture of new products. This book is devoted to the technological challenges and to one aspect in particular. Despite the sophistication of modern process technology, there are still huge benefits to be realized in many processes by more fully exploiting their fundamental physics and chemistry. The key to this puzzle is the underlying technological challenge addressed in this book: how can we combine the ability to "see inside" industrial processes with models of the fundamental phenomena, in order to improve process control?

Industrial processes tend to be highly complex and capital-intensive operations. A generic industrial process is depicted in Figure 1.1. The feedstock material, whose composition, mass flow rate, and phase distribution often vary with time, is introduced to the core process (e.g., a catalytic reactor or a flotation tank). The chemistry or physics driving the core process will generally depend upon the conditions in the process vessel. Therefore, quantities such as the distribution and flow of various phases, temperature, mixing conditions, and even the condition of the vessel itself all affect the outcome of the process. At the completion of the core process, the product must be separated from the waste stream. Clearly, the

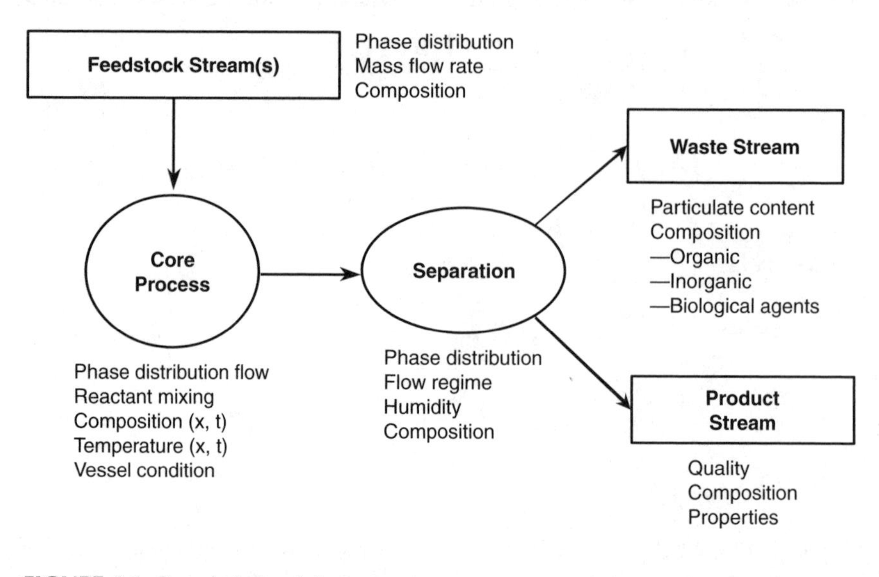

FIGURE 1.1 Generic industrial process.

separation step is affected by the composition and other characteristics of the material. The product stream must generally be assessed for quality control, and even the waste stream must often be controlled so that allowable (legally enforced) emissions levels are not exceeded.

Measurement and control technology can simplify the operation of process equipment, improve product quality and asset productivity, and minimize waste by increasing first-pass yield. Thus there are real economic incentives for improving the control of these processes. Traditionally, the feedback used for industrial control systems was based on scalar quantities such as temperature and pressure measured at single points in the process. Due to the increasingly stringent demands on the quality of products produced by increasingly complex systems, scalar sensors can no longer provide all of the necessary information. Two-dimensional and sometimes three- or four-dimensional (including the dimension of time) information is needed to determine the state of the process. Process imaging technology provides this higher dimensionality.

The term *process imaging* refers to the use of image-based sensors, data processing concepts, and display methods for obtaining information about the internal state of industrial processes. These data are used for research and development of new processes and products, process monitoring, and process control. The process generally includes equipment such as pipes, chemical reactors, or robotic systems, but it could also be a procedure, such as management of inventory. In any case, process imaging extracts information about the process based on spatio-temporal patterns in planar or volume images. These images can be obtained directly (as with a camera) or indirectly (via tomographic reconstruction from a data set of lower dimensionality).

Process imaging can be implemented by a wide variety of techniques. For this reason, the field is inherently interdisciplinary and draws from the contributions of process engineers (such as chemical engineers, mechanical engineers, and metallurgists), electronic engineers, physicists, mathematicians, chemists, computer programmers, and many others. The field has grown tremendously over the past decade due to the development of high-performance imaging systems and advances in computational power. The technologies involved include:

- Video camera systems
- Fiber optics
- Computers
- Display systems
- Optical systems
- Lasers
- Electronics
- Instrumentation development
- Image intensification
- Tomography
- Inverse problem mathematics
- Image analysis

After the desired information is extracted from the images, it is used to estimate the state of the process as part of an overall control scheme. This area has also witnessed very significant recent advances.

There are several considerations in determining how the measured data should be best processed. These include the quality of data, information required concerning the process, what the information is to be used for, and the time that can be tolerated to process the data. A more simplistic processing option that minimizes on-line computation time is often chosen to satisfy these requirements. A single-number output rather than a fully reconstructed image may be sufficient in many cases to provide the required information regarding the process. Such output is easier to feed into a control loop and reduces ambiguity for operator interpretation. In other cases a more detailed image of the system is required, or a complex model may be used to relate the state of the process to the image or to the measurements that underlie the image in the case of tomographic systems.

1.2 ROAD MAP

The first half of this book introduces the concepts and tools used to control industrial processes through process imaging, and the second half presents several applications of the technology in various industries. The intention of this "road map" section is to help the reader choose an optimal route through the book by presenting the theme of each chapter and its relation to the others. However, we believe that ultimately the content of each chapter will be of interest to a wide variety of scientists and engineers.

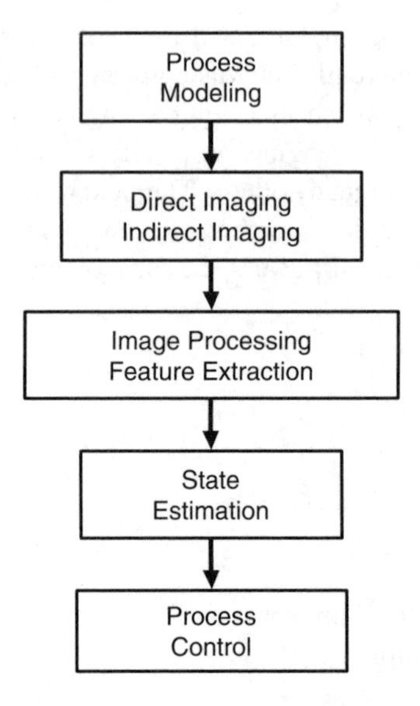

FIGURE 1.2 Steps involved in process imaging for a control application.

A complete control application can be described by the five steps shown in Figure 1.2. The actual implementation may not be as linear as suggested by the figure, but conceptually the output from each step does tend to flow into the successive step.

The ultimate goal of process imaging is improved control or operation of industrial processes. A starting point for this effort is the development of a suitable model for the process under consideration, including a description of the process, the control issues involved, and the data to be provided by the system. The model identifies the process variables and the expected outputs. Process models can be implemented in various ways, e.g., using computational fluid dynamics (CFD), neural networks, or wavelets. Chapter 2 provides an overview of process modeling techniques used to address these issues; although the emphasis is on the chemical engineering industry, the principles are widely applicable. As an example, CFD and other techniques are used to simulate fluid flow and mixing in process equipment, to help predict the impact of changes in flow rate. Due to the difficulty of correctly simulating turbulence in a multiphase system, such models must be validated with empirical data. Once validated, they provide insight into the location and type of measurements that will provide feedback about the current state of the industrial process. In practice, the model must evolve concurrently with the acquisition of empirical data, and the interpretation of the data depends to some extent on the process model.

A wide variety of technologies are used to obtain images of industrial processes. They can be classified as either direct or indirect (reconstructed) imaging. Direct imaging (such as an in-process video camera) refers to the recording of visual scenes (which may be invisible to the human eye, such as infrared or x-ray imaging).

Traditionally, this type of image was recorded directly on photographic film (sometimes with an intensifier screen). Modern applications use electronic sensors that have largely supplanted the use of film. Relevant issues involve the selection of the appropriate light source (lasers or white light sources) and sensor (CCD, intensified CCD, thermal sensor, etc.). Chapter 3 describes these aspects together with optical components such as lenses and laser scanning systems. Innovative examples of direct imaging include particle imaging velocimetry (PIV), measurement of pressure across a surface by observing color change of pressure-sensitive paint, measurement of pH and temperature profiles by fluorescence imaging, and micro-scale imaging of shear stress in fluids.

The other broad class of imaging technology is indirect imaging, in which cross-sectional two- or three-dimensional images are *calculated* through tomographic methods: measurements are made around the boundary of the measurement subject, and these data are inverted by a mathematical algorithm. Tomography itself has become familiar through various applications in the field of diagnostic medical imaging, which use x-ray CT or MRI "scanners." Chapter 4 introduces process tomography, which has been successfully applied to a large number of industrial processes. These techniques provide unique information about the internal state of the industrial process, extending even to the imaging of one chemical species as it mixes with several others, even when they are all in the same thermodynamic phase.

Many physical sensing mechanisms may be used to obtain information about the internal details of the process necessary for reconstructing a cross-sectional image. Each modality, or sensing mechanism, has its own set of strengths or weaknesses in relation to a given application. The most prominent modalities include measurement of electrical properties (by measuring capacitance, impedance, or just resistance), x-ray or gamma-ray absorption, positron emission, and optical emission or absorption. The actual reconstruction of the image can be quite difficult due to the ill-posed nature of the inverse problem. Nevertheless, a variety of reconstruction approaches have been devised, including back-projection, transform methods, iterative methods, model-based reconstruction, and heuristic approaches (which include neural networks).

Until recently, the technical complexity of process tomography has prevented most industrial users from exploring potential applications of this technology. Several companies now sell commercial turnkey process tomography systems. The increased availability of such devices at affordable prices is expected to increase the number of industrial applications and thus to augment the impressive efforts of companies that have been in the vanguard of this development.

Regardless of the source, once a digital image has been obtained, the relevant information must be extracted. It is generally necessary to perform some postprocessing on the digital images (e.g., to enhance the contrast or flatten the background intensity levels) in order to maximize the amount of information that can be extracted. Chapter 5 describes a set of image processing (as opposed to process imaging) tools commonly used to enhance digital images. The primary tasks involved in digital image processing include image enhancement (or restoration) and feature extraction. These operations are performed on digital images, which are essentially two- or three-dimensional arrays of numbers stored in computer memory. *Image enhancement* includes gray level transformations, histogram equalization, spatial filtering, and image enhancement by frequency-domain filtering. *Image restoration* is accomplished by filtering fluctuations (i.e., noise) from images, based on an understanding of the applicable noise model. Typical approaches use mean filters, adaptive filters, or frequency-domain filtering.

Feature extraction is based on segmentation and feature representation concepts. *Segmentation* refers to the operation of splitting an image into a set of regions of interest, where each region contains a particular feature of interest. Segmentation methods include edge detection, exclusion of pixels whose gray levels do not meet a predefined threshold value, and morphological methods. *Feature representation* (such as Freeman chain codes or descriptions in terms of moments) is used to describe the principal features that have been extracted from the image. Morphological operations such as dilation, erosion, opening, and closing are also used to identify the salient features in an image. These methods are used to produce a well-defined set of data from which the process state may be estimated.

Control systems are based on state-space representations of the systems that are to be controlled. The controlled variables are functions of the state variables; in several cases the controlled variables are identical to some of the state variables. Extracted features of the process images are used to estimate the state variables for the process under consideration. Chapter 6 reviews the most important state estimation methods used to optimize processes. In most cases the state variables refer to continuous time processes while the measurements occur only at certain time instants. The state-space representation of a dynamical system can be approximated by a discrete-time continuous state-space model, and the time evolution of the state can be represented as a first-order difference system. The most commonly used method for state estimation is the Kalman filter, which yields a recursive and computationally effective solution. However, other approaches such as fixed-lag and fixed-interval smoothers can in some cases give estimates that are superior to Kalman filtering or other real-time estimates.

In a typical process, the state is inherently of infinite dimension and thus cannot be estimated with any computational methods. Chapter 6 discusses the example of imaging Navier–Stokes flow with electrical impedance tomography. The state is described as a stochastic partial differential equation with partially unknown boundary values. Spatial discretization methods are used to approximate this model as a finite-dimensional first-order Markov system, and numerical results are shown.

Process control systems use the estimated state variables to determine the corrective action necessary to keep a process operating within defined specifications. In control applications, process imaging technology is unique in that it provides detailed information about distributed systems, where physical properties vary spatially as well as temporally. Chapter 7 considers several system models (mass transport, convection systems, convection-diffusion equations) from a control viewpoint. Typical control strategies are discussed, including linear quadratic regulation, model predictive control (MPC), effects of input and output constraints, and nonlinear systems.

The conversion of the partial differential equation models into state-space form generally entails an approximation to create a finite-dimensional state-space model. Control performance criteria (i.e., how "control" of a process profile is defined) are based on the concepts of controllability and observability. These two factors directly impact the number and locations of actuators and sensors needed for a particular application. Implementation issues such as limitations of actuators and speed of response are also considered in Chapter 7.

To begin the consideration of actual applications of process imaging, Chapter 8 discusses its use in the development and control of combustion systems, with a strong emphasis on internal combustion engines. There are very large economic and ecological benefits to be derived from improved control of combustion processes, and pressures both from the market and from environmental legislation have resulted in large efforts in academia and industry alike to exploit image-based process measurements. Crucial features of combustion include fuel preparation, combustion, and pollutant formation. Active imaging techniques are used to look at fuel sprays (optical imaging based on Mie scattering), the hydrocarbon content of vapors (direct and indirect imaging based on fluorescence, absorption, and Raman spectroscopy), and the flame front itself (photographic and tomographic techniques). Flow fields are studied using PIV, in which the motion of tracer particles is tracked with a high-speed camera in order to determine the velocity field. Dopant techniques (using fluorescent tracers) are often used to study fuel/air mixtures. Postcombustion imaging, based on detection of hydroxyl radicals or tracer material, provides information about the removal of waste gases, their subsequent treatment, and their eventual release into the environment. For incineration or power generation operations, the ability to image plumes of exhaust gases is critical for a scientific assessment of the impact on the surrounding community.

Transient three-dimensional multiphase flows are characteristic of many industrial processes. The experimental observation and measurement of such flows are extremely difficult, and over the past decade many tomographic methods have been developed into reliable tools for investigating these complex phenomena. Chapter 9 describes how information about flow behavior can be extracted from tomographic images, to provide valuable insight into the internal structure of flow instabilities such as plugs and slugs. The solids mass flow rate in freight pipeline systems (hydraulic or conveying systems) can also be measured.

Chapter 10 examines real-world applications of process imaging technology in the chemical process industry. A wide variety of process measurement needs have been met through either direct or indirect imaging, and this chapter includes several examples of process control schemes that rely on the technology described in this book. Additional uses include research and development of new products or manufacturing processes and process monitoring to improve fundamental understanding of the process itself. The cited examples include contamination detection and measurement of particulate size and shape, mixture uniformity, amount of fluidization, process efficiency, and various factors related to product quality.

Chapter 11 highlights the imaging methods that have been developed for industrial application in the mineral and materials processing industries. These industries deal with particulate suspensions of solids, gases, and liquids in liquids or gases or in the form of complex multiphase mixtures. Examples include measuring multiphase flow rate and auditing the operation of hydrocyclones used in mineral separation. Tomographic measurements also pertain to the design and monitoring of de-oiling cyclones and particle separation in flotation cells and columns.

The metals production industry presents a significant challenge to process control due to the severe operating conditions found in metallurgical furnaces and molten material handling processes, which include refining and casting of the final product. In many of these operations, the sensing technology will be subjected to particularly harsh environments involving high temperatures and aggressive materials. Sensing systems must be designed to withstand such challenging environments. Chapter 12 introduces applications of process imaging technology to the metals production industry. The suitability of techniques for these applications and the impact of their use are discussed, with regard to the measurement technique as well as the manner in which the measured data are processed to provide information regarding the state of the process. This state estimation, as noted above, is a prerequisite for process control. Applications of process imaging to the metals production industry include sensors for detection of entrained slag in steel, thermal imaging for monitoring flow profiles of molten materials, and hearth monitoring in blast furnaces to monitor refractory wear.

1.3 VISTA

Process imaging is already in use in a number of industries, as described in detail in this volume. The wealth of demonstrated applications of process imaging attests to the versatility of the technology and to the impact that it has already had, particularly on process and product development. The availability of commercial systems that implement the concepts described in this book will surely soon result in additional applications and extensions to other industries.

Does this book describe the pinnacle of achievement of the marriage of process engineering with imaging and control technology? Or does it establish a "base camp" from which new groups of travelers can embark on the road to establishing new and more profound applications of imaging technology and control in the process industries? Or, to return to the key question posed at the beginning: can we indeed further exploit our capability to model fundamental physical and chemical phenomena and to "see inside" industrial processes, in order to control them better and achieve much more desirable outcomes? We are sure that after reading the remainder of this book, you will agree with us that the potential is huge.

2 Process Modeling

Patrick Linke, Antonis Kokossis, Jens-Uwe Repke, and Günter Wozny

CONTENTS

2.1 INTRODUCTION

Process models are used extensively in the design and analysis of chemical processes and process equipment. Such models are either sets of differential or algebraic equations that theoretically describe the features of a process system of interest to the designer, or heuristic or self-learning models that have been developed from process data. Process models enable the prediction of the system's performance and thus enhance the understanding of the system while reducing the need for extensive experimental efforts. Models are derived to predict performance of a chemical process at steady state, dynamic behavior of a process, flow patterns inside process equipment, or even physical properties at a molecular level. The mathematical complexity of a model depends greatly upon its purpose, which determines the level of detail that is required to be captured by the model, the size of the system that is to be modeled, and the length scales to be considered. Process models can be developed with the aim of simulating the performance of a given system or of exploiting degrees of freedom to determine optimal choices for process design and operation. Optimization models offer the advantage that they incorporate decision-making capabilities, whereas simulation models enable the testing of systems for which there are no degrees of freedom, i.e., systems for which all design and operational decisions have been made by the engineer.

It is important to derive any model such that its complexity is low enough for it to be efficiently solved numerically, but at the same time detailed enough to capture realistically the system's behavior. With the rapid advances in solution algorithms and computing power, the past two decades have seen significant increases in model sizes and complexities. For instance, three-dimensional computational fluid dynamics (CFD) models are now routinely applied for reactor simulations, replacing the two-dimensional reactor models frequently used a decade ago.

The mathematical representation of systems from a molecular level through to a process or business level requires modeling across the length scales. Even with advances in computing power, it is a major challenge to solve integrated models that combine models from various levels of abstraction. This is due to the mathematical nature of the models at the individual levels of detail. Higher-level models, such as those used for simulations of entire process flow sheets, are designed as "lumped parameter" models to keep the model complexity at moderate levels. Such models assume properties to be uniform within the physical component modeled and typically consist of a set of algebraic or ordinary differential equations. On the other hand, lower-level models such as CFD models describe systems at smaller length scales. Such models give a detailed account of local effects that are neglected in the higher-level models. Lower-level models typically contain partial differential equations to describe local and dynamic effects. Multilevel modeling, the meaningful integration of models across different levels of detail, is a major research challenge.

Regardless of the level of abstraction and the type of process model to be developed, the modeling process is a systematic, well-documented activity. The derivation of a mathematical model involves the following steps:

- Problem definition, including identification of the modeling goals and the relevant chemical, physical and geometric quantities and selection of the dependent and independent variables
- Identification of the detail required to describe the phenomena of interest and availability of systems knowledge: definition of required length scale; selection of the problem boundaries; selection of physical property and reaction models; selection of conservation laws for mass, energy, or momentum that need to be considered in the model; degrees of freedom for optimization; possible approximations for problem complexities
- Derivation of the model from first principles (conservation laws and problem specifics defined in the previous steps) or by training self-learning models on process data
- Identification and checking of consistency of process data sources required for self-learning model development, if derivation of the model from first principles is not feasible (e.g., fundamental knowledge is lacking)
- Selection of an appropriate solution strategy to solve the model

- Model validation and testing; comparison of the model predictions with experimental data or comparison of the prediction from high-level models against detailed models
- Documentation of the modeling assumptions and their justification and of the derivation of the model equations

This chapter focuses on modeling issues involved in the process design, analysis, and control of multiphase systems. The next section briefly highlights the differences in modeling objectives that lead to process simulation and process optimization problems, before a number of relevant modeling techniques are reviewed in the context of process imaging and analysis. The final section of this chapter addresses practical issues in simulation and optimization for process design, diagnostics, and control on the basis of fluid separation processes.

For details on the model development procedure, the reader is referred to the textbooks by Rice and Do [1]; Luyben [2]; Biegler, Grossmann, and Westerberg [3]; Abbott and Basco [4]; and Haykin [5]. When developing a mathematical model, the engineer should always be aware that all model predictions are wrong to some extent. The engineer should always ensure that the model accuracy is sufficient to make the model useful for the given purpose while keeping the model complexity as low as possible.

2.2 SIMULATION VS. OPTIMIZATION

As mentioned above, process models are developed to support a particular aspect of process design or operation. Simulation models are developed to enable the prediction of the behavior of a particular system. Different models are developed depending upon the level of abstraction that is of interest for a particular system. For predictions at a molecular level, quantum mechanistic models, molecular dynamic models, and Monte Carlo models are frequently used. Predictions at the equipment and process level range from detailed CFD models that can predict fluid flow behavior for defined equipment geometries, via dynamic lumped parameter process models for the simulation of process control systems and process start-up behavior, to modular unit operation models as employed in commercial steady-state simulators, as well as abstract business models that enable the simulation of entire product supply chains. Simulation models are completely specified systems of equations, i.e., the models have no degrees of freedom. In other words, the modeler has made all design and operational decisions about these systems.

In contrast to the above, it is possible to make use of process models to automatically and systematically determine optimal choices for design and operational decisions. Process optimization models generally consist of a number of equality and inequality constraints (the process models and specifications) that are functions of continuous and binary variables, and of an objective function that is to be optimized by exploiting the system's degrees of freedom. Objective functions are measures of the process performance of the particular system.

Examples include cost functions, yields, and environmental impact. Conservation equations are typical examples of equality constraints, whereas product purity or equipment sizes are typical examples of inequality constraints. There are different classes of optimization problems (linear programs, nonlinear programs, mixed integer linear or nonlinear programs) and different methods for their solution, depending on the mathematical form of the objective function, the equality and inequality constraints, and the existence of continuous or discrete variables. Details of optimization techniques can be found in Diwekar [6].

Process optimization offers the advantage of decision support to the design engineer. With increasing complexity of engineering systems, it is virtually impossible for the designer to explore manually all the promising operational and design scenarios in a finite time. As a result, good choices can be easily missed, which often results in low system performance. It is therefore important to provide the engineer with optimization-based support tools that guide the selection of good candidates. The differences between simulation and optimization-based approaches to design decision-making have recently been highlighted by Rigopoulos and Linke [7], who apply optimization techniques to systematically explore design and operational candidates for a bioreaction system in waste water treatment.

Imaging information is generally used in conjunction with simulation efforts, as, for instance, in fluid flow and mixing simulations validated through process tomography. However, the application of optimization-based techniques has yielded powerful tools that speed up and improve the quality of operational and design decision-making using process models derived from first principles. Its combination with the process imaging techniques that are discussed in other chapters of this book could yield a new generation of tools to guide process design and operations and should be the focus of future research efforts.

2.3 PROCESS MODELS FOR IMAGING AND ANALYSIS

Whether a model is used for simulation or optimization, it must describe accurately the behavior of the system under investigation. In this section, we review a number of modeling approaches for process analysis that are regularly employed in conjunction with process imaging techniques. Process imaging techniques are frequently used to validate fluid flow and mixing models derived from first principles. In many cases it is not possible to derive process models from first principles. For such systems, artificial neural networks (ANNs) are often developed to model relationships between sets of process data and images.

2.3.1 FLUID FLOW AND MIXING MODELS

Chemical processing equipment design and operation require systematic tools that enable the visualization of physicochemical phenomena. The most important incentive to use computational simulation tools in equipment design is economic pressure. Such tools are more and more replacing lengthy scale-up studies and can be used to analyze and coordinate experimental efforts and support the determination of design parameters that are difficult to measure in "real life" systems [8]. CFD and cell models are frequently used to simulate fluid flow and mixing phenomena in processing equipment. Due to the fundamental difficulties in accurate modeling of turbulence phenomena in multiphase systems, it is vital to validate and assess these models by comparing the simulated flow images with those obtained from real-life experiments [9], e.g., by using tomographic sensors [8, 10–12].

Due to increased availability of computing power and advances in model accuracy, CFD models are now routinely employed in single-phase fluid flow simulations. A number of commercial CFD software packages are available (e.g., FLUENT, CFX, FEMLAB). Such packages generate and solve the partial differential equations given by the space- and time-dependent heat, mass, and momentum balances (Navier–Stokes equations). Although these model equations can generally predict accurately the behavior of single-phase systems, a number of problems in the description of multiphase phenomena have been reported [10]. These arise because the fundamentals of phase interactions are not yet properly understood, and CFD packages impose their own simplifications in the description of these phenomena. These limitations may lead to significant discrepancies between the model predictions and observations in real-life phenomena. Moreover, the complexity of the partial differential equations and the fine subdivision needed to solve them may lead to incomplete numerical convergence, and the computations are highly demanding. However, extensive research efforts in the area of computational fluid dynamics have made progress toward overcoming these problems. The CFD packages allow the inclusion of user-defined subroutines so that additional modeling detail can be added to describe multiphase phenomena more accurately. An example of such an advance is the development of a CFD model for two-phase flow in packings and monoliths and its experimental validation using capacitance tomography [11], which is described in Chapter 4.

Due to the problems associated with CFD models for multiphase systems, alternative modeling approaches have been developed. One such simplified empirical fluid mechanics modeling technique for the description of mixing phenomena is based on the "network of zones" concept [13]. In this technique, the equipment volume is divided into a large number of interacting well-mixed cells (zones). Each cell is described by a simple first-order ordinary differential equation. Network-of-zones models are therefore smaller and simpler to solve than CFD models, but even so, good accuracy has been observed in the description of mixing phenomena in single as well as multiphase (gas–liquid, liquid–solid) systems measured using electrical resistance tomography [10]. Another approach to modeling multiphase systems has recently been presented by Gupta et al. [14] for the description of gas and liquid/slurry phase mixing in churn turbulent bubble columns. Such systems cannot be accurately addressed using CFD models at present. Gupta et al. [14] decompose the overall simulation problem, i.e., they estimate the gas and liquid phase recirculation rates in the reactor with a submodel that uses a two-fluid approach in solving the Navier–Stokes equations that are the input to the mechanistic reactor model. By effectively decomposing the problem, they keep the model size at moderate levels.

2.3.2 Data- and Image-Driven Models

It is often difficult or very time consuming to establish mathematical models derived from first principles for a number of chemical processing systems. This is the case where fundamental knowledge is incomplete and cannot fully describe the system's behavior or where models become too large to be solved numerically. Examples include multiphase modeling such as describing relations between heat transfer and bubble dynamics [15], modeling of bioreaction systems [16], and the derivation of mathematical models from industrial process data for use in process control [17]. To derive models for such systems, data-driven modeling tools such as ANNs are often employed. The advantage of such supervised learning methods is the possibility of training the models with data sets.

A typical ANN is shown in Figure 2.1 (right side). It consists of a set of highly interconnected processing units that mimic the function of neurons. Each "neuron" has a number of inputs and outputs, all of which are weighted by individual multiplicative factors (as depicted on the left side of the figure). The individual neurons sum all the products of their inputs and the associated connection weights together with a set of bias values. This sum is then transformed using a sigmoidal transfer function, and the resulting signal is sent to the outputs. The biases and the connection weights are adjustable. Neural networks are trained by adjusting the weights and biases to obtain the desired mapping of

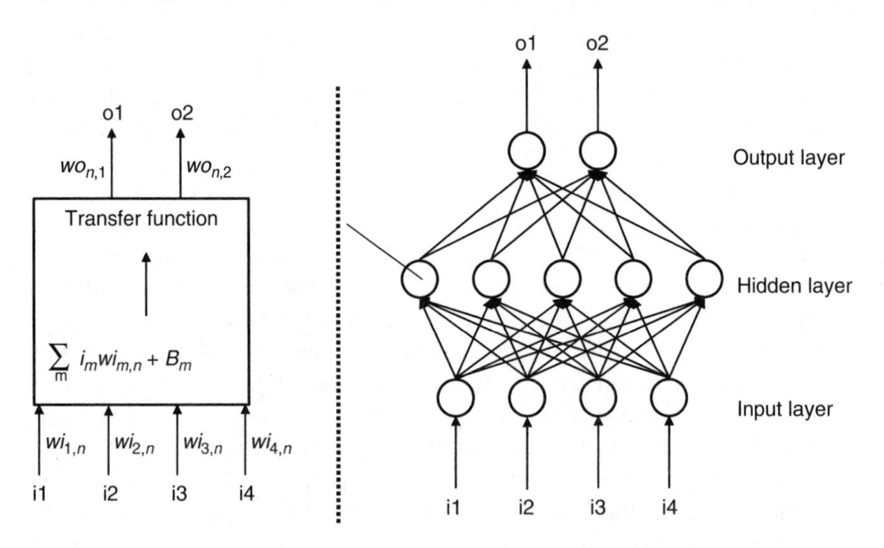

FIGURE 2.1 Artificial neural network model structure (right side). A single neuron is depicted on the left.

input data (stimuli) to output data (response), using sets of multivariate data of known system inputs and outputs. The trained ANNs are then used to predict output data from new input data (i.e., data not included in the training set). Kell and Sonnleitner [16] describe common pitfalls in applying and training ANNs and give recommendations for good modeling practice. Recommendations include ensuring the network is trained with a consistent set of data to guarantee the applicability of the model and not using the model on data outside the range of the training data. Extrapolation using ANNs is dangerous, as these models have not been derived from first principles. This lack of physical insight also makes these models difficult to interpret. ANNs and other supervised learning (heuristic) methods are often employed for reconstructing electrical tomography images, as discussed in Chapter 4. The heuristic models have the advantage that they can be implemented quickly and can relate measurements directly to the variables of interest.

Supervised learning models are useful for making process imaging information gathered using multidimensional sensors available for decision support in process operations and control. Apart from model validation using tomographic sensors for process analysis and design, such multidimensional sensors have been widely applied in process monitoring through visualization in a variety of systems including combustion systems [18] and food production [19]. Process images generally provide a more comprehensive assessment of the current state of the process than can be gained through measurement of single process variables. Recent research has focused on the development of strategies for using process images to estimate the process state variables (see Chapter 6) used in control systems. The integration of image information into control systems requires real-time processing of the image information from the sensors for data compression and pattern recognition. This is a major challenge in view of the amount of information provided by process imaging applications.

Self-learning methods such as self-organizing maps can be applied to detect intrinsic features in the images and provide compact representations of the multidimensional signals that can be integrated into control loops. Such image processing is crucial for feedback control applications, to enable the comparison of the measured signal (image) with the reference (set point) signal. Recently, Sbarbaro et al. [20] presented two strategies for the implementation of image processing in feedback process control systems. The first, a classical feedback control strategy, involves the reduction of the multidimensional information obtained from the sensors to a one-dimensional signal representing a specific characteristic of the original signal. Such a strategy involves the application of signal processing algorithms that can be difficult to apply in real time. The second strategy does not use signal processing algorithms; it avoids introducing errors into the interpretation of the multidimensional signals through the application of pattern recognition techniques. Following this strategy, the control system is designed using ANNs and finite state machines [21]. Both strategies have been successfully demonstrated for the control of fluidized bed systems. Additional control strategies are discussed in Chapter 7.

2.4 PROCESS MODELING FOR DESIGN, CONTROL, AND DIAGNOSTICS

Process design, automation, and diagnostics are based on quantitative methods and concepts of process simulation. The simulations employ process models and the balance equations for mass, components, energy, and momentum. Due to the model complexities, they are solved iteratively. Numerous types of models are published in the literature; as mentioned previously, these vary in their degree of detail or accuracy and in their application. For process design and optimization, very detailed models are required; these are termed *rigorous models*, e.g., equilibrium or nonequilibrium models [22, 23] and CFD models [24, 25]. In contrast to the rigorous models, so-called *short-cut models* (reduced models) are also frequently used; these include linear models, qualitative models [26], and trend models [27, 28]. The rigorous models are based on the balance equations for mass, energy, and momentum. In process modeling, it is always necessary to abstract from the real-world process an idealized description (in the form of equations, relations, or logic circuits) that is more amenable to analysis [29]. Hangos and Cameron [30] describe a formal representation of the assumptions in process modeling. Linninger et al. [31] and Bogusch [32] describe a modeling tool for efficient model development. Weiten and Wozny [33] describe an advanced information management system for knowledge-based documentation. Zerry et al. [34] published a method of modeling integrated documentation in MathML and automated transfer to Java-based models.

2.4.1 DEFINING THE MODEL

For the formulation of the balance equations for a chemical engineering or energy process, additional information on the properties of the deployed fluids is required. In addition to pure component properties, accurate description of the mixture properties is of vital importance to the model accuracy. The importance of the properties data and the calculation of properties are discussed by Kister [35], Carlson [36], Shacham et al. [37], and Gani et al. [38].

In Figure 2.2 a typical flow sheet of a chemical engineering process is depicted. The figure displays the various process units, such as compressors, reactors, and de-misters. The performance of a number of these units depends on the effectiveness of fluid flow, fluid contacting, and mixing. The effectiveness of the processes is linked to the internal spatial distributions of fluid inside the equipment, and it is important to model these accurately. Since the fundamental knowledge required for the accurate modeling of turbulence phenomena is still incomplete, the models need to be assessed and validated with experimental information in the form of process images. Internal spatial distributions are particularly important in reaction, mixing, heat exchange, and thermal separation equipment.

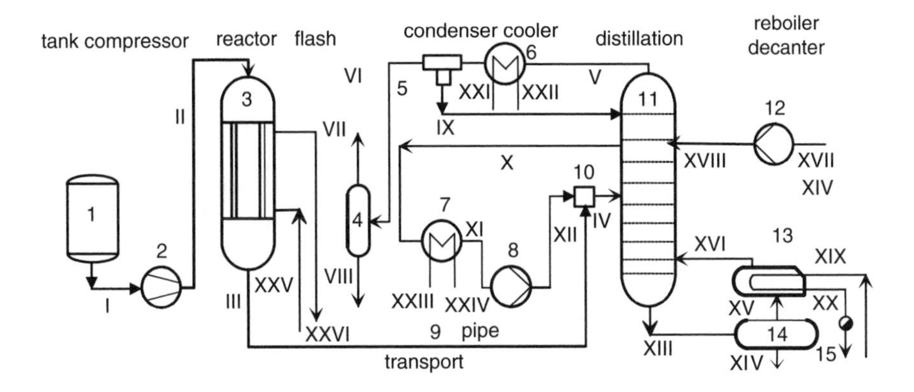

FIGURE 2.2 Flow sheet of a typical chemical engineering industrial process.

Reactors and separation columns assume particular importance for process simulation. The internal streams within the columns are in counter-current flow, causing numerical problems and requiring greater effort for the sequential-oriented solving of the model equations. The differential equations used to model reaction equipment in many cases result in boundary value problems that are difficult to solve simultaneously or sequentially with the models of the other process units. When types of pumps and valves are taken into account in the process model, the solution effort is increased further. Often these elements have to be considered in pressure-driven or closed-loop dynamic simulations. It is common to develop simulation models for single-unit operations that are solved sequentially and subsequently linked to a complete process flow sheet. Modern equation-oriented simulation software solves the models simultaneously, which offers significant advantages for dynamic simulations. As mentioned above, an important aspect in the dynamic process analysis and simulation is the estimation of thermodynamic state and transfer values. On the other hand, the dimensioning and geometry of the equipment and plant have a significant impact on the process dynamics. Consequently, the geometry has to be taken into consideration for dynamic process analysis.

For illustration purposes, Figure 2.3 and Figure 2.4 show a simulation model for a single distillation column. The design engineer has to answer a number of questions by using the model. For instance, how many trays are required to perform the separation? What is the best feed location and column pressure? How many controllers are necessary? What are the best controlled variables? What is the best pairing of the controlled variables with the manipulated variables? What is the best location of the sensors? What is the optimal set point? What are the best controller parameters?

To solve the model, it is necessary to determine its degree of freedom, i.e., the number of variables minus the number of equations. To simulate the model, the problem has to be fully specified. A number of variables equivalent to the degrees of freedom are generally specified as design parameters. The basis or

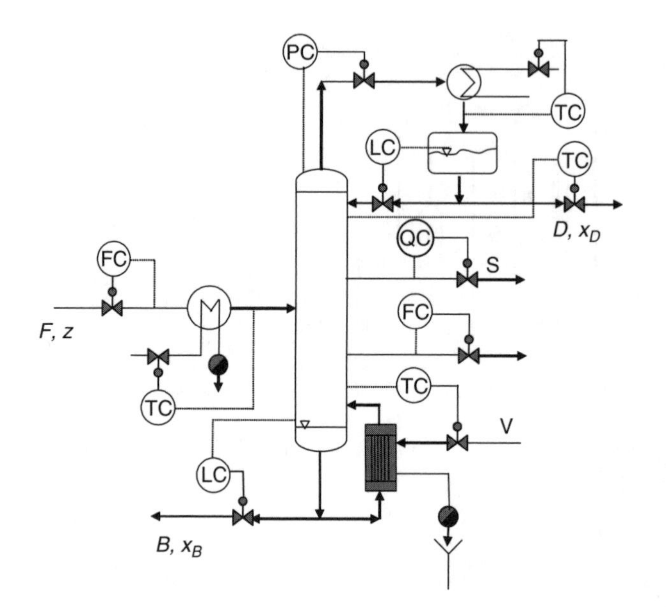

FIGURE 2.3 Flow sheet of a distillation column (F, z: feed flow rate and composition; B, x_B: bottom flow rate and composition; D, x_D: distillate flow rate and composition).

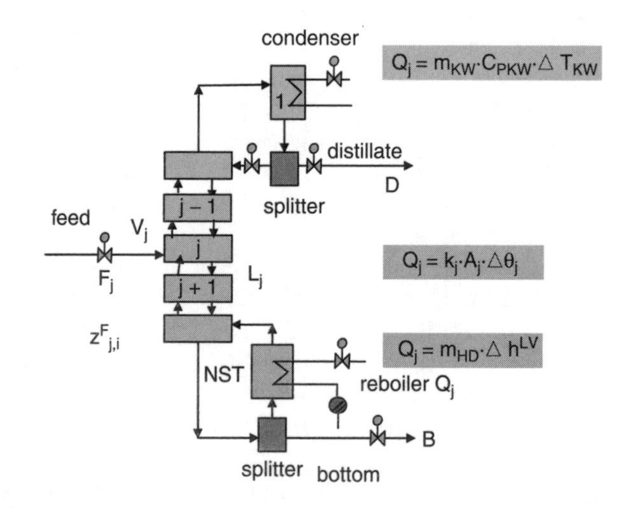

FIGURE 2.4 Schematic of a stage model for a distillation column. Indices denote j: stage number, i: component, kw: cooling water, and hd: steam. V is the vapor flow rate, z is the mole fraction, Q is the heating duty, m is the utility mass flow rate, k is the heat transfer coefficient, and A is the heat exchanger area.

default values are selected in view of market conditions, previous processes, experience, and sensitivity analysis. For the above example (Figure 2.3), the following information needs to be specified:

- Tray number 1, 2, 3, ..., $j - 1, j, j + 1$, ..., NST
- Feed tray j
- Feed flow F_j
- Feed temperature, concentration, pressure
- Geometry of the tray, area, weir height, weir length
- Condenser and reboiler area
- Design of condenser and reboiler (e.g., total, falling film, thin film; cooling water and heating medium conditions)

With this data the mathematical model is developed as shown in Figure 2.4.

For the reboiler and the condenser, the heat transfer equations are integrated into the process model. For closed-loop dynamic simulation as shown in Figure 2.5, the control structure (connection of controlled and manipulated variables) and the controller type have to be predefined. The set points are normally given as the steady-state design values, and the controller parameters have to be optimized. In some cases, the sensor dynamics and actor dynamics have to be considered. Muske and Georgakis [39] describe an optimal measurement system design procedure for chemical processes.

In the commonly used flow-driven simulation procedure, the direction of the flow is specified *a priori*. The more realistic "pressure driven" simulation procedure is more complex, and thus more physical and process data need to be considered. A detailed description of pressure drops, the valve characteristics, and the pump diagram have to be introduced in the simulation model. The basic set of equations required to simulate a column, as shown in Figure 2.3 and Figure 2.4 for the flow-driven calculation procedure, includes overall material balances, component material balances, energy balances, summation equations (mole fractions), phase equilibrium relations, control algorithms, and other functions such as hold-up correlations and pressure drop correlations.

The nonlinear differential-algebraic equation (DAE) system can be solved using a simultaneous solution procedure. The time dependency can be linearized by Euler

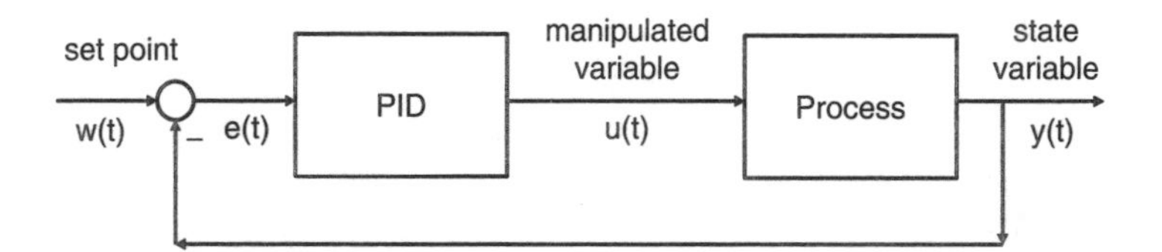

FIGURE 2.5 Layout of a closed-loop process control system ($w(t)$: set point, $e(t)$: set point error).

approximation, resulting in an equation system that can be solved by a Newton–Raphson procedure for each time step. For this, the equation system will be reformulated in a vector description where all equations are given as a vector $G(X) = 0$. A modified Gauss algorithm can be used to solve the linearized balance equation system. The method also enables the simulation of different process units such as membranes, reactors, columns and connected units, and complex flow sheets.

2.4.2 Detailed Models

To eliminate the assumption of phase equilibrium, the coupled mass and heat transfer across each boundary have to be considered in a nonequilibrium or rate-based model. For the case of three-phase distillation, the tray models shown in Figure 2.6 describe the separation process at several levels of detail.

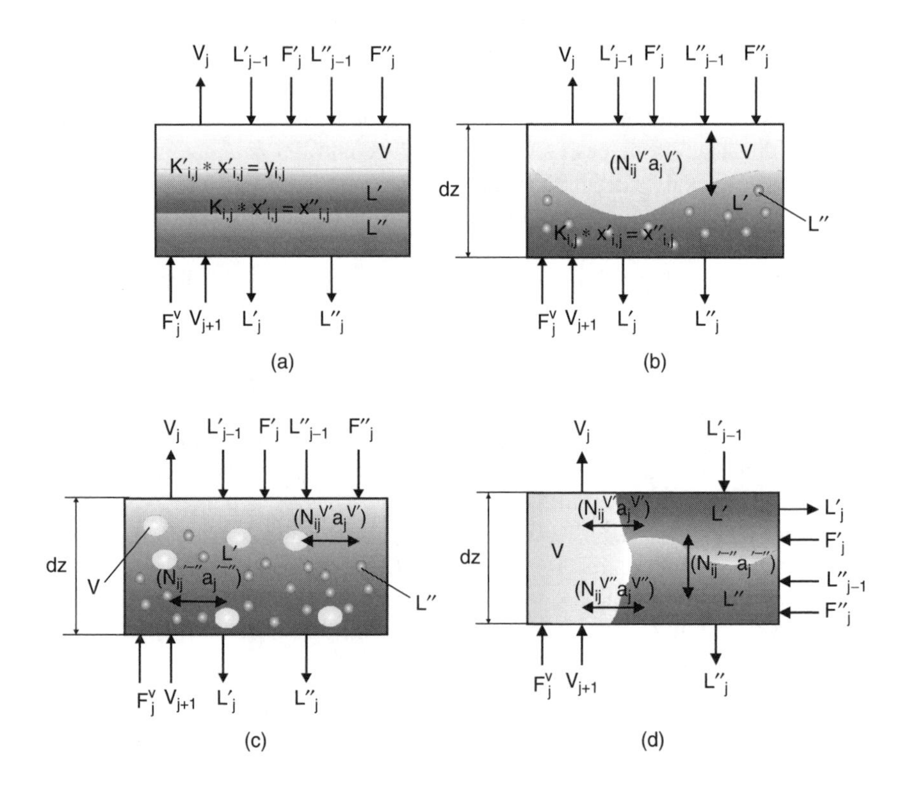

FIGURE 2.6 Possible balance regions for three-phase vapor–liquid–liquid (VLL) contacting on a distillation column stage. (a) Equilibrium model. (b) Nonequilibrium model considering V–L mass transfer. (c) Nonequilibrium model considering L–L mass transfer. (d) Full nonequilibrium model considering V–L–L mass transfer. Indices are defined in Figure 2.4; flow rates are given by F (feed), V (vapor), L (liquid 1), L (liquid 2); mole fractions are denoted as y (vapor), x (liquid 1), x (liquid 2); K is the equilibrium constant; a is interfacial area, N is specific mass transfer rate.

The equilibrium model in Figure 2.6a uses the relations for vapor–liquid (VL) and liquid–liquid (LL) equilibrium. The models in Figure 2.6b and Figure 2.6c take only one or two of the existing three mass (or heat) transfer rates into consideration. The model shown in Figure 2.6d is the generalized description of all transfer streams. The degree of accuracy desired for the description of transfer rates depends on the application and focus of the model. To calculate the mass transfer, an accurate description of the product of mass transfer coefficient (k) and mass transfer area (a) is needed. For VL mass transfer, a large number of correlations to predict this product are available from the literature. For VLL (i.e., both liquids exchanging mass with the vapor phase), the transfer coefficients and the transfer area are generally unknown. Figure 2.7 illustrates the excellent results that are possible by incorporating greater levels of detail in the model [22].

A more detailed description of the film area is possible using CFD simulation. CFD models are needed to describe wave films [40]. Figure 2.8 and Figure 2.9 show comparisons of the calculated and measured mass transfer characteristics in a packed column. These comparisons are encouraging, in a qualitative sense. However, further model development for better performance would require the use of process imaging techniques for model validation. For on-line optimization, process images could also be used to update mass transfer and interfacial area information used in the process model automatically. More research in this direction is necessary in the future.

The derived model can also be used for safety column analysis [41]. Can et al. [42] give the application of the described model equations for safety analysis

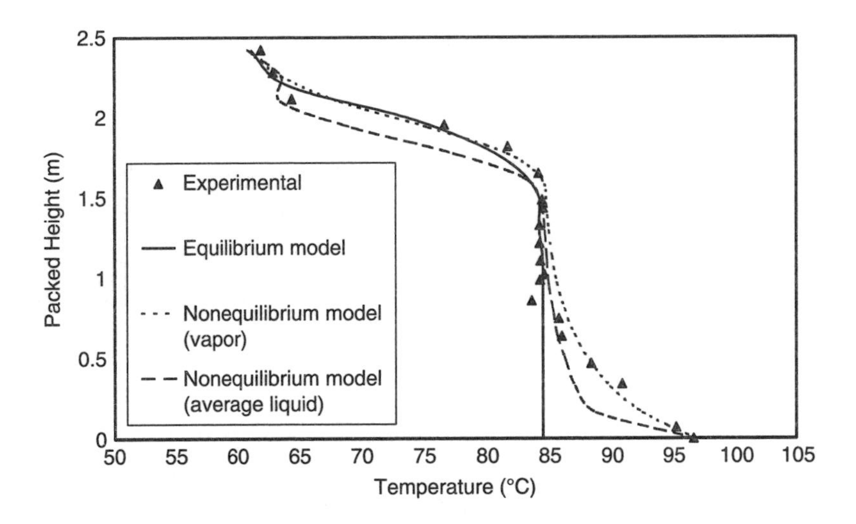

FIGURE 2.7 Temperature profile of a packed three-phase distillation column separating an acetone–toluene–water mixture at finite reflux. (From Repke, J.U., Villain, O., and Wozny, G., *Computer-Aided Chemical Engineering* 14:881–886, 2003. With permission.)

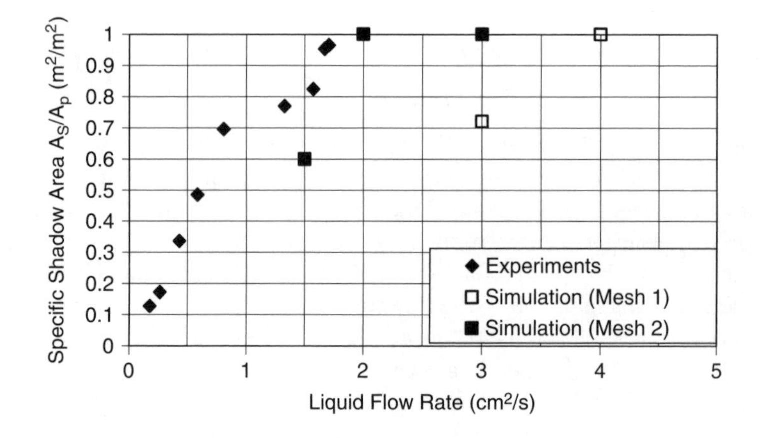

FIGURE 2.8 Comparison of experimental data and two predictions from two CFD models for the analysis of the shadow surface of a thin film in a packed distillation column. The finer MESH 2 achieves more accurate predictions of the experimental data. A_s: shadow surface area, A_p: packing surface area.

of a distillation column, including the relief system. The model describes the operational failures in a distillation column. At the top of the column, a safety valve is introduced in both the process model and the pilot plant for experimental validation (see Figure 2.10). In addition to the basic model equations described above, equations to describe the relief flow are introduced, and the model is formulated in gPROMS (Process Systems Enterprise Limited, London, U.K.). For experimental purposes a second condenser and an additional tank with a two-phase split are introduced so that the vapor relief flow and the liquid relief flow can be analyzed separately.

To simulate the system, a relief stream has to be integrated in the tray model for the first tray (Figure 2.11). Figure 2.12 shows a typical scenario of a cooling water failure for a methanol–water separation with the relief procedure. After 5 min the cooling water flow was reduced from 160 l/h to 20 l/h, to increase pressure in the system. Comparison of the theoretical results and the pilot plant experiments shows good agreement between the experimental and theoretical pressure–time dependency.

2.4.3 Start-Up and Shut-Down

Another application of the described models is the simulation of start-up and shut-down processes of distillation columns. Figure 2.13 compares the simulated and measured temperature profiles for a transesterification reactive distillation column. The selected equilibrium model accounts for chemical reaction and is

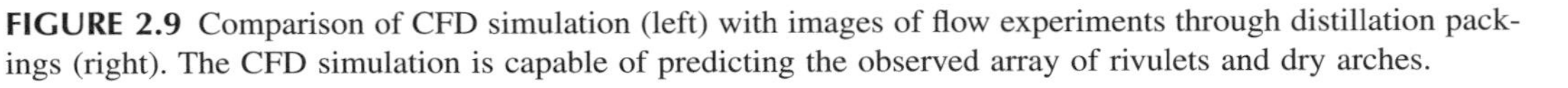

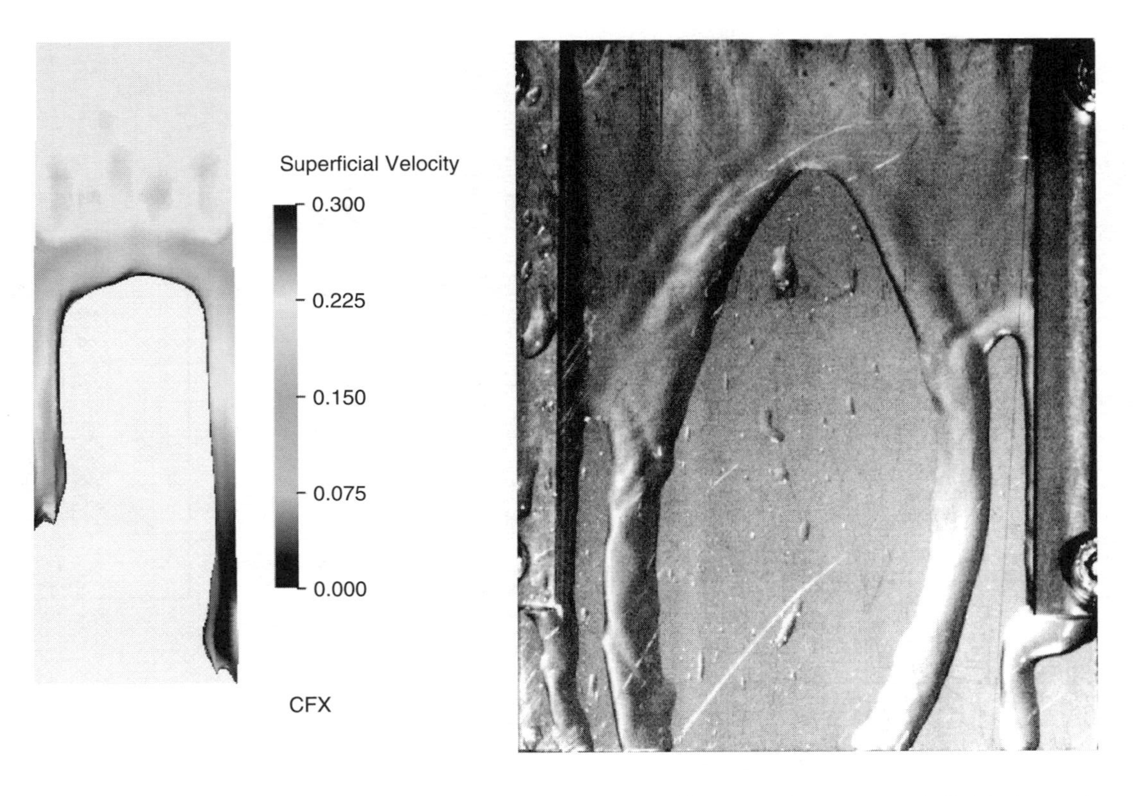

FIGURE 2.9 Comparison of CFD simulation (left) with images of flow experiments through distillation packings (right). The CFD simulation is capable of predicting the observed array of rivulets and dry arches.

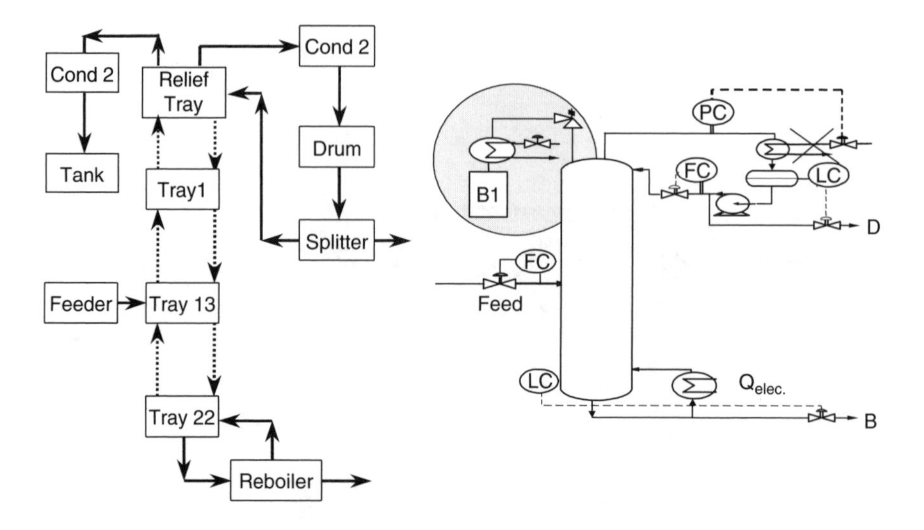

FIGURE 2.10 Structure diagram for a column relief simulation (left) and schematic plot of the pilot plant (right) with relief system including an additional condenser to analyze failure of cooling water supply.

used to describe the start-up and shut-down of a reactive distillation column, assuming:

- Reaction only takes place in the liquid phase (homogeneously catalyzed)
- Vapor and liquid phase are in equilibrium at steady state
- Vapor phase shows ideal behavior (for operation at ambient pressure)

A single set of equations is not sufficient to simulate a start-up from the cold and empty state to steady-state operation. For example, during fill-up and heating of the column, the plates are not at physical or chemical equilibrium. Therefore,

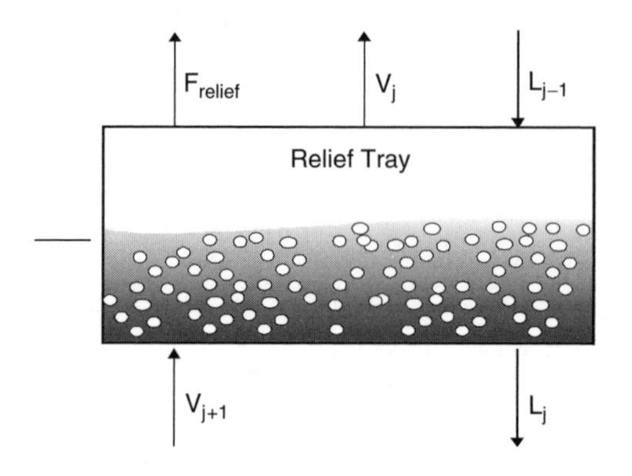

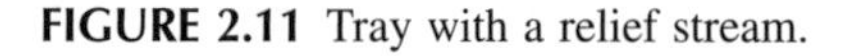
FIGURE 2.11 Tray with a relief stream.

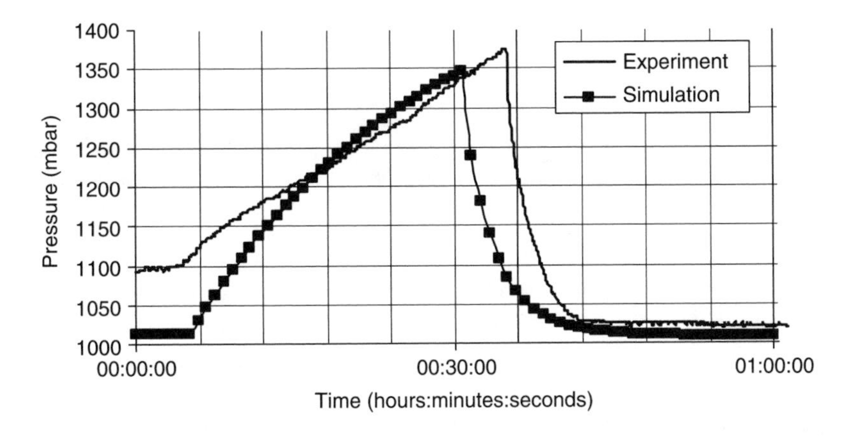

FIGURE 2.12 Comparison of simulation and experiment for a typical scenario of a cooling water failure for a methanol–water separation with the relief procedure.

additional sets of equations that are active at different times during the dynamic simulation are needed. The first set of equations is active during the fill-up and heating process of the column. Once the boiling conditions of a plate are reached, the second set is activated.

The start-up process for a single plate is depicted in Figure 2.14. In phase I, the plate is empty, cold, and at ambient pressure. The feed fills the plate until

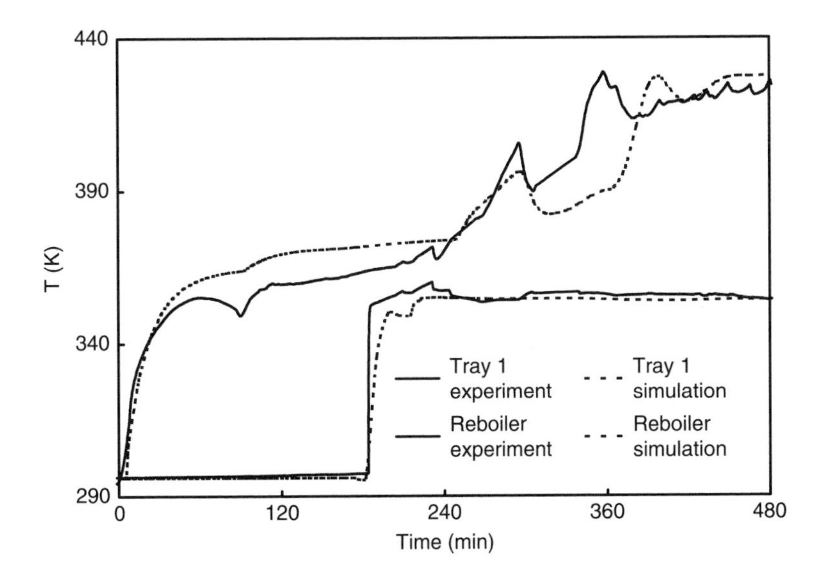

FIGURE 2.13 Dynamic validation of a reactive distillation column start-up: experiments vs. simulation for a transesterification process. (From Reepmeyer, F., Repke, J.U., and Wozny, G., *Chemical Engineering & Technology,* 26:81–86, 2003. With permission.)

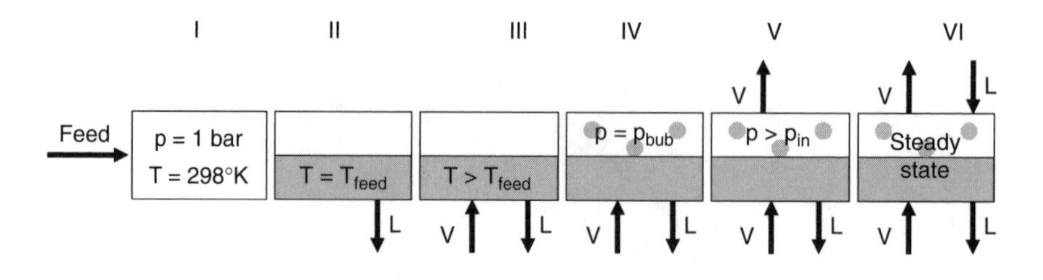

FIGURE 2.14 Different simulation phases of a sample plate during the start-up process. P_{in}: pressure at the stage above the feed stage.

liquid leaves the stage to the stage below (phase II). In phase III, vapor from the stage below is entering the stage and heating it up until, in phase IV, the mixture's bubble pressure (p_{bub}) reaches p_{set}, the set pressure (here 1 bar). In phase V the stage pressure is higher than the pressure from the stage above, so vapor is leaving the stage. In phase VI the stage is operating at steady state. In phases I to IV the first set of equations is active. The switching point is reached when $p_{bub} = p_{set}$. Then the phase equilibrium equation is applied. Comparison of the model and experimental values shows good agreement, as shown in Figure 2.13. Reepmeyer et al. [43] give details of the study.

The amount of information needed for the development of a dynamic model and for rigorous simulation of the complete start-up process is tremendous. All component and kinetic data have to be known, as well as the column, operation, and control specifications. The computational time required to complete one simulation run is long. Therefore, it is desirable to find a simple method of predicting the influence of changes in manipulated or input variables such as heating duty, reflux ratio, feed compositions, and flow. The impact of the manipulated variable on the start-up time is easier to understand on the basis of a reduced and simplified model. As an example, a reactive column can be reduced to a two-stage model consisting of a reboiler and a condenser, as depicted in Figure 2.15.

The model assumes that the condenser hold-up is negligible, the phases are in equilibrium, and the reaction takes place only in the liquid phase. From the component balance for the species X_A around the reboiler (the hold-up should be constant), the following equation results:

$$HU_B \cdot \frac{dX_A}{dt} = F \cdot X_{FA} - V \cdot Y_B + L \cdot X_D - B \cdot X_B - r_A \cdot HU_B \tag{2.1}$$

where HU_B is the reaction volume (hold-up); F, L, and B are the molar flow rates of feed, reflux, and bottom streams, respectively; Y_B, X_D, and X_B are the corresponding vapor and liquid component mole fractions in the bottom and distillate streams; and r_A is the reaction rate. Introducing the phase equilibrium equation

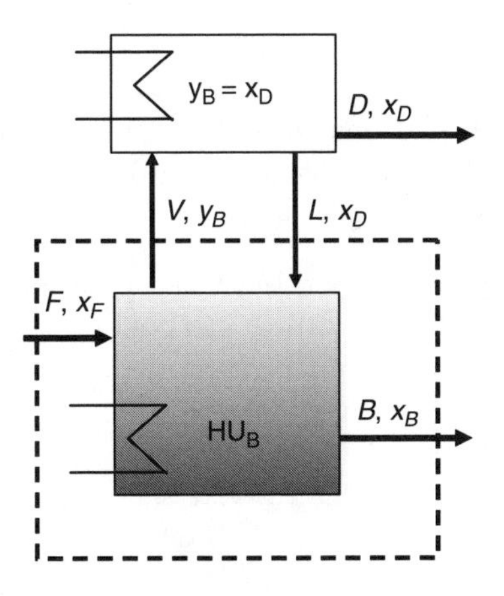

FIGURE 2.15 Reduced two-stage model. (From Reepmeyer, F., Repke, J.U., and Wozny, G., *Chemical Engineering & Technology,* 26:81–86, 2003. With permission.)

$Y_B = K \times X_B$ and the overall mass balance as well as the kinetic balance, (e.g., first-order approach to r_A), the above equation yields:

$$HU_B \cdot \frac{dX_A}{dt} = F \cdot X_{FA} + (K-1) \cdot (L-V) \cdot X_A - F \cdot X_A - (k_H \cdot X_A \cdot X_B - k_R \cdot X_C \cdot X_D) \cdot HU_B \tag{2.2}$$

where X_B, X_C, and X_D are the mole fractions of components B, C, and D in the reaction volume HU_B. To derive a time constant from this equation, as a value which indicates a trend of the start-up time, the bilinear terms involving the molar composition of all components must be eliminated. Therefore a linearization around an operating point must be applied:

$$X_A \cdot X_B \approx X_{A0} \cdot X_B + X_A \cdot X_{B0} - X_{A0} \cdot X_{B0} \tag{2.3}$$

where the subscript "0" denotes steady-state values.

Inserting the linearization of the bilinear terms in the component balance yields

$$\frac{dX_A}{dt} = \left(\frac{(K-1)(L-V)+F}{HU_B} - k_H \cdot X_{B0} \right) \cdot X_A + (-k_H \cdot X_{A0}) \cdot X_B + (k_R \cdot X_{D0}) \cdot X_C + (k_R \cdot X_{C0}) \cdot X_D + \left(\frac{F \cdot X_{FA}}{HU_B} + k_H \cdot X_{A0} \cdot X_{B0} - k_R \cdot X_{C0} \cdot X_{D0} \right) \tag{2.4}$$

TABLE 2.1
Comparison of Time Constant (Reduced Model) and Start-Up Time (Detailed Simulation) for the Ethyl Acetate Process

Strategy	Simulation (min)	Time Constant (min)
Conventional	175	118.9
Total reflux	225	122.4
Total distillate removal	183	93.2
Time optimal	191	118.1

The variables where interaction during control of the start-up process is possible are the vapor stream V (influenced using the heating power), the feed flow rate F, and the reflux stream L. Utilizing this formulated equation for all components yields a system with four components described by a 4×4 matrix where the eigenvalues of the matrix give an idea of the time constants underlying this system. The steady-state values are known in advance. Setting the manipulating variables V and L, to, for example, $V = L$ or $L = 0$ provides to the start-up strategy a total reflux condition or total distillate removal, respectively. In Reepmeyer et al. [43], the ethyl acetate process is analyzed as a reactive distillation using the reduced model. A comparison of the rigorous and the reduced model is given in Table 2.1. As can be seen, the time constant is capable of predicting the effects of variable changes on the start-up time. Here the total reflux strategy ($V = L$) shows the largest start-up time and has the highest time constant. The total distillate removal strategy ($L = 0$) shows the smallest time constant; therefore, it should deliver the fastest start-up time.

The simplifications introduced by the reduced model limit the discussion to the effect of the manipulated variables (here, heating power in the form of vapor stream V and the reflux L). Furthermore, due to the linearization (which is valid around the operating point), complex and highly non-ideal characteristics of the process are incompletely described. Nevertheless, using the reduced model, the start-up of a nonreactive [44] and heat-integrated distillation column [45] and of a reactive distillation column with "simple" reaction can be estimated in terms of the time constant to represent the trend of the start-up time. For more complex reactive distillation processes, rigorous dynamic modeling from an initially cold and empty state is necessary.

2.4.4 Control and Optimization

In process control, linearized Laplace-transformed models are often used. The advantages and disadvantages of such models are discussed elsewhere [46, 47]. Alici and Edgar [48] extend existing strategies for the solution of the nonlinear dynamic data reconciliation problem by using the process model as a constraint,

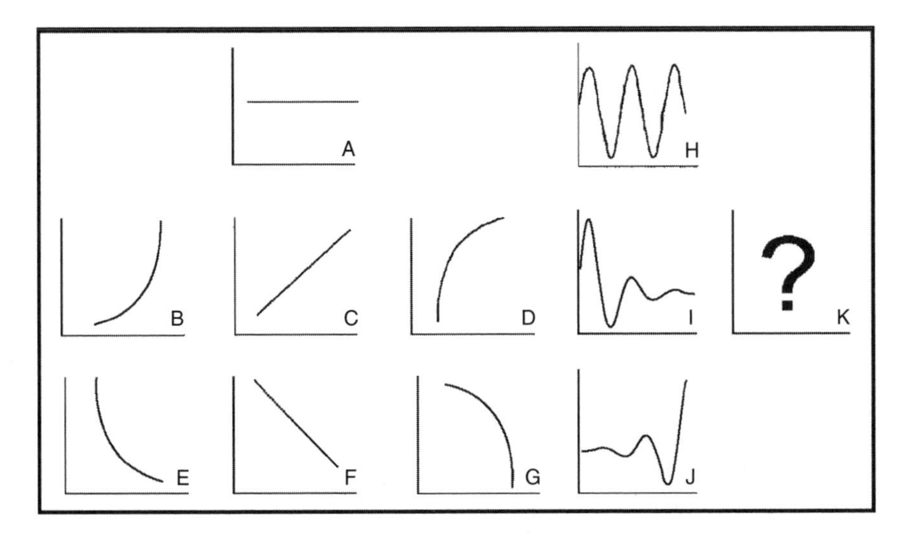

FIGURE 2.16 Trend models: different models classes to characterize the response of process variables. (From Vedam, H., and Venkatasubramanian, V.A., Proc. American Control Conference, Albuquerque, 1997; MC Kindsmüller, L Urbas, Situation Awareness in der Fahrzeug- und Prozessführung. Bonn: DGLR Bericht; 2002–04, 131–152.)

expressed above as the differential-algebraic Equation 2.4. Qualitative models are also often used in process control. Figure 2.16 shows the fundamental model behavior to depict the qualitative dynamic performance of a system. This model class describes the system's behavior for a certain time constant or oscillating behavior. Vianna and McGreavy [26] use another solution, using graph theory.

For optimization, the model equation system given above has to be expanded. The degrees of freedom will be reduced so that selected design variables such as unit number, tray number, pressure, and additional variables for operation (e.g., controller parameters, control structure) can be optimized. The expanded process model is described in the following form:

$$\begin{aligned}
&\text{Min } J(x,t,y,u,d,p,r,c,\zeta)\\
&f(x,t,y,u,d,p,\zeta) = 0 \text{ equilibrium constraints (mass, equilibrium, summation, and heat)}\\
&g(x,t,y,u,d,p,\zeta) \geq 0 \text{ nonequilibrium constraints}\\
&h(x,t,y,u,d,p,r,\zeta) = 0 \text{ controller equations}
\end{aligned}$$

where x_0, y_0, u_0 are initialization variables, c is the cost parameter, p is the model parameter, ζ is uncertainties, d is disturbances, u is manipulated variables, r is controlled variables, and t is time.

Optimization under uncertainty is often necessary for robust process design and operation. Wendt et al. [49] propose a new approach to solve nonlinear optimization problems under uncertainty, in which some dependent variables are to be constrained with a predefined probability. Such problems are called "optimization under chance constraints." The proposed approach is applied to the optimization of reactor networks and a methanol–water distillation column.

Wendt et al. [50] describe the application of the column model. A two-pressure column system modeled by the mass, equilibrium, summation, and heat (MESH) equations and solved by the algorithm described above is optimized with the sequential quadratic programming (SQP) method. In the optimization approach, the entire computation is divided into one layer for optimization and one layer for simulation. The model equations are integrated in the simulation layer, so that the state variables and their sensitivities can be calculated for given controls. The control variables, defined as piecewise constant, are calculated in the optimization layer by SQP as the decision variables. A reduction in start-up time of up to 80% was identified using this approach.

The application of the optimization method for a probabilistically constrained model-predictive controller is described by Li et al. [51]. The optimal design and control of a high-purity industrial distillation system is described by Ross et al. [52]. They developed a software implementation for the solution of the mixed integer dynamic optimization (MIDO) problem and optimized a two-pressure column system to improve operability and to identify a new process design with improved economics.

The problem of inconsistent initial values of the dependent variables is described by Wu and White [53]. Borchardt [54] describes a promising parallel approach for large-scale real-world dynamic simulation applications, such as plant-wide dynamic simulation in the chemical process industry. This approach partitions the system of differential and algebraic model equations into smaller blocks which can be solved independently. Considerable speed-up factors were obtained for the dynamic simulation of large-scale distillation plants, covering systems with up to 60,000 model equations.

Holl and Schuler [55] give an overview of process simulation in industrial application and operation. The nonlinear DAE-system mentioned in Section 2.4.1 is used for steady-state on-line optimization by Basak et al. [56], for plant-wide process automation [57], and for operator training systems [58] where the operators are able to assess the process performance.

REFERENCES

1. RG Rice, DD Do. *Applied Mathematics and Modeling for Chemical Engineers*. New York: John Wiley & Sons, 1995.
2. WL Luyben. Process Modeling, *Simulation and Control for Chemical Engineers, 2nd ed.* New York: McGraw Hill, 1990.
3. LT Biegler, IE Grossmann, AW Westerberg. *Systematic Methods of Chemical Process Design*. Upper Saddle River, NJ: Prentice Hall, 1997.
4. MB Abbott, DR Basco. *Computational Fluid Dynamics*. Singapore: Longman Scientific & Technical, 1989.

5. S Haykin. *Neural Networks.* Upper Saddle River, NJ: Prentice Hall, 1999.
6. UM Diwekar. *Introduction to Applied Optimization and Modeling.* Netherlands: Kluwer Academic Publishers, 2003.
7. S Rigopoulos, P Linke. Systematic development of optimal activated sludge process designs. *Comput. Chemical Eng.* 26:585–597, 2002.
8. H Lemonnier. Multiphase instrumentation: The keystone of multidimensional multiphase flow modeling. *Experimental Thermal Fluid Science* 15:154–162, 1997.
9. M Lance, M Lopez de Bertodano. Phase distribution phenomena and wall effects in two-phase flows. Chapter 2 in: *Multiphase Science and Technology,* Vol. 8. London, Begell House, 1996.
10. R Mann, RA Williams, T Dyakowski, FJ Dickin, RB Edwards. Developments of mixing models using electrical resistance tomography. *Chemical Eng. Science* 52:2073–2085, 1997.
11. D Mewes, T Loser, M Millies. Modeling of two-phase flow in packings and monoliths. *Chemical Eng. Science* 54:4729–4747, 1999.
12. HS Tapp, AJ Peyton, EK Kemsley, RH Wilson. Chemical engineering applications of electrical process tomography. *Sensors Actuators B* 92:17–24, 2003.
13. R Mann, AM El-Hamouz. A product distribution paradox on scaling-up a stirred batch reactor. *AIChE J.* 41:855–867, 1995.
14. P Gupta, B Ong, MH Al-Dahhan, MP Dudukovic, BA Toseland. Hydrodynamics of churn turbulent bubble columns: gas-liquid recirculation and mechanistic modeling. *Catalysis Today* 64:253–269, 2001.
15. W Chen, T Hasegawa, A Tsutsumi, K Otawara, Y Shigaki. Generalised dynamic modeling of local heat transfer in bubble columns. *Chemical Eng. J.* 96:37–44, 2003.
16. DB Kell, B Sonnleitner. GMP—Good modeling practice: an essential component of good manufacturing practice. *TIBTECH* 13:481–492, 1995.
17. MG Allen, CT Butler, SA Johnson, EY Lo, F Russo. An imaging neural network combustion control system for utility boiler applications. *Combustion Flames* 94:205–214, 1993.
18. G Lu, Y Yan, DD Ward. Advanced monitoring and characterization of combustion flames. Proceedings of IEE Seminar on Advanced Sensors and Instrumentation Systems for Combustion Processes, London, 2000, pp. 3/1–3/40.
19. PE Keller, LJ Kanngas, LH Linden, S Hashem, T Kouzes. Electronic noses and their applications. Proceedings of IEEE Northcon: Technical Applications Conference, Portland, 1995, pp. 791–801.
20. D Sbarbaro, P Espinoza, J Araneda. A pattern based strategy for using multidimensional sensors in process control. *Comput. Chemical Eng.* 27:1925–1943, 2003.
21. J Lunze. Stabilization of nonlinear systems by qualitative feedback controllers. *Int. J. Control* 62:109–128, 1995.
22. JU Repke, O Villain, G Wozny. A nonequilibrium model for three-phase distillation in a packed column: modeling and experiments. *Computer-Aided Chemical Eng.* 14:881–886, 2003.
23. JH Lee, MP Dudokovic. A comparison of the equilibrium and nonequilibrium models for a multicomponent reactive distillation column. *Comput. Chemical Eng.* 23:159–172, 1998.
24. FH Yin, CG Sun, A Afacan, K Nandakumar, KT Chuang. CFD modeling of mass-transfer processes in randomly packed distillation columns. *Industrial Eng. Chemical Res.* 39:1369–1380, 2000.

25. JB Joshi, VV Ranade. Computational fluid dynamics for designing process equipment: Expectations, current status, and path forward. *Industrial Eng. Chemical Res.* 42:1115–1128, 2003.
26. RF Vianna, C McGreavy. A qualitative modeling of chemical processes: A weighted digraph (WDG) approach. *Comput. Chemical Eng.* 19:S375–S380, 1995.
27. H Vedam, VA Venkatasubramanian. Wavelet Theory-Based Adaptive Trend Analysis System for Process Monitoring and Diagnosis. Proceedings of the American Control Conference, Albuquerque, 1997.
28. MC Kindsmüller, L Urbas. Der Einfluss von Modelwissen auf die Interpretation von Trenddarstellungen bei der Steuerung prozesstechnischer Anlagen. In: M Grandt, KP Gärtner, eds. *Situation Awareness in der Fahrzeug- und Prozessführung.* Bonn: DGLR Bericht; 2002–04, 131–152.
29. O Levenspiel. Modeling in chemical engineering. *Chemical Eng. Science* 57:4691– 4696, 2002.
30. KM Hangos, IT Cameron. A formal representation of assumptions in process modeling. *Comput. Chemical Eng.* 25:237–255, 2001.
31. AA Linninger, S Chowdhry, V Bahl, H Krendl, H Pinger. A systematic approach to mathematical modeling of industrial processes. *Comput. Chemical Eng.* 24:591–598, 2000.
32. R Bogusch. Eine Softwareumgebung für die rechnergestützte Modelierung verfahrenstechnischer Prozesse, *Fortschritt-Berichte VDI,* Nr. 705, Düsseldorf: VDI Verlag, 2001.
33. M Weiten, G Wozny. Advanced information management for process sciences: Knowledge-based documentation of mathematical models. *J. Internet Enterprise Manage.,* (2)2:178–190, 2004.
34. R Zerry, B Gauss, L Urbas, G Wozny. Web based object oriented modeling and simulation using MathML. *Computer-Aided Chemical Eng.,* 15:1171–1176, 2004.
35. H Kister. Can we believe the simulation results? *Chemical Eng. Prog.* 10:52–58, 2002.
36. EC Carlson. Don't gamble with physical properties for simulation. *Chemical Eng. Prog.* 10:35–46, 1996.
37. M Shacham, N Brauner. A dynamic library for physical and thermodynamic properties correlations. *Industrial Eng. Chem. Res.* 39:1649–1657, 2000.
38. R Gani, JP O'Connel. Properties and CAPE: From present to future challenges. *Comput. Chemical Eng.* 25:3–14, 2001.
39. KR Muske, C Georgakis. Optimal measurement system design for chemical processes. *AIChE J.* 49:1488–1493, 2003.
40. A Hoffman, I Ausner, JU Repke, G Wozny. Fluid dynamics in multiphase distillation processes in packed columns. *Computer-Aided Chemical Eng.,* 15:199–204, 2004.
41. M Jimoh, G Wozny. Simulation and Experimental Analysis of Operational Failures in a Methanol-Water Distillation Column. Proceedings (CD-ROM) of Distillation & Absorption, Baden-Baden, Germany, 2002.
42. Ü Can, M Jimoh, J Steinbach, G Wozny. Simulation and experimental analysis of operational failures in a distillation column. *Sep. Purif. Technol.* 29:163–170, 2002.
43. F Reepmeyer, JU Repke, G Wozny. Analysis of the start-up process for reactive distillation. *Chemical Eng. Technol.* 26:81–86, 2003.

44. M Flender. *Zeitoptimale Strategien für Anfahr- und Produktwechselvorgänge an Rektifizieranlagen.* Düsseldorf: VDI Verlag, ISBN 3-18-361009-5, 1999.
45. K Löwe. *Theoretische und experimentelle Untersuchungen über das Anfahren und die Prozessführung energetisch und stofflich gekoppelter Destillationskolonnen.* Düsseldorf: VDI Verlag, ISBN 3-18-367803-9, 2001.
46. FM Meeuse, AEM Huesman. Analyzing dynamic interaction of control loops in the time domain. *Ind. Eng. Chem. Res.* 41:4585–4590, 2002.
47. FG Shinskey. Process control: As taught vs. as practiced. *Ind. Eng. Chem. Res.* 41:3745–3750, 2002.
48. S Alici, TF Edgar. Nonlinear dynamic data reconciliation via process simulation software and model identification tool. *Ind. Eng. Chem. Res.* 41:3984–3992, 2002.
49. M Wendt, P Li, G Wozny. Nonlinear chance-constrained process optimization under uncertainty. *Ind. Eng. Chem. Res.* 41:3621–3629, 2002.
50. M Wendt, R Königseder, G Wozny. Theoretical and experimental studies on startup strategies for a heat-integrated distillation column system. *Trans. I. Chem. E, Part A,* 81:153–161, 2003.
51. P Li, M Wendt, G Wozny. A probabilistically constrained model predictive controller. *Automatica* 38:1171–1176, 2002.
52. R Ross, JD Perkins, EN Pistikopoulos, GLM Koot, JMK von Schijndel. Optimal design and control of high-purity industrial distillation system. *Comput. Chemical Eng.* 25:141–150, 2002.
53. B Wu, RE White. An initialization subroutine for DAEs solvers: DAEIS. *Comput. Chemical Eng.* 25:301–311, 2001.
54. J Borchardt. Newton-type decomposition methods in large-scale dynamic process simulation. *Comput. Chemical Eng.* 25:951–961, 2001.
55. P Holl, H Schuler. Simulatoren zur Unterstützung der Prozeß- und Betriebsführung. *Chemie Ingenieur Technik* 64:679–692, 1992.
56. K Basak, KS Abhilash, S Ganguly, DN Saraf. On-line optimization of a crude distillation unit with constraints on product properties. *Ind. Eng. Chem. Res.* 41:1557–1568, 2002.
57. R Lausch. Ein systematischer Entwurf von Automatisierungskonzepten für komplex verschaltete Chemieanlagen. *Fortschritt-Berichte VDI,* Reihe 3, Nr. 578, Düsseldorf: VDI Verlag, ISBN 3-18-358703-3, 1998.
58. L Urbas. Entwicklung und Realisierung einer Trainings- und Ausbildungsumgebung zur Schulung der Prozessdynamik und des Anlagenbetriebs im Internet. *Fortschritt-Berichte VDI,* Reihe 10, Nr. 614, Düsseldorf: VDI Verlag, ISBN 3-18-361410-3, 1998.

3 Direct Imaging Technology

Satoshi Someya and Masahiro Takei

CONTENTS

3.1 INTRODUCTION

This chapter describes the basic elements for direct imaging techniques and systems (i.e., systems that do not rely on mathematical reconstruction to form an image). We begin by considering the fundamental characteristics of laser and white light sources, then describe the basic properties and structure of two-dimensional image sensors, image intensifiers, and thermal sensors. We explain the properties of materials used to provide optical access to the target, such as glass materials for optics and optical fibers. We also present important current topics in applications of direct imaging, including pressure-sensitive paints and measurement of shear stresses in microelectromechanical systems (MEMS). These techniques are now seeing considerable development and advances. Finally, a brief description of machine vision will survey the future of direct imaging techniques.

3.2 LIGHT SOURCES

3.2.1 Lasers

3.2.1.1 Overview

A laser is a device that uses the principle of amplification of electromagnetic waves by stimulated emission of radiation. Lasers can operate in the infrared, visible, and ultraviolet regions of the electromagnetic spectrum, and they produce intense beams of monochromatic and coherent light. Laser light can be highly collimated to travel over great distances, or it can be focused to a very small and extremely bright spot. Because of these properties, lasers are used in a wide variety of applications, including direct imaging techniques.

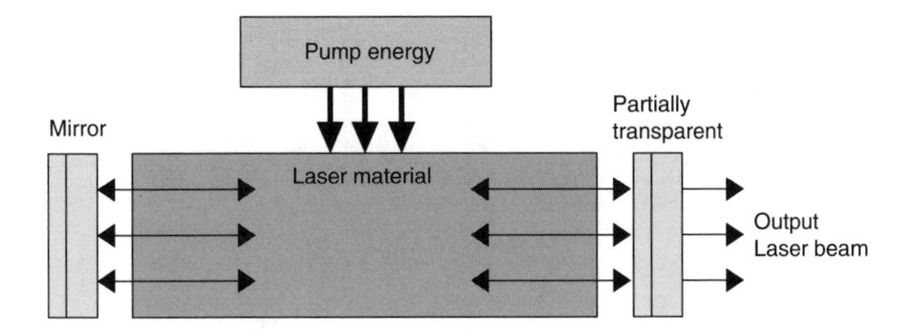

FIGURE 3.1 Schematic diagram of a laser.

Lasers generally consist of three main components: a "pumping" source or energy supply, an optical amplifier, and a resonator structure. The pumping mechanism depends on the type of optical amplifier used. Solid materials are generally pumped by electromagnetic radiation, gas lasers by electrical discharge and associated collisional processes in the ionized gas, and semiconductor lasers by electronic current. When atoms, ions, or molecules absorb energy into an excited but metastable state, they will tend to emit light only when stimulated by an external mechanism. This stimulated emission is the opposite of absorption, where absorbed photons provide the energy to cause an electronic transition to an excited state. If an optical amplifier is pumped so that there are more excited than unexcited electronic states (a situation called "population inversion"), then an incident light wave will stimulate more emission than absorption and gain intensity.

An optical amplifier can be made into a laser by arranging suitable mirrors on either end of the amplifier to form a resonator; see Figure 3.1. A photon that impinges randomly on one of the mirror surfaces is reflected into the amplifier again. Radiation bounces back and forth between the mirrors, generating an avalanche of light that increases exponentially with the number of reflections. In the amplification process, waves traveling in directions other than along the axis of the resonator soon pass off the edges of the mirrors and are lost before they are significantly amplified. Thus the output wave travels along the axis and is highly collimated.

Over the four decades since the ruby laser was first demonstrated in 1960, different kinds of lasers have found general acceptance for use in the laboratory. The helium–neon (He-Ne) laser, the helium–cadmium (He-Cd) laser, the ion laser (argon, krypton, or mixed gas), and the diode laser have become ubiquitous in laboratories and in original equipment manufacturer (OEM) applications. Recently, diode-pumped solid-state (DPSS) lasers have been used successfully in many systems, and their applications are bound to increase in the future. Commercially available DPSS lasers with output power over 30 Watts or more are used for laser machining. More than one kind of source may be used for a specific application. Lasers vary in their size, output, beam quality, power

TABLE 3.1
Wavelengths of Lasers

Laser	Wavelength of Laser (nm)
HeNe	543, 594, 612, **632.8**, 1523
Ar	351.1, 363.8, 454.5, 457.9, 465.8, 472.7, 476.5, **488.0**, 496.5, 501.7, **514.5**, 528.7, 1090.00
Krypton	406.7, 413.1, 476.2, 482.5, 520.8, 530.9, **568.2**, **647.1**, 676.4, **752.5**
CO_2	10600
N_2	337
Excimer laser	193 (ArF), 248 (KrF), 308 (XeCl), 334 (He-N_2), 337 (He-N_2)
HeCd	325, 354, 441.6
Cu vapor	511, 578
Au vapor	627.8
Ruby	694
Nd:YAG	266 (4th order harmonic), 355 (3rd order), 532 (2nd order), 1064
Nd:YLF	262, 349, 351, 523, 527, 1047, 1053, 1313
Ti-sapphire	660~ 1180
DPSS blue lasers	375, 405, 415, 440 or 430, 490 or 473 (Nd:YAG+BiBO) or 457 (Nd:YVO_4 +KTP)
Dye laser	300~ 1200

consumption, and operating life. Table 3.1 summarizes typical wavelengths of lasers.

3.2.1.2 Helium–Neon Lasers

The helium–neon laser is electrically pumped. The lasing medium is a gas mixture of about five parts of helium to each part of neon, at a pressure of a few Torr. A glow discharge established in the tube excites the helium atoms by electron impact. The energy is rapidly transferred to a neutral neon atom, which has an energy level just below that of the helium atom. This energy transfer is the main pumping mechanism. The most important laser transition occurs at 633 nm, and transitions at 1523 nm, 612 nm, 594 nm, and 543.5 nm are also possible. He-Ne lasers are continuously pumped, usually with a DC power supply. Typically, the output power ranges from 0.1 to 100 mW in the lowest transverse mode (TEM_{00}). The main properties of He-Ne lasers are the coherence of the beam and the Gaussian distribution of the beam intensity. He-Ne lasers are inexpensive and very reliable, and they are used in a wide variety of applications. He-Ne lasers are convection cooled and use relatively little power.

In many applications, the important laser properties are not just output power and wavelength but also coherence length, mode quality, beam divergence, beam-pointing stability, and output polarization. Additional factors, such as cost, reliability, lifetime, part-to-part consistency, and electrical requirements, are often also significant. Compared to other laser types, the overall performance of He-Ne lasers in most of these areas is remarkably good.

Coherence length is the distance over which phase information can be maintained, and it is an important parameter for interferometric imaging, holography, and laser-based metrology. For frequency-stabilized He-Ne lasers, this distance may be several kilometers. The extended coherence length of He-Ne lasers results directly from the narrow line width of the optical transitions. Solid-state and semiconductor laser devices have a broader continuum of electronic states than do gas lasers, resulting in a larger bandwidth and consequent shorter coherence length [1, 2]. Due to their narrow line width and wavelength stability, He-Ne lasers are widely used as wavelength-calibration standards in equipment such as Fourier-transform infrared (FTIR) spectrometers.

The rotationally symmetric cylindrical cavity of He-Ne lasers allows the resonator and mirrors to be designed so the lasers produce a Gaussian beam with more than 95% Gaussian mode content. Because of this excellent mode structure, He-Ne lasers can often be used without any spatial filtering. Even when numerous optical elements are used to improve the mode structure of the output from a laser diode, the beam quality of most laser diodes still does not compare with the beam quality of a He-Ne laser.

In addition to beam divergence, the beam-pointing stability in He-Ne lasers is also very good. A single-mode laser requires delicate temperature or length control to ensure that the cavity mode coincides precisely with the center of the spectral line. The output power will vary greatly as changes in cavity length reduce the gain by altering the resonant frequency. Thermal-expansion effects are minimized with carefully chosen tube materials and a highly symmetrical mechanical design. A well-designed He-Ne laser can easily achieve a long-term beam-pointing stability of less than 0.02 m rad.

3.2.1.3 Argon-Ion and Krypton-Ion Lasers

Argon-ion lasers are gas lasers similar to He-Ne lasers. The argon-ion laser can be made to oscillate at several wavelengths in the violet to the green end of the visible spectral region, with particularly strong lines at 488 nm and 514.5 nm. Important transitions occur between energy levels of the Ar^+ spectrum. A very high current arc discharge will produce a sufficient number of singly ionized argon atoms to bring about the required gain. Emission is produced at several wavelengths through the use of broadband laser mirrors. However, the individual wavelengths may be selected by means of Brewster prisms in the resonator structure. The individual wavelengths may be adjusted by tuning the prism. Because of the energy required to ionize an atom and then raise the ion to an excited state, the efficiencies of all arc discharge lasers are extremely low. Once a population inversion is achieved and maintained, these lasers have very high gain and can provide continuous output powers of several watts or more.

A close relative of the argon-ion laser is the krypton-ion laser, which produces a strong red line output at 647 nm, with additional lines at 568 nm and 752 nm. Mixed-gas lasers combine argon and krypton in the same laser tube to produce strong outputs across the spectrum from 457 to 647 nm. Both types of lasers exhibit good beam quality and are available in tunable as well as single-line and multiline configurations. Ion lasers typically include light stabilization electronics that effectively eliminate power drift and reduce beam noise to a small fraction of a percent. Nearly all conventional inert gas ion lasers supply a Gaussian distribution of the laser beam intensity.

3.2.1.4 Helium–Cadmium Lasers

He-Cd lasers are very good sources of ultraviolet light at 325 nm and at 354 nm with excellent beam quality. They also produce a deep blue light at 442 nm. He-Cd lasers use vaporized metallic cadmium transported down the laser bore by electrophoresis. Many of the difficulties of vaporizing sufficient cadmium metal and preventing it from plating on electrodes and cooler portions of the tube have recently been overcome. Though these lasers still have somewhat more beam noise than other gas lasers, they can be tailored to minimize this noise at specific frequencies. The He-Cd laser is the best source of pure ultraviolet light currently available.

3.2.1.5 Neodymium Lasers

The neodymium laser is optically pumped in broad energy bands and is an example of a four-level laser. The active Nd^{3+} ions may be incorporated into several host materials, such as yttrium aluminum garnet (YAG) crystals. A Nd^{3+} ion surrounds itself with several oxygen atoms that largely shield it from its surroundings. Recently, neodymium lasers (Nd:YAG, Nd:YLF, Nd:YVO4) have become more common, especially for particle image velocimetry (PIV) techniques. The Nd:YAG laser emits a near-infrared wavelength of 1064 nm. The laser can be operated in a triggered mode by including a Q switch, which alters the resonance characteristics of the optical cavity. The beam of a Q-switched laser is linearly polarized. In addition, an intra-cavity system has made it possible to generate strong pulses at several kHz [3]. The near-infrared light is frequency-doubled by using a second harmonic generation (SHG) crystal of potassium titanyl phosphate ($KTiOPO_4$ or KTP). After separation of the frequency-doubled portion, part of the original light energy is available at 532 nm. Nd:YAG lasers can be repetitively pulsed at a high rate, and the peak output power is extremely large (nearly 1 Joule for a pulse of 10 ns at 532 nm).

On the other hand, it can be difficult to obtain consistent operation with flash lamp excitation because the optical properties of a laser cavity change with temperature. Consequently, the coherence length of pulsed neodymium lasers is only a few centimeters, which is too short for many phase-sensitive imaging applications. In addition, precise laser timing and temperature control are required. A pumping system based on light-emitting diodes or a small semiconductor laser can achieve this purpose. The recently developed DPSS lasers in many ways approach the ideal. In addition, the new nonlinear optical crystals make it possible to obtain blue laser light from neodymium laser materials. A DPSS laser's output power, beam quality, stability, and repeatability approach those of a gas laser. The DPSS lasers are convection-cooled and extremely rugged, and their efficiency and size are comparable to those of a diode laser. The operating lifetime of a DPSS laser is much longer than that of a conventional ion laser.

3.2.2 White Light Sources

White light sources are frequently used in process imaging. In contrast to lasers, all conventional light sources are basically hot bodies that radiate by spontaneous emission. The tungsten filament of an incandescent lamp emits blackbody radiation due to its high temperature. In a gas lamp, an electrical discharge excites the atoms, which soon decay to the ground state by radiating the excitation energy as light. In these examples, spontaneous emission takes place independently of other emission events. The phase of the light wave emitted by one atom has no relation to the phase emitted by any other, so that the phase of the light fluctuates randomly in time and space. The light generated by a conventional light source is therefore incoherent. The white light from these conventional sources is not as easily collimated as laser light. Thus, conventional light sources clearly have some disadvantages.

However, white light sources cost much less than lasers and can be handled easily and safely. Newer white light strobe sources can be easily triggered and can supply quite short pulse duration times (10 to 100 ns) with very high repetition rates. These short pulses may be used to "freeze" the motion of dynamic processes for imaging applications (e.g., observing the shape and size of crystals in an industrial crystallizer). Moreover, fiber-optic delivery systems in which fibers are arranged in line are commercially available, considerably simplifying the formation of a light sheet. White light sources may be suitable for many applications where spatial and temporal coherence is not required. Such sources are used with a thermochromic liquid crystal (discussed in Section 3.3.3.4) to measure the temperature field. They are also necessary for investigations of absorption and emission spectral characteristics. The illumination of certain dyes and fluorescent powders requires white light sources to provide broadband excitation.

Xenon lamps have high intensity and color temperature; they provide a continuous spectrum from the ultraviolet to the infrared, making them ideal for use as a light source in a variety of photometric applications. Moreover, a mercury–xenon lamp is designed to provide high radiant energy in the ultraviolet range. This lamp offers the characteristics of both xenon lamps and high-pressure mercury lamps. For example, the spectral distribution of a mercury–xenon lamp includes a continuous spectrum from ultraviolet to infrared of the xenon gas and strong mercury

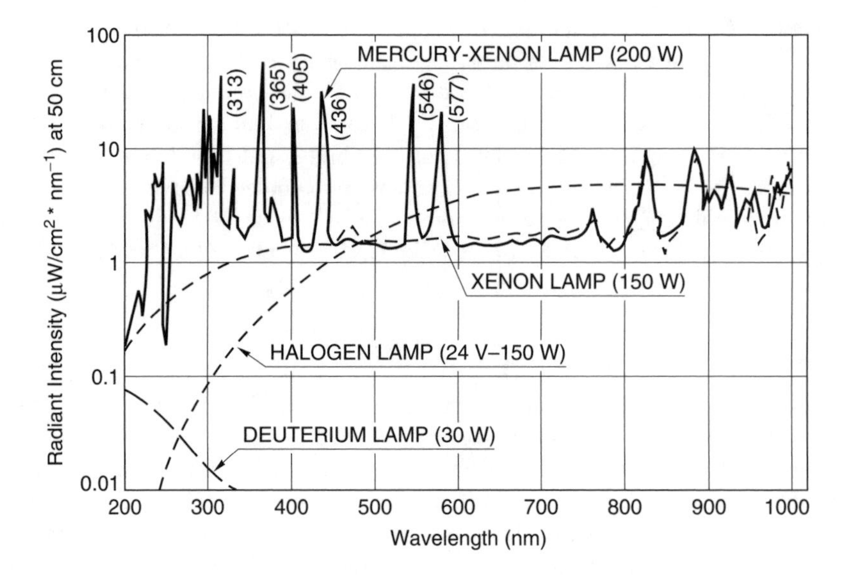

FIGURE 3.2 Spectral distribution of various lamps. (© Hamamatsu Photonics K.K. With permission.)

line spectrum in the ultraviolet to visible range; see Figure 3.2 and Figure 3.3. In comparison to high-pressure mercury lamps and xenon lamps, the radiant spectrum in the ultraviolet range, especially in the deep UV range from 300 nm downward, is higher in intensity. The mercury–xenon lamp also features instantaneous starting and restarting, which are difficult with conventional high-pressure mercury lamps; it is thus an excellent choice as an ultraviolet light source.

A deuterium (D_2) lamp is a broadband source of ultraviolet light that emits a continuous UV spectrum in the 160 to 400 nm range. The lamp consists of an anode and cathode positioned in a glass envelope filled with low-pressure deuterium gas. A small aperture centered in front of the anode constricts the flow of current. In front of this constriction, the arc is greatly intensified, producing the desired UV radiation. A deuterium lamp offers twice the stability of conventional lamps, and its typical operating life is over 4000 h. The high intensity and stability of deuterium lamps make them ideal ultraviolet light sources in a number of applications. Deuterium lamps are available from several manufacturers.

Metal halide lamps have a flash efficiency approximately four times higher than that of halogen and xenon lamps. In addition, the short-arc type is similar to a point light source, making optical design easy. As their color temperature characteristics are similar to daylight color, exact colors can be reproduced. Therefore, metal halide lamps are suitable in a number of imaging and machine vision applications where color is important.

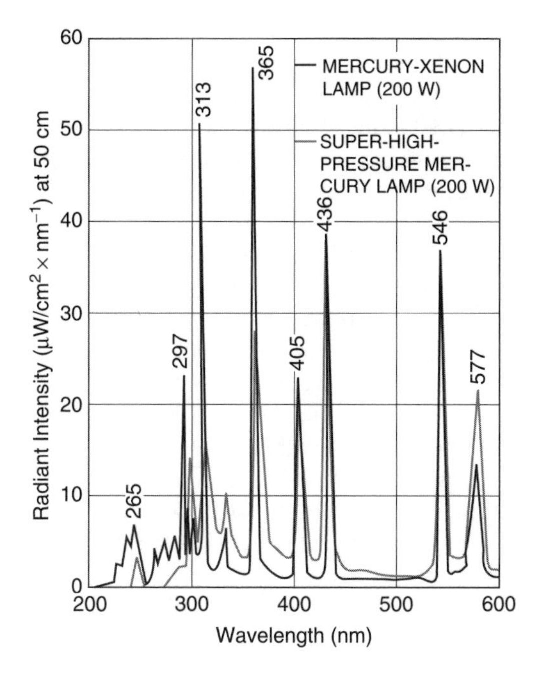

FIGURE 3.3 Spectral distribution with super-high-pressure mercury lamp (200 W). (© Hamamatsu Photonics K.K. With permission.)

3.3 SENSORS

3.3.1 Visible Image Sensors

3.3.1.1 Basic Properties

Charge coupled device (CCD) and complementary metal oxide semiconductor (CMOS) sensors are electronic devices that transform a light pattern (image) into an electronic signal. They consist of millions of individual light-sensitive elements that collect and store the pattern of energy deposited by photons impinging on the sensor. These devices have seen rapid recent development, especially in the case of CMOS sensors. Even for demanding applications such as holography, digital recording devices are now commonly used, and these technologies have forced traditional photographic methods aside. The development of these cameras has enabled a wide variety of new measurement methods, such as holographic PIV, which measures the three-dimensional spatial and temporal flow field [4]. For the purpose of understanding how a digital imaging sensor works, CCD and CMOS devices can be considered to be nearly identical. Here we will focus on CCDs, but the basic concepts also apply to CMOS sensors.

The basic mechanism of photoelectric sensors is similar to that of a solar energy cell that converts light directly into electricity. The CCD is built on a semiconductor substrate such as silicon, with metal conductors on the surface, an insulating oxide layer, an anode layer, and a cathode layer. The individual elements of the imaging device are called pixels (from "picture element"). When light strikes a pixel, some of the energy is transferred to the semiconductor. This energy promotes electrons from the valence band to the conduction band, where they become "free" to move through the semiconductor. The number of electrons in the conduction band depends on the intensity of the incident light. The charge carried by the electrons is stored by the capacitance associated with the pixel. The storage capacity is limited, but modern designs prevent the overflowing of charges to neighboring pixels. Without this feature, extra charge from saturated pixels would leak into neighboring pixels and produce "blooming" (distortion of the image in the form of bright streaks or spots). Finally, the amount of charge associated with each pixel is read out sequentially, digitized, and stored in computer memory. The digital value at each point in the array (corresponding to pixels on the CCD sensor) is the gray value of that pixel. Variations in gray value in computer memory correspond to the spatial variation of light in the original optical image. Once the digital image is in the computer, it may be processed as explained in Chapter 5.

The ratio between the maximum storage capacity and the dark current noise is the dynamic range of the sensor. Dark current arises from thermal energy within the silicon material in the CCD. Electrons that are promoted thermally instead of photoelectrically are still counted as signals, resulting in dark current. The dark current is a major source of electronic noise for the sensor. To reduce the dark current and improve the dynamic range, the sensor is often kept cool by Peltier elements. A dynamic range of 16 bits (65,535:1) has been achieved with cooled detectors. Another source of noise, read noise or shot noise, is a direct consequence of the charge-to-voltage conversion during the readout sequence.

Important characteristics of a pixel include fill factor (aperture), responsiveness, data rate, linearity, and spectral sensitivity. *Fill factor* is the ratio of the optically sensitive area to the entire area of a pixel. This factor is determined mainly by the amount of opaque area occupied by the readout electronics. However, applying an array of microlenses just above the sensor can enhance the fill factor and allow pixels to receive more photons.

The ratio of valid output signal to the total exposure under a certain illumination is defined as the *responsiveness* of a pixel; it depends on the quantum efficiency and the efficiency of the charge-to-voltage conversion. The *data rate* is the speed of the data transfer process, normally expressed in MHz. The output voltage for a pixel is theoretically proportional to the collected charge. Nonlinearity in the output signal results from overexposure or defects in the sensor, but it is typically less than 1 percent.

Sensitivity (or quantum efficiency) of a pixel is defined as the ratio of the number of collected electrons to the number of incident photons. Sensitivity depends on the sensor's design. Figure 3.4 summarizes examples of quantum efficiency of several types of CCD camera. Some CCDs can achieve high efficiencies (over 75%), but the maximum quantum efficiency of a typical CCD is about 50% at room temperature.

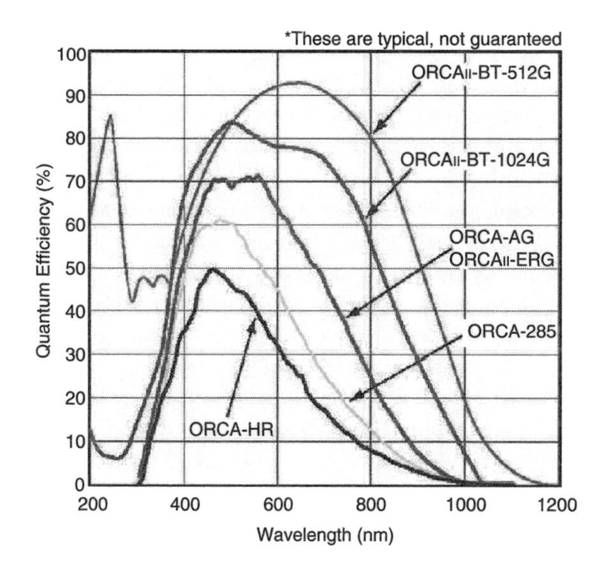

FIGURE 3.4 Examples of quantum efficiency and spectra sensitivity of CCDs manufactured by Hamamatsu Photonics K.K. (© Hamamatsu Photonics K.K. With permission.)

The *spectral sensitivity* is also an important property. Due to the band-gap of the semiconductor, a pixel can absorb only part of the electromagnetic spectrum. Some of the light that strikes a pixel does not have enough energy to promote electrons to the conduction band (i.e., the wavelength is too long). The minimum energy required to excite an electron from the valence band to the conduction band is called the band-gap energy. The band-gap energy for crystalline silicon is about 1.1 eV, which corresponds to a theoretical maximum detectable wavelength of 1.13 μm. Longer wavelengths will simply pass through the semiconductor as if it were transparent. In addition, light of too short a wavelength will not be detected. Light enters the CCD through a gate structure made of polysilicon. The gate structure is transparent at long wavelengths, but it is opaque for short wavelengths. The shorter the wavelength of incoming light, the shorter will be the penetration depth into the silicon. Therefore, light at short wavelengths is attenuated by the gate structure on the CCD.

CCD cameras are often equipped with an infrared filter in front of the sensor to reduce the susceptibility to infrared light. Other filters may also be applied to adjust the spectral sensitivity. As shown in Figure 3.4, the spectral sensitivity for green light (near a wavelength of 500 nm) is enhanced in commercial CCD cameras to correspond to the sensitivity of the human eye.

3.3.1.2 Color

Each pixel in a CCD sensor can only register the total intensity of the light that strikes its surface. It cannot sense color. A full-color imaging sensor uses filters

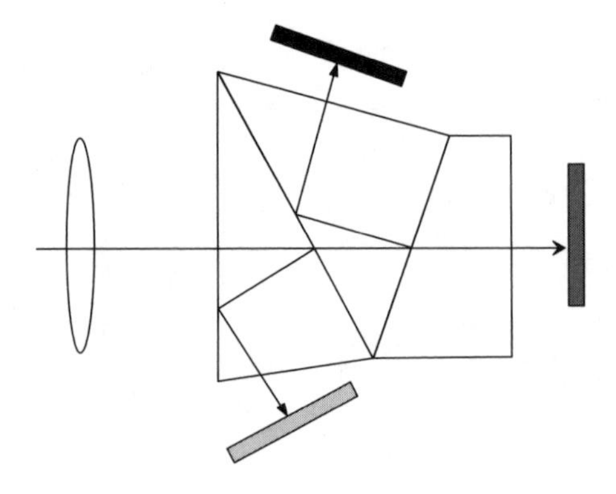

FIGURE 3.5 Arrangement of three CCDs with beam splitters and filters to obtain color images.

to register the light in its three primary colors. The recorded signals for each color can then be combined to produce a full-color image. There are several ways of recording the three colors.

The highest-quality cameras use three separate sensors, each with a different filter over it, as shown in Figure 3.5. Light is directed to the different sensors by a beam splitter in the camera. Each sensor responds only to the primary color of the filter, but all three sensors look at exactly the same view. The camera can record each of the three colors at each pixel location, giving a very high spatial resolution. This type of camera should be selected for measurement techniques based on color information such as thermochromic liquid crystal methods [5] or color-based three-dimensional PIV methods [6].

A second method is to rotate a series of red, blue, and green filters in front of a single sensor. The sensor records three separate images in rapid succession. This method also provides information on all three colors at each pixel location. However, since the three images are not taken at precisely the same moment, both the camera and the target of the photo must remain stationary for all three readings. This limitation is not practical for most industrial measurements.

A more economical and common way to record the three primary colors from a single image is to position a composite filter over the sensor, as shown in Figure 3.6. This filter divides the sensor into smaller areas composed of red-, blue-, and green-sensitive pixels. It is possible to gather information from vicinal pixels to reconstruct a full-color image at that location. The advantages of this method are that only one sensor is required, and all the color information is recorded at the same moment. The most common filter pattern is the Bayer pattern (shown in Figure 3.6) used in early handheld video cameras. This pattern alternates a row of red (colored black in the figure) and green (white) filters with a row of blue (gray) and green filters. Due to the green sensitivity of the eye, it is necessary to include extra green pixels to create an image that the eye will perceive as being true.

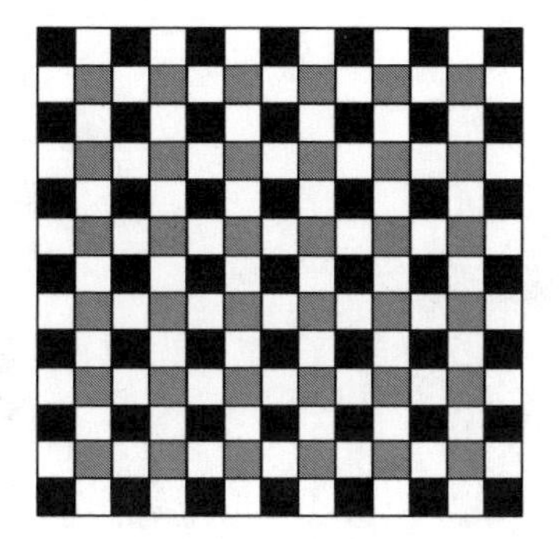

FIGURE 3.6 Bayer filter for color imaging.

3.3.1.3 Differences between CCD and CMOS Sensors

The electrical charges stored in each pixel of the sensor must be read out before they can be digitized. The major difference between a CCD-type and a CMOS-type sensor is the method for transferring and reading out the stored charges. A CCD sensor can be considered to contain a two-dimensional array of pixels linked together. In a CCD device, the charge from a given pixel is transported across the chip, passing from one pixel to another, and read at one corner of the array. On the other hand, in most CMOS devices, each pixel has its own buffer amplifier and can be addressed and read individually. Figure 3.7a shows a schematic structure of a CMOS sensor.

The advantage of CCD sensors is that they tend to have a larger fill factor than CMOS sensors. The exposure time of all pixels is uniform, and it is independent of the readout time.

The advantage of CMOS sensors is that they are extremely inexpensive and consume little power compared to CCDs. CMOS devices can be designed to be much simpler and smaller. The most remarkable advantage is that CMOS sensors can operate at very high speeds. The most recent commercial high-speed cameras with CMOS sensors can record 2000 images per second at a resolution of 4 million pixels with a dynamic range of 12 bits. Each pixel area is enlarged to counter the relatively low light sensitivity.

3.3.1.4 Architectures

CCD cameras use a variety of architectures. The most important ones, shown schematically in Figure 3.7, are described below.

3.3.1.4.1 Interline Transfer

A typical consumer digital camera uses what is called an interline transfer CCD, in which each pixel has both a photodetector and a charge storage area. The storage area is formed by shielding or masking part of the pixel from light and using it only for the charge transfer process (see Figure 3.7b). The masked areas

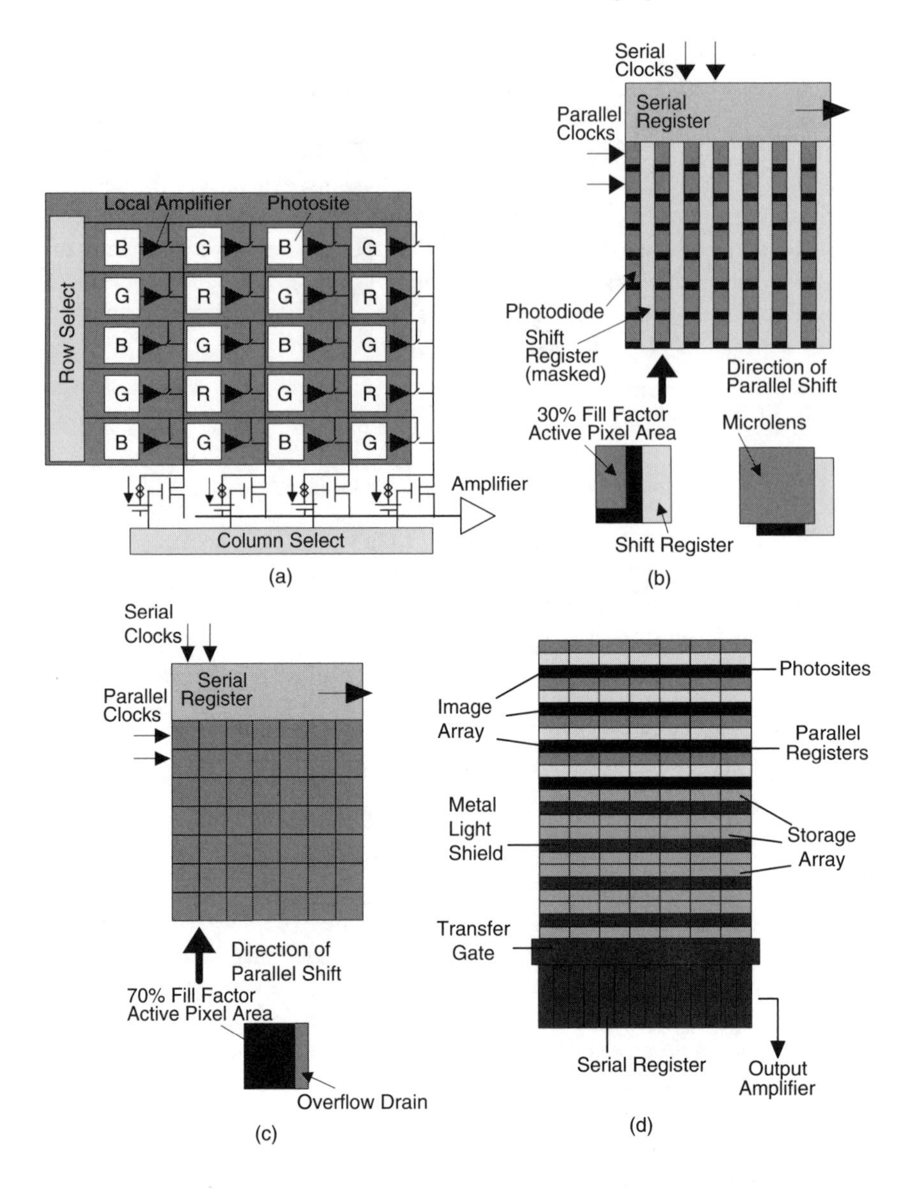

FIGURE 3.7 Schematics showing architecture of pixel structure. (a) CMOS sensor. (b) Interline transfer CCD. (c) Full-frame CCD. (d) Frame transfer CCD.

for each pixel form a vertical charge transfer channel that runs from the top of the array to the horizontal shift register. The area that remains exposed to light is called the *aperture.*

The interline design allows the pixel's electric charge to be quickly shifted to an adjacent masked storage area, where the charge can be moved row by row to the horizontal shift register. Since the signal is transferred in microseconds, image blurring is undetectable for typical exposures. In digital cameras, this quick availability of the pixel aperture to accept the next frame of image data is what enables the capture of video. Interline transfer CCDs can be controlled by software and do not require a mechanical shutter. The disadvantage of the design is that it exhibits relatively poor sensitivity to light, because a large portion of each pixel is covered by the extra electronics and is not photosensitive. To counter this small fill factor, high-quality interline transfer CCDs are equipped with microlenses to focus the light onto the photosensitive area of the pixel. This approach improves the fill factor from about 30% to over 70%.

3.3.1.4.2 Full Frame

Full-frame CCD sensors have the simplest architecture among CCDs and are preferred for professional use. The pixels are typically square (as shown in Figure 3.7c), so there is no image distortion inherent to the detector. The classical full-frame CCD design employs a single parallel register for photon exposure, charge integration, and charge transport. It does not have a shift register. Therefore, a mechanical shutter is required to control the exposure and block light from striking the CCD during the read process. The shutter is not necessary when the duration and amount of light are controlled externally, as with pulsed lasers or strobe lights.

In the full-frame format, the total area of the CCD is available for sensing incoming photons during the exposure period. Full-frame sensors have a high fill factor (around 70%) and do not require microlenses. The active pixel area of a full-frame sensor is 1.5 times that of an interline sensor of equal size. For some special designs, the fill factor can approach 100%.

3.3.1.4.3 Frame Transfer

The pixel architecture of the frame transfer sensor (see Figure 3.7d) is essentially equivalent to that of a full-frame sensor, but half of the array is masked out to provide temporary storage for the integrated electric charges prior to readout. The frame transfer CCD has its parallel register divided into two distinct areas, the storage array and the image array (where the image is focused). Typically, the storage array is identical in size to the image array. Once the integration period ends and the charges have been stored in each pixel, they are quickly transferred to the storage array, which operates without shutter delay. While the masked storage array is being read, the image array can integrate charges for the next image. The frame transfer sensor has two sets of parallel register clocks that independently shift charge on either the image or storage array. Therefore, a frame transfer sensor can operate continuously without a shutter at high frame rates.

The sensor can be used in conjunction with a mechanical shutter to acquire two images in rapid succession. This technique is one of the most interesting aspects of the frame transfer system, as it allows the capture of two images with an extremely short time delay. This double-exposure mode is used especially for PIV measurements, as explained in Section 3.5.2. It can also be used to acquire images at different excitation or emission wavelengths, as in certain biomedical experiments.

3.3.1.4.4 Interlaced and Progressive Scanning

Interline CCD sensors use one of two methods for reading data out: interlaced or progressive scan. Those selecting sensors for scientific use must carefully consider the differences between these two methods. These differences relate to the order in which the CCD columns of data are fed to the horizontal transfer register and off the sensor, similar to the scan modes of NTSC and PAL systems for video signals. Progressive CCDs read each line in the order in which it appears in the image. Interlaced CCDs first read the even lines, then read the odd lines, and reintegrate them later through image processing. The interlaced scan sensors cannot be applied to measurements when the duration and amount of light are controlled externally, such as by using flash lamps and lasers. They also cannot be used to observe time-dependent behaviors.

3.3.2 Image Intensifiers

3.3.2.1 Basic Principles

Image intensifiers were primarily developed for nighttime viewing and surveillance. They amplify low-light images into bright, high-contrast images. Image intensifier applications have spread from nighttime viewing to various fields including industrial product inspections and scientific research, especially when used in conjunction with CCD cameras (called intensified CCD).

The basic principle of operation is identical for all intensifiers. A low-intensity image is projected onto the transparent window of a vacuum tube, as shown in Figure 3.8. The vacuum side of this window carries a sensitive layer called the photocathode. Incident radiation causes the emission of electrons from the photocathode into the vacuum. The flux of electrons from a given spot on the photocathode is proportional to the incident light intensity at that location. These electrons are accelerated by an applied electric field that focuses them onto a luminescent phosphor screen situated opposite the photocathode. The phosphor

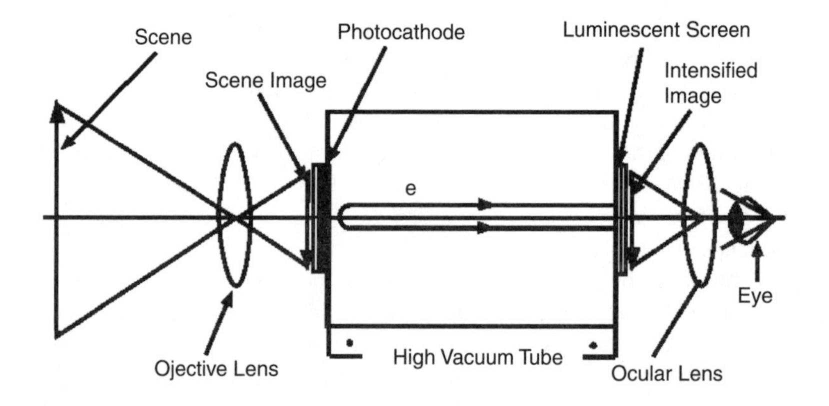

FIGURE 3.8 Basic principle of image intensification.

in turn converts high-energy electrons back to photons, which correspond to the distribution of the input image radiation but with a flux amplified many times. The image intensifier functions not only as an amplifier of light intensity but also as an image converter from an invisible to a visible spectral range.

3.3.2.2 Image Intensifier Designs

Image intensifiers are classified in three categories: first, second, and third generation. Each generation has specific advantages and disadvantages. First-generation image intensifier tubes feature high image resolutions, a wide dynamic range, and low noise. These tubes use a single potential difference to accelerate electrons from the cathode to the anode (screen). They possess moderate gain in the range of some hundreds of Lumens per Lumen (Lm/Lm). Focusing is achieved by placing the screen in close proximity to the photocathode or by using an electron lens to focus electrons originating from the photocathode onto the screen.

Second-generation tubes use electron multipliers so that not only the energy but also the number of electrons is significantly increased. Multiplication is achieved by the use of a device called a microchannel plate (MCP), consisting of an array of millions of thin glass tubes or channels bundled in parallel and sliced to form a disk. Each channel has an internal diameter of 6 to 25 mm and works as an independent electron multiplier. As Figure 3.9 illustrates, when a large voltage is applied across the ends of the MCP, a potential gradient builds up along the channel direction. When an incident electron strikes an inner wall on the input side, a number of secondary electrons are emitted. Each of these secondary electrons is accelerated by the potential gradient and collides with the opposing wall surface, causing another group of secondary electrons to be emitted. In this manner, the electrons collide repeatedly within the channel as they pass toward the output side. The result is a large multiplication of the incident electrons. The achievable image resolution and dynamic range of second-generation intensifiers are less than those of first-generation designs, but the gain in luminosity is significantly higher. The luminous gain ranges from 10^4 Lm/Lm for a single-stage MCP up to 10^7 Lm/Lm for intensifiers with two microchannel plates.

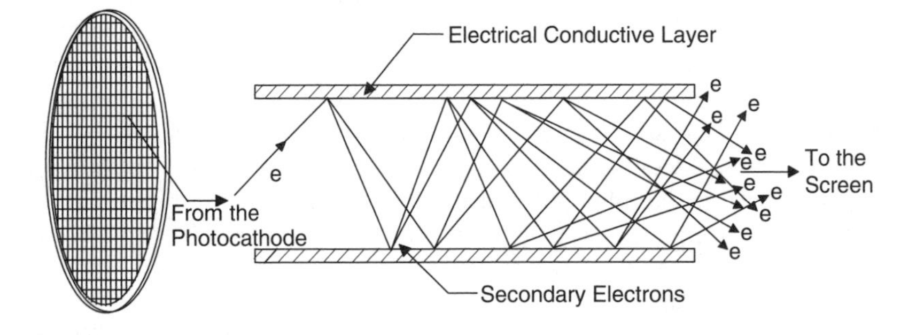

FIGURE 3.9 Microchannel plate (MCP).

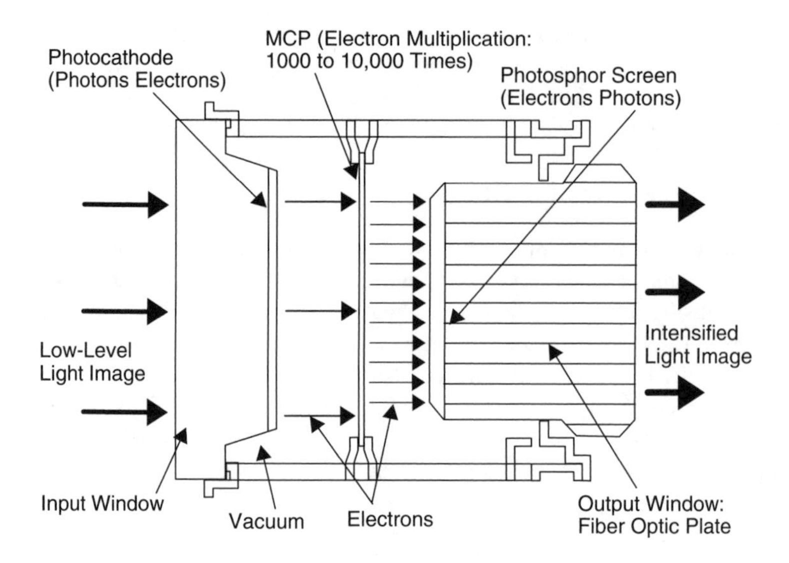

FIGURE 3.10 Structure of a typical third-generation image intensifier with proximity focus MCP.

Third-generation image intensifier tubes (see Figure 3.10) employ proximity focus MCP intensifiers with photocathodes made of semiconductor crystals such as GaAs (gallium arsenide) and GaAsP. These devices are very small, and their design allows gated operation (explained in the next section). The photocathodes offer extremely high sensitivities, typically four times those found in the multi-alkali photocathodes used in first- and second-generation intensifiers. In addition, the use of proximity focus (instead of an electron lens) provides an image with no geometric distortion, even at the periphery.

3.3.2.3 Gated Intensifiers

An image intensifier can be gated (turned on and off to provide an optical shutter) by varying the potential between the photocathode and the MCP input. Most photocathodes have a high electrical resistance and are not suitable for gate operation when used separately. To allow for gate operation at a photocathode, a low-resistance photocathode electrode, such as a metallic film, is usually deposited between the photocathode and the incident window. Gate operation can be performed by applying a high-speed voltage pulse to the low-resistance photocathode electrode. Metallic thick films or mesh-type electrodes may be provided rather than metallic thin films, since they offer an even lower surface resistance. To turn the gate operation on, a high-speed, negative-polarity pulse of about 200 V is applied to the photocathode while the MCP input potential is fixed. The pulse width is the gate time.

When the gate is on, the photocathode potential is lower than the MCP input potential, so the electrons emitted from the photocathode are attracted toward the

MCP and multiplied there. An intensified image can then be obtained on the phosphor screen. When the gate is off, the photocathode has a higher potential than the MCP input, so the electrons emitted from the photocathode are prevented from reaching the MCP by this reverse-biased potential. In the gate off mode, no output image appears on the phosphor screen even if light is incident on the photocathode.

The gate function is very effective for analyzing high-speed optical phenomena (e.g., high-speed moving objects, fluorescence lifetimes, bioluminescence and chemiluminescence images). Gated intensifiers and intensified CCDs incorporating this gate function are capable of capturing instantaneous images of high-speed optical phenomena while excluding extraneous signals.

3.3.2.4 Intensified CCDs

An intensified CCD camera is easily handled and very useful for capturing images in low light levels. Figure 3.11 shows an intensified CCD camera schematically. A CCD sensor is coupled with a proximity focused image intensifier by a bundle of optical fibers. The resolution of an intensified CCD depends on both the intensifier and the CCD. Since image intensifiers can be rapidly gated (turned off or on within a few nanoseconds), relatively bright objects can be visualized by a reduction in the exposure time. Intensifier gain may be rapidly and reproducibly changed to accommodate variations in scene brightness, thereby increasing the dynamic range. A gated, variable-gain intensified CCD camera is commercially available with a 12 order-of-magnitude dynamic range. Intensified CCD cameras have very fast response times (limited only by the time constant of the output phosphor), and often the CCD camera readout is the slowest step in image acquisition. Gated, intensified CCD cameras are required for most time-resolved

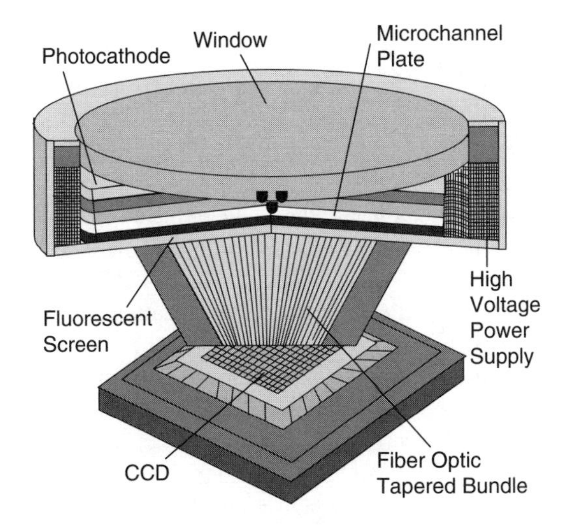

FIGURE 3.11 Schematic of intensified CCD camera.

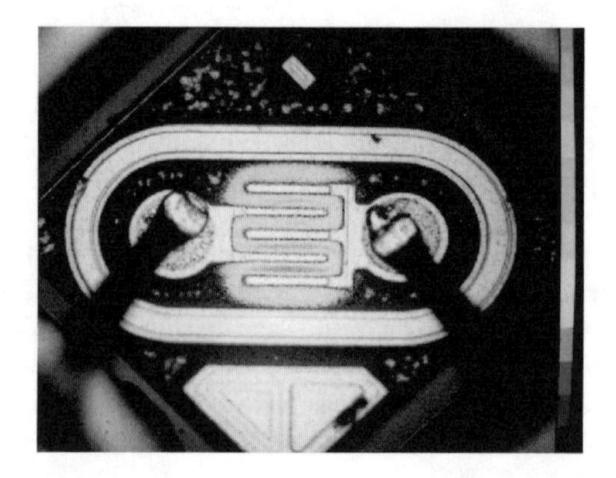

FIGURE 3.12 An example image captured by a hot electron analyzer. (© Hamamatsu Photonics K.K. With permission.)

fluorescence microscopy and spectroscopy applications because the detector must be turned on and off in nanoseconds or its gain rapidly modulated synchronously with the light source. In addition, simultaneous or near-simultaneous acquisition of two images at different excitation or emission wavelengths is required for ratio imaging, and intensified CCD cameras have the requisite speed and sensitivity.

Thermal noise from the photocathodes and electronic noise from the microchannel plates reduce the signal to noise (S/N) ratio in an intensified CCD camera. The contribution of these components to the noise created by the statistical nature of the photon flux depends on the gain of the device and the temperature of the photocathode. Generally the gain of the intensification stage is reduced to limit this noise, although intensified CCD cameras are available with a cooled photocathode.

Figure 3.12 is a sample of the type of images visualized by new high-sensitivity imaging devices. This figure is an image of the emission from recombination of minority carriers in a bipolar transistor. It was captured with the Phemos-50 (from Hamamatsu Photonics K.K.), a hot electron analyzer used as a high-resolution emission microscope. A cooled CCD camera is used to detect faint emissions caused by various phenomena inside the semiconductor device.

A hybrid of the image intensifier and the CCD camera is the recently introduced electron-bombarded CCD (EBCCD). The EBCCD is a high-sensitivity imaging device that employs the electron bombardment (EB) effect for image intensification. The EBCCD consists of a photocathode and a CCD chip arranged in parallel in a vacuum tube. Photons are detected by a photocathode, similar to an image intensifier. An optical image is converted at the photocathode into an electron image. The released electrons are accelerated across a gap and impact on the back side of a CCD. These energetic electrons generate multiple charges in the CCD, resulting in a high gain of up to 1200. Even a low-light image can be brought into view with a high S/N ratio. The advantages of this device over a cooled, slow-scan CCD are the additional gain and accompanying speed; the main disadvantages are a diminished dynamic range and a limited gain adjustment range. Compared to an intensified CCD, the EBCCD usually has a higher spatial resolution and a better S/N at moderate light levels.

3.3.2.5 X-Ray Image Intensifiers

X-ray image intensifiers are large-diameter imaging tubes that convert low-contrast x-ray images into visible light images. They are widely used in medical diagnostics and nondestructive inspection systems. For a significant improvement in the transmittance of x-rays at low energies (down to several keV), a beryllium (Be) input window is sometimes used. This type of intensified CCD is ideal for nondestructive inspection of light-element materials and radiation imaging at low-energy x-ray levels. A process imaging example (for polymer compounding) is described in Chapter 10.

A fiber optic plate is an optical plate composed of millions of glass fibers with diameters of several microns, bundled parallel to one another. Such plates are capable of transmitting an optical image from one surface to the other without image distortion. Fiber optic plates may be fabricated from x-ray scintillator material, so that radiation entering the plate is converted to visible light, which is channeled directly to the output plane of the plate. This design prevents loss of light and degradation of image quality. A fiber optic plate with x-ray scintillator offers high sensitivity and high resolution, and it allows real-time digital radiography when directly coupled to a commercially available CCD. The plate has excellent x-ray absorption characteristics, so that radiation that is able to penetrate the scintillator and directly enter the CCD is minimized. This feature protects the CCD from device deterioration caused by irradiation, assuring a long CCD service life.

3.3.3 Thermal Sensors

3.3.3.1 Pyrometers

All objects emit blackbody radiation, the spectral distribution of which is determined by temperature. The simplest and oldest noncontact method for estimating the temperature of a radiating body is to observe its color. Table 3.2 summarizes

TABLE 3.2
Relationship between Temperature and Color

Temperature (°C)	Color	Temperature (°C)	Color
550–650	Black or purple	830–880	Dark carmine
650–750	Purple	880–1050	Orange
750–780	Dark carmine	1050–1150	Orange yellow
780–800	Carmine	1150–1250	Yellow
800–830	Orange carmine	1250–1320	White yellow

the relationship between temperature and color. Using this method, experienced observers can estimate temperatures over about 700°C with a precision sufficient for simpler heat treatment processes. Pyrometers, infrared thermometers, and radiation thermometers are noncontact thermometers that measure the temperature of a body based on its emitted thermal radiation, thus extending the ability of the human eye to sense heat. No disturbance of the existing temperature field occurs with this noncontact method. The most important radiation wavelengths are from 0.4 to 20 μm in the visible and infrared radiation bands. A typical pyrometer is composed of an optical system concentrating the radiation onto a detector and the radiation detector itself, which consists of a thermal or photoelectric sensor, a signal converter, and a signal output channel or a display. Several types of optical pyrometers are available.

Thermal radiation detectors are heated by the incident radiation. The wavelength band is about 0.2 to 14 μm. This type of pyrometer can measure a wide range of temperatures (wavelength bands). The radiation from bodies is concentrated on a thermal radiation detector by an optical light guide, fiber, lens, or mirror system. Heating of the thermal detector by the concentrated incident thermal radiation gives a detector output signal in proportion to the detector's temperature and also to the measured body's temperature.

Photoelectric pyrometers are used in selected wavelength bands; the signal is generated by photons bombarding a photoelectric detector. Due to the thermal inertia of thermal radiation detectors, they are not well suited for the measurement of rapidly changing temperatures. The time constant of a thermal radiation detector (such as a bolometer) is about 1 ms, and that of a thermopile is much longer. The wavelength band of a photoelectric pyrometer depends on the spectral sensitivity of the photo-element and on the spectral transmission of the lens. Pyrometers with narrow working wavelength bands are called monochromatic pyrometers; others are called band pyrometers. Band (or multiwavelength) pyrometers are used for measuring the temperatures of bodies with low emissivity.

3.3.3.2 Detector Elements for Infrared Light

Generally, when light is directed on a material, the electrons in the material obtain energy from the light and become excited. The excited electrons in a semiconductor move freely when they exceed a minimum energy called the *band-gap* (the energy difference between the valence and conduction bands in a solid). A semiconductor-type infrared detector gathers these excited electrons to gauge the light intensity. The choice of detector material depends on the wavelength of the observed light, because the type of semiconductor determines the band-gap. Generally, the detector needs to be cooled to suppress electrical noise caused by thermal motion of the electrons inside the detector. Table 3.3 lists typical detectors and their wavelength ranges. In general, infrared detectors are of four principal types: photovoltaic (PV), photoconductor (PC), impurity band conductor (IBC), and bolometer (B).

TABLE 3.3
Types of Semiconductors for Infrared Detectors

Materials	Type	Wavelength Range
Si	PV	<1.1 μm
Ge	PV	<1.8 μm
HgCdTe	PV	0.9–2.5 μm
PtSi	SD	1–4 μm
InSb	PV	0.9–5.6 μm
Si:As	IBC	6–27 μm
Si:Sb	IBC	14–38 μm
Ge:Be	PC	30–50 μm
Ge:Ga	PC	40–200 μm
Ge or Si	BM	200–1000 μm

The photovoltaic detector measures the light intensity as a voltage by accumulating electron-hole pairs in the depletion layer at a junction between *n* and *p* types of intrinsic semiconductor. The PV detector is used as a near infrared detector. Generally, a reverse bias voltage (positive to *n* type, negative to *p* type) is applied before an exposure to increase sensitivity.

The photoconductor detector measures the light intensity as an electric current, and it is used to measure the far infrared region of the spectrum. PC detectors are made from extrinsic semiconductors (whose band-gaps have been shifted by doping the crystal lattice with impurities). A disadvantage of these detectors is that recombination of the electron-hole pairs reduces the quantum efficiency.

In the case of the impurity band conduction (IBC) detector, an insulator is added to the photoconductor detector to accumulate electric charges as a capacitor. The IBC is similar to the PC in terms of operating principle; however, the physical structure is similar to the PV. It is possible to produce a large number of elements, and the IBC is used as an intermediate infrared detector using extrinsic semiconductors.

The bolometer measures the temperature when an object cooled to cryogenic temperatures warms by the absorption of a photon. The quantum efficiency is 100%; however, the stability is not good because the bolometer must be maintained at a given cryogenic temperature with a very high accuracy. It is also extremely difficult to produce a sensor with large numbers of elements. The bolometer is mainly used for far-infrared astronomical telescopes and for submillimeter wave telescopes.

3.3.3.3 Thermal Cameras

Infrared thermal imaging of temperature fields has become an extremely versatile and popular method of real-time measurement in numerous industrial and research applications. The development of this technology has been rapid, and it is finding new applications in thermal measurement by offering the benefit of a two-dimensional view of temperature profiles. Example applications are discussed in Chapter 12.

Among the first commercially available infrared systems were the two-dimensional opto-mechanical scanning systems, which were appropriately called thermovision systems. In these systems, all of the target surface points are scanned sequentially with rotating or oscillating mirrors. Detectors of InSb, which are cooled either thermoelectrically or by liquid nitrogen, ensure a temperature resolution of 0.1°C and an accuracy of ±2%. The measurement range is from –20 to 1500°C. The resulting series of signals is transformed in the detector into electrical signals and displayed on a monitor screen as a visible image of the temperature field. The local emissivity is an important factor in determining the temperature at each point in the thermal image. The emissivity is defined as the ratio of radiant energy emitted by a surface to that emitted by a classical blackbody at the same temperature. This factor can be determined from a handbook or from a calibration.

Systems based on a two-dimensional array sensor are now the most popular ones, since they have no moving parts. They are available in two types. The first type includes one-photon detectors based on photovoltaic (InSb, HgCdTe) or photoconductive (PbSe, PbS) cells. The second type includes thermopiles or microbolometers, the operation of which is based on the absorption of thermal energy. The signals from individual detectors, corresponding to different pixels in the image, are scanned in a manner similar to that of CCDs. Detectors using monolithic PtSi microbolometers are now the most popular since they allow the integration of many detectors into one matrix. These microbolometers do not require cooling, a considerable benefit. Small variations in ambient temperature are monitored by temperature sensors inside the camera and used to correct the results.

3.3.3.4 Thermochromic Liquid Crystals

Encapsulated liquid crystals for use in heat transfer measurements or for visualization of flow patterns are available either as a water-based slurry (encapsulated liquid crystals dispersed in water) or as microcapsules containing powder. Thermochromic liquid crystals show a reversible color effect and respond to changes in ambient temperature. There are two types of thermochromic liquid crystals. The first, known as cholesteric liquid crystals, is made from cholesteryl esters. The second type, chiral nematic liquid crystals, is made from organic compounds. Both have a similar color play (amount of shift in color space), but chiral nematic liquid crystals give a sharper and more intense color play with a quicker response to a change in ambient temperature, relative to the cholesteric liquid crystals. The specific gravity of encapsulated liquid crystal capsules is typically about 1.01 g/cm^3. Table 3.4 summarizes the thermochromic liquid crystals.

TABLE 3.4
Thermochromic Liquid Crystals. R Denotes Cholesteric Liquid Crystals; K Denotes Chiral Nematic Liquid Crystals

Type	Product form	Capsule Membrane	Capsule Diameter, μm	Temp. Available, °C
R	powder	gelatin	20–30	−10 ~ +60
K	powder	gelatin	20–30	−20 ~ +100
K	slurry	urea formalin resin	20–30	−20 ~ +49 +50 ~ +100
R	water-based	gelatin	~20	−10 ~ +60
K	ink	gelatin	~15	−20 ~ +49 +50 ~ +100
R	LC-coated	gelatin	~20	−10 ~ +60
K	polyethylene sheet	gelatin	~15	−20 ~ +49 +50 ~ +100
R	LC-coated	gelatin	~20	−10 ~ +60
K	polyethylene sheet (laminated)	gelatin	~15	−20 ~ +49 +50 ~ +100

3.3.3.5 Fiber Optic Thermometers

Although fiber optic thermometers are not inherent imaging devices, they can be used in conjunction with process imaging applications (see, for example, Chapter 12). These devices use many of the same concepts found in thermal imaging systems. The most popular devices use an optical fiber to transmit light to a sensor, where it is modulated through some temperature-dependent interaction outside the optical transmission system. The modulated light is read by a remote readout system, where the change in light intensity, color, or phase shift is used to determine the local temperature at the sensor. Typical examples of such sensors include:

- GaAs semiconductors, whose absorption coefficient depends upon their temperature and the frequency of the incident radiation
- Thermochromic thermometers, which are based on the temperature dependence of the reflectance of liquid crystals
- Fluorescent thermometers, in which a fluorescent material is excited by pulsed radiation and the temperature is inferred from the decay time of the fluorescence signal
- Fabry–Perot sensors, which are based on the temperature-dependent spectral reflection coefficient of a thin monocrystalline silicon film

In the case of the thermochromic and fluorescent thermometers, it is easy to extend measurements to two dimensions by using a multifiber system and a CCD sensor.

Other thermometers are based on temperature-dependent changes in the optical properties of the optical fiber itself. They indicate the average temperature along the length of the fiber. Some are based on Raman scattering, while others are based on changes in the refractive index of the fiber.

3.4 OPTICAL COMPONENTS

3.4.1 Glass Materials for Optics

Most of the well-known glass types had already been developed by the 19th and early 20th century. Many of these are summarized in Table 3.5. The names of the glasses are based more on historical circumstances than on their chemical composition. The Schott company in Germany was a leading center in the creation of new glass types. Hence, many of these glass types originally had German names.

Glasses are often summarized in an Abbe number glass map (see Fig. 13), which shows the Abbe number versus the refractive index. The Abbe number (also

TABLE 3.5
Types of Glass

Type of Glass	Abbreviation
Fluorkron (FluorCrown)	FK
Phosphatkron (Phosphate Crown)	PK
Phospatschwerkron (Phosphate Dense Crown)	PSK
Borkron (Borosilicate)	BK
Kron (Crown)	K
Zinkkron (Zinc Crown)	ZK
Baritleichtkron (Light Barium Crown)	BaK
Schwerkron (Dense Crown)	SK
Schwerstkron (Extra Dense Barium Crown)	SSK
Lanthankron (Lanthanum Crown)	LaK
Lanthanschwerkron (Dense Lanthanum Crown)	LaSK
Kurzflint (Antimony Flint)	KzF
Kronflint (Crown Flint)	KF
Baritleichtflint (Light Barium Flint)	BaLF
Doppelleichtflint (Extra Light Flint)	LLF
Baritflint (Barium Flint)	BF
Leichtflint (Light Flint)	LF
Flint (Flint)	F
Baritschwerflint (Dense Barium Flint)	BaSF
Schwerflint (Dense Flint)	SF
Tiefflint	TF
Lanthanflint (Lanthanum Flint)	LaF
Lanthanschwerflint (Dense Lanthanum Flint)	LaSF

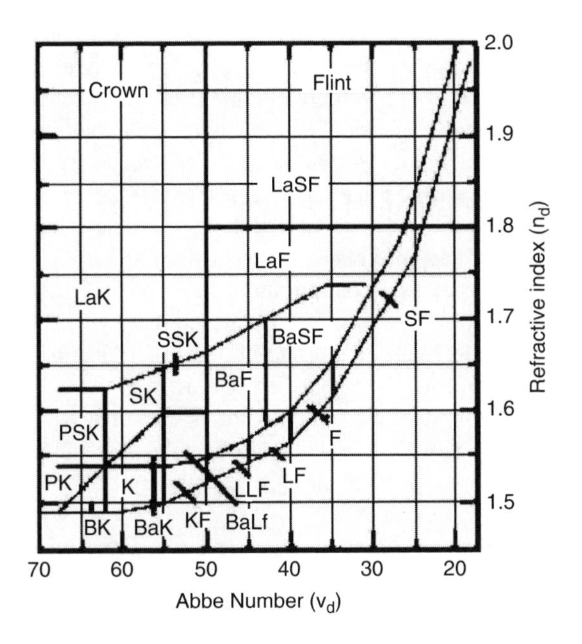

FIGURE 3.13 Abbe number glass map.

called V number) expresses the dispersion of an optical medium, which is directly related to the chromatic quality of a lens. Its value is given by the expression

$$V = (n_d - 1)/(n_F - n_c) \tag{3.1}$$

where n_d, n_F, and n_c are the refractive indices at the wavelengths of the Fraunhofer lines 587.6, 486.1, and 656.3 nm, respectively. The Abbe number of a glass mixture is carefully monitored and controlled by manufacturers to achieve target dispersion values.

More than 200 glass formulations exist, with refractive indices ranging from 1.46 to 1.97 and Abbe numbers from 20 to 85 (although the glass map in Figure 3.13 ends at $V = 70$). By using glass maps, optical designers can determine the range of Abbe number formulations available for a particular refractive index. In the subregions of the glass map, individual formulations have designations that end in K (referring to a crown glass) or F (Flint glass). The border between crown and flint glasses for the entire refractive index spectrum is at $V = 50$.

Only a fraction of all known glasses are employed in optical design on a routine basis. The choice is based on cost, ease of optical fabrication, thermal stability, and optical property requirements. The most important types of optical glass for lenses, fiber optics, and sight glasses are introduced below.

Pyrex®—Pyrex is a borosilicate glass with a low coefficient of thermal expansion. It is mainly used for nontransmissive optics, such as mirrors, due to its low homogeneity and high bubble content.

Sapphire—Artificial sapphire is frequently used as a sight glass in high-pressure vessels. The hardest of the oxide crystals, sapphire retains its high strength at high temperatures, and it has good thermal properties and excellent transparency. Its coefficient of thermal expansion is very high. It is chemically resistant to common acids and alkalis at temperatures up to 1000°C, as well as to HF below 300°C. These properties encourage its wide use in hostile environments where optical transmission in the range from visible to the near-infrared is required. The refractive index of sapphire glass is typically about 1.75.

BK7—BK7 is one of the most common borosilicate crown glasses used for visible and near-infrared optics. Its high homogeneity, low bubble content, and ease of manufacture make it a good choice for transmissive optics. The transmission range for BK7 is 380 to 2100 nm. It is not recommended for temperature-sensitive applications such as precision mirrors.

UV Grade Fused Silica—UV grade fused silica is synthetic amorphous silicon dioxide of extremely high purity. This noncrystalline, colorless silica glass combines a very low thermal expansion coefficient with good optical qualities and excellent transmittance in the ultraviolet. Transmission and homogeneity exceed those of crystalline quartz without the problems of orientation and temperature instability inherent in the crystalline form. Fused silica is used for both transmissive and reflective optics, especially where a high laser damage threshold is required.

Quartz Crystal—Quartz crystal is a positive uniaxial, birefringent single crystal grown using a hydrothermal process. It shows good transmission from vacuum UV to near infrared. Because of its birefringent nature, quartz crystal is commonly used for wave plates.

CaF_2—Calcium fluoride is a cubic single crystal material, grown using the vacuum Stockbarger technique, with good vacuum UV to infrared transmission. Calcium fluoride's excellent UV transmission, down to 170 nm, and nonbirefringent properties make it ideal for deep UV transmissive optics. Material for infrared use has been grown using naturally mined fluorite, at a much lower cost. CaF_2 is sensitive to thermal shock, so care must be taken during handling.

MgF_2—Magnesium fluoride is a positive birefringent crystal grown using the vacuum Stockbarger technique with good vacuum UV to infrared transmission. It is often oriented with the c axis parallel to the optical axis to reduce birefringent effects. High vacuum UV transmission (down to 150 nm) and its proven use in fluorine environments make it ideal for lenses, windows, and polarizers for excimer lasers. MgF_2 is resistant to thermal and mechanical shock.

3.4.2 Basic Optical Elements

3.4.2.1 Lens Design

Ray tracing (geometrical optics) is an important tool in the design of optical components for direct imaging systems. Figure 3.14a shows a basic example of an ideal thin lens and three light rays. Any two of these three rays fully determine

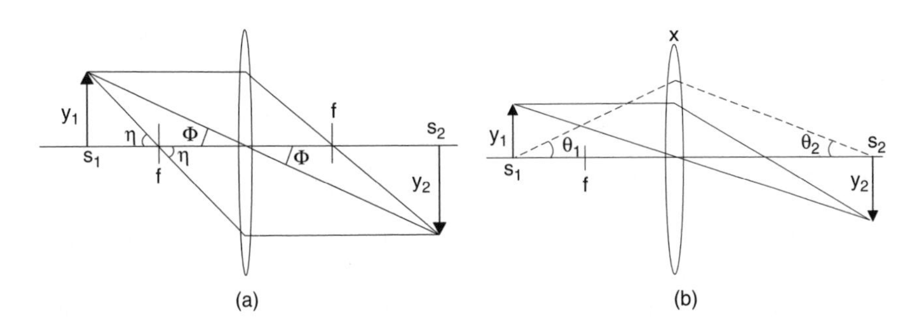

FIGURE 3.14 Optical ray tracing. (a) Derivation of lens equation. (b) An arbitrary ray to define the Lagrange invariant.

the size and position of the light (image). The figure here shows an object of height y_1 at a distance s_1 from an ideal thin lens of focal length f. The lens produces an image of height y_2 at a distance s_2 on the far side of the lens. One ray extends from the object parallel to the optical axis of the lens and is refracted by the lens through the optical axis at a distance f on the far side of the lens. A second ray passes through the optical axis at a distance f in front of the lens. This ray is then refracted into a path parallel to the optical axis on the far side of the lens. The third ray passes through the center of the lens. Since the surfaces of the lens are normal to the optical axis and the lens is very thin, the deflection of the ray is negligible as it passes through the lens.

Because we assume an ideal thin lens, aberrations and thick-lens effects can be neglected here. The thickness does not contribute to the focal length, and the lens only changes the angle of the ray passing through it.

The third ray mentioned above intersects the optical axis at an angle ϕ, and of course the opposite angles of two intersecting lines are equal. It follows from the congruency of the triangles formed in Figure 3.14a that

$$y_1/s_1 = y_2/s_2 \tag{3.2}$$

The magnification factor, M, is therefore defined as

$$M = y_2/y_1 = s_2/s_1 \tag{3.3}$$

If a particular magnification of the image is desired, there is only one position of any given lens that will satisfy the requirement.

The second ray, which goes through the front focus, intersects the optical axis at an angle of η. From the sine of this angle, we obtain

$$y_2/f = y_1/(s_1 f) \tag{3.4}$$

and

$$M = f/(s_1 f) \tag{3.5}$$

Substituting Equation 3.3 into Equation 3.5, the Gaussian lens equation is obtained:

$$1/f = 1/s_1 + 1/s_2 \quad (3.6)$$

This equation provides the fundamental relationship between the focal length of the lens and the size of the optical system. A specification of the required magnification and the Gaussian lens equation form a system of two equations with three unknowns: f, s_1, and s_2. When the focal length of the lens, f, or the distance from the object to image, $s_1 + s_2$, is given, all three variables are then fully determined.

Another important constant is the Lagrange invariant H. Figure 3.14b shows an arbitrary ray from the object or a point source. The ray has an angle θ_1 with the optical axis and passes through the lens inside of its clear aperture. This arbitrary ray goes through the lens at a distance x from the optical axis. In the paraxial approximation (where $\sin\theta \approx \theta$), we obtain

$$\theta_1 = x/s_1 \quad (3.7)$$

$$\theta_2 = x/s_2 = (x/s_1)(y_1/y_2) \quad (3.8)$$

and then

$$\mathrm{H} = (n_2)y_2\theta_2 = (n_1)y_1\theta_1 \quad (3.9)$$

Here, n_1 and n_2 are refractive indices in each side of the lens. This is a fundamental law of optics. H is constant in any optical system containing only lenses. It is valid for any number of lenses, as could be verified by tracing the ray through a series of lenses. Since the total flux collected by an optical system from a uniformly radiating source is proportional to H^2, its invariance is a consequence of conservation of energy.

3.4.2.2 Source Optics

Focusing a laser beam to a small spot is a simple but important requirement for direct imaging. Figure 3.15a shows an example where a laser beam with radius y_1 and divergence θ_1 is focused to a spot by a lens of focal length f. From the geometry it is evident that

$$\theta_2 = (y_1/f) \quad (3.10)$$

Substitution of the Lagrange invariant from Equation 3.9 leads to the conclusion that

$$y_2 = \theta_1 f \quad (3.11)$$

Equation 3.11 places a fundamental limitation on the minimum size of the focused spot. Because we have already assumed an aberration-free lens, diffraction will

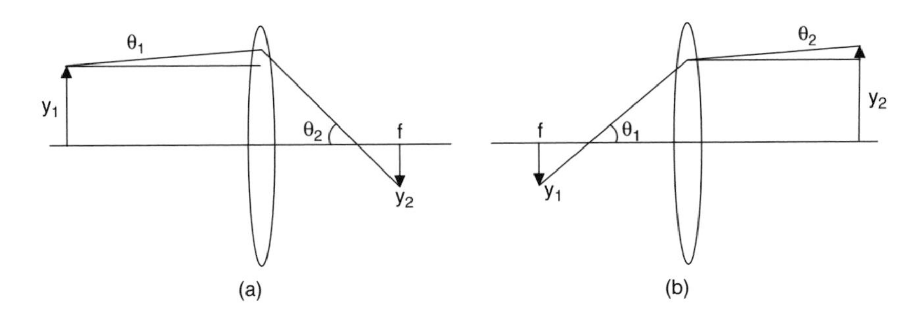

FIGURE 3.15 (a) Focusing a collimated beam. (b) Collimating light from a small source.

tend to broaden the spot to an even larger size. The only way to make the spot size smaller is to use a lens of shorter focal length or to expand the beam before it goes through the lens.

Similar limitations apply when collimating light from a point source (such as a light-emitting diode or fiber optic). When light from a small source of radius y_1 and a maximum aperture angle of θ_1 is collimated with a lens of focal length f (see Figure 3.15b), the radius of the collimated beam and its divergence angle become $y_2 = \theta_1 f$ and $\theta_2 = y_1/f$, respectively.

Many imaging applications, such as optical metrology, laser scanning, spectroscopy, PIV, and laser induced fluorescence (LIF), require a thin sheet of light rather than a narrow ray. Cylindrical lenses, which focus or expand light along one axis only, are often used in such light sources. In the case of lasers with sufficiently small beam diameter and divergence (e.g., HeNe), one cylindrical lens is sufficient to form a thin light sheet. For example, when a collimated laser beam of radius r_0 is incident upon a cylindrical plano-concave lens of focal length $-f$, as shown in Figure 3.16a, the laser beam will expand with a half-angle θ of r_0/f. The laser beam will appear to be expanding from a virtual source placed a distance f behind the lens. At a distance z ahead of the lens, a sheet will be formed with thickness $2r_0$ (ignoring expansion of the Gaussian beam) and a height

$$L = 2(r_0/f)(z + f) \tag{3.12}$$

For other lasers, however, a combination of different lenses is required to form a thin light sheet with less energy loss. One option for a configuration of lenses is shown in Figure 3.16b, where one lens is used for expanding the beam, the second one is used for making the beam sufficiently thin, and the third one is used to keep the height of the sheet constant. The position of minimum thickness is given by the beam divergence of the light source and the focal length of the right-side cylindrical lens.

The combination of a cylindrical lens with two telescopic lenses, as shown in Figure 3.16c, makes the system more versatile. The height of the light sheet

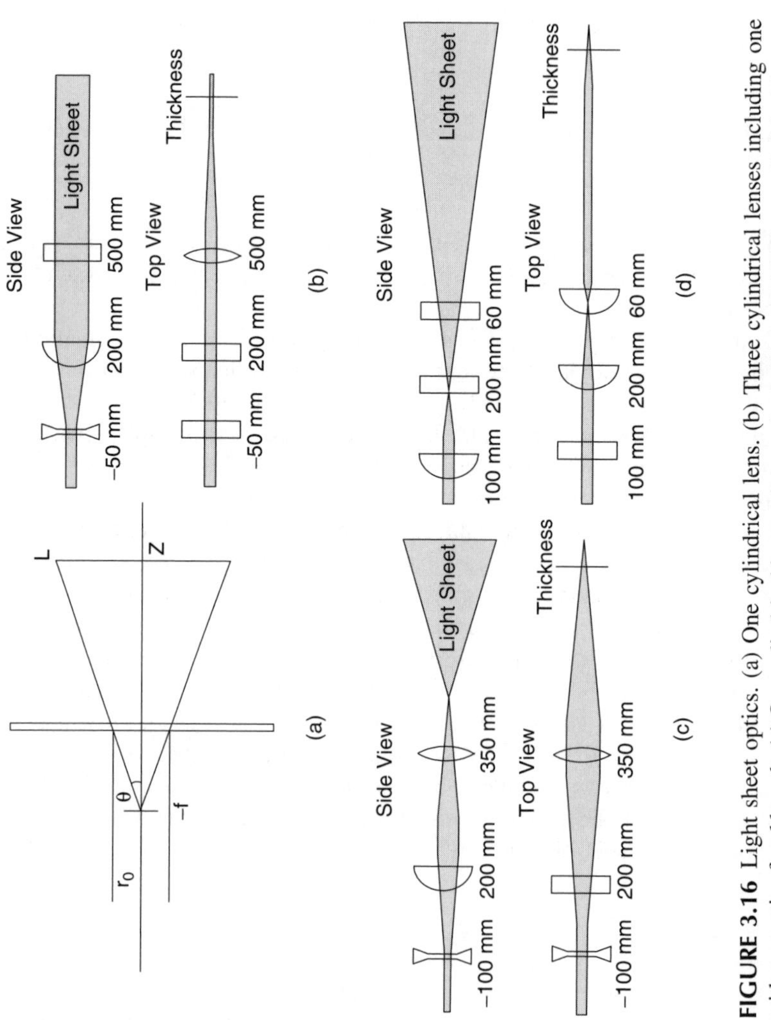

FIGURE 3.16 Light sheet optics. (a) One cylindrical lens. (b) Three cylindrical lenses including one with a negative focal length. (c) One cylindrical lens and two spherical lenses. (d) Three cylindrical lenses.

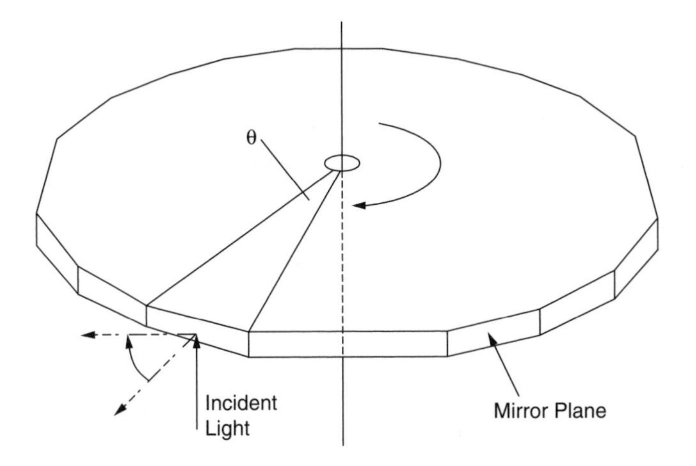

FIGURE 3.17 A polygon mirror used to sweep a light beam.

is mainly dictated by the focal length of the cylindrical lens in the middle and may be adjusted by changing the cylindrical lens. The thickness of the light sheet is adjusted by shifting the spherical lenses relative to each other. Note that the light sheet is not square or rectangular. Applying spherical lenses generally does not allow for the height and the thickness of a light sheet to be varied independently. The most useful configuration is shown in Figure 3.16d. This system generates a light sheet that is thinner than the laser beam diameter.

Light sheets can also be created with the beam sweep illumination method, where the laser scans the space with high speed. A motor-driven polygon mirror is typically used for scanning, although electro-optic or acousto-optic modulators are also widely used. The polygon mirror, shown in Figure 3.17, is made from aluminum or glass into a polyhedron with 8 to 48 mirrored surfaces. As the polygon rotates, a laser beam reflecting off the surfaces is scanned at an angular rate twice that of the rotation frequency. The scanning frequency increases with the number of surfaces, but the range of scanning angle decreases.

Galvano scanners are also frequently used to scan a laser beam. These devices use an electromagnetic drive to provide oscillatory motion of a plane mirror. The scanning angle is controlled by the amount of drive current, and the scan frequency can reach hundreds of Hz.

Two-dimensional scanning is achieved by using a combination of lenses, a polygon mirror, and a galvano scanner. In this case, the imaging speed and the time constant of the process must be considered. Typically, the polygon provides the high-frequency horizontal sweep while the galvano provides the vertical deflection.

3.4.2.3 Optical Fiber

Optical fibers are widely used in direct imaging and other optical measurement techniques, such as PIV and laser Doppler velocimetry (LDV). Single-mode

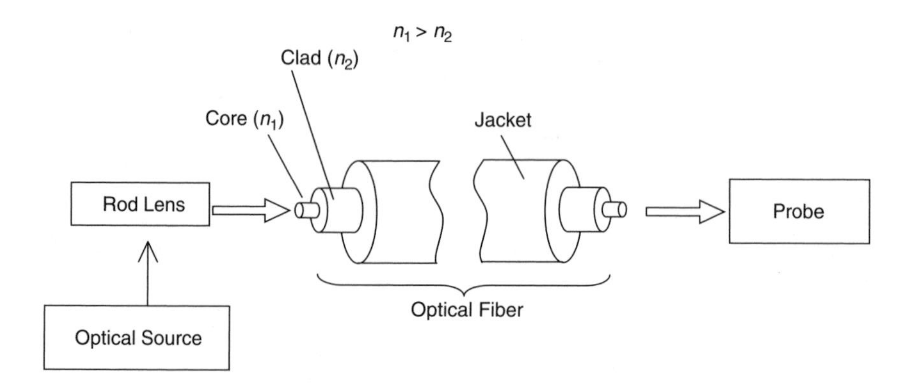

FIGURE 3.18 Structure of an optical fiber.

optical fibers typically consist of a dielectric core roughly 5 to 10 μm in diameter, cladding, and a jacket as shown in Figure 3.18 [7]. The refractive index of the core (n_1) is higher than that of the cladding (n_2). Because of the difference in refractive index, light rays in the fiber experience total internal reflection at the boundary and stay within the core. The jacket serves as protection to increase the mechanical strength of the fiber assembly. The propagation of light in optical fibers occurs in modes defined by solutions to the wave equation. For fibers with a step change in refractive index, it is found that each mode travels with a different group velocity. This condition leads to pulse broadening and waveform distortion, so these fibers are operated as single-mode waveguides to avoid modal dispersion [8].

Graded index fibers, in which the refractive index changes smoothly with radius from the center of the fiber, are also available. This design helps to decrease optical dispersion (distortion of the waveform), and it supports multiple propagation modes. The core diameter ranges from 50 to 100 μm for multimode fibers.

A typical LDV uses a single polarization, single-mode fiber for the transmitting fiber and a graded index multimode fiber for the receiving fiber. A multifiber bundle is widely used for a fiberscope camera. From a materials viewpoint, optical fibers can be made from quartz glass, fluoride glass, chalcogenide glass, and plastic as the dielectric core. Among these, quartz is popular because the transmission loss is very small.

3.5 APPLICATIONS

3.5.1 Particle Imaging Velocimetry

Particle imaging velocimetry is a planar measurement technique wherein a pulsed laser light sheet is used to illuminate a flow field seeded with tracer particles small enough to accurately follow the flow. The positions of the particles are recorded on digital CCD cameras at each instant the light sheet is pulsed. The data processing consists of determining either the average displacement of the particles over a small region of interest in the image or the individual particle displacements between pulses of the light sheet. Precise knowledge of the time intervals between light sheet pulses then permits computation of the flow velocity. In a typical PIV experiment, a pair of pulsed Nd:YAG lasers are used to provide a doubly pulsed light sheet illumination. The positions of particles entrained in the flow are recorded by a CCD camera, which is oriented at 90° to the plane of the light sheet. Depending on the type of CCD camera used and the particle concentration, either particle tracking or correlation processing can be used to produce the processed velocity vector map. In low particle concentration cases, so-called particle tracking velocimetry (PTV) can determine the individual particle displacements. In high particle concentration cases, correlation processing is usually the technique of choice. Particle tracking may be used after correlation processing to provide "super resolution" particle velocity maps.

Traditionally, both exposures are recorded on a single frame, creating a "double exposure." The double-exposed frame is then processed using auto-correlation techniques. However, this leads to a directional ambiguity because the double-exposed frame contains no information about which set of particle images was recorded from the first laser pulse and which from the second. Image shifting using a rotating or spinning mirror can be used to overcome this ambiguity, but it does increase experimental complexity.

A better alternative when using CCD cameras is to record each of the two exposures on a separate frame and to compute the cross-correlation of the two frames. Recording on separate frames preserves the time sequence of the pulses and avoids directional ambiguity. Also, cross-correlation processing provides an improved dynamic range for the velocity compared to auto-correlation of double exposures, because the time interval between two exposures depends on the separation time of double-pulsed lasers, which can be changed by the user. This technique of positioning the two laser pulses on sequential CCD frames is known as frame-straddling.

The cameras used for PIV are standard or high-resolution full-frame CCD imagers (or interline transfer progressive-scan CCDs) running at about 30 frames/sec. Full-frame CCD cameras, described in Section 3.3.1.4, have both a light-sensitive frame integration area and a frame storage area. After frame integration (light detection from the pulsed laser), the entire image is quickly shifted into the frame storage area. The image in the storage area is read out while a new frame is being collected in the light sensitive region of the CCD. The laser pulses are timed so two successive single-exposure image frames are obtained. The first laser pulse occurs just at the end of one video frame integration period, and the second laser pulse at the beginning of the next frame integration period; see Figure 3.19. The minimum time separation between laser pulses is limited by the camera's frame transfer period (the time it takes to shift a video frame to the camera's storage register). By using the frame-straddling technique, commercial PIV camera systems offer interframe exposure intervals of $^1/_{30}$ sec down to less than 1 μs. This means that cross-correlation imaging can be used

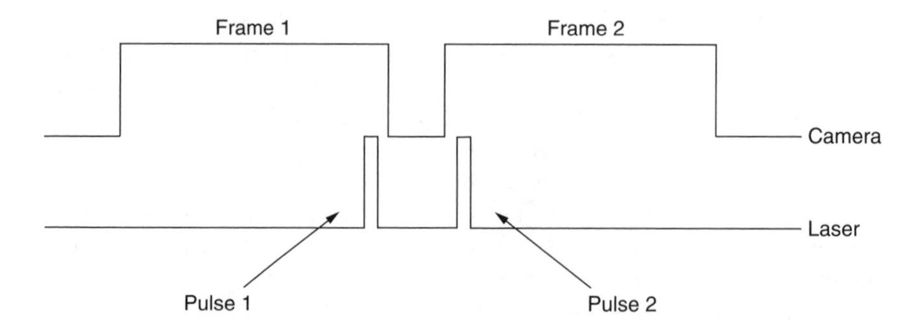

FIGURE 3.19 Timing of frame-straddling laser pulses used in PIV.

for a variety of applications, even when the flow velocity is several hundreds of meters per second.

3.5.2 Pressure Measurements

Pressure-sensitive paints (PSP) have been developed for directly visualizing pressure distributions on a surface in a high-speed flow of air [9–12]. This noncontact method is especially beneficial for the aerospace industry, significantly reducing development costs by eliminating the need for more complicated sensors and providing global images of aerodynamic data.

These paints are composed of a chemical dye and a special polymer binder [13, 14]. The chemical dye is a porphyrin or metal complex with photochemical features. Such dyes emit luminescence at specific wavelengths. The pressure sensitivity of these dyes arises from the characteristic that the luminescence is quenched by the adsorption of oxygen when exposed to air. The degree of luminescence loss is directly related to the partial pressure of oxygen, allowing the air pressure on a painted surface to be visualized in terms of the dimness of luminescence at any point.

In the conventional form of a PSP, the pressure-sensitive dye is distributed evenly in a binder. The polymer binder functions to immobilize the chemicals on the solid surface and is hyperpermeable to allow the full partial pressure of oxygen to act on the dye. The oxygen partial pressure is then related to a pressure value by Henry's law. Though the oxygen permeability and hence luminescence intensity vary with the type of binder employed, this composition is useful for painting on solid surfaces. The choice of an appropriate binder is a key factor in the performance of PSPs. The binder is an intermediate substance in this system, and it complicates the mechanism of dye quenching by the adsorption of oxygen molecules. Recently, a number of binderless systems, called porous PSP systems, have been proposed for a faster response due to direct exposure of the dye. The dye molecules are trapped in a porous structure formed from silica gel. Thin-layer chromatography PSP is an example of a porous technique [12].

Luminescence lifetimes may be measured instead of intensity [15, 16], although it remains necessary to verify the lifetime characteristics of the proposed system. The relationship between the concentration of oxygen and the lifetime or intensity of luminescence is represented by the Stern–Volmer equation:

$$I_0/I = t_0/t = 1 + K_{SV}[O_2] \tag{3.13}$$

where I_0 (t_0) and I (t) are luminescence intensity (triplet lifetime) in the absence and in the presence of oxygen, respectively; K_{SV} is the Stern–Volmer quenching constant; and $[O_2]$ is the concentration of oxygen. Intensity-based PSP techniques require the intensity to be measured against a reference image captured using the same exact measurement setting and illumination conditions. Much more information is available on the Web site of the Molecular Sensors for Aero-thermodynamIC Research (Mosaic) project available at http://www.ista.jaxa.jp/fluid/eng/mosaic/index.html.

Abe et al. applied this concept to make pressure-sensitive particles for use in PIV and measured the oxygen density distribution and velocity distribution simultaneously [17]. Their PIV–PSP hybrid system provides a visualization of oxygen motion, which could prove useful in many industrial situations. An example of an important application is in the analysis of fire outbreak mechanisms, in which the behavior of oxygen plays a crucial role. This hybrid system is in essence a lifetime-based sensing technique. However, the authors used a CCD camera to obtain a global image, in contrast to point measurements using a photomultiplier as in conventional lifetime measurement techniques. They selected particles with a porous microstructure. The dye used was a ruthenium complex, tris(2,2'-bipyridine) ruthenium(II) chloride, or $[Ru(bpy)_3^{2+}]Cl_2$. The ruthenium complex $[Ru(bpy)_3^{2+}]$ is a well-investigated dye and a well-known photosensitizer in organic and inorganic chemistry, with many applications in PSPs [13, 14]. Figure 3.20 shows an example of the luminescence decay of $[Ru(bpy)_3^{2+}]$-based dye, which is excited by a pulsed YAG laser with a 532-nm wavelength. The luminescence intensity reaches an immediate peak and then begins to decay due to oxygen quenching. When the oxygen fraction is increased, the lifetime becomes shorter.

Abe et al. captured two images by using a cooled CCD with double-shutter operation. In luminescence lifetime measurements, the peak intensity before the luminescence starts to decay is considered to be affected only by the intensity of the exciting light and by the density of dye particles, independent of oxygen concentration. The first image captured is therefore the reference image, representing the distribution of the excitation intensity and the particle density. On the other hand, the intensity after the onset of luminescence decay is a function of oxygen concentration. The second image (just 400 ns after the first one) has a slightly longer exposure time and shows the intensity after onset of decay. The second image contains information about oxygen density, and the ratio between these two images gives the correct distribution of oxygen concentration.

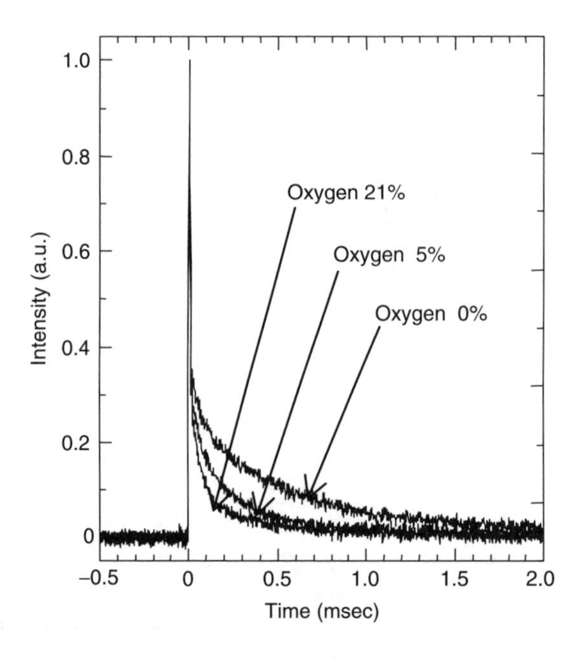

FIGURE 3.20 The luminescence decay of $[Ru(bpy)_3^{2+}]$-based dye.

3.5.3 Measurement of pH and Temperature by Fluorescence

The laser induced fluorescence technique is a useful and powerful tool for measuring a distribution of chemical compatibility (typically pH) or temperature. Generally, applications of LIF fall into two categories. One is the measurement of the pH or temperature field in which a molecular probe, usually a fluorescent dye, is added in the system in advance. This type of approach is discussed below. The second category is the measurement of radical ions produced in a combustion field without any addition of dye. Applications of this nature are discussed in Chapter 8.

A noncontact measurement method such as LIF is appropriate for studying dynamic behavior, especially in systems where it is difficult to insert a contact probe or to seed relatively large particles. Many kinds of molecular probes are readily available, and LIF is becoming more useful in wider applications, including measurement of thermo-physical properties. LIF directly yields temporal and spatial distribution of scalars in the flow field, as well as velocity distributions. By adding a few different dyes one can determine the distribution of several properties simultaneously, which is a major advantage.

On the other hand, a few problems are associated with the LIF technique, aside from the addition of impurities. These include high sensitivity to fluctuations in excitation intensity and dye concentration, and a susceptibility to quenching. If a very stable excitation light source is used under good conditions, fluctuations are not an issue. In many cases, however, it is difficult to avoid instabilities in intensity,

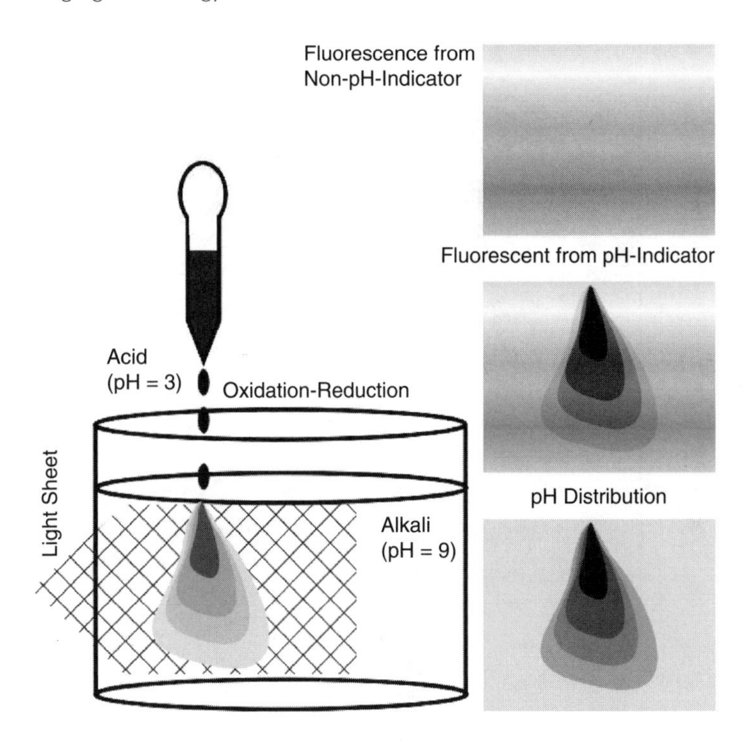

FIGURE 3.21 The DeLIF technique.

irregular reflection, and refraction of incident laser light, especially when using a pulsed laser. These noise sources decrease the reliability of measurement [18]. The dual emission (two-color) LIF (DeLIF) technique is one way to cancel these effects [19–21]. In DeLIF, the fluorescence from two kinds of dye is simultaneously measured. One dye is an indicator of the target scalar, while the other is not sensitive to the target scalar but is affected by the same noise sources as the indicator dye. Therefore, the ratio between the intensities of the two dyes gives a highly reliable value of the scalar distribution, as Figure 3.21 shows schematically. Using this technique, Sakakibara and Adrian obtained precise temperature distributions, with a random error of 0.17 K as a standard deviation [20].

The problem of quenching is more difficult to overcome. Quenching is caused by oxygen infiltration into the ambient structure and subsequent thermal deterioration of the dye. This process is not well understood at present, and it generally requires an empirical correction. If the quenching properties of a dye are known in advance, that effect could be used in the measurement.

3.5.3.1 Dependency on pH and Temperature

In the case of a pH-dependent dye, the molar absorptivity ε depends on the pH. In the case of a temperature-dependent dye, the quantum efficiency Φ depends

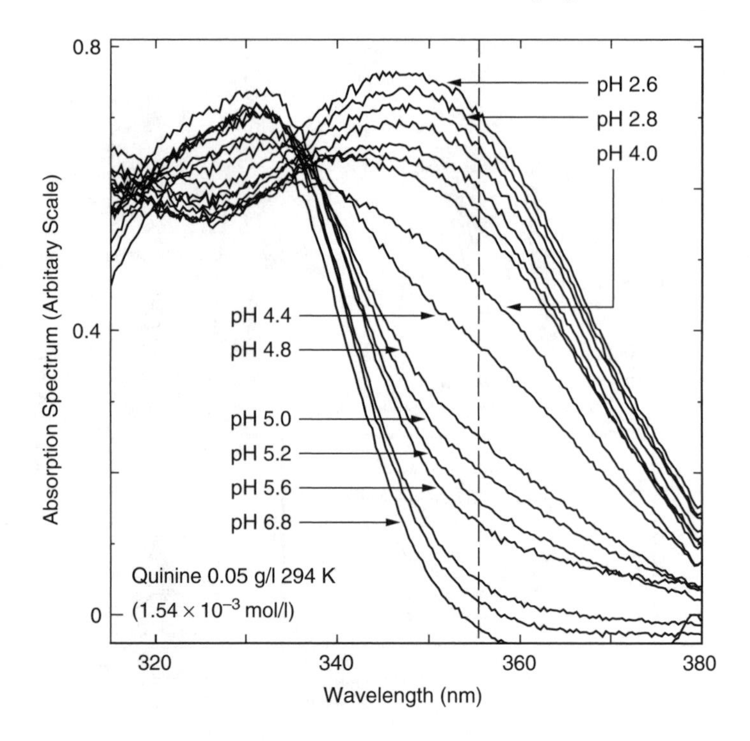

FIGURE 3.22 Absorption spectra of quinine, a pH indicator.

on temperature. The fluorescence intensity at each point, $I(pH, T)$, is expressed by the following equation, where I_0 is the excitation intensity, C is the molar concentration of dye, l is the length of the sampling region along the path of the incident light, and F is a fraction of the light coming into the camera:

$$I(pH, T) = I_0 e^{-\varepsilon(pH)lC} \Phi(T) \varepsilon(pH) CF \tag{3.14}$$

The dependency on oxygen concentration can be treated similarly to the dependency on temperature. Therefore, the ratio between the values of $I(pH, T)$ for each dye is not affected by the excitation intensity for constant molar concentration of dyes and gives reliable results. Note that while pH affects the absorptivity, quenching affects the emissivity. Figure 3.22 shows an example of absorption spectra of the pH-indicator quinine for different pH conditions at constant temperature (294 K). Coppeta and Rogers have summarized absorption and emission spectra for many kinds of dye [19].

3.5.3.2 Spectral Conflicts

As indicated by Coppeta and Rogers [19], there are spectral conflicts which must be minimized for the DeLIF technique to be effective. One is an overlap between emission bands of each dye within the measured spectral bands. Another is an

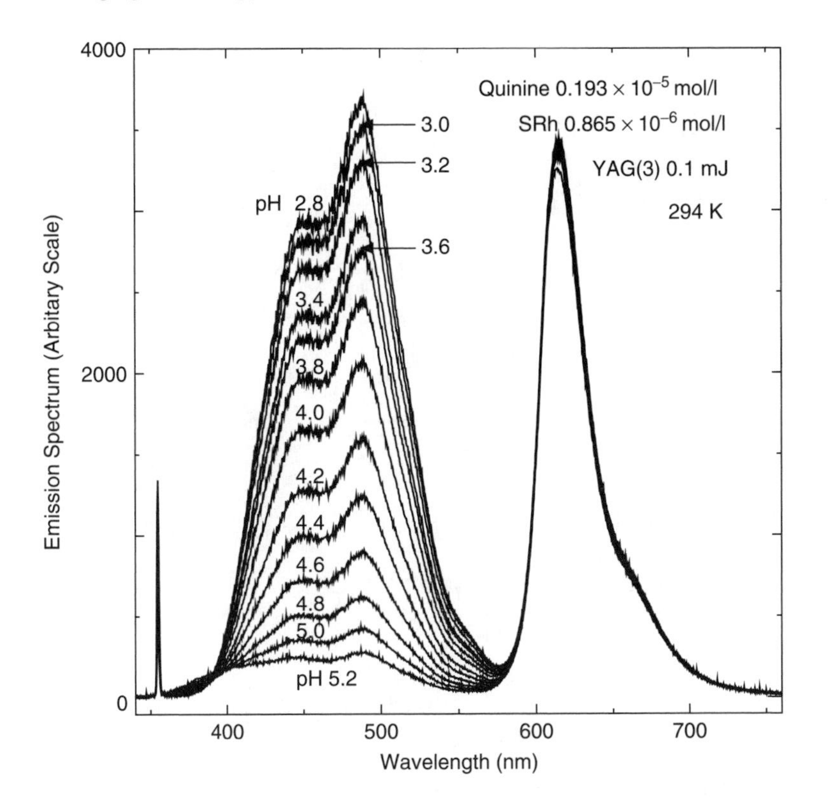

FIGURE 3.23 Emission spectra of mixed dye solution.

overlap between the emission band of one dye (A) and the absorption band of another dye (B). When B is not affected by the target scalar, the measured fluorescence of A is attenuated, corresponding to the path length from the measured point to the outside, along the optical axis of the camera lens. When B is affected by the target scalar, pH, this combination of dyes should not be selected.

Figure 3.23 shows an example of emission spectra of mixed dye solutions for different pH conditions at a constant temperature (294 K), where the excitation wavelength is 355 nm. The pH-indicator dye is quinine and the other, non-pH-indicator dye is Sulforhodamine 101 (SRh). The solvent of the required pH is a buffer solution composed of $KHC_8H_4O_4$ and NaOH. The fluorescence intensity from quinine (at wavelengths under 570 nm) was affected by pH, but the intensity from SRh (at wavelengths over 580 nm) was independent of pH at least from 5.2 to 2.8 of pH. It seems relatively easy to measure the fluorescence from each dye separately (due to the large peak shift) by using a dichroic mirror.

3.5.3.3 Measurement Example

This section describes an example of measurement using the DeLIF technique. Someya et al. [21] applied the DeLIF technique to the measurement of dissolution

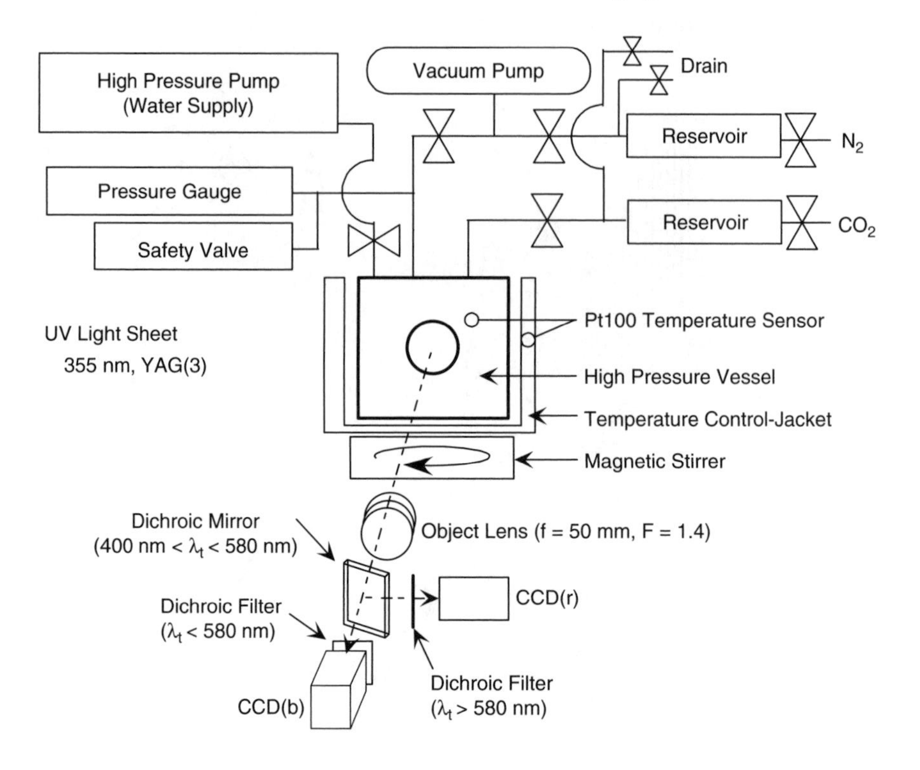

FIGURE 3.24 Experimental apparatus for DeLIF measurements of Someya et al. (From Someya, S., Bando, S., Sakakibara, J., Okamoto, K., and Nishio, M., *Proc. of FLUCOME '03*, Sorrento, Italy, 2003.)

behavior of liquid CO_2 droplets and CO_2 solubility into pure water at 9.81 MPa and 288 K. In such a high-pressure condition, it is difficult to measure the dissolution behavior and the solubility by other methods. The experimental apparatus is schematically shown in Figure 3.24. The optical system was composed of a light source and two cameras. A light sheet generated by cylindrical lenses from the third-order harmonic of a Nd:YAG laser (355 nm, 30 Hz, 0.1~0.2 mJ/pulse) was used as the excitation light. The thickness of the light sheet was less than 0.5 mm. In the receiving optics, the fluorescence was collected by a lens (f = 50 mm, F1.4) and split by a dichroic mirror that transmitted wavelengths from 400 to 580 nm at an incident angle of 45°. The transmitted and reflected fluorescences were refiltered by dichroic filters and simultaneously recorded by each CCD camera. The positions of the CCD cameras were adjusted to capture images of approximately the same spot just below the CO_2 injection nozzle. It was not necessary for the cameras to be positioned exactly since the physical and image coordinates were calibrated by using a regular grid pattern to determine a third-order polynomial function relating the two image spaces. This procedure compensated for minor misalignment of the cameras and the distortion due to refraction along the optical path of fluorescence from the target to the cameras.

FIGURE 3.25 Measured pH distribution (at 288°K and 9.81 MPa).

A pH distribution as shown in Figure 3.25 was obtained. The pH distribution shows the dissolution behavior of liquid CO_2, and the pH at the interface (at the edge of the concentration boundary layer) gives the solubility. The measured pH value was 3.12, and the estimated value of solubility was 0.0286 mole fraction. This estimation required only the ion product [22] and the dissociation constants of CO_2 [23, 24]. The measured value is close to the CO_2 solubility (0.0283 mole fraction) as measured by a conventional method, which needs 24 hours or more for the acquisition of only one data point [25]. The DeLIF technique is thus very useful for measuring dynamic behavior of chemical reactions and some kinds of thermo-physical properties.

3.5.4 Micro Shear-Stress Imaging Chip

3.5.4.1 Basic Structure and Calibration

Not all direct-imaging applications are based on optical sensors. For example, Figure 3.26 shows a micro shear-stress imaging chip, which is composed of multiple thermal-type sensors [26]. The imaging chip has three rows of microsensors, which contain an array of 25 sensors each. Figure 3.27 shows the top view and a cross-section of the micro shear-stress sensor. Each microsensor consists of a 150-μm long, 3-μm wide, and 0.45-μm thick polysilicon resistor wire and a 1.2-μm thick and 200×200 μm² silicon nitride diaphragm that seals off a 2-μm deep vacuum cavity underneath. The cavity is necessary to reduce heat transfer from the resistor wire to the substrate and to increase the sensitivity of the sensor. The sensors are connected to external constant temperature mode circuits through gold bonding wires, which are used in a hot-wire anemometer driven at a 1.1 overheat ratio. The output from the anemometer circuits is digitized by a 64-channel Keithley Metrabyte ADC board in a Pentium-based PC. The sensitivity of the shear-stress sensor is about 1.0 V/Pa for a gain of 10. The heating power of a

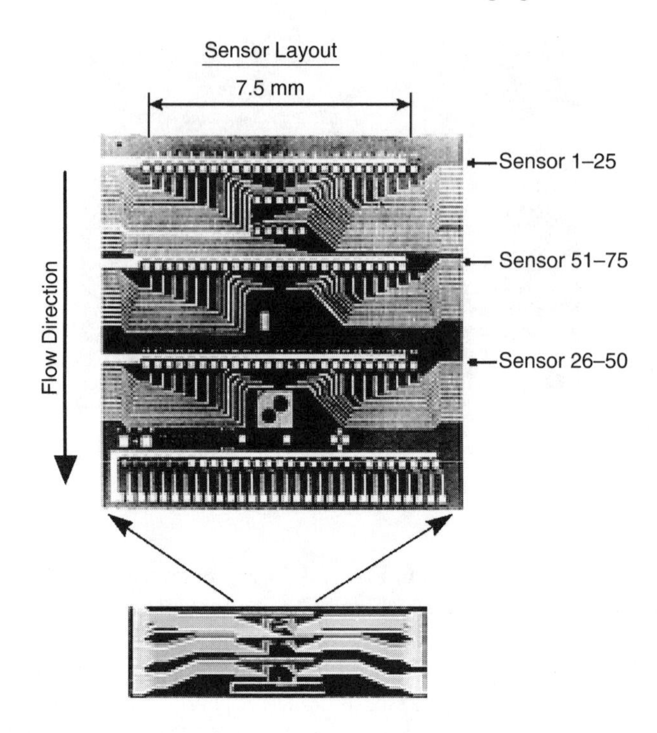

FIGURE 3.26 A surface shear-stress imaging chip.

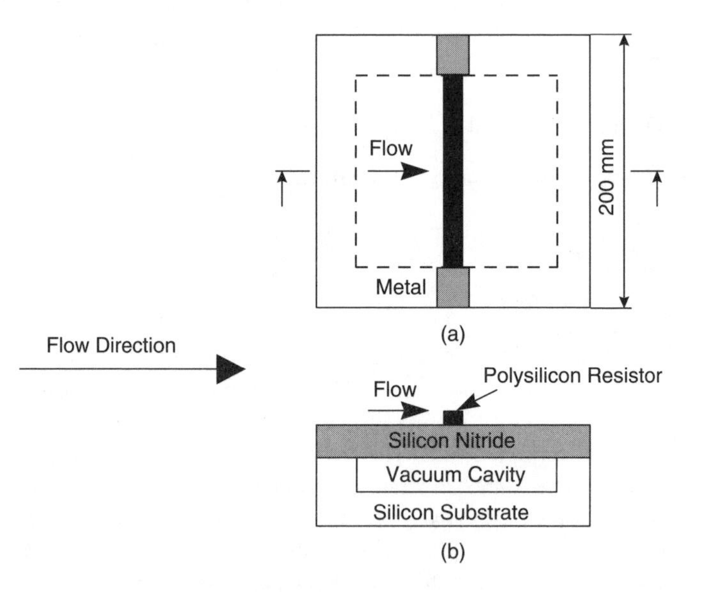

FIGURE 3.27 Schematic showing (a) top and (b) cross-sectional views of the micro shear-stress sensor.

shear-stress sensor operating in steady state can be correlated with wall shear stress τ as follows [27]:

$$i^2R^2 = (T_f T)(A + B\tau^{1/3}) \tag{3.15}$$

Here T_f and T are the temperatures of the heated sensor and the measured fluid, R is the resistance of the sensor, i is the heating current through the sensor, and A and B are calibration constants.

To correlate the output voltage with the wall shear stress τ, the sensor must be calibrated. In a fully developed turbulent flow, τ is related to the stream-wise pressure gradient:

$$dP_x/dx = -\tau/h \tag{3.16}$$

where P_x is the local pressure, x is the stream-wise coordinate, and h is the half-height of the wind channel. The pressure drop and output voltage of the sensor are measured at different center velocities of the channel ranging from 8 to 20 m/s. If the temperature of the measured fluid is constant, the wall shear stress can be directly related to the output voltage E_0 by a sixth-order polynomial with the data

$$t = a_0 + a_1E_0 + \cdots + a_6E_0^6 \tag{3.17}$$

where a_0, a_1, a_2, ..., a_6 are calibration constants. These constants are calibrated in the channel flow in a downstream region where the turbulent flow is fully developed. This chip senses the stream-wise pressure gradient and calculates the corresponding wall shear stress.

3.5.4.2 Measurement Example

Figure 3.28 shows contours of two-dimensional shear-stress distributions measured by the shear-stress imaging chip [28]. The horizontal axis covers a distance of 7.5 mm transverse to the flow area, and the vertical axis represents 51.2 ms of time for three different Reynolds numbers. The experimental conditions are as follows. The channel, constructed of 13-mm Plexiglas, is 610 mm × 25.4 mm in cross-section and 4880 mm long. An axial blower generates the air flow in the channel. The imaging chip was flush mounted into the wall at 4267 mm from the channel inlet where a fully developed turbulent channel flow exists. The Re number (defined by hu_∞/v, where h is the half width of the channel and u_∞ is the centerline velocity) ranged from 6,960 to 17,400. Each shear stress is normalized by the mean shear stress and root mean square shear stress for each Reynolds number, as shown in Figure 3.28. The gray levels in the images represent the amount of shear stress. Figure 3.29 shows the original data (before normalization) in a vertical direction at the point where the maximum value appears. The streaks are narrower and packed more densely as the Re number increases. They also

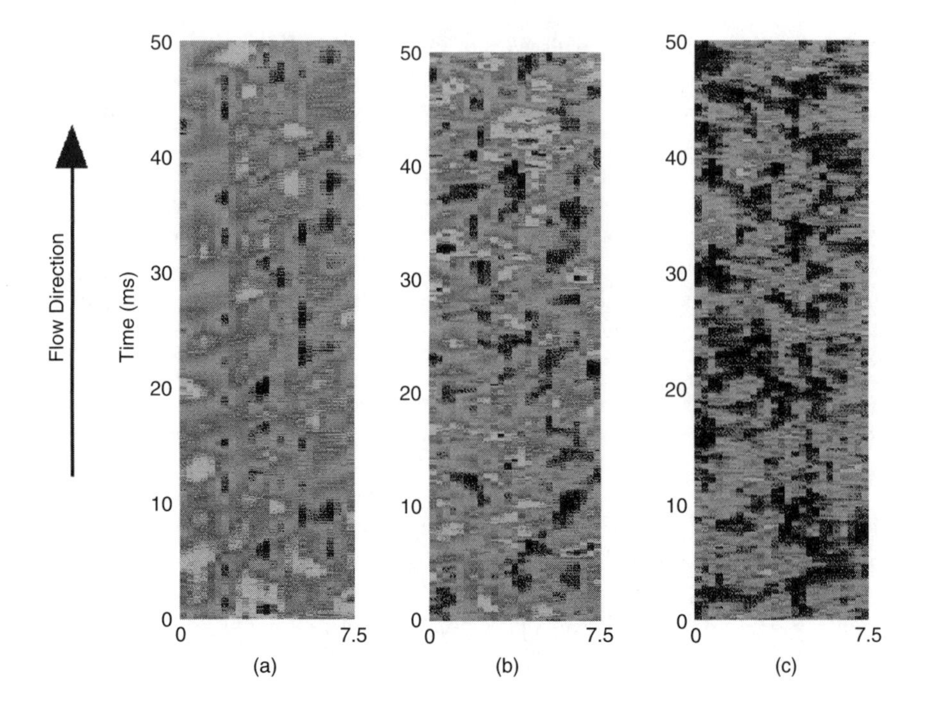

FIGURE 3.28 Shear stress measured with the imaging chip. (a) Re = 6,960. (b) Re = 12,180. (c) Re = 17,400.

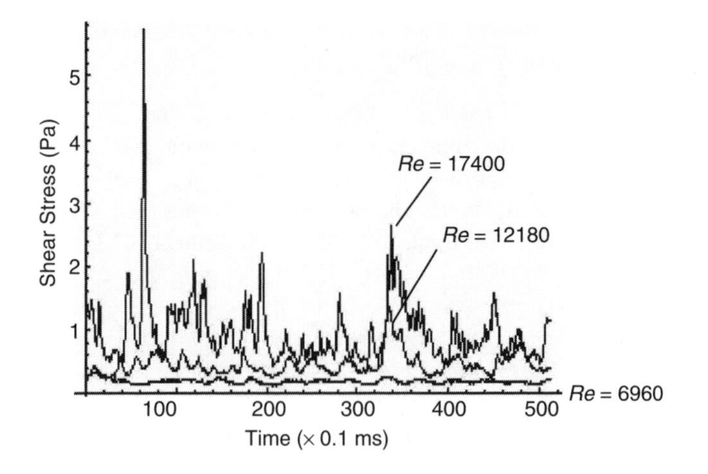

FIGURE 3.29 One-dimensional data from stress imaging chip.

appear at shorter time intervals as the Re number increases. The shear-stress fluctuation near the wall increases rapidly, resulting in a steep peak in a turbulence boundary layer, because a velocity gradient rapidly increases due to the influence of the eddy guided by bursting events.

3.6 MACHINE VISION

Machine vision techniques are used for determining the relative position of objects and precise three-dimensional shape measurement. Shape determination techniques are classified as either active systems or passive systems. J.Y. Zheng and A. Murata have proposed an active system that incorporates rotating the object and illuminating it with fixed ring-shaped light sources [29]. Their system can handle specularly reflecting surfaces. However, the applicability of any active system is limited since these systems tend to be complicated and large.

Passive stereovision systems, on the other hand, are relatively free from limits in usage and applicability. They do, however, require very precise calibration of the cameras. In stereo matching, the disparity or parallax r is usually measured as the displacement between corresponding points of two images. As Figure 3.30 shows, the disparity is related to the distance z by the equation

$$z = fR/r \tag{3.18}$$

where R and f are the baseline and the focal length, respectively. This equation indicates that for a given distance, the baseline length R acts as a magnification

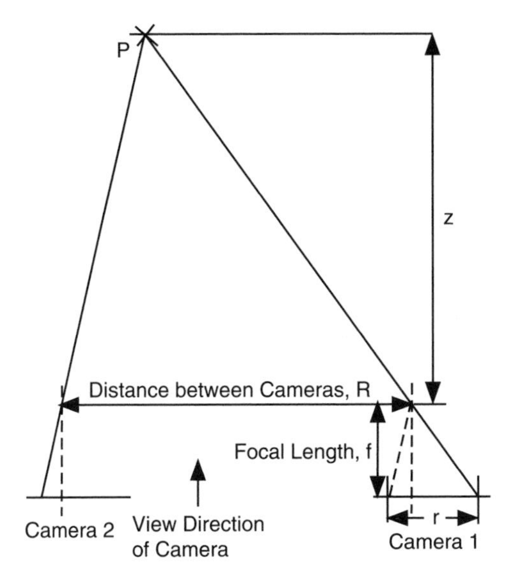

FIGURE 3.30 Basic geometry of stereo vision.

factor for measuring r to obtain z. In other words, the estimated distance is more precise if we set the two cameras farther apart from each other, which means a longer baseline.

A multiple-vision system needs a longer disparity range, though a trade-off must be made between precision (disparity) and correctness in pattern matching. Current three-dimensional shape measurement techniques used in computer vision cannot deal with specular surfaces successfully. Even laser range finders are not able to get correct shape information from a surface with strong specular reflectance. Only magnetic range finders can get the information from specular surfaces. Passive systems also need high textures at the surface and similar brightness (no shadow).

Another major problem with stereovision systems is associated with occlusion. Occlusion is a critical and difficult phenomenon to be dealt with by stereo algorithms. M. Okutomi and T. Kanade adopted a multiple baseline stereo system to clear the trade-off problem by using many pairs of stereo images with varying amounts of disparity [30]. C.L. Zitnick and T. Kanade proposed a stereo algorithm for obtaining disparity maps with occlusions explicitly detected, with assumptions of uniqueness and continuity [31].

More recently, three-dimensional images have been measured using stereo systems with an active camera system as proposed for the multibaseline system [30]. Figure 3.31 schematically shows an example of a three-dimensional measurement system [32]. A mirror is set in front of the camera at an angle of 45°. Another mirror is set parallel to the first mirror, separated at an interval R. This periscope is rotated coaxially with the fixed camera. The angle of another mirror is arranged according to the object. A number of pairs of stereo images are

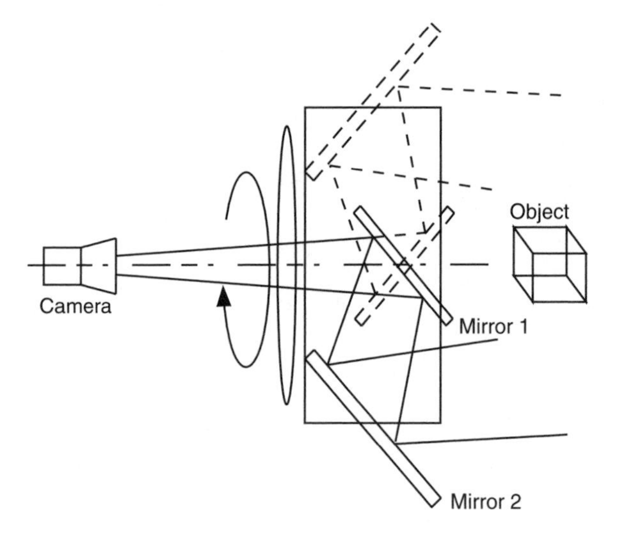

FIGURE 3.31 Schematic of three-dimensional shape measurement system.

captured at intervals of small angles (e.g., 2°). The rotating camera can also capture equivalent images, while the periscope-type mirror is used for downsizing and stabilizing the system. The distance, parallel to the axis of rotation of the periscope, from the focal point of the camera system to the corresponding point at the object surface is inversely proportional to the disparity. In this system, the baseline R is the radius of the periscope rotation or the interval between the mirrors. The disparity, r, can be calculated from the radius of the circular streak, which is obtained by tracing the corresponding points on each image.

REFERENCES

1. W Koechner. *Solid-State Laser Engineering.* Springer Series in Optical Sciences, Vol. 1, 3rd ed. Heidelberg: Springer Verlag, 1992.
2. E Kapon. *Semiconductor Lasers I: Fundamentals (Optics and Photonics).* New York: Academic Press, 1998.
3. LA Coldren, SW Corzine. *Diode Lasers and Photonic Integrated Circuits.* New York: Wiley-Interscience, 1995.
4. K Ikeda, K Okamoto, H Madarame. Measurement of three-dimensional density and velocity distributions using HPIV (evaluation of simultaneous recording and filtering extraction). *JSME Int. J., Ser. B* 43:155–161, 2000.
5. I Kimura, T Hyodo, M Ozawa. Temperature and velocity measurement of 3d thermal flow using thermo-sensitive liquid crystals. *J. Visualization* 1:145–152, 1998.
6. C Brucker. 3-D PIV via spatial correlation in a color-coded light-sheet. *Experiments in Fluids* 12:312–314, 1996.
7. NS Kapany. *Optical Fibers, Principles and Applications.* New York: Academic Press, 1967.
8. A Yariv. *Optical Electronics,* 3rd ed. New York: Holt, Rinehart and Winston, 1985, pp. 57–83.
9. JI Peterson, VF Fitzgerald. New technique of surface flow visualization based on oxygen quenching of fluorescence. *Rev. Sci. Instrum.* 51:133–136, 1980.
10. T Liu, BT Campbell, SP Burns, JP Sullivan. Temperature and pressure-sensitive luminescent paints in aerodynamics. *Appl. Mech. Rev.* 50:227–246, 1997.
11. B McLachlan, JH Bell. Pressure-sensitive paint in aerodynamic testing. *Experimental Thermal and Fluid Science* 10:470–485, 1995.
12. AE Baron, JDS Danielson, M Gouterman, JR Wan, B McLachlan. Submillisecond response times of oxygen-quenched luminescent coatings. *Rev. Sci. Instrum.* 64:3394–3402, 1993.
13. JN Demas. Luminescence spectroscopy and bimolecular quenching. *J. Chem. Edu.* 52:667–679, 1975.
14. B Durham, JV Casper, JK Nagle, TJ Meyer. Photochemistry of $Ru(bpy)_3^{2+}$. *J. Am. Chem. Soc.* 104:4803–4810, 1982.
15. JN Demas. Luminescence decay times and bimolecular quenching. *J. Chem. Edu.* 53:657–663, 1976.
16. JW Holmes. Analysis of radiometric, lifetime and fluorescent lifetime imaging for pressure sensitive paint. *Aeronautical J.* 2306:189–194, 1998.

17. S Abe, K Okamoto, H Madarame. Paper No. 3227, *5th International Symposium on Particle Image Velocimetry,* Busan, Korea, September 22–24, 2003.
18. S Someya, M Nishio, B Chen, H Akiyama, K Okamoto. Visualization of the dissolution behavior of a CO_2 droplet into seawater using LIF. *J. Flow Visualization Image Process.* 6:243–259, 1999.
19. J Coppeta, C Rogers. Dual emission laser induced fluorescence for direct planar scalar behavior measurements. *Experiments in Fluids* 25:1–15, 1998.
20. J Sakakibara, RJ Adrian. Whole field measurement of temperature in water using two-color laser induced fluorescence. *Experiments in Fluids* 26:7–15, 1999.
21. S Someya, S Bando, J Sakakibara, K Okamoto, M Nishio. DeLIF measurement of pH distribution induced by CO_2 dissolution in high pressure vessel. *Proc. FLUCOME '03,* Sorrento, Italy, 2003.
22. WL Marshall, EU Franck. Ion product of water substance, 0-1000C, 1-10000 bars new international formulation and its background. *J. Phys. Chem. Ref. Data* 10:295–304, 1981.
23. RE Zeebe, D Wolf-Gladrow. *CO_2 in Seawater: Equilibrium, Kinetics, Isotopes.* Amsterdam: Elsevier Science, 2003, p. 55.
24. K Saruhashi. In: S Horibe et al., eds., *Kaisui-no-Kagaku.* Tokyo: Tokai Univ. Press, 1970 (*in Japanese*), Sec. 3.2.
25. S Someya et al. CO_2 solubility into water with clathrate hydrate. *National Conf. of JSME,* Tokushima, Japan, 2003.
26. JB Huang, S Tung, CH Ho, C Liu, YC Tai. Improved micro thermal shear-stress sensor. *IEEE Transactions on Instrumentation and Measurement* 45:570, 1996.
27. HH Bruun. *Hot-Wire Anemometry: Principles and Signal Analysis.* Oxford: Oxford University Press, 1995, pp. 272–286.
28. M Kimura, M Takei, CM Ho, Y Saito, K Horii. Visualization of shear stress with micro imaging chip and discrete wavelets transform. *J. Fluid Eng. ASME* 124:1018–1024, 2002.
29. JY Zheng, A Murata. Acquiring a complete 3D model from specular motion under the illumination of circular-shaped light sources. *IEEE Trans. PAMI* 22:913–920, 2000.
30. M Okutomi, T Kanade. A multiple-baseline stereo. *IEEE Trans. PAMI* 15:353–363, 1993.
31. CL Zitnick, T Kanade. A cooperative algorithm for stereo matching and occlusion detection. *IEEE Trans. PAMI* 22:675–684, 2000.
32. S Someya, K Okamoto, G Tanaka. 3-D shape reconstruction from synthetic images captured by a rotating periscope system with a single focal direction. *J. Visualization* 6:155–164, 2003.

4 Process Tomography

Brian S. Hoyle, Hugh McCann, and David M. Scott

CONTENTS

4.1 INTRODUCTION

The challenge of the efficient design and flexible operation of many industrial manufacturing processes is coupled closely to the availability of high-quality information concerning their *actual* internal state. Many process systems are incompletely understood or are designed for operation under specific conditions or with particular grades of raw materials. Process upsets occur when unexpected variations in the quality of feed stock or other critical parameters cause the process to operate outside its specified range. The end result is typically a product of lower quality (and therefore of lower value), lower productivity, or at worst a process that no longer produces useful product. Real-time information about the internal state of the industrial process enables control schemes that adapt to changing conditions. In addition, such information allows a more flexible use of the process to meet new markets or tighter environmental standards.

Process imaging from a generic standpoint can monitor an element or facet of a process if a convenient portal is available. For example, the light from a flame that is the key element in a combustion process can be monitored with a simple optical fiber. Where such portals can be provided to expose critical facets of a process, a vision-based process imaging system is possible. Where a more pervasive and general view of the process is required, an approach based on multiple process sensors can be employed. Current-generation process sensors typically provide single-point indication of a process parameter (for example, a thermocouple that estimates the temperature at a point). Obviously, these sensors are placed at carefully selected points to provide general information. However, when operating conditions change, perhaps away from initial norms, single-point information is likely to be reduced in value or may even be misleading.

Generally, it is better to have knowledge of the internal state as a two-dimensional (2D) cross-sectional distribution of the key parameter of interest, or in terms of the three-dimensional (3D) distribution. When the need for process information justifies the cost, an array of sensors can provide the information needed to reconstruct the 2D or 3D distribution of the parameter of interest, which is generally shown as an image. The estimation of such internal distributions within industrial processes is called *process tomography* (PT).

This chapter reviews the background to this technology, its major operating modes, and some examples of current applications. References in general cite either key articles that first describe an innovation or review articles that offer a comprehensive set of further references. A large collection of literature now exists in the general field, and the references are intended to assist readers to access more focused information. Readers seeking a general introduction will find a textbook on this topic of interest [1].

4.1.1 Tomography and Tomographic Reconstruction

Tomography refers to the reconstruction of the internal distribution of 2D and 3D objects from multiple external viewpoints, thus providing cross-sectional slices through the object. The term originates from the Greek words *tomos* (for slice) and *graph* (for image). The term is commonly found in medical diagnostics, where computerized tomography (CT) x-ray scanning systems are well known. The medical applications of tomography provide a useful illustration of the principle, although this review is of course concerned primarily with industrial processes.

In a conventional x-ray imaging system, a cone-shaped beam of x-rays is used to illuminate the subject on one side. In passing through the subject, the radiation is attenuated (either absorbed or scattered) in proportion to the local electronic density; thus, variations in density or composition determine the intensity of the radiation as it leaves the subject. The final intensity pattern is captured as an image on a photographic plate mounted on the opposite side. The pixels of the resulting photograph (radiograph) represent the attenuation integrated through the object along the rays emanating from the source. This integration in effect removes depth information from the image, so if for example a tumor were seen on a radiograph one could not determine (on the basis of one image) which of several overlying organs was affected.

In contrast, early tomography systems [2] used several different views to build up an image of a slice at a predetermined depth in the subject. Most x-ray tomography systems in recent times have been designed to obtain cross-sectional slice images perpendicular to some axis, around which the sources and detectors are rotated. This process is known as axial tomography; for more details, see Webb [2]. In a first-generation CT system, a narrow x-ray beam is transmitted through the subject, and a collimated detector is placed on the other side of the subject. This measures the line-integrated attenuation of the beam through the subject along the path so defined. Figure 4.1 illustrates this arrangement in general form. The example subject has a circular cross-section and contains further contrasting circular features. An x-ray transmission source U and receiver V are shown. A beam is transmitted from the source along the line shown, displaced by length s from the origin and at angle ϕ, creating the line of integration, l. The measurement system measures the path integral of the subject's density distribution along this line, and we call this quantity the path density integral (PDI). If the angle is held constant and the beam is displaced across the subject (in direction s), a "shadow" of the subject can be seen in the attenuation data. This composite set of data is called the *projection*.

A set of projections can be obtained by repeating this process for a range of values of the angle ϕ, in effect providing an observation of the subject's shadow from multiple viewpoints. From an intuitive perspective, these projections are the prerequisite for tomographic imaging. In fact, from a mathematical perspective, the individual PDI values are the basic requirement.

Clearly, the measured PDI values will depend on the attenuation distribution, depicted in Figure 4.1 as $f(x, y)$. If sufficient PDI data are obtained, these may be used to estimate this distribution. In mathematical terms, this is a classic inverse problem, and the process of solving it to generate the estimated distribution is commonly called *image reconstruction*. The quality of the resulting estimate depends on the quality and number of PDI data and on the reconstruction algorithm.

In the above discussion, it has been a defined condition that the x-ray beam attenuation is measured along a collimated straight path through the subject. Such a system is called a hard-field tomography system, since an attenuation measurement can only be influenced by material that lies directly along the straight collimated

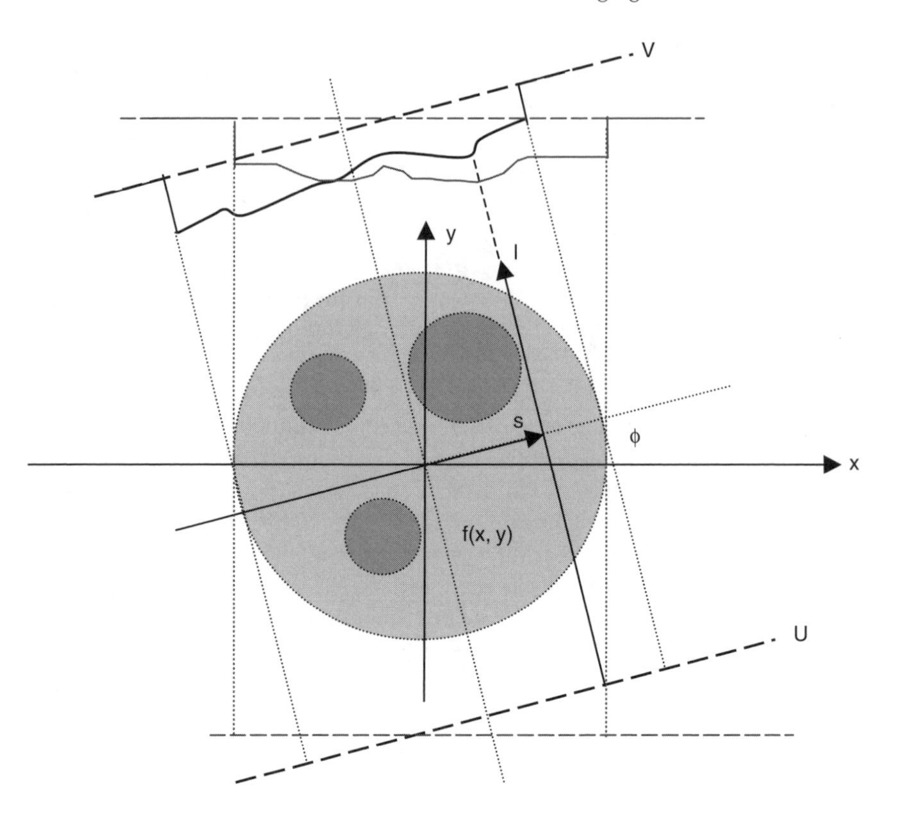

FIGURE 4.1 X-ray tomographic projection.

path from source to detector. This helps enormously in setting out the mathematical basis for image reconstruction. We shall see below that many process tomography systems do not share this feature, and the reconstruction task is complicated as a result.

4.1.2 Background on Tomography

The field of x-ray tomography is nearly as old as radiography itself, and a recital of the history of its development would be outside the scope of this chapter. Therefore, only a few highlights are mentioned here; for a more thorough account, the reader is referred to the book by Webb [2].

In 1917 Radon provided the basic mathematical description of projections and tomographic reconstructions [3]. Taking the example illustrated in Figure 4.1, Radon defined a projection as:

$$p(s, \varphi) = \int_{l} f(x, y)\,\mathrm{dl}, \quad \text{for all } s \tag{4.1}$$

The mapping defined by the projection along all possible angles ϕ is the (2D) *Radon transform*, R. A limited set of projections, as illustrated by multiple discrete applications of Equation 4.1, is known as a *sample* of the Radon transform.

Radon showed that, in general, an inverse transform exists such that

$$f = R^{-1}\{R\{f\}\} \tag{4.2}$$

The proof showed in essence that an N-dimensional subject can be reconstructed from an infinite number of (N – 1)-dimensional projections. In a physical implementation, naturally there are only a finite number of projections, so the reconstruction will have a certain amount of error in it [4]. However, research into practical tomography proceeded for 50 years without any knowledge of the Radon transform.

Given the limitations of technology in the early 20th century, the first functional x-ray tomography machines were based on radiographic film and mechanical contrivances. In 1914, K. Mayer, working in Poznan, showed that it was possible to erase the overlying shadows in a radiographic image by moving the x-ray source to various positions [2]. Another early tomography method collimated the x-ray beam to create a sheet of radiation slicing through the subject and placed the radiographic film in the plane of the beam [5]. When the object and film were synchronously rotated, a tomographic image was produced on the film. This second approach is an analog implementation of an unfiltered backprojection reconstruction, which will be discussed further in Section 4.3.1. Hundreds of schemes were devised for generating tomographic images, and by the late 1930s several European and U.S. companies were manufacturing commercial instruments for medical applications. Although these machines produced useful images, the quality was limited due to the dynamic range limitations of film and the relatively broad point spread function caused by backprojection, as discussed below.

An approximate inverse function is needed to estimate the original attenuation function in Equation 4.2, and this demands computational power and flexibility. Pioneering work on computed tomography was done in the early 1960s by Cormack at Tufts University; he was apparently unaware of Radon's original work but calculated a solution to the inverse problem and demonstrated it on real projection data [6, 7]. Hounsfield, working at EMI in the United Kingdom, is generally credited with creating (in the late 1960s) the first practical x-ray tomography system where the reconstruction was accomplished with a digital computer [8, 9]. This device was soon commercialized as the EMI Head Scanner. Cormack and Hounsfield shared the 1979 Nobel Prize in Medicine and Physiology for their contributions to this important diagnostic tool.

It should be noted that the selection of the inversion process is inevitably an engineering compromise that also takes into account the resources and time available for computation. In the case of the patient whose body is the object under examination, the objective is to provide an estimate of the internal distribution to allow the detection of any abnormal features. This examination is typically carried out by a clinician, perhaps assisted or advised by a specially trained radiographer. In this case, the associated computer system could realistically be allowed to work for several minutes, or even hours, to achieve an image of the highest practicable quality. In modern medical imaging systems, the clinical team can typically call upon the assistance of a range of image interpretation techniques to highlight particular features.

For an industrial process, internal state estimation is likely to have specific data resolution and real-time constraints. If the objective is model verification for process optimization, then obtaining an adequate resolution to observe the features of interest will be more important than real-time performance. If the objective is to offer state information for control, then data estimates must be at a rate consistent with the relevant process dynamics. In some cases, this time scale may be of the order of hours; in others it may be milliseconds. In this latter case the spatial resolution will be limited to that achievable within budget and technology bounds.

In order to review the implementation of process tomography systems, it is useful to define the generic architecture shown in Figure 4.2. The *sensor array* is the hardware that both defines the mode of tomography that will be employed (commonly called the *modality*) and implements the interface of the following tomography system modules. The *sensor data acquisition module* must provide the appropriate control and energy source to the sensor array and, where necessary, marshal the sensor signals into logical projection data. These data are then processed by the *reconstruction processing module* to yield an image that estimates the internal process distribution under the chosen modality. Many systems that aim to offer some diagnostic or experimental internal view of the process may not go further. However, a process sensor system also requires a further module, the *interpretation processing module*, to extract relevant process information from the tomographic image. For example, the 2D or 3D distribution can be used to provide a *mixing index* or *homogeneity index* that in turn may be used to determine the endpoint of a mixing process. Similarly, the cross-sectional distribution of a flow process may be used to estimate component composition. In general, the various modules of Figure 4.2 may be implemented in a mixture of hardware, firmware, and software.

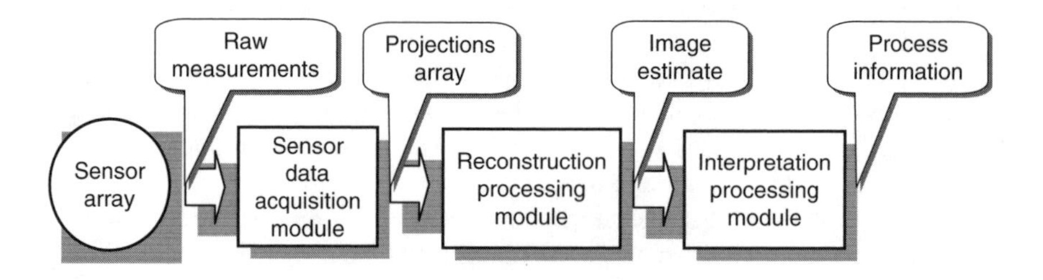

FIGURE 4.2 Generic process tomography system architecture.

4.1.3 Links with Medical Tomography

As we have noted, process tomography has specific constraints compared to medical applications, where image detail may be more critical than real-time reconstruction. However, medical tomography was clearly an inspiration for process applications. Indeed, additional tomographic modalities (besides x-ray) have been borrowed from medical imaging and adapted to industrial processes, as discussed in the following sections.

In the early 1970s medical scientists began to realize the potential of electrical impedance tomography (EIT), as a method for imaging soft tissues such as the lung and stomach more effectively than with x-rays. The technique is based on taking conductivity measurements by passing small currents through the body via contact electrodes, as illustrated in Figure 4.3. Cross-sectional images of the variations in resistivity (corresponding to organs or other structures) can then be reconstructed by means analogous to CT. This work was presaged by early work on electrical measurements in the human body [10–12] and also by the work on impedance imaging of geological structures [13, 14]. Pioneering work in biomedical EIT was performed at a number of institutions, including University of Wisconsin–Madison, Rensselaer Polytechnic Institute, and Sheffield University (Royal Hallamshire Hospital) in the United Kingdom. [15–17]. The imaging system developed at Sheffield has been widely applied for both medical and industrial purposes.

The success of this early work encouraged the setting up in 1988 of a European Union Concerted Action on EIT for medical applications. The original network group is still active. This program's final report provides a useful review of the early work in EIT for medical applications [18]. Holder has edited a comprehensive collection of papers covering a range of medical applications of EIT [19]. Additional information can be found in the book by Webster [20].

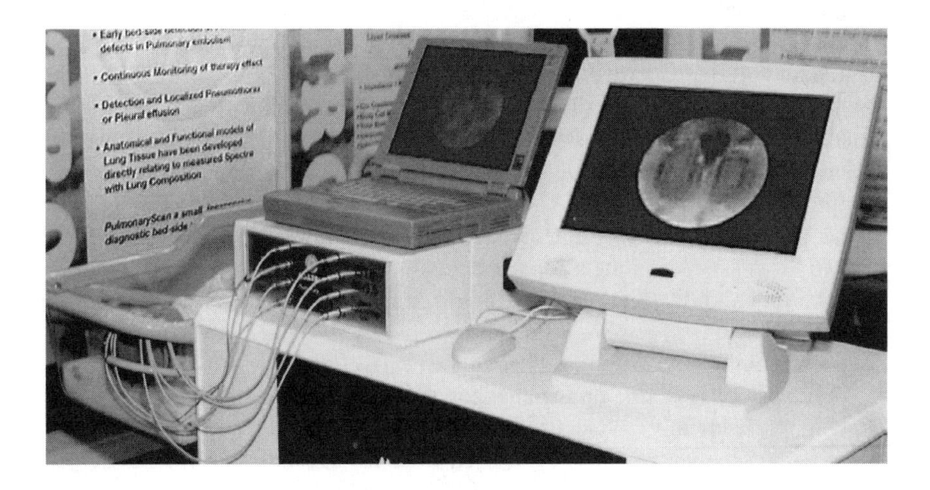

FIGURE 4.3 The Sheffield EIT system (courtesy of Maltron International Ltd.).

Links between researchers with interests in tomography in the medical and industrial communities have stimulated cross-fertilization. Brown has provided a useful review of the interaction between medical- and process-targeted research in the field of EIT [21]. Despite the potential benefits of this synergy, it must be noted that there are two clear differences that set process tomography apart from its medical counterpart. The first is the wide range of time scales that characterize industrial processes. The second is their broad range of discontinuity and inhomogeneity. In contrast, a medical tomography system need only deal with relatively slow dynamics (for almost all applications) and weak inhomogeneity.

4.1.4 Historical Developments in Process Tomography

It is useful to classify progress in terms of the modules of the generic PT system illustrated in Figure 4.2. Much work has been carried out on sensor arrays of various modalities, in effect covering a wide range of potential candidate applications. The elements of Figure 4.2 that deal with reconstruction and interpretation have also been explored, with increasing emphasis in the last five years or so, but to a lesser degree. Here we give a brief chronological overview of major developments in each key module area as a historical preamble to the detailed sections on each topic in this chapter. We begin at the front end with the development of a range of sensor arrays and their sensor data acquisition modules.

A number of applications of tomographic imaging of process equipment were described in the 1970s, but generally these involved using ionizing radiation from x-ray or isotope sources. Although of interest for research, these approaches were unsatisfactory for the majority of process applications at the routine level because of the high cost involved and the limitations due to safety constraints. Most of the radiation-based methods also required long data collection periods, which meant that dynamic measurements of the real-time behavior of process systems were not feasible.

In the mid-1980s work started that led to the present generation of process tomography systems. At the University of Manchester Institute of Science and Technology (UMIST) in the United Kingdom, a research investigation was initiated on electrical capacitance tomography (ECT) [22]. The objective was to image the multicomponent flows in oil wells and pneumatic conveyors, via the permittivity contrast within the subject. At about the same time, a research group at the U.S. Department of Energy's Morgantown Energy Technology Center (METC) was also designing a capacitance tomography system. Their aim was to measure the void distribution in gas fluidized beds [23]. The capacitance transducers used in these systems were suitable only for use in an electrically nonconducting process. Figure 4.4 illustrates an early system.

Work also was started on the application of EIT for imaging process vessels containing electrically conducting fluids. EIT was already well known in the geophysics community, where it was used to survey subsurface anomalies during prospecting studies [24, 25]. Medical EIT results also stimulated progress, and early UMIST work employed the sensor data acquisition module developed for

(a)

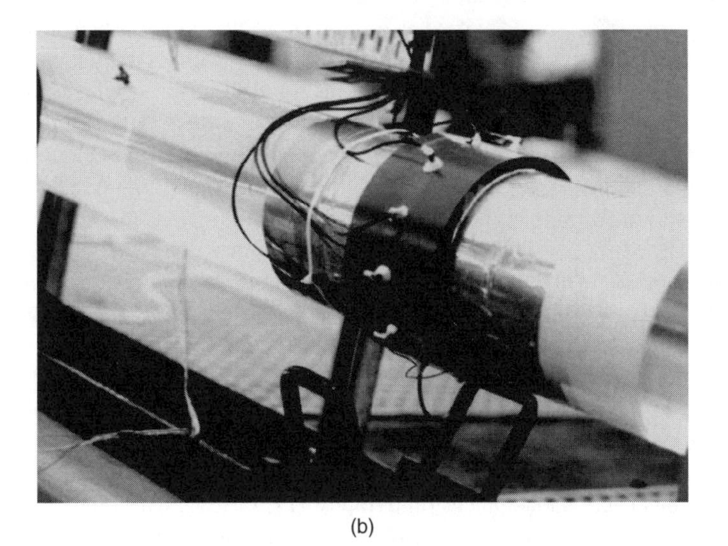

(b)

FIGURE 4.4 An early ECT system: (a) Parallel computing unit and display showing individual electrode measurements and reconstructed image; (b) eight-electrode sensor with electrostatic screen.

medical EIT at Sheffield. A specific unit suited to process applications was developed later [26].

The above electrical modalities are distinguished from x-ray techniques in a particularly fundamental respect: the measurement between any pair of electrodes is sensitive to the distribution of impedance (EIT) or permittivity (ECT) throughout the entire subject. These modalities are examples of soft-field tomography, the name arising from the fact that the electric field is changed everywhere in the subject due to a change of impedance or permittivity in any one location. This is in contrast to the hard-field nature of x-ray CT. This has the consequence that image reconstruction for EIT and ECT is more complex than for x-ray CT. Accordingly, the progress of electrical tomography has gone through a number of stages where hardware has led software and vice versa.

Other PT methodologies were also investigated in the late 1980s and early 1990s. For example, an ultrasound-based tomography system was developed at the University of Leeds in the United Kingdom [27], microwave imaging technology at École Supérieure d'Électricité in France [28], radioactive particle tracking and visualization at Washington University at St. Louis in the United States [29], positron-emission tomography at the University of Birmingham in the United Kingdom [30], electromagnetic induction tomography at UMIST [31], and gamma-ray tomography at the University of Bergen in Norway [32]. The use of twin sets of electrodes to estimate flow parameters was also demonstrated during that time [33].

The obvious implementation of the reconstruction processing module of Figure 4.2 was to deploy an appropriate small computer. Most early developments were critically constrained by the available resources for reconstruction processing. To set this in perspective, in 1988, the UMIST team employed a British Broadcasting Corporation Microcomputer for 8-electrode ECT sensor data acquisition and reconstruction processing. This microcomputer was chosen for its simple and flexible external interfacing facilities. The 8-bit, 6502 processor required about 20 minutes to compute the estimated image using the simplest possible reconstruction algorithm.

The viability of PT in most dynamic applications is critically dependent on the availability of computational power to deliver reconstruction and interpretation processing within the real-time constraints of the process of interest. For example, to realize simple flow imaging, about 100 cross-sectional estimates per second are required. This in turn demands that processing for each set of projections is completed in less than 10 ms. In terms of the historical perspective of the early UMIST prototype system, a speed increase of five orders of magnitude was required.

It was clear that specialized hardware was required for the reconstruction processing module. In the late 1980s, low-cost parallel processors were first harnessed, at the University of Leeds, to provide high-speed implementations of tomographic image reconstruction for ultrasound tomography [27, 34]. INMOS Transputer processing elements were selected for their scalability and flexible interconnection links. These elements could be connected in multiple instruction, multiple data (MIMD) topologies to yield powerful processing systems. These systems were later applied to ECT to produce the first high-speed real-time tomography system in a joint program undertaken at UMIST and Leeds [35]. This system was able to generate several hundred frames of tomographic frames per second (fps), sufficient for the real-time visualization of flow images in oil and gas pipelines [36]. As the power of digital signal processor (DSP) devices

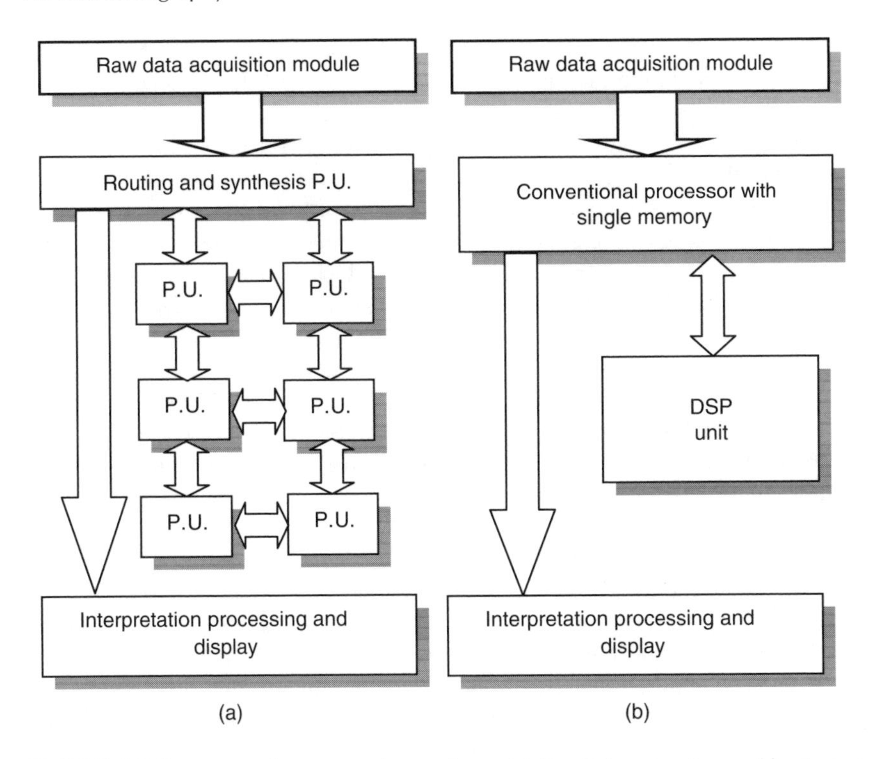

FIGURE 4.5 Reconstruction processing options: (a) Parallel processing architecture: a supervisory processor unit (PU) distributes data to subordinate units and integrates their results; (b) DSP-based architecture: a conventional processor shares a bus with a DSP that performs complex computations on the supplied data.

has increased, the provision of such firmware-based processing resources has been embedded in systems [37], and parallel solutions are now reserved for problems that require very sophisticated reconstruction processing. Figure 4.5 illustrates these contrasting approaches. Where process dynamics are modest, the sensor data acquisition module may be limited in function to sensor control, data stacking, and communication with a computer host.

A key aspect of the reconstruction processing module is the algorithm implemented. The most widely used algorithms are described in Section 4.3. As noted, early work was critically limited by the computing resources available. The simplest algorithm, known as linear backprojection (LBP), was typically employed. Here, tomographic data projections are simply linearly weighted by a matrix set to yield the estimated image (more details are presented in Section 4.3.1). The matrix set characterizes the sensitivity of each point in the object space to nonuniform process features, for each possible sensor measurement. The resulting image is generated by an overlay of all the linear backprojected weightings through the sensitivity matrix set. As a simple linear operation this is relatively fast, but the resulting image will typically be corrupted with artifacts. The image can be corrected to some extent with thresholding operations or by limiting the pixel content within physically reasonable bounds [38]. Early work explored more sophisticated algorithms, but in general it was limited to theoretical and off-line studies due to the large processing burden.

Most of the early research focused on the core technology, with the creation of an image seen as the endpoint in itself. Some early work did go further and attempt to encompass the generic interpretation processing module of Figure 4.2 [39]. The final part of this chapter deals with applications and will revisit this important topic.

The historical development of PT has gained much from an active community of researchers, aided by a number of networking initiatives. By 1990, it was felt that PT was progressing strongly as a potentially useful technique for application to industrial process design and operation, although the effort was somewhat fragmented. The European Commission provided funding for networking support under its Brite-Euram program to a consortium led by Professor Maurice Beck at UMIST, in a 4-year "European Concerted Action on Process Tomography" (ECAPT). Its first workshop was held in March 1992 in Manchester, United Kingdom [40]. Three further annual workshops were held in Karlsruhe, Germany [41]; Oporto, Portugal [42]; and Bergen, Norway [43]. A notable outcome from this series of workshops was the textbook noted above, which contain contributions from most of the active groups [1]. The ECAPT initiative was important in building a research and development community that clearly advanced the technology [44]. Although it had a European focus, the community also embraced key researchers in the United States, Japan, Australia, and elsewhere. Two further meetings, sponsored by the Engineering Foundation, were held in Shell Beach, California, in 1995 [45], and Delft, The Netherlands, in 1997 [46].

Following the early-stage ECAPT activity, the European Commission offered further potential support for specific networks having a more coherent structure and specific cross-collaboration goals. Approximately 30 research organizations (mainly universities) and companies agreed to take part in the resulting Advanced Tomographic Systems Network (ATS-NET) coordinated by Professor Brian Hoyle at Leeds University. This network sponsored plenary meetings and visits between research partners. Specific meetings were organized on Measurement and Optimization of Separation Processes in late 1999 and on Tomography in Mixing Processes in early 2000 [47].

The considerable impetus by research groups at Leeds and UMIST led to a large project funded under the U.K. government's "Technology Foresight Challenge" initiative. This program had specific goals to develop advanced multimodal PT systems and demonstrate their application in a number of commercial processes. The university team members resolved to create an ongoing scientific society dedicated to advancing the development of PT technology and its applications. The Virtual Centre for Industrial Process Tomography (VCIPT) was instigated to fulfill this mission on a nonprofit basis. A key objective was to continue the valuable interaction of plenary community meetings.

The research and development community has continued to expand since the mid-1990s. The VCIPT has succeeded in the aim noted above and provided a key dissemination platform in mounting a biennial series of World Congresses on Industrial Process Tomography (WCIPT). These meetings have been held in Buxton, United Kingdom (1999); Hanover, Germany (2001); and Banff, Canada (2003). The next congress, WCIPT-4, is planned for Aizu, Japan, in 2005. This event has grown to offer a world platform for the dissemination of the core technologies and applications of industrial process tomography. The proceedings of these meetings provide a valuable compendium of research and development since 1997. The VCIPT Web site (www.vcipt.org) provides further information. Developments published in the Congress proceedings and elsewhere are typical of a midcycle emergent technology, consisting of much useful but rather incremental progress (particularly in sensor development) and a small number of more novel breakthroughs.

4.2 TOMOGRAPHIC SENSOR MODALITIES

4.2.1 Introduction

For a specific application, the sensor modality will be the most critical design choice, since it must be sensitive to the variations of interest in the process and offer suitable noise properties. A relatively large number of candidate sensors are available, including infrared, optical, x-ray and gamma-ray, electrical capacitance, resistance and more generally impedance, positron emission, magnetic resonance, and acoustic or ultrasonic sensors. Each of these techniques has its advantages, disadvantages, and limitations. The choice of a particular technique is usually dictated by many, often contradictory factors. These include the physical properties of the constituents of the process, the desired spatial and temporal resolution of imaging, cost of the equipment, its physical dimensions, human resources needed to operate it, and the potential hazards to the personnel involved (e.g., radiation). Many university and industrial research groups have contributed to the development of the above range of modalities for a very wide set of applications; space limitations in the present volume prevent the exhaustive presentation of each group's contributions. We hope that the interested reader will pursue more detail in the references cited at the end of this chapter.

4.2.2 CT and Gamma-Ray Tomography

As discussed in the first section of this chapter, CT uses measurement of x-ray attenuation to obtain the PDI and projection data needed for tomographic reconstruction. X-rays are energetic photons resulting from deep electronic transitions within atoms, and they are readily produced by directing a high-energy electron beam onto a metal target. Clinical imaging applications typically call for x-ray energies of the order of 100 to 300 keV, but large-scale industrial uses sometimes require several MeV. Gamma rays result from nuclear transitions within atoms and (due to the higher binding energies of the nucleons) therefore have much higher energies than x-rays. Gamma rays are produced by radioactive sources (which may be natural or created in a nuclear reactor). Due to their high energy and short wavelength, x-rays and gamma rays can penetrate the steel casing, insulation, and piping typically found in process equipment. However, an effect known as beam hardening occurs as the radiation traverses large, dense objects. The lower-energy rays are preferentially absorbed or scattered. This effect shifts the overall contrast in certain regions of the reconstructed image. In addition, Compton scattering and fluorescence create a diffuse background that can also decrease the contrast. Compared to the electrical sensing modalities, image capture can be relatively slow (several seconds to several minutes), which can be problematic for control applications. In spite of these potential difficulties, the spatial resolution of the images produced by CT and gamma-ray tomography can be very attractive.

4.2.3 Electrical Modalities

Sensors based on the measurement of electrical properties of target processes promise relative simplicity and therefore low cost compared, for example, with those that involve high-energy radioactive sources. As noted, considerable work has been carried out in this area. A number of commercial products are available based on these sensor modalities; several of these are described below. A review of the process applications of electrical tomography has been given by York [48].

Developments have progressed in polarized directions, to suit electrically nonconducting and conducting processes respectively. The front-end design is typically challenging in all cases due to the small quantities involved. Considerable development work has been done to enhance performance in terms of speed of data acquisition, sensitivity, and noise immunity. For example, a fundamental approach by Yang et al. aims to explore optimal electronic circuit designs for tomographic sensors [49].

Where the process contents of interest are nonconducting, an ECT instrument can image the 2D or 3D permittivity (ε) distribution by sensing the capacitance between electrodes mounted on the boundaries of the subject. Typical examples of measurement subjects here include minerals in liquid and solid forms, such as oils and powders. Specific designs have been proposed to enhance performance in terms of speed of data acquisition, sensitivity, and low-noise [50, 51]. The excitation frequency applied in these systems is typically of the order of 1 MHz, which is low enough to allow the use of electrostatic models of the subject.

For processes in which contents are conducting, the sensor data acquisition module is designed to sense resistance values in a 3D resistivity distribution; hence the description electrical resistance tomography (ERT). Typical applications include a wide range of processes based on aqueous mixtures, such as pharmaceuticals and foodstuffs. (Alternatively, this instrument can be considered as imaging the conductivity [σ] distribution.) Recent innovations have enabled in-process measurement of the complex impedance, justifying the term electrical impedance tomography. New designs have attempted to push forward the specification to offer very high speeds and increased flexibility [52–54]. In these systems, current injection frequencies vary from about 1 kHz to about 100 kHz.

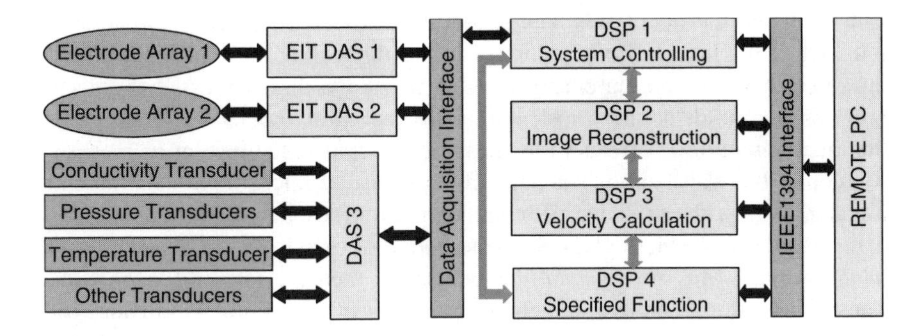

FIGURE 4.6 High speed EIT System Block Diagram (courtesy of M Wang, University of Leeds).

Figure 4.6 illustrates the complexity of a novel EIT design for high-speed data acquisition and processing to suit flow applications. In this case the system is designed to deliver 1000 reconstructed fps.

As illustrated in Figure 4.2, following the specific front-end sensing electronics, all systems typically employ a multiplexing arrangement enabling their sensor data acquisition module to deliver a complete set of tomographic data for display and interpretation.

Some applications exhibit complex electrical characteristics; for example, insulating mineral liquids such as oil will also exhibit a "loss angle" corresponding to their resistive (conductive) component in the field. Corresponding developments in both ECT and ERT instrumentation technologies offer a true impedance measurement. A comprehensive literature exists on both ECT and ERT methods. In addition to the paper by York [48], the proceedings of the VCIPT World Congresses in Industrial Process Tomography offer a major set of papers.

Electromagnetic induction tomography (EMT) is a further low-frequency electrical variant. Although it can be considered in principle to complete the trio of the standard R, C, L electrical parameters, its applications are limited in practice to those that exhibit significant sensitivity in inductive terms. Of relevance are those that center on processes dealing with some minerals, metals, and magnetic materials and ionized liquids (including water). This range includes the important class of ferrous metal ores, the production of iron and steel raw material stock, and corresponding parts under manufacture. A detailed review of basic sensor design and corresponding reconstruction software is given by Peyton et al. [55] and Borges et al. [56]. Tapp and Peyton have provided a recent updated review of developments [57].

4.2.4 Microwave Tomography

The term *microwave* is used loosely here to mean the frequency region of about 1 to 100 GHz. These frequencies correspond to wavelengths ranging from 30 cm down to 3 mm in air, and from approximately 3 cm down to 300 μm in water.

In this region, the interaction between matter and electromagnetic radiation is relatively complicated for our purposes, and this offers both advantages and disadvantages. For measurement subjects that have dimensions of the same order of magnitude as the wavelength of the radiation, the near-field nature of the interaction allows effects to be measured with a spatial resolution that is a fraction of the wavelength. The penetrating nature of microwave radiation and its interaction with the material's complex permittivity $\varepsilon^* = \varepsilon - j\sigma/\omega$ (where ω is the angular frequency) gives access to both the conductivity and permittivity of the subject. Microwaves can thus be used to measure materials with wide ranges of these parameters, which are functions of frequency. Disadvantages are that for application to process engineering subjects, very advanced high-frequency low-noise electronic circuits are required to measure these effects, and the near-field sensitivity can lead to complicated interactions of multiple antennae.

Microwave tomography can be implemented at a fixed frequency by sampling the wavefront with an array of antennae or by sweeping the frequency and using spectral time-domain analysis [58]. Both planar and circular arrays have been demonstrated; one antenna in an array can be used to broadcast while the others are used to sample the diffracted wavefront [59]. The data acquisition times for such systems are about 1 sec. The best image contrast comes from objects that contain water, but this modality has also been applied to dry conveyed products [58], and a 32-antennae system has been demonstrated with free-space coupling [60].

4.2.5 Optical Modalities

Early work on tomographic imaging systems using optical radiation simply exploited the subject's bulk attenuation in a manner that was not dependent on wavelength. These methods are discussed in the text edited by Williams and Beck [1]. With the advent of small, robust lasers emitting a very narrow spectrum of light around a well-defined wavelength, much progress has been made in recent years toward tomography systems that provide information on the distribution of chemical species within a subject. This type of information can now be provided with very high temporal resolution and, in principle, for a wide range of chemical species. By using modern laser diodes, optical fibers, wavelength mixers, multiplexers, and photodiodes, these systems can be very robust. The principles of optical tomography systems for chemical species tomography (CST) are therefore discussed here in some detail.

Many atomic or molecular species have characteristic absorptions at UV or visible wavelengths that are strong, due to transitions of electrons between different quantum states. Subsequent photon emission by the excited electronic state (fluorescence) is generally very specific and is isotropic. Figure 8.6 illustrates the fluorescence process. The need for high levels of transmission through the subject by both the stimulation beam and the emitted fluorescence photons has led to an emphasis on gas-phase applications, with occasional examples of liquid-phase imaging. The use of fluorescence from high-energy pulsed laser sheets for species imaging is very advanced for processes where large-scale optical access is available, as discussed at length in Chapter 3 and Chapter 8.

The UMIST group has demonstrated the concept of fluorescence imaging in systems with optical access restricted to a single plane [61], with a technique called optical fluorescence auto-projection tomography (OFAPT). With current laser technology, OFAPT could be applied to liquid-phase subjects or to slow imaging (about 1 fps) in gas-phase subjects. Application to fast imaging in gas-phase subjects awaits the development of high-powered UV laser diodes.

When light absorption is achieved by exciting vibrational or rotational transitions of molecules (i.e., at wavelengths in the infrared [IR] spectral region), it is highly characteristic of the molecular species involved and can be very strong, particularly in the mid-IR above 2.5 μm. However, the technology available today for robust systems is much more mature in the near-IR (from about 700 nm to 2.5 μm), and the discussion here is restricted to that case.

Photon scattering within the subject under study complicates the above mechanisms, but it can be exploited when there is still some transmission through the subject. Photons that are not scattered are termed *ballistic* and can be distinguished from scattered photons by their time of flight through the subject. The principal elastic scattering mechanism is described by Mie theory [62]. In the case where the diameter of the scattering centers is significantly less than the wavelength of light, λ, the scattering intensity varies as λ^{-4}. Hence, for longer wavelengths incident on a given subject, the attenuation due to scattering is much less than that experienced by shorter wavelengths. This situation arises in many gas-phase reaction systems such as combustors.

In-situ chemical sensing systems are widely used in industrial application, using single-path IR absorption techniques. In recent years, near-IR chemical sensing by single-path near-IR absorption has received a great deal of attention, principally due to the availability of a growing range of laser diodes producing light at wavelengths absorbed by various target species. (See the discussion on this subject in Chapter 8 for a range of references.)

A compact all-optoelectronic chemical species tomography system is shown in Chapter 8 [63]. This system images hydrocarbons (HCs) mixing with air at rates up to 3500 fps. It uses a probe wavelength of 1700 nm and a reference wavelength of 1546 nm. Optical fibers and graded index (GRIN) lenses are used to launch these wavelengths simultaneously along 32 paths (arranged in four projections) through a gas-phase subject or a subject containing a volatile liquid spray. This work was initially targeted at combustion applications (example images are presented in Chapter 8), but application to other HC processing systems is feasible. In principle, by substituting into this system laser diodes yielding appropriate wavelengths, different target species can be imaged. Although different light mixing, multiplexing, and detection components may be required, the concept is clearly flexible. Research is now under way to apply this technique to a four-cylinder gasoline engine. Among several engineering challenges, the task of mounting the optical elements in the cylinder wall is particularly demanding.

With only 32 path concentration integrals (PCIs) available, careful attention has been paid to image reconstruction issues, which are discussed at length in Section 4.3. Early work in IR absorption tomography [63] used the algebraic reconstruction technique (ART). The gas-phase images shown in Chapter 8 have been reconstructed by an algorithm that uses interpolation both between beams within a projection and between different angular projections, followed by filtered backprojection and Landweber iteration [64]. While this latter approach is applicable to an object with smoothly varying spatial concentration (as would be expected for a gas cloud), it is not acceptable for objects with sharply defined concentration boundaries, such as the liquid distribution within sprays. In general, a spatial resolution of about 5% of the field of view has been obtained with this system. It should be noted that the spray case discussed above may be better analyzed by a model-dependent fit to the data (see Section 4.3.4) with a relatively small number of model parameters.

The Hannover group [65] has also demonstrated a laser diode-based CST system and used it to image water distributions above a packed-bed adsorber column. Some example images from a 50 mm-diameter column are shown in Figure 4.7. The probe wavelength used in this case is 1396 nm, and reference data are obtained by wavelength modulation of the laser. The light is fanned out to three sheets, and absorption is measured along 128 paths in each of the three projections, using photodiode array detectors. Image reconstruction from the resulting 484 PCIs is carried out using ART.

With the combination of chemical specificity and good temporal resolution, the IR absorption modality is sure to find other applications in a number of academic and industrial systems. Even in the case where the subject has a high concentration of scattering centers, recent advances in medical systems [66] have shown the feasibility of tomographic imaging based on time-domain detection. For many process applications of IR CST, the ability to deal with high levels of scattering will be crucial.

4.2.6 Acoustic and Ultrasonic Methods

Ultrasonic imaging of the human body is commonplace, and it is tempting to assume that techniques may transfer simply to industrial applications. However, medical systems are designed to deal with weakly inhomogeneous objects; in contrast, most process applications are likely to present strongly inhomogeneous objects. The primary property relating to ultrasound propagation through materials is *acoustic impedance*, which is defined as the product of density and sound speed. For inhomogeneous objects, the results are related to the difference in acoustic impedance and the size and shape of the interface. For small particles, diffraction effects may also occur where the wavelength of the sound energy is of the order of the particle size. Measurements may be classified into three basic methods:

1. Transmission mode (where attenuation of the subject is sensed)
2. Reflection mode (where backscatter from the subject is sensed)
3. Diffraction mode (where diffraction by the subject is sensed)

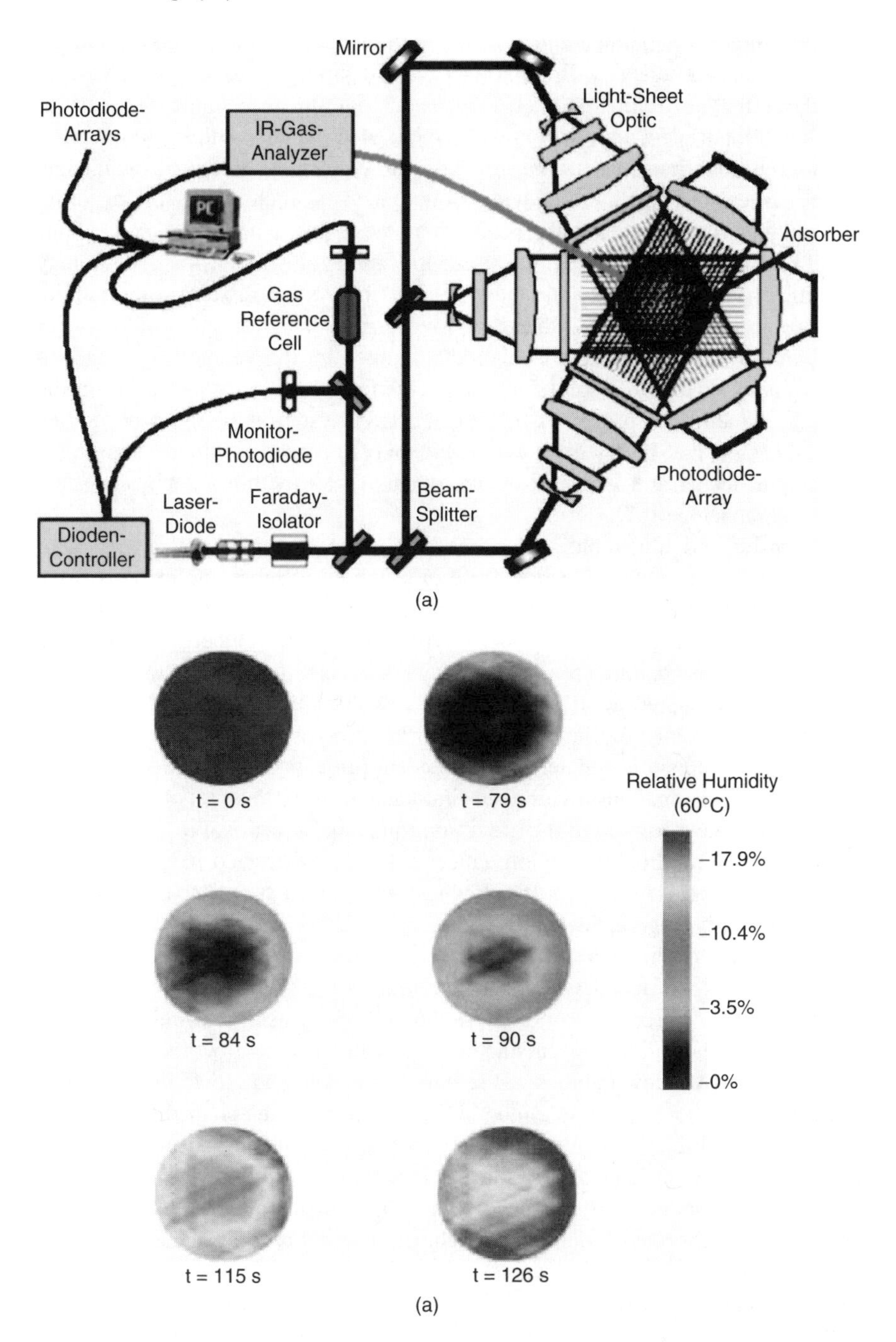

FIGURE 4.7 (a) Schematic of near-IR absorption tomography system. (b) Example images for water breakthrough above a packed bed absorber. (From Salem, K., Mewes, D., and Tsotsas, E., *Proc. 3rd World Congress on Industrial Process Tomography,* Banff, Alberta, Canada, pp. 199–206, 2003. With permission.)

Ultrasound is generated and detected with piezoelectric transducers, and the acoustic energy interacts with objects of interest through scattering and absorption. For ultrasonic tomography (UST) in industrial processes, multiple measurements will be needed, and, in contrast to medical applications, this will probably require multiple transducers. The time taken to collect multiple projections may not be critical in applications having slow dynamics, such as the mixing of an injected material in a chemical reactor where the requisite number of projections could be obtained sequentially. In other applications, the dimension of time may be critical; for example, on-line measurements in flowing mixtures may require short projection intervals so that flow evolution is insignificant. Measurement times in ultrasonic systems are inherently limited by the velocity of sound. In water, this is of the order of 1.5 mm/μs. This fact places an upper limit on the intrinsic dynamics of processes for which a UST system is likely to be usable. Thus, it is clear that UST will be worthwhile only in processes where a significant interaction occurs and where real-time dynamics are within the physical data collection envelope [67].

Acoustic and ultrasonic transducers for process tomography are in their infancy relative to their widespread use in a variety of other applications, such as nondestructive testing and particle size measurement. Hence the realization of UST depends critically on the design of existing transducers. Transducers designed for those mature applications may not be useful in process tomography. Most nontomographic applications demand a narrowly focused beam. A UST system must insonify the interfaces within the process of interest from several viewpoints, to obtain a number of independent projections. A wide-angle beam offers poor directional information but simultaneous insonification of a wide area. Norton and Linzer suggested the use of multiple omnidirectional transducers and illustrated how multiple projection data could be reconstructed [68]. Moshfeghi has demonstrated a transducer employing a fan-shaped beam, able to simultaneously insonify large areas [69].

Experience of practical ultrasonic process tomography is limited. Wolf has described an experimental system that determines the spatial distribution of the gas bubbles in the cross-section of a gas–liquid flow by using transmission mode sensing [70]. The results are satisfactory only when the gas flow rate is low and there are relatively few bubbles in the liquid. Wiegand and Hoyle demonstrated a real-time ultrasonic process tomography system using fan-shaped beam transducers with multiple segments, employing both reflection and transmission mode data. This system utilizes all transducers as simultaneous receivers while transmitted pulses sequentially insonify the subject from a number of angular positions [71]. Processing demand is high, and parallel processing has been used to maximize real-time performance [72].

Interpretation based on bubble identification has also been explored, coupled with a Hough transform to automatically fit circular cross-sections and estimate centers [73]. Further applications have been demonstrated both in pipelines [74, 75] and in cyclonic separators whose internal air-core can be detected [76, 77].

4.2.7 Other Modalities Borrowed from Medicine

A range of other tomographic modalities have been demonstrated, of which two in particular are well developed for both medical and process engineering application, albeit with the disadvantage (for process engineering) of nonportability. These are magnetic resonance imaging and positron emission tomography.

4.2.7.1 Magnetic Resonance Imaging

Nuclear magnetic resonance imaging (NMRI or MRI) exploits the interaction of electromagnetic radiation, at particular radio frequencies, with the magnetic dipole moment of certain nuclei when they are subject to a spin-aligning magnetic field. The principles of this technique are well known [78], and process engineering applications have been demonstrated by several authors [79, 80, 81]. NMRI offers a very high spatial resolution, and it has made a spectacular impact on medical science. Lauterbur [82] and Mansfield [83] received the Nobel Prize in 2003 for the invention and initial demonstration of NMRI.

The most frequently used nucleus for MRI is that of hydrogen (H), a single proton, although several nuclei have usable magnetic dipole moments. We illustrate the technique briefly here by reference to the H case. It is important to note, however, that the frequency of the H resonance is sensitive to the particular molecular species within which it is incorporated [79]. In general, therefore, MRI is chemically specific; i.e., it provides images of the concentration distribution of particular chemical species.

In simple terms, MRI of H in H_2O is obtained by the following sequence:

- A large, static, spatially uniform magnetic field $\mathbf{B}_0$ is applied to the subject to be imaged. The magnetic dipoles of the H nuclei in the subject align with $\mathbf{B}_0$.
- A radio-frequency (RF) electromagnetic field of angular frequency ω is applied to the subject.
- An additional magnetic field $\Delta\mathbf{B}(x, y)$, varying linearly with space in one direction (a gradient field), is applied to the subject to give the total magnetic field $\mathbf{B}_{tot} = \mathbf{B}_0 + \Delta\mathbf{B}(x, y)$. Suppose that the gradient is in the x direction. The gradient field is chosen so that the nuclear spin-flip condition $\omega = \gamma\mathbf{B}_{tot}$, where γ is the gyro-magnetic ratio, which is specific to any given nucleus) is satisfied in a very small spatial region along the gradient (x) direction. Along the line of constant x, in the y direction, every point satisfies the nuclear spin-flip condition. Hence, absorption of the RF field and changes of nuclear spin orientation will occur only along this line.
- By applying the gradient field with a temporal ramp, the "critical line" is swept through the subject in the x direction, while RF absorption measurements are made or another means of detecting the local spin state is used.

- By rotating the direction of the gradient field, different projections of RF absorption (or local spin state) are obtained.
- Classical image reconstruction techniques yield the distribution of the absorbing species.

The above sequence is an oversimplification of the process, intended to convey the basic principles of the technique and to illustrate their elegance. Various versions of the pulse sequence are used. In medical applications, the frame acquisition time for a single cross-section is typically a few seconds. Exploitation of different pulse sequences in process engineering applications yields imaging rates of tens of frames per second [84]. Spatial resolution depends on the uniformity of $\mathbf{B}_0$ and the steepness of the $\Delta\mathbf{B}$ ramp; in medical application, spatial resolution of 40 μm is typical, and this has been achieved in process engineering studies also, although faster imaging can result in some degradation of this performance. Many successful process engineering studies have used medical MRI systems [80]. Additional flexibility has been obtained from systems customized for process application, e.g., incorporating vertical-bore magnets [84].

Typically, MRI systems require large cryogenically cooled superconducting magnets and considerable shielding from metallic objects in the vicinity. They are therefore implemented as static facilities, to which the subject is brought. For that reason, they are not considered at greater length here. Nevertheless, many ingenious schemes have been applied to insert various models of process engineering systems into MRI scanners [79, 80]; vertical-bore systems offer the best scope for this [84]. The fundamentals of a variety of engineering processes are being opened up for study by this technique.

4.2.7.2 Positron Emission Tomography

Positron emission tomography (PET) and its derivatives suffer from the same limitations in terms of portability as outlined for MRI above, but for different reasons. However, PET offers the scope to "follow" different labeled molecules or particles through a process, and thus provides a similar, but complementary, window on process fundamentals to MRI. Simplistically, PET imaging is performed as follows:

- A sample of a positron-emitting isotope of the required element is prepared at a facility such as a small cyclotron. The positron (e^+) is the anti-particle of the electron (e^-). Nuclear radioactive decay by positron emission is more commonly known as β radioactivity, and these artificially produced radioactive samples typically have half-lives of minutes to hours.
- The radioisotope is chemically incorporated into a molecule or into a particle.
- The "labeled" molecules or particles are introduced into the process.

- Each time a radioactive nucleus decays, the ejected positron almost immediately annihilates with a local electron to produce two photons of energy 511 keV that are emitted in colinear but opposite directions.
- The photons have sufficient energy to penetrate large amounts of material in the host process and its containment, and they travel outward to be detected by specialized photon detectors.
- The detection positions of the photons (on opposite sides of the process) define a line through the host process along which, at any point, there is an equal probability that the radioactive nucleus decayed. A large number of these data can be used to perform image reconstruction of the concentration of labeled molecules or particles.

The data acquisition phase of PET is typically of the order of a few minutes.

In a frequently used derivative of this process, only a single labeled particle is placed into the host process to perform so-called positron emission particle tracking (PEPT), which provides dynamic process information. Both PET and PEPT have been used for a wide range of process engineering applications [85].

4.2.8 Multimodal Systems

As in the medical case, the use of a single tomographic modality "sees" the process in terms of the "contrast" to the form of energy deployed. Major breakthroughs have been made in the medical sphere through the fusion of information relating to the hard structures of the human body illuminated by x-ray CT and that of the soft functional parts of the body illuminated by (nuclear) MRI. To fuse such datasets clearly requires precise physical scaling and registration of the image. As always, the process case is more complex, because even colocated process sensors will tend to view slightly different 3D volumes. In some cases, it will not be possible to colocate sensors, and the data fusion may need to use a model to compensate for a process delay.

A dual-mode system employing gamma-ray and ECT modalities, illustrated in Figure 4.8, was devised by Johansen and colleagues at Bergen [86]. This dual-mode system is analogous to the medical x-ray/MRI combination. In this case, the integration needed for the gamma-ray element makes the system useful only for relatively slow processes. The target application was the assessment of flow regime in three component oil–gas–water flows. The integration time for the gamma-ray element limited the imaging rate to about 10 fps, and hence the system could not provide information on transient events.

A further key step forward has demonstrated a system designed intrinsically to support extensive multimodal data capture and fusion. It comprises modular hardware and features a layered data protocol to facilitate highly flexible multimodal configurations. Data streams are time registered, allowing data to be fused in four dimensions. The system also permits on-line streaming video and data from non-tomographic sensors to be recorded and time synchronized with the tomographic

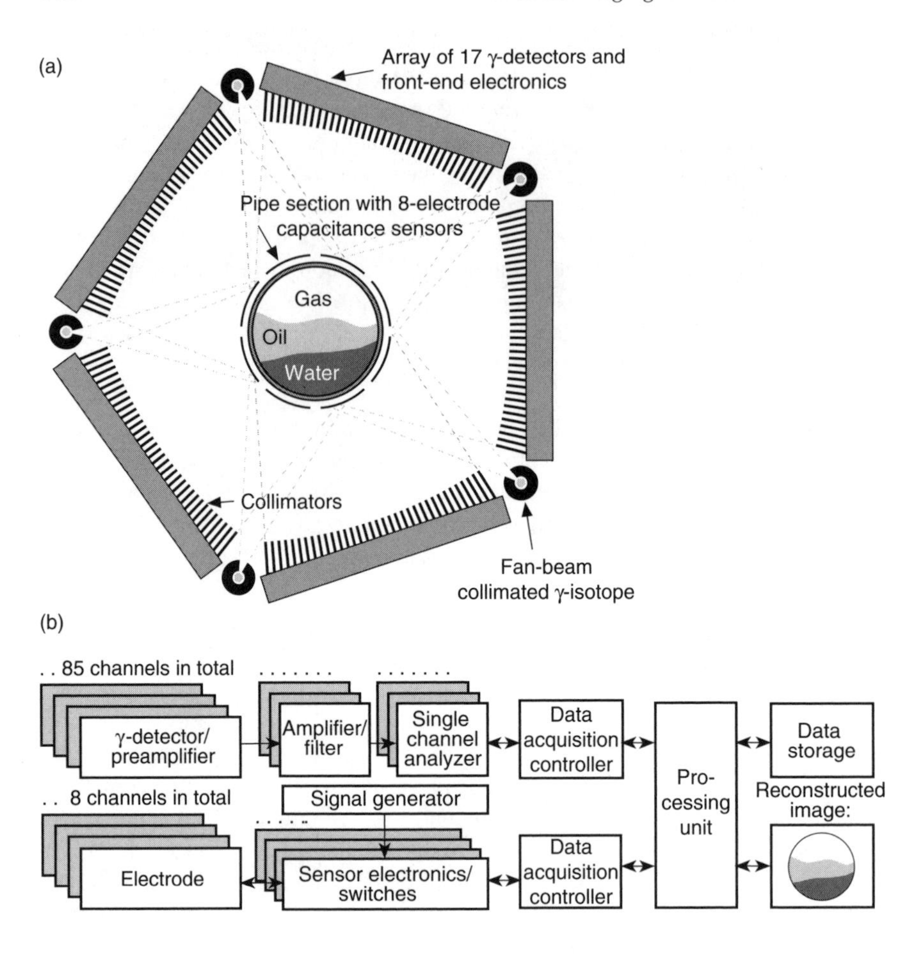

FIGURE 4.8 Dual-mode gamma-ray/ECT system: (a) dual-mode sensor; (b) data acquisition and processing units (courtesy of GA Johansen, University of Bergen).

data [87]. Figure 4.9 shows the system in use to perform a multicomponent flow study in which tomographic flow estimates obtained by fusing twin datasets were verified by the time synchronized recorded video.

4.3 IMAGE RECONSTRUCTION

For any of the sensor types listed above, the sensor output (i.e., the path integral and projection data) depends on the physical state and distribution of material in the industrial process under scrutiny. The task of tomographic reconstruction is to determine the material distribution that would give the observed sensor responses. Typically, the spatial variation (of the parameter of interest) is represented as a cross-sectional image. This process is known as *image reconstruction.*

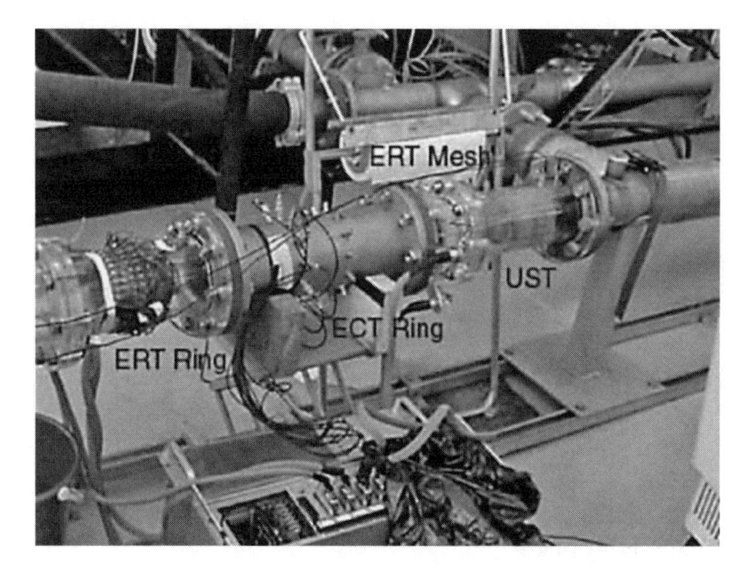

FIGURE 4.9 Multimodal PT system used for multiphase flow studies.

In mathematical terms, tomographic reconstruction is an "ill-posed" problem, which means that one or more of the following statements must be true:

- There is no solution to the problem.
- If a solution does exist, it is not unique (other solutions exist).
- The solution does not have a continuous dependence on the projection data.

A discussion of ill-posedness in this context is presented by Bertero and Boccacci [88], an excellent introductory text in tomographic image reconstruction.

It is assumed that there is a solution to the problem (the desired cross-sectional image), so reconstruction algorithms must be able to cope with nonunique or wildly varying solutions. In practice, the undesired alternative solutions (images other than the one that actually describes the process) can be eliminated on physical grounds. If the ill-posed nature is manifested as an erratic solution, it may be necessary to explore thoroughly some well-chosen test cases of the problem at hand in order to understand the connection between projection data and reconstructed image.

A wide variety of techniques are available for reconstructing cross-sectional images from the path integral and projection data. The main techniques are briefly described here. In most process tomography systems, only a few dozen path integral measurements are obtained. This has stimulated a great deal of research into image reconstruction techniques in the PT community, along with a similarly large effort in the community using EIT for medical applications.

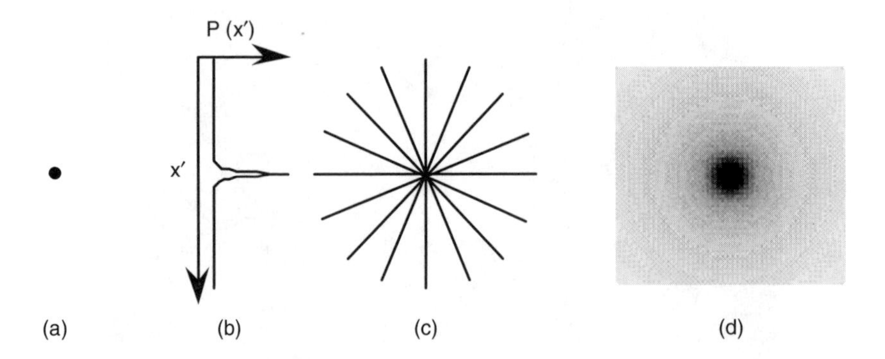

FIGURE 4.10 (a) A point object; (b) the projection of a point; (c) the backprojection of a delta function at many observation angles; (d) the reconstructed image of a point object.

4.3.1 BACKPROJECTION

As mentioned briefly in Section 4.1.2, one way to reconstruct an image of the original object is to backproject (replicate on the subject space) each path integral measurement along the direction of y' (see Figure 4.1) and to sum all the backprojections together, in the process of linear backprojection. Backprojection is represented in terms of integration over all projection angles:

$$f_{FB}(x, y) = \int_0^{\pi} p(x\cos\varphi + y\sin\varphi, \varphi)\, d\varphi \tag{4.3}$$

This process is illustrated in Figure 4.10, which shows the impulse response of the backprojection reconstruction. Each one of the projections of a point object (a) is a delta function (b), which is nonzero only at a single point of the projection. The backprojection is a line, and the summation over many angles is seen in Figure 4.10(c). Since these projections are one-dimensional (1D) objects, backprojection can be considered as a "smearing" of the projections into a second dimension. The final reconstructed image of the point object is shown in Figure 4.10(d). This image illustrates the point-spread function of the reconstruction.

It is evident that images tend to be blurred in this method of reconstruction. Due to the linearity of the imaging process, the reconstructed image of a more complicated object will simply be the superposition of a large number of blurred points. Although the point-spread function for backprojected images is known [89] to have a relatively long-range $1/r$ dependence (where r is the distance from the point), a convolution (filtering) operation performed on the projections before reconstruction improves the image clarity [90]. This approach is the basis of the filtered backprojection (FBP) method. In medical CT studies, it is typical to record 0.5 million path integral data for one image. In that case, the above techniques, along with the transform techniques discussed in Section 4.3.2, are able to generate excellent images.

In one form or another, backprojection (as described above) is one of the most common reconstruction algorithms. Initially it was used to reconstruct single cross-sectional images of an object from 1D projections. However, direct algorithms have also been developed to reconstruct 3D volume images from radiographic 2D projections [91].

It is generally assumed that the reconstruction process is carried out on a digital computer, but analog methods of calculating the backprojected image were developed several decades before the invention of computed tomography. For example, one of the early tomography systems (mentioned previously) produced an unfiltered backprojection image by placing film in the plane of an x-ray fan beam [5]. As the object and film were rotated synchronously, a tomographic image was produced on the film. At each orientation angle, the 1D projected radiation intensity pattern exposed the entire film in a way that backprojected it into the second dimension. Since film exposure is additive, as the film rotated it accumulated a latent tomographic image. More recently, incoherent optical processors based on optical transfer function synthesis have demonstrated the ability to reconstruct cross-sectional images directly from projection data stored on x-ray film [92]. Other optical devices reconstruct the images directly from projection data provided by x-ray image intensifiers.

Backprojection calculations (both filtered and unfiltered) can also be performed with analog electronic circuits so that the tomographic image forms directly on a display screen in real time [93]. In the early days of tomography, analog methods were faster and less expensive to implement than computer-assisted tomography. However, their image quality is inferior to that of computer-generated images. Given the computation speed attainable with present-day computers, analog methods have generally fallen into disuse.

Many process tomography methods (most notably the electrical methods) require modifications to the reconstruction procedure just outlined. Electrode arrays that circle pipes generate curved electric fields, rather than the straight-line propagation of x-rays in CT. Moreover, the soft-field nature of electrical modalities, discussed in Section 4.1.4, complicates the reconstruction process still further. The implementation of LBP in this case is by means of the sensitivity matrix **S**. Consider the case of ECT:

Each element S_{ij} of **S** defines the expected change of capacitance between electrodes i and j when the permittivity content in pixel k in the subject space is changed. The whole set of capacitance measurements can then be calculated as:

$$C = \mathbf{S}G \tag{4.4}$$

where C is the vector of all capacitance measurements, **S** is the sensitivity matrix, and G is the vector describing the distribution of permittivity in the subject space. Typically, we attempt to reconstruct many more pixels than the number of measurements. Hence, **S** is not square, and its inverse does not exist. LBP can be expressed as:

$$\bar{G} = \mathbf{S}^T C \tag{4.5}$$

where $\bar{G}$ is an estimate of the true distribution G, and $\mathbf{S}^T$ is the transpose of **S**, used as an approximation to the inverse of **S**. This gives, in general, a crude approximation to the solution, and it is often necessary to implement further stages to achieve an acceptable solution. As discussed in Section 4.1.4, such further stages may consist of simply limiting the pixel content of the image within known bounds or iterating the process (Section 4.3.3).

4.3.2 Transform Methods

Transform methods are based on the pioneering work of Radon [3] and the mathematical properties associated with the Radon transform (Equation 4.1). The most familiar example is the central slice method used extensively in computed tomography [94]. Lewitt has reviewed additional methods of this type [95].

As an example, we consider here the Central Slice Theorem, which assumes parallel rays and no soft-field effect. It states that at any angle the (1D) Fourier transform of the projection of a 2D function (i.e., the cross-section of interest) is the central slice (through the origin) of the Fourier transform of that function. Therefore, a complete transform of the cross-section can be assembled by appropriately sampling the transforms of the projections. The final image $f(x, y)$ is obtained through an inverse 2D Fourier transform (where P is the Fourier transform of the projection):

$$f(x, y) = \int_0^{2\pi} d\varphi \int_0^{\infty} d\nu \nu P(\nu, \varphi) e^{-i2\pi\nu(x\cos\varphi + y\sin\varphi)} \tag{4.6}$$

This approach assumes that a full set of projection data is available. If the projection set is incomplete or noisy, the final image will be distorted by artifacts of the reconstruction process [96]. A systematic study of such artifacts was undertaken by Müller [97], who showed that multiple truncated projections (those not "wide enough" to span the width of the object) could be combined to give an error-free reconstruction. This synthetic aperture approach is clearly applicable to the tomography of large industrial vessels. Given the small number of measurements typically obtained in process tomography, the interpolation techniques explored by Yau and Wong [98], for example, are of considerable interest.

4.3.3 Iterative Methods

An alternative strategy to cope with small numbers of measurements is to employ iterative algorithms. This strategy has been heavily used in medical applications of EIT and in process tomography over recent years. Iterative reconstruction techniques solve the inversion problem by comparing the projections of an assumed solution to the projections actually measured, adjusting the assumed solution for a better fit, and repeating the process until the solution converges. A finite element method is often employed for the forward calculations in such algorithms. These methods include the optimization approach [99] and variations on the algebraic reconstruction technique [89, 100], the Landweber technique [101], and the Newton–Raphson method [102]. Some important examples are discussed in detail here.

The Newton–Raphson method is often employed in EIT applications [102]. As in any iterative technique, there is an error signal (the "objective function") that is to be minimized. In this case, the objective function $\Phi(\rho)$ is defined to be:

$$\Phi(\rho) = (1/2)(\mathbf{f}(\rho) - \mathbf{V}_0)^T(\mathbf{f}(\rho) - \mathbf{V}_0) \tag{4.7}$$

where $\mathbf{V}_0$ is the measured voltage and $\mathbf{f}(\rho)$ represents the estimated voltages based on a resistivity distribution ρ. To minimize $\Phi(\rho)$, its first derivative is expanded in a Taylor series about a point ρ^k and set equal to zero, keeping only the linear terms. Solving for the change in ρ^k, one obtains the update equation [18]:

$$\Delta\rho^k = [\mathbf{f}'(\rho^k)^T\mathbf{f}'(\rho^k)]^{-1}[\mathbf{f}'(\rho^k)]^T[\mathbf{f}(\rho^k) - \mathbf{V}_0] \tag{4.8}$$

After the resistivity distribution is updated according to Equation 4.8, the process is repeated until the convergence criterion is met. This algorithm is often used for nonlinear problems.

A more physically inspired form of iterative algorithm is the Landweber technique, which was initially employed for ECT by Yang et al. [103]. The updated estimate of the permittivity image $\bar{G}_{p+1}$ is obtained from the current estimate $\bar{G}_p$ as follows:

$$\bar{G}_{p+1} = f\left[\bar{G}_p + \alpha\mathbf{S}^T(C - \mathbf{S}\bar{G}_p)\right] \tag{4.9}$$

where $\mathbf{S}$ is the sensitivity matrix, C is the vector of capacitance measurements as described in Section 4.3.1, α is an arbitrary gain factor, and f is a function that ensures each pixel's permittivity value lies between physically determined upper and lower limits.

The quantity $\mathbf{S}\bar{G}_p$ is simply the forward calculation of the capacitance values that would be expected, given the present image estimate $\bar{G}_p$. The residual vector between that calculation and the actual capacitance measurements is calculated and subjected to LBP by multiplication with $\mathbf{S}^T$. After the gain factor is applied, the image is updated and physically "regularized" by the function f. This last step is a crude form of applying bound constraints and is critically important in many electrical tomography systems. In most cases, the permittivity contrast is sufficiently small to allow the linear form of forward calculation given above. However, it is possible to use a full electric field calculation, if necessary, for the forward calculation step [103].

Recent years have seen a very healthy and robust interaction on the subject of image reconstruction algorithms between the process tomography and medical EIT communities. Considerable progress has been achieved at a fundamental level, as reflected in two recent Ph.D. dissertations [104, 105]. In the case of low-contrast imaging, where the linear approximation can be used, iterative methods can all be seen to fall into the class of Tikhonov regularization schemes [88, 104, 106]. These types of analyses have revealed new insights into the image reconstruction process. In particular, the cessation of the iteration process is understood to be equivalent to applying a low-pass filter to the spatial frequencies in the image. However, the process of iteration can continue only as far as is justified by the signal-to-noise ratio in the measured data, and a robust criterion for stopping the iteration procedure is necessary [105, 107]. The joint effort of these two communities has yielded an excellent public-domain software package called EIDORS [108] that is highly recommended for new workers in the field. The package also includes reconstruction software for diffuse optical tomography.

4.3.4 Model-Based Reconstruction

It is clear that many implicit and explicit forms of incorporating *a priori* knowledge of the subject are used in process tomography. Specifically, model-based methods assume a particular physical model of the process to be imaged. From the physical model, which includes parameters that describe the spatial distribution of material, one can calculate the interaction of the sensor (x-ray, electrical current, etc.) with the contents of the process. One can reconstruct the image by first using an optimization routine to find the parameters that give the best agreement between the calculated and measured projections. Once the parameter values have been established, the tomographic image can be reconstructed from the model [99]. The optimization component of this approach requires a number of iterations, so these methods may be seen as a subset of the iterative methods.

A model-based reconstruction may be used effectively even when the projection data are noisy or incomplete, because the model allows the substitution of observed (or assumed) information for some of the projection data. Of course, this technique can be applied only if sufficient *a priori* knowledge is available about the system; therefore, the approach may work well only for certain applications. When the temporal variation of the process being measured is similar to the data acquisition period for the tomography system, it is necessary to model the process in the reconstruction algorithm. The group at Kuopio University in Finland has played a leading role in this development [109].

4.3.5 Heuristic and Neural Net Methods

Neural networks offer a heuristic method of relating input signals to output signals for complex systems that cannot otherwise be modeled. These networks are based on neural structures found in simple organisms, and they may be either hard-wired or virtual (existing as a program in a computer). Many of the concepts can be traced to the work on perceptrons and to previous research on self-organizing systems [110].

A typical neural network is built from interconnected layers composed of single processing elements ("neurons"). A neuron has several inputs but only one output, which may be connected to other neurons through weighted connections. Thus, the output of a given neuron is based on a nonlinear transformation of the weighted sum of its inputs. The output values of the neurons in the final (output) layer of the network depend on the pattern of values presented at the input layer. To train the neural network, one presents a series of input patterns and compares the output to the "correct" response for that particular input. The weights interconnecting the neurons are changed according to a "learning rule"; the network adapts to its environment by modifying the weights according to this learning rule [111]. The "knowledge" contained in the network is stored in the values of the weights interconnecting the neurons. Therefore, knowledge can be transferred from one net to another simply by copying the values of the weights.

The application of neural networks to process tomography is based on the concept that useful information about a process (such as void fraction or multiphase flow velocities) can be obtained directly from the projection data without a full reconstruction of the cross-sectional image. This approach has the benefit of exploiting faster sampling rates and thus achieving higher throughput. In addition, the hardware needed to extract pertinent information may be less complex than that needed to reconstruct images.

A neural network of this type has been used to identify flow regimes in sand flow [112]. Recent developments have expanded these simple approaches in terms of both methods and applications [113–116].

4.4 CURRENT TOMOGRAPHY SYSTEMS

4.4.1 Current Status

Development of process tomography has accelerated progressively since its inception in the late 1980s. Research and development has grown across the range in hardware and software dimensions. A considerable number of well-resourced research and development teams are working on new innovations. In general, work has focused on improving (a) hardware, to improve system operation (for example, to reduce noise) and stretch operational range, and (b) software, to enhance the robustness and spatial resolution of images and the quality and speed of processing. There has also been considerable research work into new modalities, for example in the application of near-IR measurements to gain new insight into combustion processes, as discussed in Section 4.2.5.

Considering these developments against a normal technology development lifecycle, the early electrical methods have now received considerable integrated effort, and the resulting knowledge delivers very stable and broad-ranged tomography platforms. For example, early ERT systems were limited to a narrow range of "intermediate" conductivity and were therefore equally limited in their application domain. Current systems can operate with processes having a very wide range of properties, down to liquids with very high conductivity values. Similarly, early reliance on crude reconstruction techniques has been overtaken by a range of methods where current-generation computer processing can support a choice of algorithms in which resolution can be traded for time.

A great deal of work is under way in many processing engineering laboratories to exploit process tomography systems to reveal more fundamental knowledge about process phenomena, to validate and help quantify physical process models, and to optimize the design of process equipment.

At the time of writing, no examples exist where a process tomography sensor subsystem is routinely deployed as part of a standard process or product. Some of the underlying reasons are related to the relative complexity and specificity of the sensor elements and to their need for close integration into the process, but there are also issues in the lack of generic access and interface standards for software and postsensor hardware [117]. Recent proposals for standardization of smart sensor networks (IEEE Standard P1451) may add impetus to this avenue of development. It is certainly also true that incorporation of tomography systems into process control loops is both very attractive and a major hurdle to overcome. Hence, much R&D work is anticipated in this topic, to follow up the pioneering work discussed in other chapters of this volume.

4.4.2 Commercial Instruments

Although no mass-produced applications of process tomography technology exist, a number of small companies have entered the market to supply experimental and pilot-scale equipment. Hence, a small number of general-purpose process tomography instruments are now available.

Industrial Tomography Systems Ltd. (www.itoms.com) manufactures a range of ERT and multimodal process tomography instruments, including a retrofit ERT Linear Probe, offering an insight into an existing process. Their standard ERT instrument can also be supplied in a form that has gained intrinsic safety certification, allowing its use in a range of difficult processes involving explosive, flammable, and similar mixtures. The instrument is shown in Figure 4.11.

Supported applications are classified by process type and industrial sector. A wide range of process types are supported: solid–liquid, including crystallization; switch from batch to continuous; control of in-line mixing; gas–liquid, including optimization of high-intensity gas mixing; and liquid–liquid, including investigation of static mixers. Application sectors supported are the food, chemical, and pharmaceutical industries. In the food industry, applications include investigation of new mixing processes, on-line measurement detection and identification of foreign bodies, concentration measurement, dough proving, sugar systems, food quality during processing, and an in-line food grade sensor. In the chemical and pharmaceutical industries, applications noted are experimental verification of CFD calculations, experimental testing of mixing parameters (such as residence times and mixing quality), on-line measurement,

FIGURE 4.11 A commercial ERT instrument (courtesy of Industrial Tomography Systems Ltd.).

batch mixing, crystallization, polymerization, pressure filtration, chromatography, bubble columns, and packed beds.

Process Tomography Ltd. (www.tomography.com) manufactures ECT instruments (see Figure 4.12). The ECT image estimates the permittivity distribution of the vessel contents averaged over the length of the sensor electrodes. Spatial resolution achievable depends on the size and radial position of the target object, together with its permittivity difference relative to that of the other material in

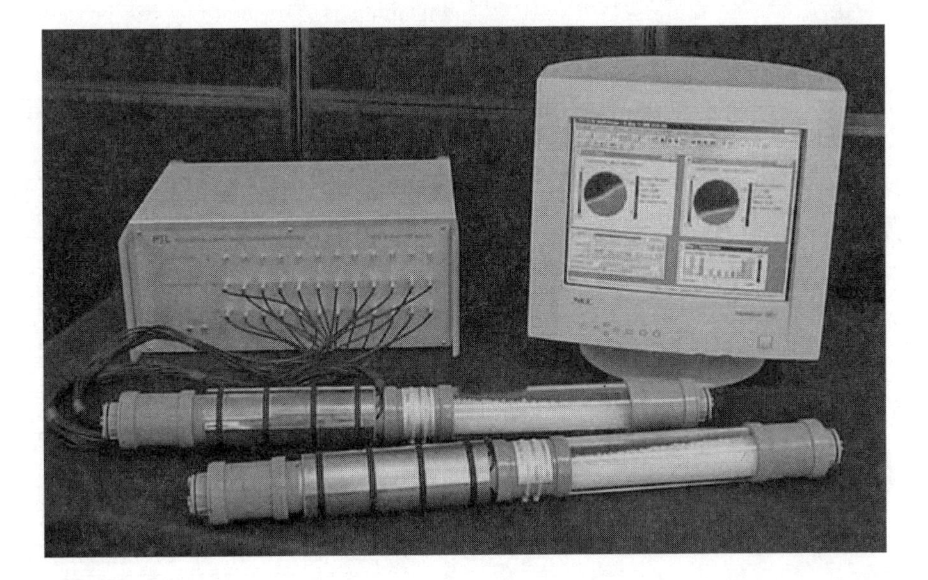

FIGURE 4.12 A commercial ECT instrument (courtesy of Process Tomography Ltd.).

the pipe. Typically, target objects with a diameter 5% of that of the pipe or vessel are detectable, subject to adequate contrast between the permittivity of the target and the surrounding media. Applications are noted in fluidized beds, gas in oil, flames, and velocity and flows in pneumatic conveying.

Tomoflow Ltd. (www.tomoflow.com) offers a system based on twin-plane correlation techniques from ECT tomographs. Their system supports a range of applications: image flows of fluid–fluid and fluid–solid mixtures, velocity measurement at any point or over any zone in the flow field, investigation of the details of flow structures, measurement of flow rate of a dispersed phase as a function of time and space, and measurement of the volume of moving structures within the section.

TABLE 4.1
Recent Multiphase Process Applications

Application	PT Sensing Technique	Key Attributes	Ref.
Pneumatic conveyors	Capacitance	Measures solids flow in sliding bed on production plant. 100 fps	118
Oilfield pipelines	Capacitance	Measures void fraction and flow regime in oil well riser at 200 fps	119
	Ultrasound	Real-time tomography for two-component flows	120
Extrusion	Resistance	Paste extrusion imaging	121
Fluidized beds	Gamma	Risers and standpipes (<1 m diameter) on production plants	122
	Capacitance	Bubble rise velocity, emulsion phase expansion	123
		Cross-correlation of images	124
Media Milling	Capacitance	Quantification of fluidization in mills	125
Trickle beds	X-ray	Packing morphology in 60 cm diameter laboratory bed, 512 × 512 pixels	126
	Capacitance	High speed liquid pulse imaging in 12 cm diameter lab bed, 32 × 32 pixels	127
Hydrocyclones	Resistance	Solids concentration profile and stability of air core in 20 mm separator	128
	X-ray	CT in industrial air sparged hydrocyclone	129
Colloidal suspensions	Resistance	Platelet settlement and alignment rate	130
	Magnetic resonance	Velocity estimation of fluids in motion	131
Environmental (large scale sites)	Resistance	Leakage detection from storage tanks	132
		Ionic transport through soil	133
		Gasoline plume imaging	134
Polymerization	Capacitance and resistance	Polymerization reaction in pilot plant	135
Mixers	Resistance	Multiplane imaging of mixer at 25 fps	136

4.5 APPLICATIONS

The potential value of early PT developments in the design, evaluation, monitoring, and control of industrial processes has become more evident. Developments in the availability of sophisticated computer aids have also changed design capabilities in the chemical process industries. Design strategies are now based firmly on a variety of modeling techniques and simulations that may incorporate thermodynamics, reaction kinetics, and hydrodynamics. For instance, the increased dependence on computational fluid dynamics models must be accompanied by independent validation of each model against experimental results. It is in this context that the various sensing methods of process tomography offer a convenient means of model verification in an industrial environment, rather than in a simplified "model" reactor using conventional laser or optical tracer techniques.

In addition to model validation, process tomography has the ability to measure concentration profiles and to identify phase boundaries within vessels and pipelines. These data are relevant to the elucidation of fundamental reaction kinetics and the optimum design of large-scale equipment. It is envisaged that process tomography will eventually be used for the purposes of plant control, and that this will become a strong R&D theme in the discipline.

A range of proven applications, varying from laboratory demonstrations to pilot plant evaluations, is listed in Table 4.1. This list is not exhaustive, but it conveys the breadth and potential impact of process tomography. Several of these applications are discussed in more depth later in this volume.

REFERENCES

1. RA Williams, MS Beck, eds. *Process Tomography: Principles, Techniques and Applications.* Oxford: Butterworth-Heinemann, 1995.
2. S Webb. *From the Watching of Shadows: The Origins of Radiological Tomography.* New York: Adam Hilger, 1990.
3. J Radon. *Berichte Verh. Sachsische Akad. Wiss.,* Leipzig, Math. Phys. Kl. 69:262–267, 1917.
4. GT Herman. *Image Reconstruction from Projections.* New York: Academic, 1980.
5. W Watson. X-ray apparatus. U.S. Patent No. 2,196,618 (1940).
6. AM Cormack. Representation of a function by its line integrals, with some radiological applications. *J. Appl. Phys.* 34:2722–2727, 1963.
7. AM Cormack. Representation of a function by its line integrals, with some radiological applications II. *J. Appl. Phys.* 35:2908–2913, 1964.
8. GN Hounsfield. Computerised transverse axial scanning (tomography). Part 1: Description of system. *Br. J. Radiol.* 46:1016–1022, 1973.
9. J Ambrose, G Hounsfield. Computerised transverse axial scanning (tomography). Part 2: Clinical applications. *Br. J. Radiol.* 46:1023–1047, 1973.
10. H Fricke, S Morse. The electric capacity of tumors of the breast. *J. Cancer Res.* 10:340–376, 1926.
11. RA Goldensohn, L Zablow. An electrical spirometer. *J. Appl. Physiol.* 14:463–464, 1959.

12. LE Baker, LA Geddes, HE Hoff. Quantitative evaluation of impedance spirometry in man. *Am. J. Med. Elect.* 4:73–77, 1965.
13. JR Inman, Jr., J Ryu, SH Ward. Resistivity inversion. *Geophysics* 38:1088–1108, 1973.
14. WM Telford, LP Geldart, RE Sheriff, DA Key. *Applied Geophysics: Resistivity Methods.* Cambridge: Cambridge University Press, 1976.
15. RP Henderson, JG Webster. An impedance camera for spatially specific measurements of the thorax. *IEEE Trans. Biomed. Eng.* BME-25:250–254, 1978.
16. D Isaacson. Distinguishability of conductivities by electric current computed tomography. *IEEE Trans. Medical Imaging* MI-5:91–95, 1986.
17. DC Barber, BH Brown. Applied potential tomography. *J. Phys. E: Sci. Instrum.* 17:723–733, 1984.
18. K Boone, D Barber, BH Brown. Imaging with electricity: Report of the European Concerted Action on Impedance Tomography. *J Med. Eng. Technol.* 21:201–232, 1997.
19. DS Holder, ed. *Clinical and Physiological Applications of EIT.* London: UCL Press, 1993.
20. JG Webster. *Electrical Impedance Tomography.* Bristol: Adam Hilger, 1990.
21. BH Brown. Medical impedance tomography and process impedance tomography: a brief review. *Meas. Sci. Technol.* 12:991–996, 2001.
22. SM Huang, A Plaskowski, CG Xie, MS Beck. Tomographic imaging of two-component flow using capacitance sensors. *J. Phys. E* 22:173–177, 1989.
23. GE Fasching. 3-D capacitance density imaging of fluidized beds. U.S. Patent No. 4,926,112 (1990).
24. EE Bliamptis. Plural electrode system for determining the electrical parameters of large samples of material *in situ.* U.S. Patent No. 3,975,676 (1976).
25. JO Parra, TE Owen, BM Duff. A synthetically focused resistivity method for detecting deep cavities. *Geophysics* 52:388, 1987.
26. M Wang, FJ Dickin, MS Beck. Improved electrical impedance tomography data collection system and measurement protocols. In: MS Beck et al., eds. *Tomographic Techniques for Process Design and Operation (Proceedings of 1st ECAPT Workshop, Manchester, UK 1992).* Southampton, U.K.: Computational Mechanics Publications, 1993, pp. 75–88.
27. BS Hoyle, F Wiegand. Real-time parallel tomographic ultrasound imaging using Transputers. *Electronics Lett.* 24:605–606, 1988.
28. JC Bolomey. Recent European developments in microwave imaging techniques for ISM applications. *IEEE Trans.* MTT-37:777–781, 1989.
29. N Devanathan, D Moslemian, MP Dudukovic. Flow mapping in bubble-columns using CARPT. *Chem. Eng. Sci.* 45:2285–2291, 1990.
30. DJ Parker, MR Hawksworth, TD Benyon, J Bridgewater. Process engineering studies using positron-based imaging techniques. In: MS Beck et al., eds. *Tomographic Techniques for Process Design and Operation (Proceedings of 1st ECAPT Workshop, Manchester, UK 1992).* Southampton, U.K.: Computational Mechanics Publications, 1993, pp. 239–250.
31. ZZ Yu, AJ Peyton, MS Beck, LA Xu. Imaging system based on electromagnetic tomography (EMT). *Electronics Lett.* 29:625–626, 1993.
32. GA Johansen, T Froystein. Gamma detectors for tomographic flow imaging. *Flow Meas. Instrum.* 5:15–21, 1994.

33. DG Hayes, IA Gregory, MS Beck. Velocity profile measurement in two-phase flows. In: MS Beck et al., eds. *Tomographic Techniques for Process Design and Operation (Proceedings of 1st ECAPT Workshop, Manchester, UK 1992).* Southampton, U.K.: Computational Mechanics Publications, 1993, pp. 369–380.
34. F Wiegand, BS Hoyle. Development and implementation of real-time ultrasound process tomography using a transputer network. *Parallel Computing* 17:791–807, 1991.
35. CG Xie, SM Huang, BS Hoyle, R Thorn, C Lenn, D Snowden, MS Beck. Electrical capacitance tomography for flow imaging: System model for development of image reconstruction algorithm and design of primary sensors. *Proc. IEEE Part G* 139:89–98, 1992.
36. CG Xie, BS Hoyle, DG Hayes. Parallel processing approach using Transputers. In: RA Williams, MS Beck, eds. *Process Tomography: Principles, Techniques and Applications.* Oxford: Butterworth-Heinemann, 1995, pp. 251–280.
37. GM Lyon, JP Oakley. A digital signal processor based architecture for EIT data acquisition. In: MS Beck et al., eds. *Tomographic Techniques for Process Design and Operation (Proceedings of 1st ECAPT Workshop, Manchester, UK 1992).* Southampton, U.K.: Computational Mechanics Publications, 1993, pp. 89–94.
38. CG Xie. Image reconstruction. In: RA Williams, MS Beck, eds. *Process Tomography: Principles, Techniques and Applications.* Oxford: Butterworth-Heinemann, 1995, p. 309.
39. J Bond, I Faulks, J Xiaodong, K Ostrowski, RM West, RA Williams. Optimization of solid liquid mixing using three-dimensional resistance tomography. *Proc. Frontiers Industrial Process Tomography II,* Delft, 1997, pp. 343–348.
40. MS Beck, RA Williams, E Campogrande, MA Morris, RC Waterfall, eds. *Tomographic Techniques for Process Design and Operation (Proceedings of 1st ECAPT Workshop, Manchester UK, 1992).* Southampton, U.K.: Computational Mechanics Publications, 1993.
41. MS Beck, RA Williams, E Campogrande, MA Morris and RC Waterfall, eds. *Proceedings of 2nd ECAPT Workshop,* Karlruhe, Germany, 1993.
42. MS Beck, RA Williams, EA Hammer, E Campogrande, MA Morris, RC Waterfall, eds. *Proceedings of 3rd ECAPT Workshop,* Oporto, Portugal, 1994.
43. MS Beck, BS Hoyle, MA Morris, RC Waterfall, RA Williams, eds. *Proceedings of 4th ECAPT Workshop,* Bergen, Norway, 1995.
44. MS Beck, RA Williams. Process tomography: A European innovation and its applications. *Meas. Sci. Technol.* 7:215–224, 1996.
45. DM Scott, RA Williams, eds. *Frontiers in Industrial Process Tomography.* New York: Engineering Foundation, 1995.
46. BS Hoyle, DM Scott, eds. *Proceedings of Frontiers in Industrial Process Tomography II.* Delft, 1997.
47. Commission of the European Communities. Advanced tomographic sensors network. In: *A Road to European Cooperation,* 1999, pp. 40–41.
48. T York. Status of electrical tomography in industrial applications. *J. Electronic Imaging* 10:608–619, 2001.
49. D Georgakopoulos, WQ Yang, RC Waterfall. Best value design of electrical tomography systems. *Proc. 3rd World Congress on Industrial Process Tomography,* Banff, Canada, 2003, pp. 11–9.
50. WQ Yang. Hardware design of electrical capacitance tomography systems. *Meas. Sci. Technol.* 7:225–232, 1996.

51. WQ Yang, TA York. New AC-based capacitance tomography systems. *IEE Proc. Sci. Meas. Technol.* 146:47–53, 1999.
52. Y Ma, N Holliday, Y Dai, M Wang, RA Williams, G Lucas. A high performance online data processing system. *Proc. 3rd World Congress on Industrial Process Tomography,* Banff, Canada, 2003, pp. 27–32.
53. Y Ma, Y Dai, and M Wang. A specific signal conditioner for electrical impedance tomography. *Proc. 3rd World Congress on Industrial Process Tomography,* Banff, Canada, 2003, pp. 45–49.
54. T Savolainen, LM Heikkinen, M Vauhkonen, JP Kaipo. A modular adaptive electrical impedance tomography system. *Proc. 3rd World Congress on Industrial Process Tomography,* Banff, Canada, 2003, pp. 50–55.
55. AJ Peyton, MS Beck, AR Borges, JE de Olivera, GM Lyon, ZZ Wu, MW Brown, J Ferrerra. Development of electromagnetic tomography (EMT) for industrial applications, Part 1: Sensor design and instrumentation. *Proc. 1st World Congress on Industrial Process Tomography,* Hannover, Germany, 1999, pp. 306–312.
56. AR Borges, JE de Olivera, J Velez, C Taveres, F Linhares, AJ Peyton. Development of electromagnetic tomography (EMT) for industrial applications, Part 1: Image reconstruction and software framework. *Proc. 1st World Congress on Industrial Process Tomography,* Hannover, Germany, 1999, pp. 219–225.
57. HS Tapp, AJ Peyton. A state of the art review of electromagnetic tomography. *Proc. 3rd World Congress on Industrial Process Tomography,* Banff, Canada, 2003, pp. 340–346.
58. JC Bolomey. Microwave sensors. In: RA Williams and MS Beck, eds. *Process Tomography.* Oxford: Butterworth-Heinemann, 1995, pp. 151–164.
59. JM Rius, M Ferrando, L Jofre, E de los Reyes, A Broquetas. Microwave tomography: an algorithm for cylindrical geometries. *Electron. Lett.* 23:564–565, 1987.
60. A Boughriet, Z Wu, AT Nugroho, H McCann, LE Davis. Free-space imaging with an active microwave tomographic system. *Proc. SPIE* 4129:59–66, 2000.
61. FP Hindle, H McCann, KB Ozanyan. First demonstration of optical fluorescence auto-projection tomography. *Chem. Eng. J.* 77:127–135, 2000.
62. CF Bohren, DR Huffman. *Absorption and Scattering of Light by Small Particles.* New York: John Wiley & Sons, 1983, pp. 82 ff.
63. FP Hindle, SJ Carey, KB Ozanyan, DE Winterbone, E Clough, H McCann. Measurement of gaseous hydrocarbon distribution by a near-infrared absorption tomography system. *J. Electronic Imaging* 10:593–600, 2001.
64. CA Garcia-Stewart, N Polydorides, KB Ozanyan, H McCann. Image reconstruction algorithms for high-speed chemical species tomography. *Proc. Third World Congress on Industrial Process Tomography,* Banff, Canada, 2003, pp. 80–85.
65. K Salem, W Kwapinski, E Tsotsas, D Mewes. Novel experimental techniques for packed bed adsorber investigation. *Proc. AChEMA,* Frankfurt, 2003.
66. FEW Schmidt, ME Fry, EMC Hillman, JC Hebden, DT Delpy. A 32-channel time-resolved instrument for medical optical tomography. *Rev. Sci. Inst.* 71:256–265, 2000.
67. BS Hoyle. Real-time ultrasonic process tomography of flowing mixtures. *Part. Part. Syst. Charact.* 12:81–86, 1995.
68. SJ Norton, M Linzer. Ultrasonic reflectivity tomography: reconstruction with circular transducer arrays. *Ultrasonic Imaging* 1:154–184, 1979.

69. M Moshfeghi. Ultrasonic reflection mode tomography using fan-shaped beam insonification. *IEEE Trans. Ultrason. Ferroelec. Freq. Control* UFFC-33:299–314, 1986.
70. J Wolf. Investigation of bubbly flow by ultrasonic tomography. *Part. Syst. Charact.* 5:170–173, 1988.
71. F Wiegand, BS Hoyle. Simulations for parallel processing of ultrasound reflection-mode tomography with applications to two-phase flow measurement. *IEEE Trans. Ultrason. Ferroelec. Freq. Control* UFFC-36:652–660, 1989.
72. M Yang, HI Schlaberg, BS Hoyle, MS Beck, C Lenn. Parallel image reconstruction in real-time ultrasound process tomography for two-phase flow measurements. *Real-time Imaging J.* 3:295–303, 1997.
73. M Yang, HI Schlaberg BS Hoyle. Parallel implemented Hough transform for pattern recognition in ultrasound process tomography. *Proc. Frontiers in Industrial Process Tomography II,* Delft, 1997, pp. 349–354.
74. HI Schlaberg, M Yang, BS Hoyle. Ultrasonic reflection tomography for industrial processes. *Ultrasonics* 36:297–303, 1998.
75. M Yang, HI Schlaberg, BS Hoyle, MS Beck, C Lenn. Real-time ultrasonic tomography for two-phase flow imaging using a reduced number of transducers. *Trans. Ultrason. Ferroelec. Freq. Control* UFFC-46:492–501, 1999.
76. FJW Podd, HI Schlaberg, BS Hoyle. Model-based parameterization of a hydrocyclone air-core. *Ultrasonics* 38:804–808, 2000.
77. HI Schlaberg, FJW Podd, BS Hoyle. Ultrasound process tomography system for hydrocyclones. *Ultrasonics* 38:813–816, 2000.
78. MA Brown. *MRI: Basic Principles and Applications.* New York: Wiley-Liss, 1999.
79. LF Gladden. Nuclear magnetic resonance in chemical engineering: principles and applications. *Chem. Eng. Sci.* 49:3339–3408, 1994.
80. LD Hall, MHG Amin, S Evans, KP Nott, L Sun. Magnetic resonance imaging for industrial process tomography. *J. Electronic Imaging* 10:601–607, 2001.
81. DF Arola, GA Barrall, RL Powell, KL McCarthy, MJ McCarthy. Use of nuclear resonance imaging as a viscometer for process imaging. *Chem. Eng. Sci.* 52:2049–2057, 1997.
82. PC Lauterbur. Image formation by induced local interactions: examples employing nuclear magnetic resonance. *Nature* 242:190–191, 1973.
83. P Mansfield, PK Grannell. NMR ‘Diffraction’ in solids. *J. Phys. C: Solid State Phys.* 6:L422–426, 1973.
84. LF Gladden. Recent advances in MRI studies of chemical reactors: ultrafast imaging of multiphase flows. *Topics in Catalysis* 24:19–28, 2003.
85. DJ Parker, RN Forster, P Fowles, PS Takhar. Positron emission particle tracking using the new Birmingham positron camera. *Nucl. Instrum. Meth. Phys. Res. A* 477:540–545, 2002.
86. GA Johansen, T Froystein, BT Hjertaker, O Olsen. A dual-mode sensor flow imaging tomographic system. *Meas. Sci. Technol.* 7:297–307, 1996.
87. BS Hoyle, X Jia, FJW Podd, HI Schlaberg, M Wang, RM West, RA Williams, TA York. Design and application of a multi-modal process tomography system. *Meas. Sci. Technol.* 12:1157–1165, 2001.
88. M Bertero, P Boccacci. *Introduction to Inverse Problems in Imaging.* Bristol: IOP Publishing, 1998.
89. A Rosenfeld, AC Kak. *Digital Picture Processing* (2nd ed.). New York: Academic Press, 1982.

90. HH Barrett, W Swindell. Analog reconstruction methods for transaxial tomography. *Proc. IEEE* 65:89–107, 1977.
91. LA Feldkamp, LC Davis, JW Kress. Practical cone-beam algorithm. *J. Opt. Soc. Am. A* 1:612–619, 1984.
92. JE Greivenkamp, W Swindell, AF Gmitro, HH Barrett. Incoherent optical processor for x-ray transaxial tomography. *Applied Optics* 20:264–273, 1981.
93. DM Scott. Non-computed tomography. *Proc. European Concerted Action on Process Tomography Workshop,* Karlsruhe, Germany, 1993, pp. 124–127.
94. AC Kak, M Slaney. *Principles of Computerized Tomographic Imaging.* New York: IEEE Press, 1987.
95. RM Lewitt. Reconstruction algorithms: Transform methods. *Proc. IEEE* 71:390–408, 1983.
96. M Soumekh. Image reconstruction techniques in tomographic imaging systems. *IEEE Trans. Acoustics Speech and Signal Processing ASSP.* 34:952–962, 1986.
97. M Müller. Tomographic imaging algorithms and synthetic detector arrays. Ph.D. dissertation, University of Delaware, 1993.
98. SF Yau, SH Wong. A linear sinogram extrapolator for limited angle tomography. *Proc. ISCP,* pp. 386–389, 1996.
99. Ø Isaksen, JE Nordtvedt. A new reconstruction algorithm for process tomography. *Meas. Sci. Technol.* 4:1464–1475, 1993.
100. D Mishra, K Muralidhar, P Munshi. A robust MART algorithm for tomographic applications. *Numerical Heat Transfer* B35:485–506, 1999.
101. L Landweber. An iterative formula for Fredholm integral equations of the first kind. *Am. J. Math.* 73:615–624, 1951.
102. TJ Yorkey, JG Webster, WJ Tompkins. An optimal impedance tomographic reconstruction algorithm. *Proc. 8th Annual Conference of the IEEE Engineering in Medicine and Biology Society,* Fort Worth, Texas, 1986, pp. 339–342.
103. WQ Yang, DM Spink, TA York, H McCann. An image-reconstruction algorithm based on Landweber's iteration method for electrical capacitance tomography. *Meas. Sci. Technol.* 10:1065–1069, 1999.
104. M Vauhkonen. Electrical impedance tomography and prior information. Ph.D. dissertation, University of Kuopio, 1997.
105. N Polydorides. Image reconstruction algorithms for soft-field tomography. Ph.D. thesis, University of Manchester Institute of Science and Technology, 2002.
106. AN Tikhonov, VY Arsenin. *Solutions of Ill-Posed Problems.* Washington: Winston/Wiley, 1977.
107. N Polydorides, H McCann. Electrode configurations for improved spatial resolution in Electrical Impedance Tomography. *Meas. Sci. Technol.* 13:1862–1870, 2002.
108. N Polydorides, WRB Lionheart. A Matlab toolkit for three-dimensional electrical impedance tomography: a contribution to the EIDORS project. *Meas. Sci. Technol.* 13:1871–1883, 2002.
109. A Seppänen, M Vauhkonen, PJ Vauhkonen, E Somersalo, JP Kaipio. State estimation with fluid dynamical evolution models in process tomography: An application to impedance tomography. *Inverse Problems* 17:467–483, 2001.
110. DE Rumelhart, JL McClelland. *Parallel Distributed Processing, Vol.* 1. Cambridge, MA: MIT Press, 1986.
111. DO Hebb. *The Organization of Behavior.* New York: Wiley, 1949.

112. AR Bidin, RG Green, ME Shackleton, RW Taylor. Neural networks for flow regime identification with dry particulate flows. *Part. Syst. Charact.* 10:234–238, 1993.
113. PM Duggan, TA York. Tomographic image reconstruction using RAM-based neural networks. *Proc. European Concerted Action on Process Tomography Workshop,* Bergen, Norway, 1995, pp. 411–419.
114. J Mohamad-Saleh, BS Hoyle, FJW Podd, DM Spink. Direct process estimation from tomographic data using artificial neural systems. *J. Electronic Imaging* 10:646–652, 2001.
115. J Mohamad-Saleh, BS Hoyle. Determination of multi-component flow process parameters based upon electrical capacitance tomography data using artificial neural networks. *Meas. Sci. Technol.* 13:1815–1821, 2002.
116. W Warsito, LS Fan. Development of 3-dimensional electrical capacitance tomography based on neural network multi-criterion optimization image reconstruction. *Proc. 3rd World Congress on Industrial Process Tomography,* Banff, Canada, 2003, pp. 391–396.
117. BS Hoyle, X Jia, XZ Wang, RA Williams. Generic reality visualisation models through fusion of process tomography sensor data. *Proc. 3rd World Congress on Industrial Process Tomography,* Banff, Canada, 2003, pp. 565–570.
118. T Dyakowski, RB Edwards, CG Xie, RA Williams. Application of capacitance tomography to gas-solid flows. *Chem. Eng. Sci.* 52:2099–2110, 1997.
119. CG Xie, WQ Yang, MS Beck. Electrical capacitance tomography: from design to applications. In: DM Scott, RA Williams, eds. *Frontiers in Industrial Process Tomography.* New York: Engineering Foundation, 1995, p. 292.
120. HI Schlaberg, M Yang, BS Hoyle. Real-time ultrasonic process tomography for 2-component flows. *Electronics Lett.* 32:1571–1572, 1996.
121. RM West, DM Scott, G Sunshine, J Kostuch, L Heikkinen, M Vauhkonen, BS Hoyle, HI Schlaberg, R Hou, RA Williams. *In situ* imaging of paste extrusion using electrical impedance tomography. *Meas. Sci. Technol.* 13:1890–1897, 2002.
122. JR Bernard, L Desbat, P Turlier. Gamma ray imaging of industrial beds. In: DM Scott, RA Williams, eds. *Frontiers in Industrial Process Tomography.* New York: Engineering Foundation, 1995, pp. 197–206.
123. JS Halow. Capacitance imaging of fluidized beds. In: RA Williams, MS Beck, eds. *Process Tomography: Principles, Techniques and Applications.* Oxford, U.K.: Butterworth-Heinemann, 1995, pp. 447–486.
124. T Dyakowski. Electrical capacitance tomography for fluidized bed analysis. In: DM Scott, RA Williams, eds. *Frontiers in Industrial Process Tomography.* New York: Engineering Foundation, 1995, pp. 185–196.
125. DM Scott, OW Gutsche. ECT studies of bead fluidization in vertical mills. *Proc. of 1st World Congress on Industrial Process Tomography,* Buxton, U.K., 1999, pp. 90–95.
126. D Toye, P Marchot, M Crine M, G L'Homme. Analysis of a liquid flow distribution in a trickle bed reactor. In: DM Scott, RA Williams, eds. *Frontiers in Industrial Process Tomography.* New York: Engineering Foundation, 1995, pp. 185–196.
127. N Reinecke, D Mewes. Calculation of interface velocities from single plane tomographic data. *Proc. of European Concerted Action on Process Tomography Workshop,* Bergen, Norway, 1995, pp. 41–50.

128. RA Williams, FJ Dickin, A Gutierrez, MS Beck, M Wang, OM Ilyas, T Dyakowski. Monitoring cyclone performance using resistance tomography. In: DM Scott, RA Williams, eds. *Frontiers in Industrial Process Tomography.* New York: Engineering Foundation, 1995, pp. 261–274.
129. JD Miller, CL Linn. Quantitative x-ray computed tomography and its application in the process engineering of particulate systems. In: DM Scott, RA Williams, eds. *Frontiers in Industrial Process Tomography.* New York: Engineering Foundation, 1995, pp. 207–221.
130. RA Williams, T Dyakowski, CG Xie, SP Luke, PJ Gregory, RB Edwards, FJ Dickin, LF Gate. Industrial measurement and control of particulate processes using electrical tomography. *Proc. of European Concerted Action on Process Tomography Workshop,* Bergen, Norway, 1995, pp. 3–15.
131. SJ Gibbs, B Newling, LD Hall, DE Haycock, WJ Frith, S Ablett. Magnetic resonance velocity. In: DM Scott, RA Williams, eds. *Frontiers in Industrial Process Tomography.* New York: Engineering Foundation, 1995, pp. 47–58.
132. W Daily, A Ramirez. Environmental process tomography in the United States. *Chem. Eng. J.* 56:159–165, 1995.
133. LJ West, DI Stewart, AM Bailey, CBC Shaw. Electrical resistance tomography of ionic transport through soil in an electric field. In: DM Scott, RA Williams, eds. *Frontiers in Industrial Process Tomography.* New York: Engineering Foundation, 1995, p. 315.
134. W Daily, A Ramirez, D LaBrecque, A Binley. Detecting leaks in hydrocarbon storage tanks using electrical resistance tomography. In: DM Scott, RA Williams, eds. *Frontiers in Industrial Process Tomography.* New York: Engineering Foundation, 1995, pp. 173–183.
135. T Dyakowski, T York, M Mikos, D Vlaev, R Mann, G Follows, A Boxman, M Wilson. Imaging nylon polymerisation processes by applying electrical tomography. *Chem. Eng. J.* 77:105–109, 2000.
136. R Mann, RA Williams, T Dyakowski, FJ Dickin, RB Edwards. Development of mixing models using resistance tomography. In: DM Scott, RA Williams, eds. *Frontiers in Industrial Process Tomography.* New York: Engineering Foundation, 1995, p. 324.

5 Image Processing and Feature Extraction

Dongming Zhao

CONTENTS

5.1 INTRODUCTION

Digital images are discrete two-dimensional (2D) representations of real visual scenes, typically comprising an array of gray-level values that represent light intensity and color. Images are encountered in a wide variety of disciplines including art, human vision, astronomy, and industrial engineering applications (the focus of this book). Digital image processing describes operations carried out on digitized images. Common image processing operations include filtering, sampling, coding, feature extraction, pattern recognition, motion analysis, and coloring processes. This processing is generally accomplished electronically with advanced electronic hardware and computing software.

The imaging process itself compresses the original physical information into a gray- level array of pixels (picture elements). In this respect, the image is a collection of degenerate transformations. The information is not irrevocably lost, because there is much spatial redundancy: neighboring pixels in an image have the same or nearly the same physical parameters. Several techniques described in this chapter (such as filtering) exploit this redundancy to undo the degeneracies in the image acquisition. These techniques reflect the spatial properties of the scene. Common measures for such properties are surface discontinuities, range, surface orientation, and geometry. Pertinent operations include *image enhancement,* such as gray-level transformations, histogram processing, smoothing and sharpening spatial filters, and image enhancement by frequency domain methods, and *image restoration* through the use of various filters, including noise reduction by frequency domain filtering.

In a vision system, a systematic approach to extracting information from processed images is *feature extraction.* This procedure groups parts of a generalized image into recognizable units and classifies them into a set of one or more characteristics. The results in such a process become segmented images. The features, which in general also represent a class of patterns, are based on internal domain-dependent models of the objects. Feature extraction methods include *segmentation* by means of edge detection, thresholding, region-based segmentation, and adaptive thresholding, and *feature representation* using Freeman chain code, object size filtering, moment invariants, and skeletonization.

Morphological image processing in machine vision and process control is a commonly used tool for extracting image components that can represent a region's shape, such as boundaries, skeletons, and the convex hull. Many morphological image processing algorithms are inherently parallel and have been successfully implemented in real-time processors. Machine vision applications extensively employ this type of image processing. Relevant topics in this chapter include basic operations (dilation, erosion, opening, and closing) and feature representation using morphological methods (boundaries, skeletons, watersheds, hit-or-miss transformation).

As a convention throughout this chapter, a digital image is defined as a 2D function, $f(x, y)$, where x and y are spatial coordinates, and the intensity of the image is represented by amplitude of f. The notation $f(x, y)$ is called intensity or gray-level of the image at (x, y), and the values of f are finite and discrete quantities.

This chapter consists of five sections, each of which deals with one distinct area in image processing and analysis. Section 5.2 presents basic concepts in image enhancement, and it covers image processing in the spatial and frequency domains. Fundamental operations such as spatial filtering, Fourier transform and filtering, spatial transformations, and histogram are discussed. Section 5.3 covers topics in image restoration, including noise, spatial filtering of noise, and noise smoothing in frequency domain. Section 5.4 includes topics in image segmentation: edge detection, Hough transform, thresholding, boundary detection, and segment labeling. In Section 5.5, feature representation for images is investigated. The topics include chain code, Fourier descriptor, moments, shape feature and representation, medial axes, and skeletons. Section 5.6 covers the basic concepts in morphological image processing, as well as basic operators, binary and grayscale images, and some useful transformations such as hit-or-miss transformation and morphological skeletonization.

5.2 IMAGE ENHANCEMENT

Approaches to image enhancement may be classified into two broad categories: spatial domain approaches and frequency domain approaches. Spatial domain refers to the image plane itself, and the approaches in this category are based on direct manipulation of image pixels. Frequency domain processing is based on modifying the frequency components in the Fourier transform of an image.

5.2.1 Basic Gray-Level Transformations

The term *spatial domain* refers to the geometrical distribution of image pixels. The operations of spatial domain processing function directly on these image pixels. The spatial domain processes are denoted by the expression

$$s(x, y) = Tr[f(x, y)] \tag{5.1}$$

where $f(x, y)$ is the input image, $s(x, y)$ is the transformed image, and Tr is an operator. An operator is typically applied to a sub-block of the image around a pixel location (x, y). The center of the sub-block is moved from pixel to pixel. The operator Tr is applied to each pixel at (x, y) to yield the output s. A sub-block is usually a circle or a rectangular array. Image contrast stretching is a simple transformation where Tr only applies to a single pixel at a time. The gray levels of $f(x, y)$ and $s(x, y)$ have a one-to-one correspondence. For example, $s = Tr[f]$ where Tr is shown in Figure 5.1a. In this operation, the values of f around m are stretched into a wider range of s, and conversely the values of f far from m are compressed into narrow ranges. The equation for this operation is

$$s = \frac{1}{1+\left(\frac{m}{f}\right)^{E}} \tag{5.2}$$

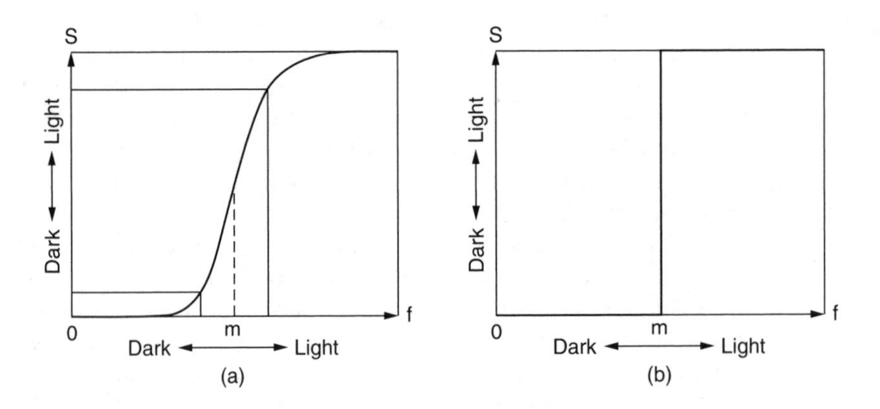

FIGURE 5.1 Gray-level transformation for contrast enhancement. (a) Conversion from a narrow band to full range output. (b) Binary transition, or thresholding at m.

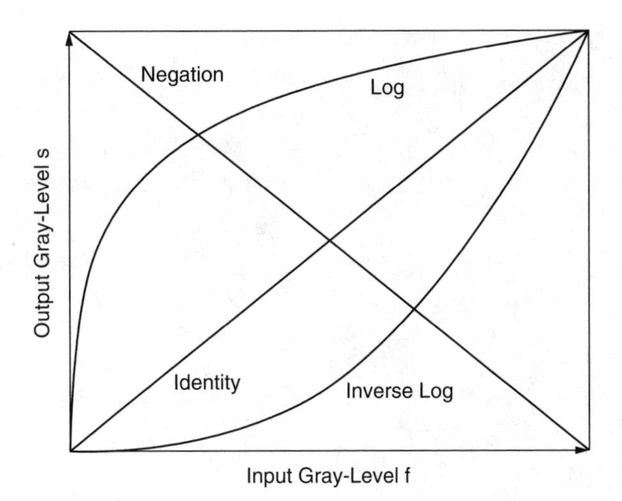

FIGURE 5.2 Some basic gray-level transformation functions for image enhancement.

A limiting case, where the parameter E approaches infinity, is shown in Figure 5.1b. The operation generates a binary image. A mapping of a gray-level image into a two-level image is called *thresholding*.

Basic gray-level transformations are illustrated in Figure 5.2, which shows three types of transformations used frequently for image enhancement: linear, logarithmic, and power-law.

5.2.1.1 Image Negation

The negative of an image with gray levels in the range from 0 to $L - 1$ is obtained through the negative transformation shown in Figure 5.3 as

$$s = L - 1 - f \tag{5.3}$$

Reversing the intensity levels of an image produces the equivalent of a photographic negative. This type of transformation does not enrich information in an image; rather, it is for ease of human inspection.

5.2.1.2 Log Transformations

The log transformation as shown in Figure 5.2 is

$$s = c\log(1 + f) \tag{5.4}$$

where c is a constant, and $f \geq 0$. The log curve in Figure 5.2 shows that this transformation maps a narrow range of low gray-level values in the input image into a wide range of output levels, and a wider range of high gray-level values in the input image into a narrow range output. Therefore, the transformation

FIGURE 5.3 Negation sample. (a) Original image of roses. (b) Negative image obtained using the negative transformation in Equation 5.3.

expands the values of dark pixels in an image while compressing the higher-level pixels.

The log function has the important characteristic that it compresses the dynamic range of images with large variations in pixel values. Figure 5.4 shows the effects of a log transformation that expands the dark areas of a gray-level image into a broader range of brighter intensities.

5.2.1.3 Power-Law Transformation

The output image of a power-law transformation is related to its input image by

$$s(x, y) = c \cdot [f(x, y)]^{\gamma} \tag{5.5}$$

where c and γ are constant. The value of γ determines the degree to which the intensity range is compressed or expanded. In a power-law transformation, each pixel of the original image is raised to a specified exponent value. By selecting the exponent values appropriately, one may boost either high or low luminance values. The slope of the transformation is chosen to be greater than 1 in the region of stretch and less than 1 in the region of compression; see Figure 5.5. The exponent in the power-law equation is referred to as gamma, and the process is called gamma correction. Figure 5.6 shows two samples with different histograms.

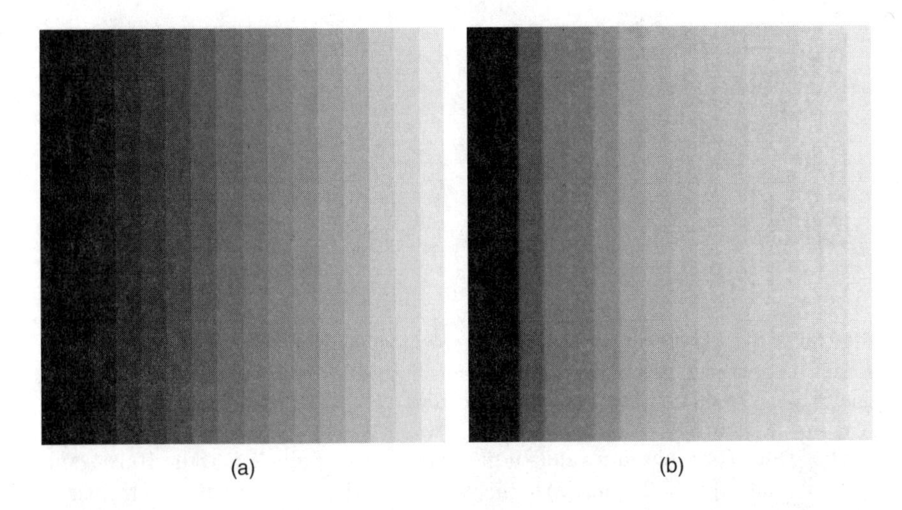

FIGURE 5.4 Log transformation. (a) Original gray-scale image. (b) Result of applying the log transformation.

5.2.1.4 Histogram Processing

Frequently, an image is acquired in such a way that the resulting brightness values do not make full use of the available dynamic range. This can be easily observed in the histogram of the brightness values in Figure 5.7. This situation is corrected by stretching the histogram over the available dynamic range. If the image is intended to go from brightness 0 to brightness $L - 1$, then one generally maps

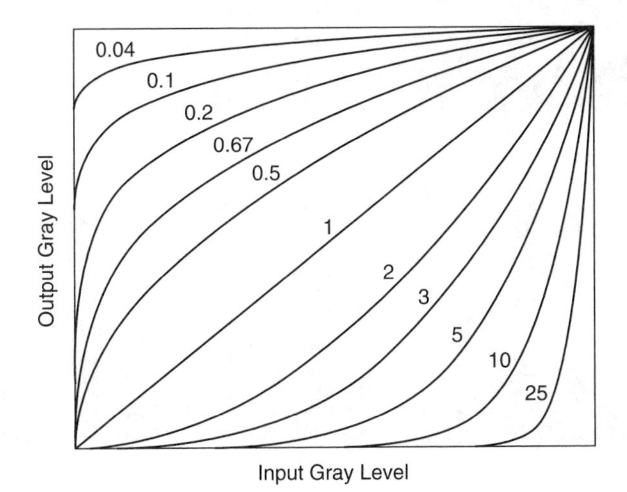

FIGURE 5.5 Power-law transformation. Plots of Equation 5.5 for various values of γ and $c = 1$.

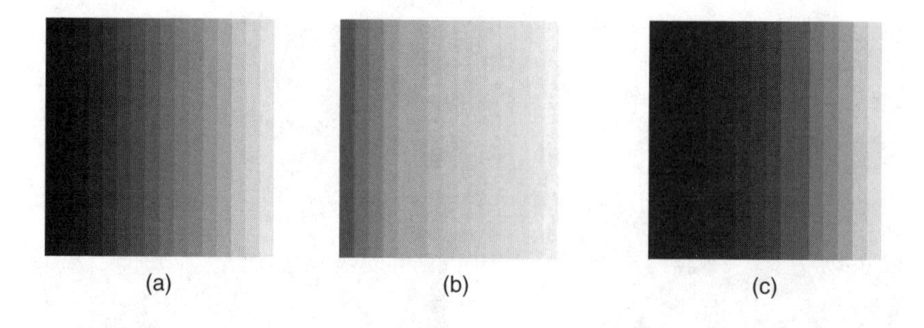

FIGURE 5.6 Examples of power-law transformation. (a) Original gray-scale image. (b, c) Results of applying the transformation in Equation 5.3.2 and Equation 5.3.3 with $\gamma < 1$ and $\gamma > 1$, respectively, and $c = 1$ in both cases.

the 0% value (or minimum value of the image) to the value 0 and the 100% value (or maximal value of the image) to the value $L - 1$. The appropriate transformation is given by

$$s(x, y) = (L-1)\frac{f(x, y) - f_{\min}}{f_{\max} - f_{\min}} \tag{5.6}$$

where $f_{\min}$ and $f_{\max}$ are the minimal and maximal values over the entire image region.

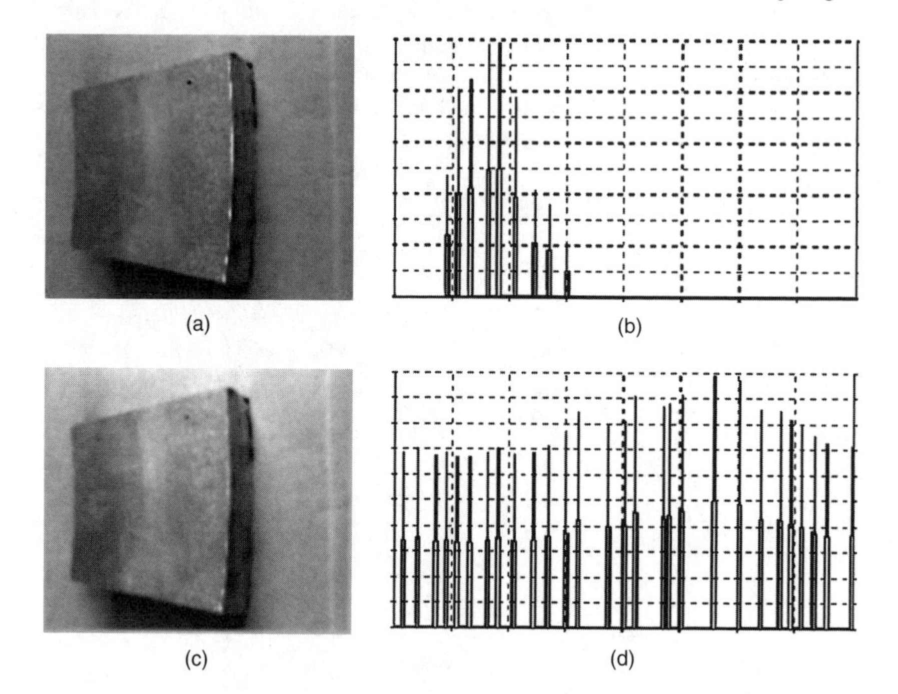

FIGURE 5.7 Histogram. (a) Original image of a brick on the wall. (c) Result of histogram equalization. (b) and (d) are the corresponding histograms.

Equation 5.6, however, can be somewhat sensitive to outliers; a less sensitive and more general version is given by

$$s(x, y) = \begin{cases} 0, & f(x, y) \le p_{\text{low}}\% \\ (L-1)\dfrac{f(x, y) - f_{\min}}{f_{\max} - f_{\min}}, & p_{\text{low}}\% < f(x, y) < p_{\text{high}}\% \\ L-1, & f(x, y) \ge p_{\text{high}}\% \end{cases} \tag{5.7}$$

In this version one might choose the 1% and 99% values for $p_{\text{low}}\%$ and $p_{\text{high}}\%$, respectively, instead of the 0% and 100% values represented by Equation 5.6. It is also possible to apply the contrast-stretching operation on a regional basis by using the histogram from a region to determine the appropriate limits for the algorithm. Note that it is possible to suppress the term $L - 1$ and simply normalize the brightness range to $0 \le s(x, y) \le 1$.

Histogram modification processes can be considered as a monotonic transformation $s_k = T\{f_j\}$ where the input image intensity range $f_1 \le f_j \le f_J$ is mapped into an output variable $s_1 \le s_k \le s_K$ such that the output probability distribution $\Pr\{s_k = b_k\}$ derives from a given input probability distribution $\Pr\{f_j = a_j\}$ while a_j and b_k are reconstruction values of the j-th and k-th levels. *Histogram equalization* refers to a mapping of gray levels f into gray levels s such that the distribution of gray levels s is uniform. This mapping stretches contrast (expands the range of gray levels) for gray levels near histogram maxima and compresses contrast in areas with gray levels near histogram minima. Since contrast is expanded for most of the image pixels, the transformation usually improves the detectability of many image features.

The histogram equalization mapping may be defined in terms of the cumulative histogram for the image. To see this, consider Figure 5.8. To map a small

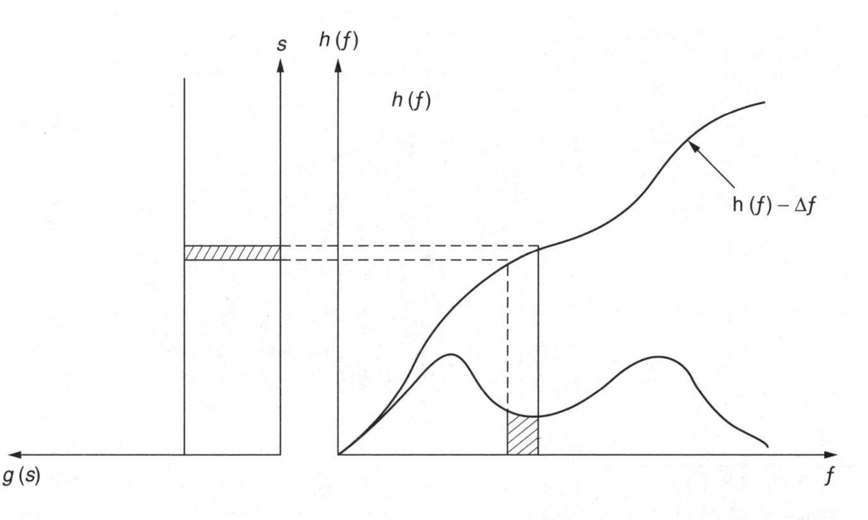

FIGURE 5.8 Illustration of a histogram equalization technique.

interval of gray levels Δf onto an interval Δs in the general case, it must be true that

$$g(s) \cdot \Delta s = h(f) \cdot \Delta f \tag{5.8}$$

where $g(s)$ is the new histogram. In histogram equalization case, $g(s)$ should be uniform; then

$$g(s) = \frac{N}{M} \tag{5.9}$$

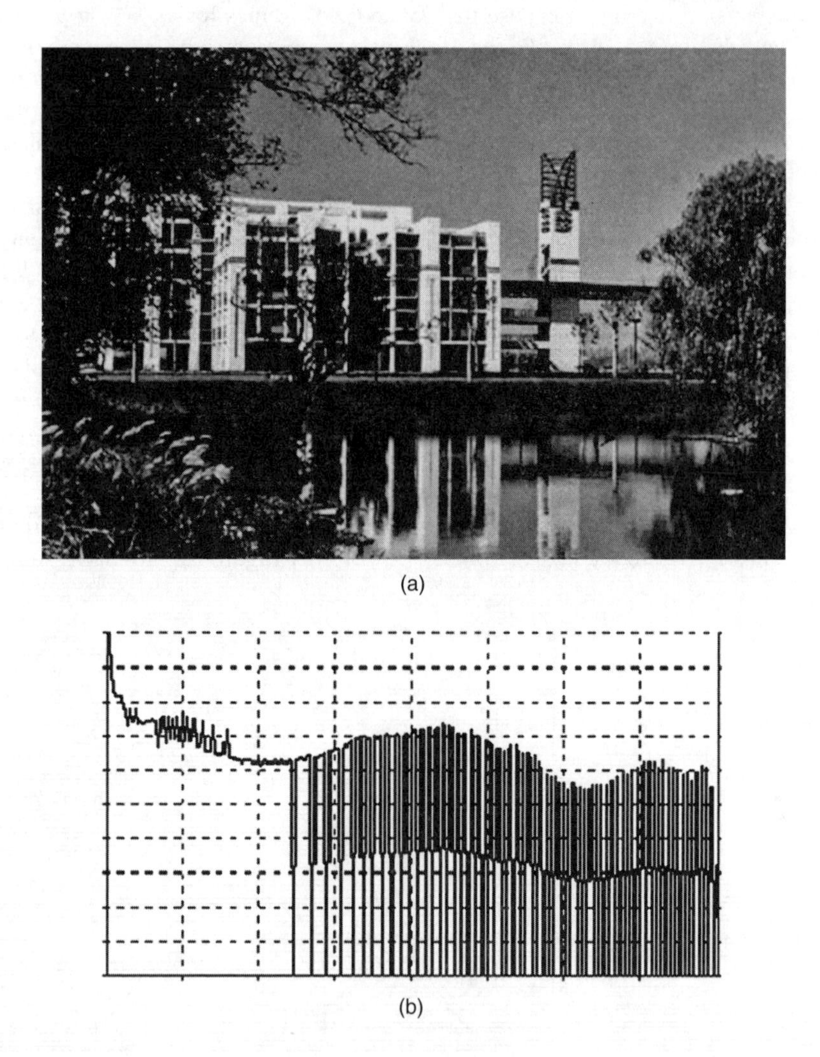

FIGURE 5.9 Histogram comparison prior to and after histogram equalization. (a) Original image. (b) Histogram of (a). (c) Image after histogram equalization of (a). (d) Histogram of (c).

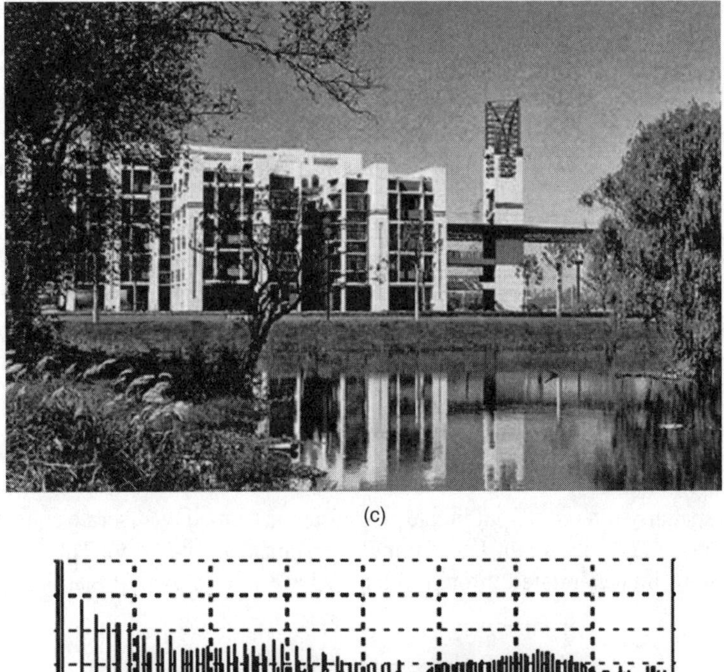

(c)

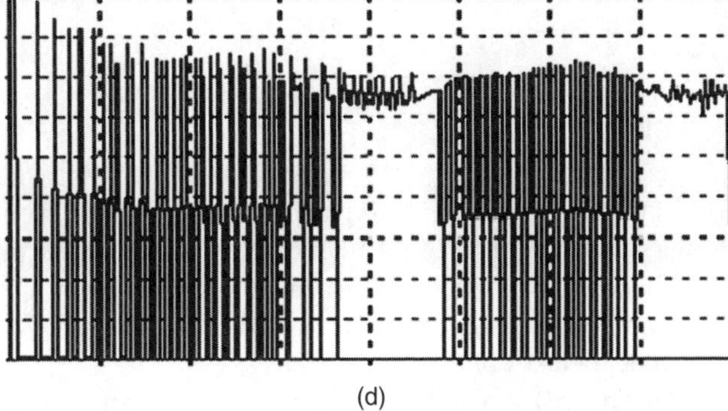

(d)

FIGURE 5.9 (Continued).

where N is the total number of pixels in the image and M is the number of gray levels. Combining Equation 5.8 and Equation 5.9 and integrating, we have

$$s(f) = \frac{M}{N}\int_0^f h(x)\,dx \tag{5.10}$$

Figure 5.9 shows a comparison of histograms, prior to and after histogram equalization.

5.2.2 Spatial Filters

5.2.2.1 Smoothing

An image may be subject to noise and interference from several sources including electronic sensor noise, photographic grain noise, and errors from media transfers. These noise effects can be reduced through smoothing filtering techniques.

Image noise arising from a noisy sensor or channel transmission errors usually appears as discrete isolated pixel variations that are not spatially correlated. Pixels that are in error often appear visually to be markedly different from their neighboring pixels. Here we discuss several filtering techniques that have proved useful for noise reduction.

Figure 5.10 contains three test images. Figure 5.10a is the original image. Figure 5.10c was obtained by adding uniformly distributed noise to the original image of Figure 5.10a. In the impulse noise example in Figure 5.10e, maximum amplitude pixels replace original image pixels in a spatially random manner. The corresponding histograms are shown in the figure.

Noise in an image generally has a higher spatial frequency spectrum than the normal image components. Hence some simple low-pass filtering can be effective for cleaning noise. We discuss spatial filtering here and will later deal with Fourier domain methods.

A spatially filtered output image $g(x, y)$ can be formed by discrete convolution of an image $f(x, y)$ with an $L \times K$ impulse response array $h(x, y)$. Then the input and output images related through filter $h()$ have the following relation:

$$g(x, y) = \sum_{m=-W}^{W} \sum_{n=-H}^{H} h(m, n) f(x+m, y+n) \tag{5.11}$$

where $W = (L - 1)/2$ and $H = (K - 1)/2$. Equation 5.11 shows that an output image pixel value is the sum of products of the filter coefficients with the corresponding pixels directly under the filter. Note that when ($m = 0$, $n = 0$), $f(x, y)$ and $g(x, y)$ correspond with $h(0, 0)$; the filter center is always at (x, y) when the computation of the sum of products takes place.

For noise cleaning, $h()$ should be a low-pass filter with all positive elements. Typical ones are 3×3 masks as shown below.

Mask 1

$$h = \frac{1}{9}\begin{bmatrix} 1 & 1 & 1 \\ 1 & 1 & 1 \\ 1 & 1 & 1 \end{bmatrix} \tag{5.12a}$$

Mask 2

$$h = \frac{1}{10}\begin{bmatrix} 1 & 1 & 1 \\ 1 & 2 & 1 \\ 1 & 1 & 1 \end{bmatrix} \tag{5.12b}$$

(a)

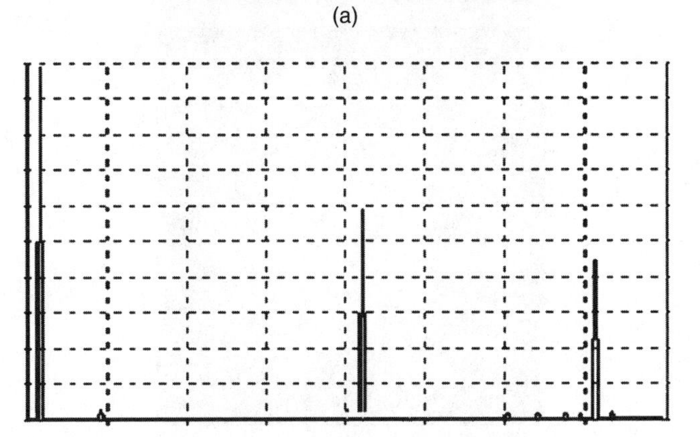

(b)

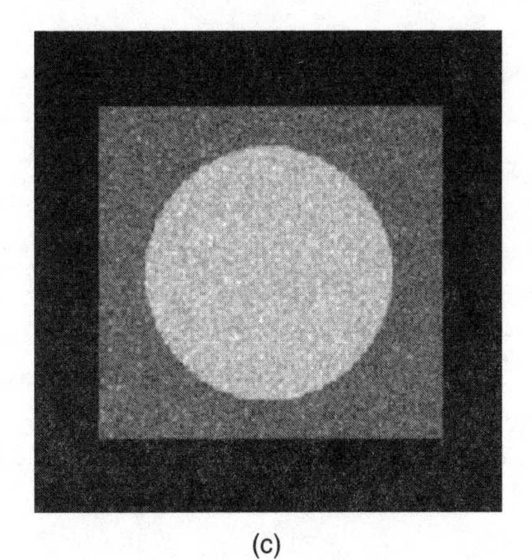

(c)

FIGURE 5.10 (a) Original image. (b) Histogram of (a). (c) Image of (a) corrupted by Gaussian noise. (d) Histogram of (c). (e) Image of (a) corrupted by impulse (pepper and salt) noise. (f) Histogram of (e).

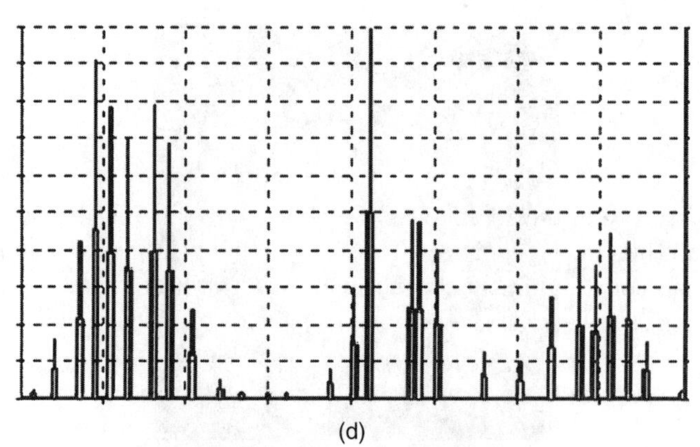

(d)

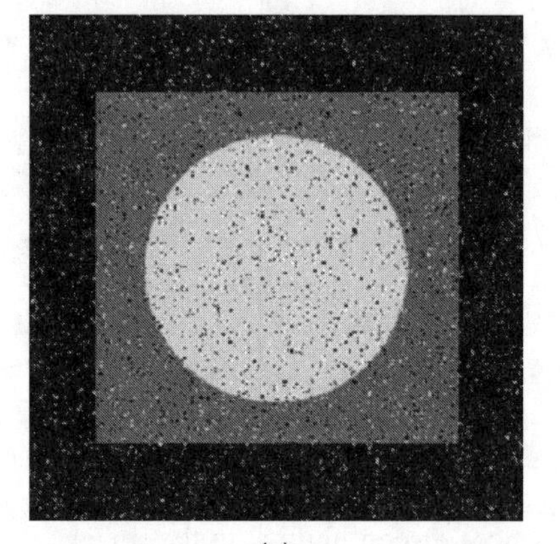

(e)

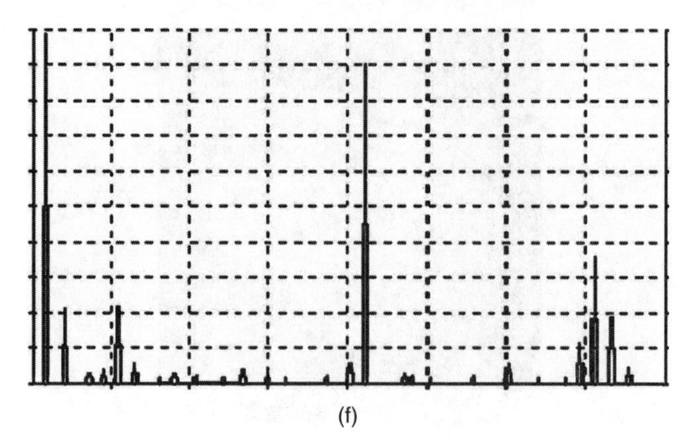

(f)

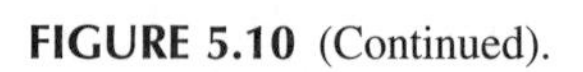
FIGURE 5.10 (Continued).

Mask 3

$$h=\frac{1}{16}\begin{bmatrix}1 & 2 & 1\\ 2 & 4 & 2\\ 1 & 2 & 1\end{bmatrix} \tag{5.12c}$$

These arrays are normalized to unit weighting so the filtering does not introduce an amplitude bias in output images. The effect of noise cleaning with the arrays on the Gaussian noise and impulse noise images is shown in Figure 5.11. Masks 1 and 3 are a special case of a 3×3 parametric low-pass filter whose impulse response is defined as

$$h=\frac{1}{[b+2]^2}\begin{bmatrix}1 & b & 1\\ b & b^2 & b\\ 1 & b & 1\end{bmatrix} \tag{5.13}$$

The concept of low-pass filtering can be extended to larger impulse response arrays. When noise is a little more than impulse or pepper-and-salt type, a larger array, say 5×5 or 7×7, provides better noise reduction, but filters with large masks destroy the fine details in an image.

5.2.2.2 Sharpening

A sharpening or high-pass filter has the effect of augmenting fine detail in an image. Image sharpening is particularly useful in industrial inspection. Sharpening can be accomplished by Fourier domain methods that emphasize the higher-frequency components or by spatial differentiation. Here we discuss the spatial approach, which enhances discontinuities in the image.

Sharpening filters are normally first-order or second-order derivatives. A first-order derivative operator is of the form

$$\frac{\Delta f}{\Delta x}=f(x+1)-f(x) \tag{5.14}$$

Note that in an image that is two-dimensional, the derivatives are taken with respect to the two directions x and y. They can be one of these options:

$$\left.\begin{aligned}\frac{\Delta f(x,y)}{\Delta x}&=f(x,y)-f(x-1,y)\\ \frac{\Delta f(x,y)}{\Delta y}&=f(x,y)-f(x,y-1)\\ \frac{\Delta f(x,y)}{\Delta (x,y)}&=f(x,y)-f(x-1,y-1)\\ \frac{\Delta f(i,j)}{\Delta (x,-y)}&=f(x,y)-f(x-1,y+1)\end{aligned}\right\} \tag{5.15}$$

First-order derivatives generate thick edges in the image.

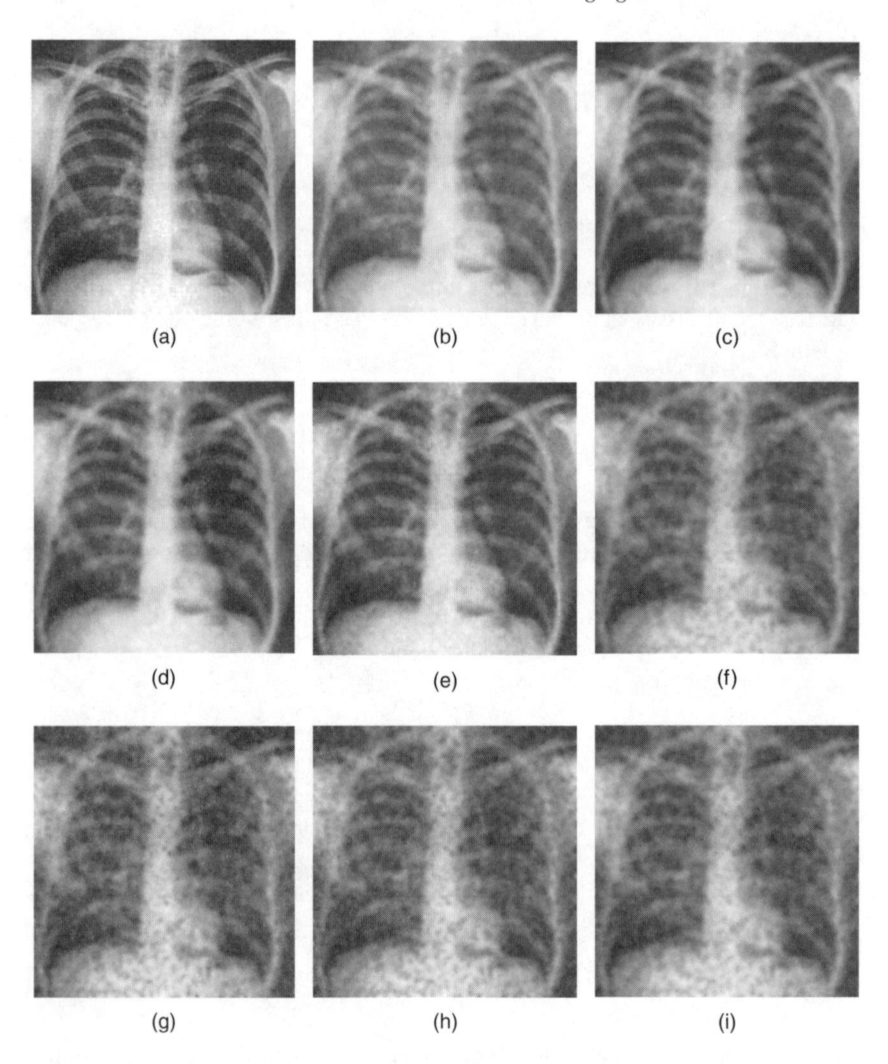

FIGURE 5.11 (a) Original image. (b) Image of (a) corrupted by Gaussian noise with a probability of 20%. (c) Result of applying Mask 1 on (b). (d) Result of applying Mask 2 on (b). (e) Result of applying Mask 3 on (b). (f) Image of (a) corrupted by impulse (pepper and salt) noise with a probability of 20%. (g) Result of applying Mask 1 on (f). (h) Result of applying Mask 2 on (f). (i) Result of applying Mask 3 on (f).

The second-order derivative operator (with respect to x) is of the form

$$\frac{\Delta^2 f}{\Delta^2 x} = f(x+1) + f(x-1) - 2f(x) \tag{5.16}$$

Second-order derivatives produce thin double edges in the image. Because images are two-dimensional, Laplacian operators are used as a high-pass filter. A Laplacian operator is defined as

$$\nabla^2 f = \frac{\partial^2 f}{\partial x^2} + \frac{\partial^2 f}{\partial y^2} \tag{5.17}$$

The implementation of the Laplacian operator is the sum of the second-order derivatives along x and y directions, respectively:

$$\nabla^2 f = f(x+1, y) + f(x-1, y) + f(x, y+1) + f(x, y-1) - 4f(x, y) \tag{5.18}$$

Equation 5.18 is equivalent to the following spatial filter:

$$L = \begin{bmatrix} 0 & 1 & 0 \\ 1 & -4 & 1 \\ 0 & 1 & 0 \end{bmatrix} \tag{5.19}$$

Other variations of Laplacian filters are

$$L = \begin{bmatrix} 1 & 1 & 1 \\ 1 & -8 & 1 \\ 1 & 1 & 1 \end{bmatrix} \quad L = \begin{bmatrix} 0 & -1 & 0 \\ -1 & 4 & -1 \\ 0 & -1 & 0 \end{bmatrix} \quad L = \begin{bmatrix} -1 & -1 & -1 \\ -1 & 8 & -1 \\ -1 & -1 & -1 \end{bmatrix} \tag{5.20}$$

Laplacian filters are derivative operators, and they highlight gray-level discontinuities in an image. This tends to produce images that have enhanced edges and other discontinuities on a dark background. One may preserve the background by subtracting the output image from the original image with the following operators:

$$g(x, y) = \begin{cases} f(x, y) - \nabla^2 f(x, y) \\ f(x, y) + \nabla^2 f(x, y) \end{cases} \tag{5.21}$$

where the difference is used if the center coefficient of Laplacian mask is negative and the addition is for the center coefficient positive. Figure 5.12 shows an image enhanced by a Laplacian filter. Figure 5.12b is the resulted edge from Laplacian operator, and Figure 5.12c shows the enhanced details where the details in (a) are not as apparent. For detailed discussion, see Andrews and Hunt [1], Bovik et al. [4], Oppenheim and Schafer [46], Prewitt [54], Preston et al. [56], and Rosenfeld [57].

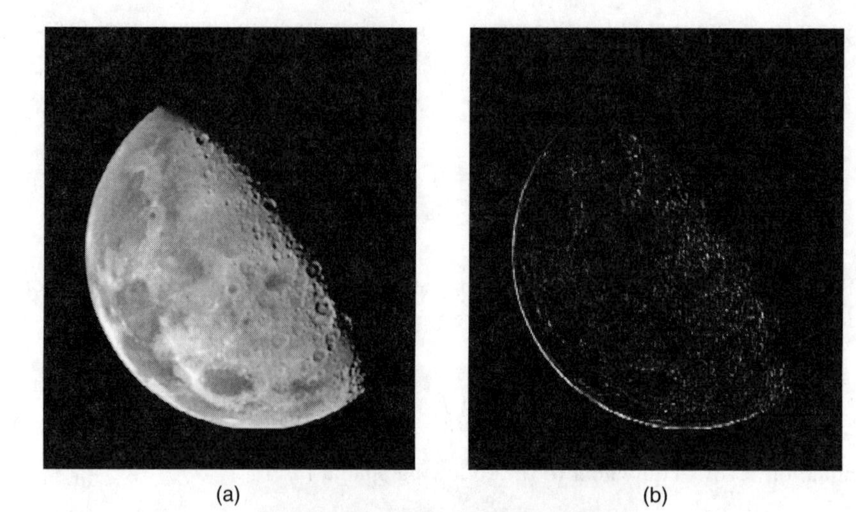

(a) (b)

(c)

FIGURE 5.12 (a) Image of North Pole of the moon. (b) Laplacian filtered image. (c) Image enhanced by using Equation 5.21.

5.2.3 Frequency Domain Processing

Frequency domain image analysis provides quantitative and intuitive insight into the nature of noise. The high-frequency noise effects can be reduced by frequency domain filtering, which employs the properties of Fourier transform. Discrete Fourier transform of a 2D array, or image, $f(x, y)$ of size $M \times N$ is defined as

$$F(u, v) = \frac{1}{MN} \sum_{i=0}^{M-1} \sum_{j=0}^{N-1} f(x, y) e^{-j2\pi(ux/M + vy/N)} \tag{5.22}$$

where (u, v) are spatial frequencies (i.e., cycles per unit length instead of cycles per unit time) of the image $f(x, y)$, $u = \{0, 1, 2, \ldots, M-1\}$ and $v = \{0, 1, 2, \ldots, N-1\}$. The inverse Fourier transform is defined as

$$f(x,y) = \frac{1}{MN}\sum_{v=0}^{N-1}\sum_{u=0}^{M-1} F(u,v)e^{j2\pi(ux/M+vy/N)} \tag{5.23}$$

where $x = \{0, 1, 2, \ldots, M-1\}$ and $y = \{0, 1, 2, \ldots, N-1\}$. Most images have a large value around $(u, v) = (0, 0)$ where

$$F(0,0) = \frac{1}{MN}\sum_{i=0}^{M-1}\sum_{j=0}^{N-1} f(x,y) \tag{5.24}$$

The frequency domain of an image normally shows the primary information within low spatial frequency regions where u and v are small. Figure 5.13 and Figure 5.14 show the images and their corresponding Fourier spectra. Noise, on the other hand, has high spatial frequency properties. Therefore, reducing the high-frequency components produces a less noisy image. The discontinuities and edges have high-frequency components. A low-pass filter thus reduces the sharpness of an image, whereas a high-pass filter enhances the edges and discontinuities. Figure 5.15 illustrates the effects of low-passed filtering (a) and high-pass filtering (b); the original image is shown in Figure 5.14a. Figure 5.15c shows the low-passed spectrum, and Figure 5.15d shows the high-pass spectrum. Detailed discussions may be found in frequency analysis in Levialdi [36], Meyer [45], and Oppenheim and Schafer [46, 47].

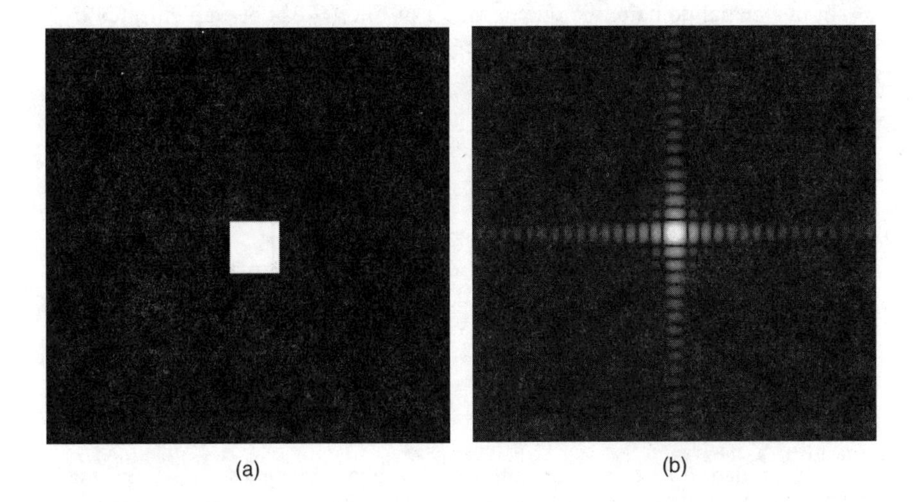

FIGURE 5.13 (a) Image of a 2D box on a black background. (b) Centered Fourier spectrum of (a).

FIGURE 5.14 Fourier spectrum. (a) Original image. (b) Centered Fourier spectrum of (a).

Frequency filters can be equivalently implemented in the spatial domain. Let $H(u, v)$ be a frequency filter mask applied to $F(u, v)$. For example, let $H(u, v)$ be zero for all except for $H(0, 0) = 1$; then the outcome is the DC component of $f(x, y)$. The frequency filter mask H is added on $F(u, v)$ directly, and the output of the filter is

$$g(x, y) = \frac{1}{MN}\sum_{v=0}^{N-1}\sum_{u=0}^{M-1}[F(u, v)H(u, v)]e^{j2\pi(ux/M+vy/N)} \tag{5.25}$$

In the spatial domain, the output image of filter H can be implemented by a convolution of $f(x, y)$ with $h(x, y)$, where $h(x, y)$ is the inverse Fourier transform of $H(u, v)$. They are related by the following equations:

$$F(u, v)H(u, v) \Leftrightarrow f(x, y) * h(u, v) \tag{5.26}$$

and

$$f(x, y) * h(x, y) = \frac{1}{MN}\sum_{j=0}^{N-1}\sum_{i=0}^{M-1} f(i, j)h(x-i, y-j) \tag{5.27}$$

5.3 IMAGE RESTORATION

5.3.1 Types of Noise

Image degradation and noise are unavoidable and sometimes natural in image acquisition. Image restoration techniques model the degradation and apply the inverse process to recover the original image. The degradation function h can be superficially modeled for any physical process such as photographic recording,

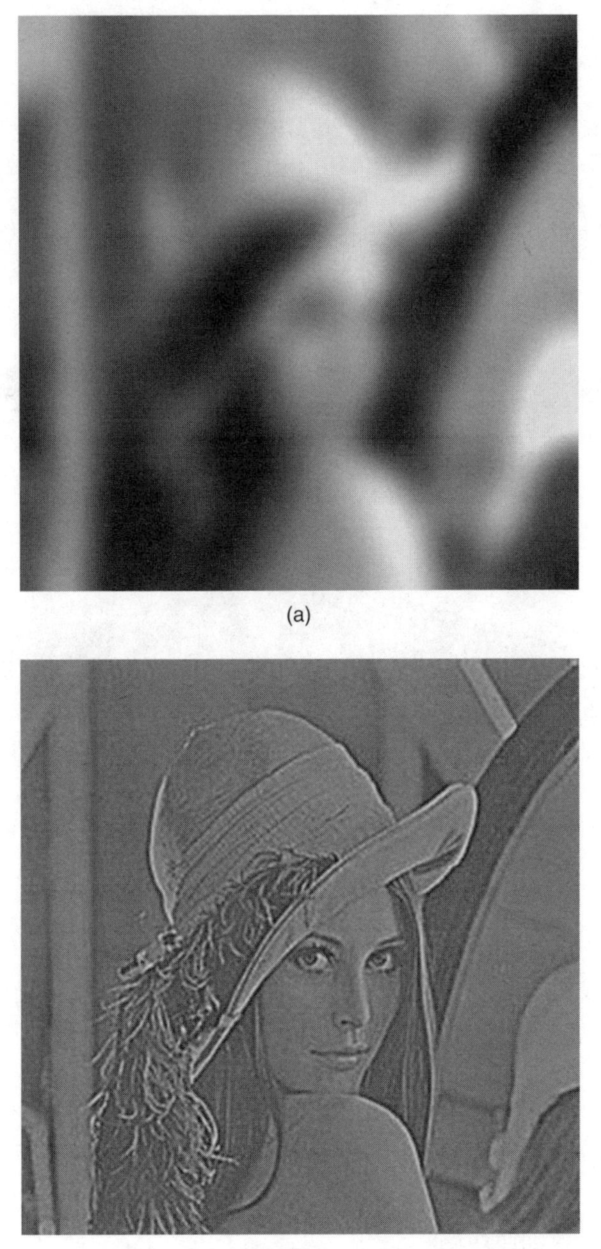

(a)

(b)

FIGURE 5.15 Low-pass and high-pass filtering. (a) Result of low-pass filtering of the image shown in Figure 5.13 (a). (b) Result of high-pass filtering of the image in Figure 5.13a. (c) A 2D low-pass filter function. (d) A 2D high-pass filter function.

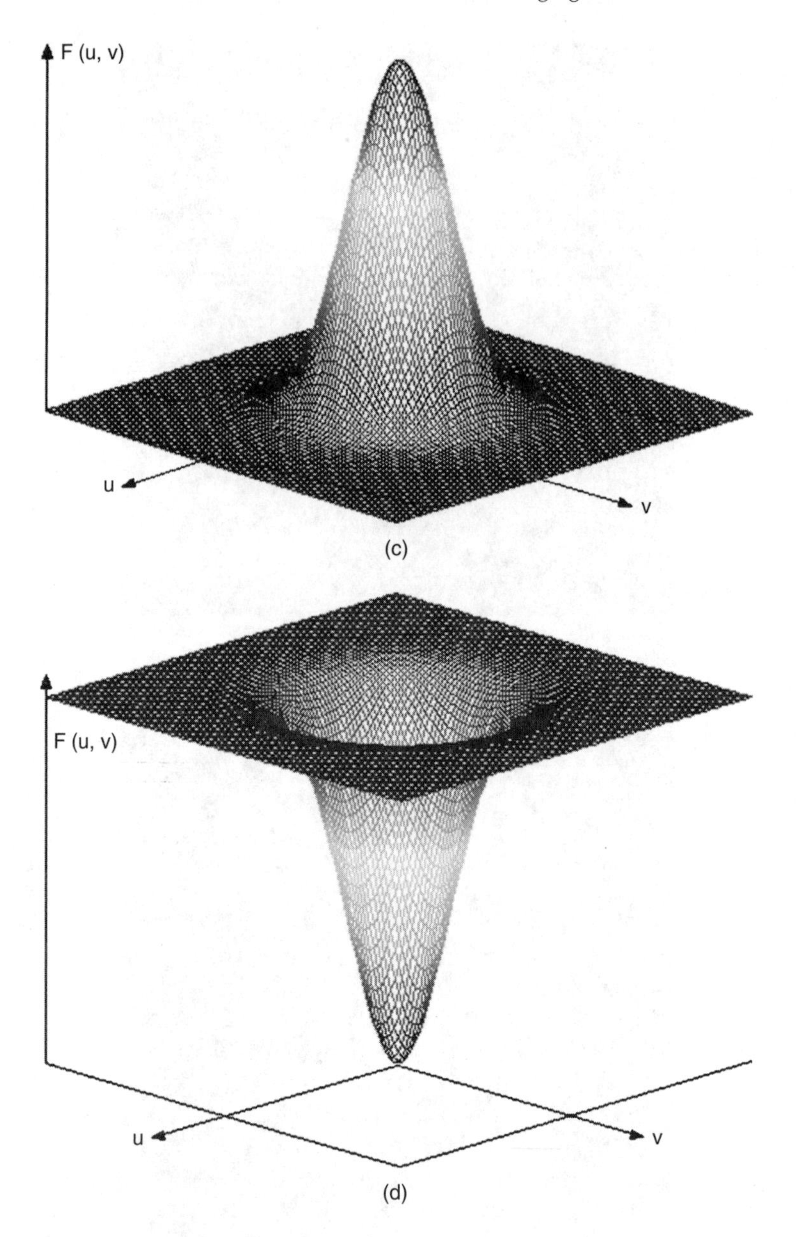

FIGURE 5.15 (Continued).

digitizing, or display. This modeling h assumes a linear property such that an inverse operation can be approximated. The result of degradation and addition of noise is an image $g(x, y)$:

$$g(x, y) = h(x, y) * f(x, y) + n(x, y) \tag{5.28}$$

where $h(x, y)$ is the spatial function for degradation and the symbol * denotes convolution.

Noise is statistically characterized as a random variable and mathematically illustrated by a probability density function (pdf). Some commonly modeled noise representations are used for their ease of handling in mathematical terms and closeness to the underlying nature of various noise types.

Gaussian noise is perhaps the most frequently cited form of noise because of its mathematical tractability. The pdf of a Gaussian random variable n is defined as

$$p(n) = \frac{1}{\sqrt{2\pi}\sigma} e^{-(n-\mu)^2/2\sigma^2} \tag{5.29}$$

where n is noise gray level, μ is the mean, and σ is the standard deviation of variable n.

Rayleigh noise has a pdf $p(n)$ equal to

$$p(n) = \frac{1}{b}(n-a)e^{-(n-a)^2/b}, \quad \text{for } n \geq a \tag{5.30}$$

where its mean $\mu = a + \sqrt{\pi b/4}$ and variance $\sigma^2 = b(4 - \pi)/4$. The pdf $p(n) = 0$ for $n < a$.

For Gamma noise, the pdf is defined as

$$p(n) = \frac{a^b n^{b-1}}{(b-1)!} e^{-an}, \quad \text{for } n \geq 0 \tag{5.31}$$

where $a > 0$, b is an integer. The pdf $p(n) = 0$ for $z < 0$. Figure 5.16 illustrates the pdf functions for Gaussian, Rayleigh, and Gamma types of noise.

Impulse noise is the most common type of noise in industrial applications because it is generated by electrical interference caused by operating equipment. This type of noise resembles salt-and-pepper granules randomly distributed over the image. The noise appears to be either completely dark points (gray level 0) or extremely bright (gray level 255 for an 8-bit image). Although the impulse noise appears to be annoying, it can be cleaned out most effectively through a number of nonlinear filtering techniques.

Estimation of noise types is done by inspection using Fourier spectrum of the image. A practical way to reduce noise is to test and establish a suitable noise model for the system, not merely the sensors themselves. Given the known probability mass distribution, there is always a proper way to smooth out noise while keeping most original object information in an image. In this section, we discuss some useful filtering techniques for noise reduction.

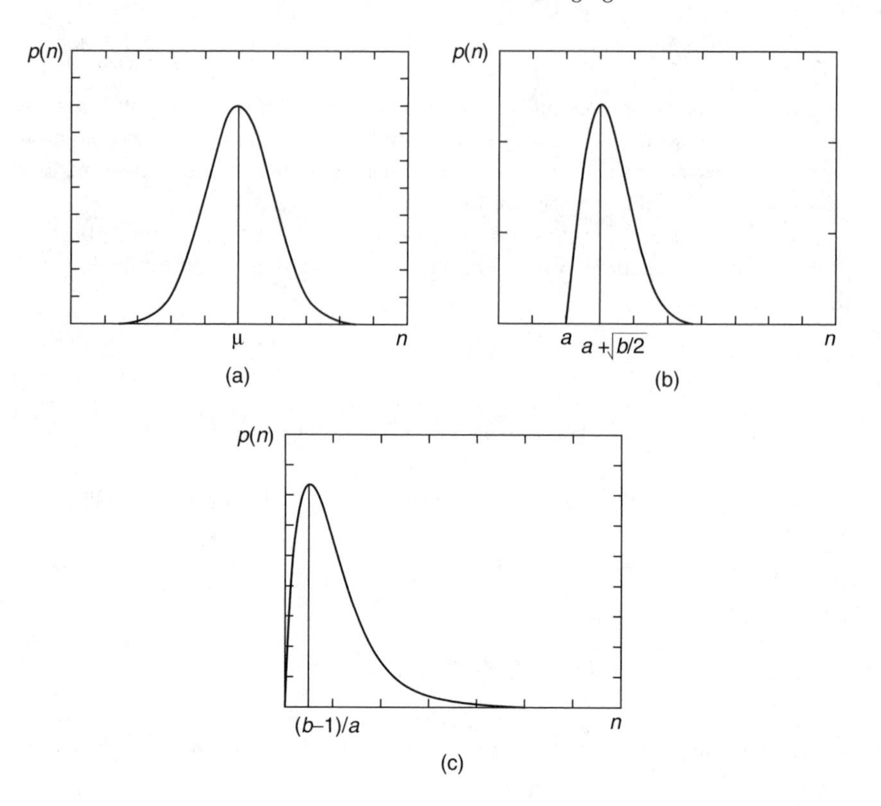

FIGURE 5.16 Noise PDF plots. (a) Gaussian. (b) Rayleigh. (c) Gamma.

5.3.2 Spatial Filtering

Spatial filtering of noise is a quick and effective solution. Many types of such filtering techniques are available. We introduce some of the most frequently used ones.

5.3.2.1 Mean Filters

Mean filters work well in noise smoothing. The output of a mean filter is the average of a block of the image. Let $f(x, y)$ be the input image; then the output image $\hat{f}(x, y)$ is a mean value of $f(x, y)$ on block B. A typical example is the arithmetic mean filter, which is defined by the arithmetic mean of an image on block B of size $m \times n$. The output $\hat{f}(x, y)$ is defined as

$$\hat{f}(x, y) = \frac{1}{mn} \sum_{(i,j) \in B} f(i, j) \tag{5.32}$$

This mean filter simply smoothes local variations. Since salt-and-pepper noise would be factored into the filter output, this mean filter does not produce desired results for filtering out spots.

5.3.2.2 Median Filters

The median filter considers each pixel in the image in turn and looks at its nearby neighbors to decide whether or not it is representative of its surroundings. Instead of simply replacing the pixel value with the mean of neighboring pixel values, it replaces it with the median of those values. The median is calculated by first sorting all the pixel values from the surrounding neighborhood into numerical order and then replacing the pixel being considered with the middle pixel value. If the neighborhood under consideration contains an even number of pixels, the average of the two middle pixel values is used. Median filters are popular in noise cleaning, particularly for cases where salt-and-pepper noise is present. They provide excellent noise reduction without blurring the image as much as a mean filter might.

5.3.2.3 Other Filters

Maximum and minimum (max, min) filters output the top or the bottom values within a filtering mask; therefore, they are also order-statistics filters. A max filter's output image is the brightest pixels that are taken from filtering mask B moving throughout the input image $f(x, y)$:

$$\hat{f}(x, y) = \max_{(i,j)\in B}\{f(i, j)\} \tag{5.33}$$

A min filter produces the darkest pixel covered under the filtering mask B of the input image $f(x, y)$:

$$\hat{f}(x, y) = \min_{(i,j)\in B}\{f(i, j)\} \tag{5.34}$$

A midpoint filter outputs the average of the max and the min from a filtering mask B and reduces noise types like Gaussian or uniform.

$$\hat{f}(x, y) = \frac{1}{2}\left[\max_{(i,j)\in B}\{f(i, j)\} + \min_{(i,j)\in B}\{f(i, j)\right] \tag{5.35}$$

Alpha-trimmed mean filtering uses the concept of adaptive filtering. The filtering algorithm selects certain pixels to be removed from the mean filtering as in Equation 5.32. Let a stack, in a descending order, hold pixels from a masked block of the input image. The top $d/2$ pixels and the bottom $d/2$ pixels are removed from the stack. The rest of the pixels are averaged as the filter output. Let $f_S(x, y)$ stand for the image pixels on the remaining coordinates; then the filter is defined as

$$\hat{f}(x, y) = \frac{1}{mn - d}\sum_{(i,j)\in B} f_s(i, j) \tag{5.36}$$

5.3.3 Frequency Domain Filtering

Periodic noise can be reduced using frequency domain filters. The frequency domain approaches are very powerful if the noise components in an image can be estimated. Frequency domain filters include bandreject filters, bandpass filters, and notch filters.

5.3.3.1 Bandreject Filters

Bandreject filters reduce or remove a band of frequencies corresponding to the frequency spectrum contained in the noise. A simple bandreject filter $D(u, v)$ can be expressed as

$$H(u,v)=\begin{cases} 1 & \text{if } D(u,v) < D_0 - \frac{W}{2} \\ 0 & \text{if } D_0 - \frac{W}{2} \le D(u,v) \le D_0 + \frac{W}{2} \\ 1 & \text{if } D(u,v) > D_0 + \frac{W}{2} \end{cases} \tag{5.37}$$

where $D(u, v)$ is the distance in frequency domain from the center frequency, W is the width of the rejection band, and D_0 is its center.

Two other frequently adopted bandreject filters are the Butterworth and Gaussian filters. The Butterworth filter is

$$H(u,v)=\frac{1}{1+\left[\frac{D(u,v)W}{D^2(u,v)-D_0^2}\right]^{2n}} \tag{5.38}$$

while the Gaussian filter is

$$H(u,v)=1-\exp\left\{-\frac{1}{2}\left[\frac{D^2(u,v)-D_0^2}{D(u,v)W}\right]^2\right\} \tag{5.39}$$

Figure 5.17 shows the effects of bandreject filtering, where (a) shows the original image and (b) is the Fourier spectrum of (a). The image in (a) is corrupted by sinusoidal noise, as shown in (c), along with its spectrum in (d). The bandreject filter, shown in (e), is applied to the image in (c), and the image resulting after filtering is shown in (f). In this case, the Butterworth filter of Equation 5.38 was used.

5.3.3.2 Bandpass Filters

Bandpass filters are the opposite of bandreject filters in that they pass only the selected spatial frequencies and suppress all others. The transfer function of a

(a)

(b)

(c)

(d)

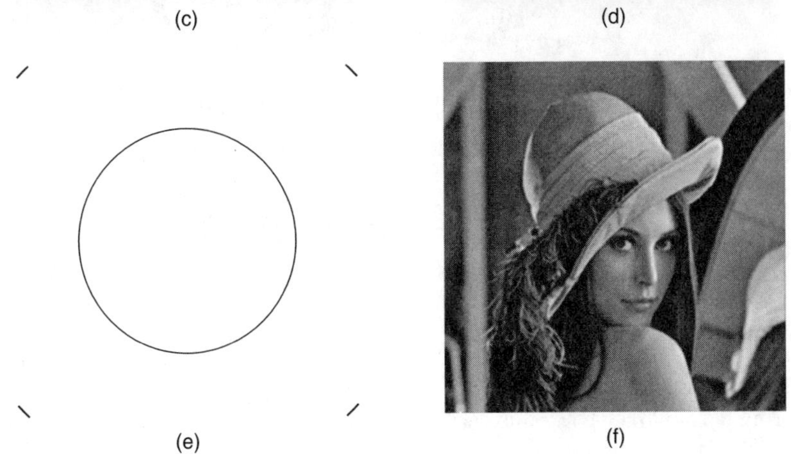

(e)

(f)

FIGURE 5.17 Effect of bandreject filtering. (a) Original image. (b) Spectrum of (a). (c) Image (a) corrupted by sinusoidal noise. (d) Spectrum of (c). (e) Butterworth bandreject filter. (f) Result of filtering.

bandpass filter $H_{bp}(u, v)$ can be obtained by subtracting the transfer function of a bandreject filter $H_{br}(u, v)$ from unity over (u, v), where the bandreject filters are given by Equation 5.37, Equation 5.38, and Equation 5.39:

$$H_{bp}(u, v) = 1 - H_{br}(u, v) \tag{5.40}$$

5.3.3.3 Notch Filters

A notch filter performs functions similar to a bandpass or bandreject filter, but with greater flexibility. The perspective plots of notch filters with ideal, Butterworth, and Gaussian properties are shown in Figure 5.18. Note that the frequency symmetry property requires that the filter functions be designed symmetrically about the origin. For example, the transfer function of an ideal notch reject filter of radius D_0 with centers at (u_0, v_0) and $(-u_0, -v_0)$ can be designed as

$$H(u, v) = \begin{cases} 0 & \text{if } D_1(u, v) \le D_0 \quad \text{or} \quad D_2(u, v) \le D_0 \\ 1 & \text{otherwise} \end{cases} \tag{5.41}$$

where

$$D_1(u, v) = \left[(u - M/2 - u_0)^2 + (v - N/2 - v_0)^2 \right]^{1/2} \tag{5.42}$$

and

$$D_2(u, v) = \left[(u - M/2 + u_0)^2 + (v - N/2 + v_0)^2 \right]^{1/2} \tag{5.43}$$

This equation assumes that the center of the frequency response has been shifted to the point $(M/2, N/2)$.

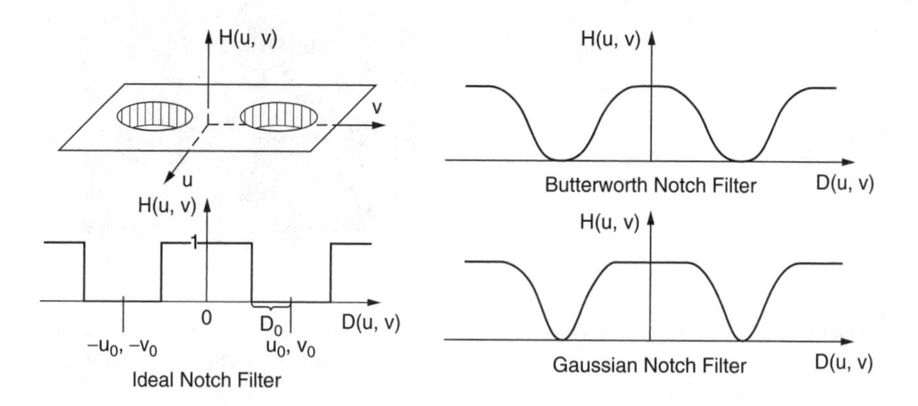

FIGURE 5.18 Notch filters: ideal, Butterworth, and Gaussian.

Likewise, with the same definition of D_1 and D_2, a Butterworth notch reject filter can be designed as

$$H(u, v) = \frac{1}{1 + \left[\frac{D_0^2}{D_1(u, v) D_2(u, v)}\right]^n} \quad (5.44)$$

and a Gaussian notch reject filter can be designed as

$$H(u, v) = 1 - \exp\left\{-\frac{1}{2}\left[\frac{D_1(u, v) D_2(u, v)}{D_0^2}\right]\right\} \quad (5.45)$$

Notch bandpass filters work the same way as bandpass filters. The transfer function of notch bandpass filters is given as

$$H_{np}(u, v) = 1 - H_{nr}(u, v) \quad (5.46)$$

where $H_{np}(u, v)$ is the transfer function of the notch pass filter. Figure 5.18 shows the plots of ideal Butterworth, and Gaussian notch filters. Additional information on this topic may be found in Glassner [25], Levialdi [36], Oppenheim and Schafer [46], and Prewitt [54].

5.4 SEGMENTATION

Image segmentation is one of the first tasks in image analysis and machine vision applications. Segmentation subdivides an image into subregions or objects from which the essential information is extracted. The accuracy of the segmentation step determines the eventual success or failure of the subsequent image analysis.

Image segmentation techniques in general are based on discontinuity and similarity. Discontinuity is typically represented by abrupt changes in intensity, such as edges. Similarity is used to partition an image into regions whose properties are the same according to a predetermined set of criteria.

5.4.1 Edge Detection

Edges may very well be the first feature to be detected in a typical image analysis. The edges of objects tend to appear as intensity discontinuities (local, abrupt changes in gray level). An *edge operator* is a mathematical operator designed to detect the presence of edges in an image.

Edge operators fall into three main classes: operators that calculate the mathematical gradient, template matching operators, and operators that fit local intensities with parametric edge models. The most common edge operator is the Laplacian already introduced in Equation 5.17, Equation 5.18, and Equation 5.19. The Laplacian is essentially a high-pass filter that can detect edges and discontinuities in the image. An example of this operation appears in Figure 5.12.

Discrete implementations of the transform are given by Equation 5.19 and Equation 5.20 (in conjunction with Equation 5.11).

5.4.2 Hough Transform

The Hough transform for curve detection may be used to locate a boundary if its shape can be described by a parametric curve (which could in fact be a straight line). The transform is relatively unaffected by gaps in the curves and by noise. Consider the problem of detecting straight lines in images. Assume that by some process image points have been selected that have a high likelihood of being on separate linear boundaries. The Hough transform organizes these points into straight lines and rates each one on how well it fits the data. More discussion on the Hough transform can be found in Baxes [3], Gonzalez and Woods [26], Haralick and Shapiro [28], and Jain [33].

5.4.2.1 Line Detection

Figure 5.19 shows the detection of a line. Consider the point x' in (a) and the equation for a line $y = mx + c$. All lines with m and c satisfying $y' = mx' + c$ would be the answer. Regarding (x', y') as fixed, the last equation is that of a line in (m, c) space, or parameter space. Repeating this reasoning, a second point (x'', y'') will also have an associated line in parameter space and, furthermore, these lines will intersect at a point (m', c') which corresponds to the line AB connecting these points. In fact all points on the line AB will yield lines in parameter space which intersect at the point (m', c'), as shown in (b). The relation between image space x and parameter space suggests the following algorithm for detecting lines:

```
Quantize parameter space between appropriate
maximum and minimum values for m and c.

Form an accumulator array A(c,m) whose elements
are initially zero.

For each point (x, y) in gradient image such that
the strength of the gradient exceeds some
threshold, increment all points in the accumulator
array along the appropriate line, i.e., A(c,m) :=
A(c,m) + 1 for m and c satisfying c = -mx + y
within the limits of the digitization.

Local maxima in the accumulator array now
correspond to collinear points in the image array.
The values of the accumulator array provide a
measure of the number of points on the line.
```

Since m may be infinite in the slope-intercept equation, a better parameterization of the line is $x\cos\theta + y\sin\theta = r$. This produces a sinusoidal curve in (r, θ) space for fixed (x, y), but otherwise the procedure is unchanged.

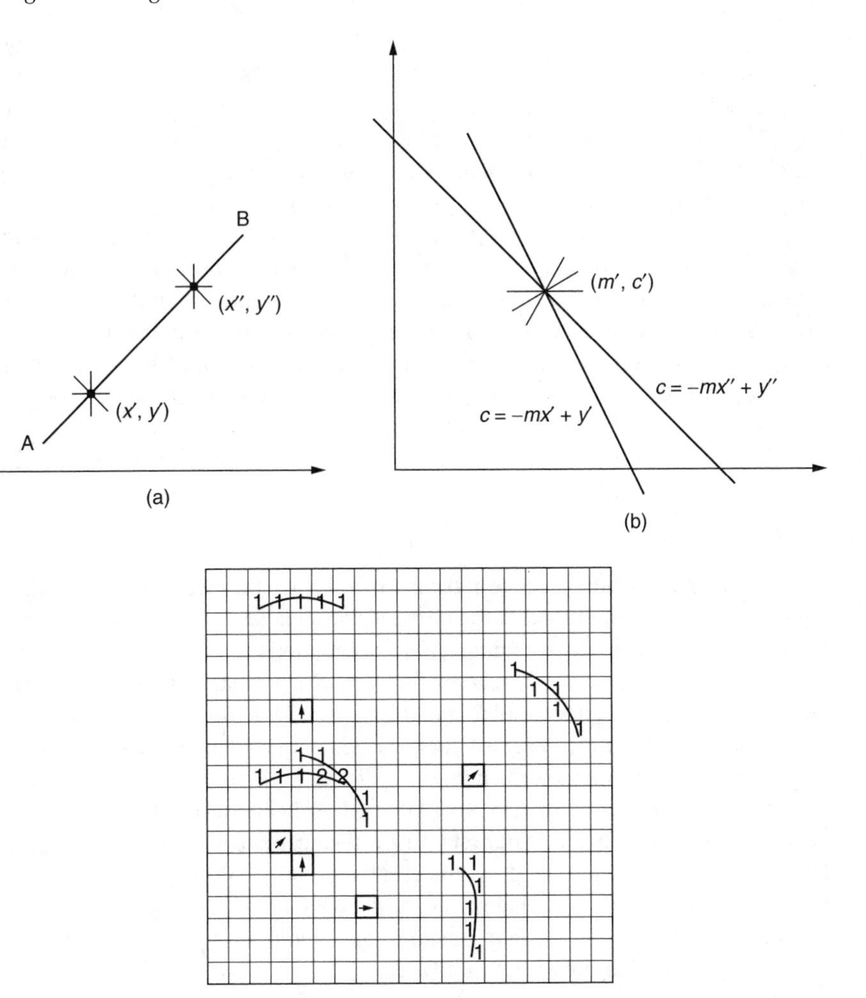

FIGURE 5.19 Detecting a line: a line in image space (a) and in parameter space (b). (c) Gradient information computation in a Hough transform algorithm.

The generalization of this technique to other curves is straightforward, and this method works for any curve $f(\vec{x}, \vec{a}) = 0$ for fixed image space $\vec{x}$ and parameter $\vec{a}$. In the case of a circle parameterized by

$$(x-a)^2 + (y-b)^2 = r^2 \tag{5.47}$$

for fixed $\vec{x}$, the Hough transform algorithm increments values of $\{a, b, r\}$ lying on the surface of a cone. The computation and the size of the accumulator array increase exponentially as the number of parameters, making this technique practical only for curves with a small number of parameters. The Hough transform is an efficient implementation of a generalized matched filtering strategy. For instance, in the case of a circle, imagine a template composed of a circle of 1s at a fixed radius (x, y) and 0s everywhere else. If this template is convolved with the gradient image, the result is the portion of the accumulator array $A(a, b, r)$.

In the Hough transform, dramatic reductions in the amount of computation can be achieved if the gradient direction is integrated into the algorithm. For example, consider the detection of a circle of fixed radius R. Without gradient information, all values a, b lying on the circle given by Equation 5.47 are incremented. With the gradient information, fewer points need be incremented, as shown in Figure 5.19c—only those on an arc centered at (a, b),

$$a = x + r\cos\phi \tag{5.48a}$$

$$b = y + r\sin\phi \tag{5.48b}$$

where $\phi(x)$ is the gradient angle returned by an edge operator. Implicit in these equations is the assumption that the circle is the boundary of a disk that has gray levels greater than its surroundings. These equations may also be derived by differentiating Equation 5.47, recognizing that $dy/dx = \tan\phi$, and solving for a and b between the resultant equation and Equation 5.47. Similar methods can be applied to other conics. In each of such cases, the use of the gradient saves one dimension in the accumulator array.

The gradient magnitude can also be used as heuristic in the incrementing procedure. Instead of incrementing by unity, the accumulator array location may be incremented by a function of the gradient magnitude. This heuristic can balance the magnitude of brightness changes across a boundary with the boundary length, but it can also lead to detection of phantom lines indicated by a few bright points or to missing dim but coherent boundaries.

5.4.2.2 Parameter Space and Image Space

Consider the example of detecting ellipses that are known to be oriented so that a principal axis is parallel to the x axis. These can be specified by four parameters. Using the equation for the ellipse together with its derivative, and substituting for the known gradient as before, one can solve for two parameters. In the equation

$$\frac{(x-x_0)^2}{a^2} + \frac{(y-y_0)^2}{b^2} = 1 \tag{5.49}$$

x is an edge point and $\{x_0, y_0, a, b\}$ are parameters. The equation for its derivative is

$$\frac{(x-x_0)}{a^2} + \frac{(y-y_0)}{b^2}\frac{dy}{dx} = 0 \tag{5.50}$$

where $dy/dx = \tan\phi(x)$. The Hough transform algorithm becomes:

For each discrete value of x and y, increment the point in parameter space given by x_0, y_0, a, b, where

$$x = x_0 \pm \frac{a}{(1 + b^2/a^2 \tan^2 \phi)^{\frac{1}{2}}} \tag{5.51a}$$

$$y = y_0 \pm \frac{b}{(1 + a^2 \tan^2 \phi/b^2)^{\frac{1}{2}}} \tag{5.51b}$$

That is,

$$A(a, b, x_0, y_0) := A(a, b, x_0, y_0) + 1 \tag{5.52}$$

Consider all pair-wise combinations of edge elements. This introduces two additional equations like Equation 5.49 and Equation 5.50, and now the four-parameter point can be determined exactly:

$$\frac{(x_1 - x_0)^2}{a^2} + \frac{(y_1 - y_0)^2}{b^2} = 1 \tag{5.53a}$$

$$\frac{(x_2 - x_0)^2}{a^2} + \frac{(y_2 - y_0)^2}{b^2} = 1 \tag{5.53b}$$

$$\frac{(x_1 - x_0)}{a^2} + \frac{(y_1 - y_0)}{b^2}\frac{dy}{dx} = 0 \tag{5.53c}$$

$$\frac{(x_2 - x_0)}{a^2} + \frac{(y_2 - y_0)}{b^2}\frac{dy}{dx} = 0 \tag{5.53d}$$

where

$$\frac{dy}{dx} = \tan\phi \tag{5.54}$$

5.4.2.3 Generalized Hough Transform

The Hough transform can be generalized as follows. Suppose the object has no simple analytic form but has a particular silhouette. Since the Hough transform is so closely related to template matching, and template matching can handle this case, it is not surprising that the Hough transform can be generalized to do so. Suppose the object appears in the image with known shape, orientation, and scale. Pick a reference point in the silhouette and draw a line to the boundary. At the boundary point, compute the gradient direction and store the reference point as a function of this direction. Thus it is possible to precompute the location of the reference point from the boundary points given the gradient angle. The set of all

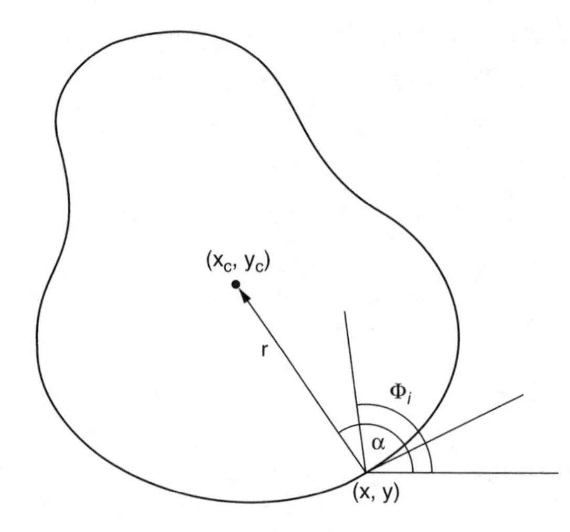

FIGURE 5.20 Geometry used to form the *R*-table.

such locations, indexed by gradient angles, makes up a table termed the R-table. The basic strategy of the Hough technique is to compute the possible loci of reference points in the parameter space from edge point data in image space and increment the parameter point in an accumulator array. Figure 5.20 shows the relevant geometry, and Table 5.1 shows the form of the R-table. For the moment, the reference point coordinates (x_c, y_c) are the only parameters, assuming that rotation and scaling have been fixed. Thus an edge point (x, y) with gradient orientation ϕ constrains the possible reference points to be at $\{x + r_1(\phi) \cos[\alpha_1(\phi)], y + r_1(\phi) \sin[\alpha_1(\phi)]\}$, and so on.

The corresponding generalized Hough transform algorithm can be described as follows.

```
Step 0:  Make a table (like Table 5.1) for the shape
         to be located.
```

TABLE 5.1
Increments in Generalized Hough Transform Case

Angle Measured from Figure Boundary to Reference Point	Set of radii$\{\mathbf{r}^k\}$ where $\mathbf{r} = (r, \alpha)$
ϕ_1	$\mathbf{r}_1^1, \mathbf{r}_2^1, \ldots, \mathbf{r}_{n1}^1$
ϕ_2	$\mathbf{r}_1^2, \mathbf{r}_2^2, \ldots, \mathbf{r}_{n2}^2$
.	.
.	.
.	.
ϕ_m	$\mathbf{r}_1^m, \mathbf{r}_2^m, \ldots, \mathbf{r}_{nm}^m$

```
Step 1:  Form an accumulator array of possible
         reference points A(xc min:xc max, yc min:yc max)
         initialized to zero.

Step 2:  For each edge point, do the following:

         Step 2.1: Compute φ (x).

         Step 2.2: Calculate the possible centers;
                   for each table entry for φ
                   compute
```

$$x_c := x + r\phi \cos[\alpha(\phi)]$$

$$y_c := y + r\phi \sin[\alpha(\phi)]$$

```
         Step 2.3: Increment the accumulator array:
```

$$A(x_c, y_c) := A(x_c, y_c) + 1$$

```
Step 3:  See the maxima in array A for possible
         locations for the shape.
```

5.4.3 Thresholding

Image thresholding has intuitive properties and is simple to implement; it therefore enjoys a central position in applications of image segmentation. One approach in thresholding an image is to employ the histogram. Suppose the grayscale histogram shown in Figure 5.21a corresponds to image $f(x, y)$, composed of bright objects on dark background. The objects and background have grayscale values that can be grouped into two dominant modes. One obvious way to extract the objects from the background is to select a threshold T that separates these two modes. Against this threshold T, any pixel (x, y) with $f(x, y) > T$ is taken as an object pixel; otherwise, it is a background point.

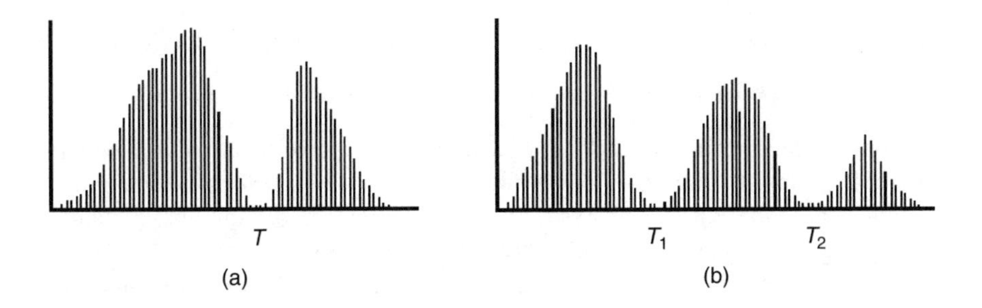

FIGURE 5.21 (a) Grayscale histograms by a single threshold; (b) multiple thresholds.

Figure 5.21b shows a slightly more general case, where three dominant modes characterize the image histogram. Here, multilevel thresholding classifies a pixel (x, y) as belonging to one object class if $T_1 < (x, y) \leq T_2$, to the other object class if $f(x, y) > T_2$, and to the background if $f(x, y) \leq T_1$.

The simple thresholding approach does not work in general for objects with varied background. In general, the threshold is a function, instead of a single value. This leads to adaptive thresholding. Adaptive thresholding is a process of decomposing a scene into its components, where each of the components has its own threshold similar to the simple thresholding discussed above. Section 5.6 introduces a practical adaptive algorithm called *moving ball thresholding,* where the local background of a small region is averaged into a curved surface that is then used as the threshold function. The objects that stick out of the curved surface are picked up, while small variations such as pepper–salt noise are smoothed out. Adaptive thresholding is much like a process of moving a window over an image. The window is considered to be an image of rather a small size, and all pixels in the window are treated via a simple thresholding. Note that in many applications, each pixel has its windowed output, and the output is usually either foreground or background—a binary decision: object or background. There are various versions of thresholding and adaptive thresholding techniques; for a quick reference, see Ballard and Brown [2], Dougherty [12], Gonzalez and Woods [26], Haralick and Shapiro [28], Jain [33], Pitas [53], and Winkler [76].

5.4.4 Edge Linking

Image analysis often requires that one segment an image into regions of common attribute by detecting the boundary of each region where an attribute changes significantly across the boundary. Boundary detection can be accomplished by means of edge detection (see Section 5.2.2 and Section 5.4.1). Figure 5.22a illustrates the segmentation of an object from its background. In Figure 5.22b, a derivative of the Gaussian edge detector is used to generate the edge map. Here, the region separation is apparent. If an image is noisy or if its region attributes differ by only a small amount between regions, a detected boundary may be broken. Edge-linking techniques can be employed to bridge short gaps in such a region boundary.

5.4.4.1 Curve Fitting

In some instances, one can link edge map points of a broken segment boundary to form a closed contour by curve fitting methods. If *a priori* information is available as to the expected shape of a region in an image, for example a rectangle or a circle, the fit may be made directly to that closed contour. For more complex shape regions, as illustrated in Figure 5.23, it is necessary to break up the supposed closed contour into chains with broken links. One such chain, shown in Figure 5.23 starting at point A and ending at point B, contains a single broken

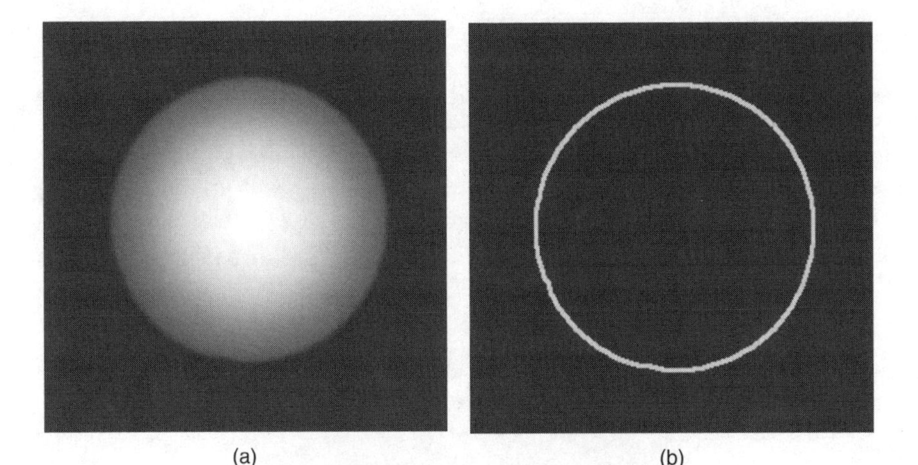

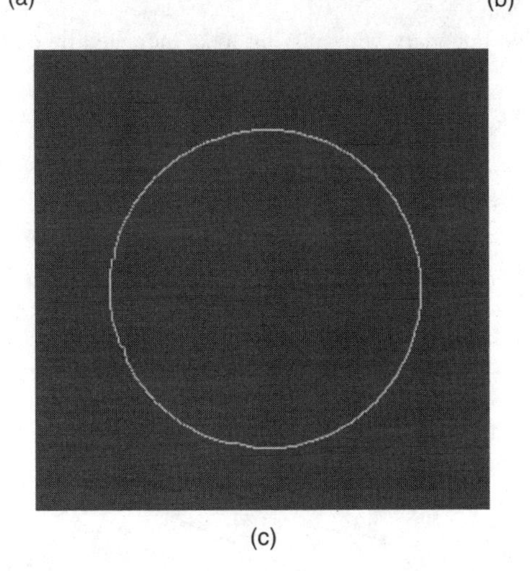

FIGURE 5.22 (a) Grayscale image; (b) edge map; (c) thinned edge map.

link. Classical curve fitting methods such as Bezier polynomial or spline fitting can be used to fit the broken chain.

The following is a simple piecewise linear curve fitting procedure called the *iterative endpoint fit.* In the first stage of the algorithm, illustrated in Figure 5.24, data endpoints A and B are connected by a straight line. The point of greatest departure from the straight line (point C) is examined. If the separation of this point is too large, the point becomes an anchor point for two straight-line segments (A to C and C to B). The principal advantage of the algorithm is its simplicity; its disadvantage is susceptibility to error caused by incorrect data points. The curve fitting approach is reasonably effective for simply structured objects. Difficulties occur when an image contains many overlapping objects, and its corresponding edge map contains branch structures [49].

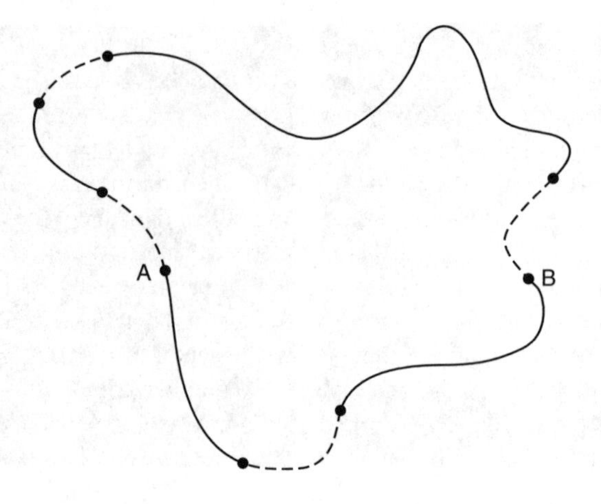

FIGURE 5.23 Region boundary with missing links indicated by dashed lines.

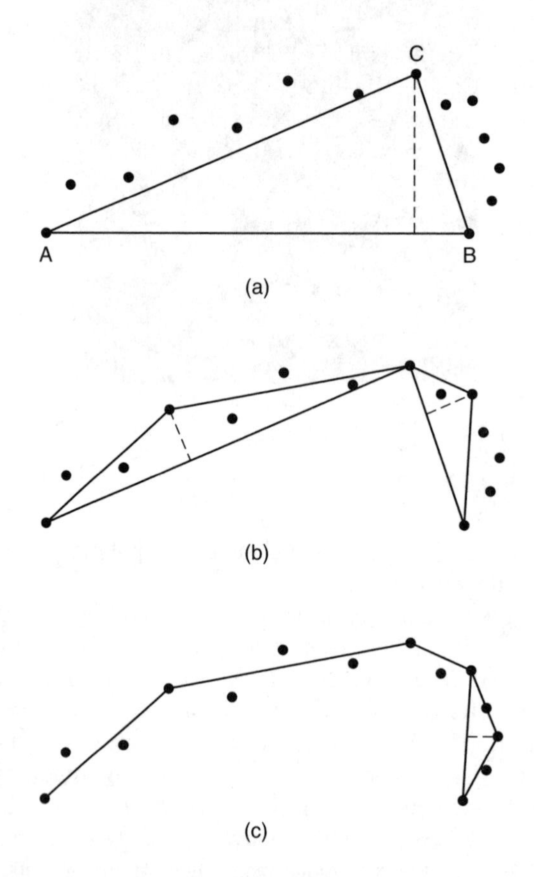

FIGURE 5.24 Iterative end point curve fitting.

5.4.4.2 Roberts Method

Edge segmentation by Roberts [60] is one of the heuristic edge linking methods. In Roberts' method, one examines edge gradients in 4×4 pixels blocks. The pixel whose magnitude gradient is largest is declared a tentative edge point if its magnitude is greater than a threshold value. Then north, east, south, and west oriented lines of length five are fitted to the gradient data about the tentative edge point. If the ratio of the best fit to the worst fit, measured in terms of the fit correction, is greater than a second threshold, the tentative edge point is declared valid, and it is assigned the direction of the best fit. Next, straight lines are fitted between pairs of edge points if they are in adjacent 4×4 blocks, and if the line direction is within 23° of the edge direction of either edge point. Those points failing to meet the linking criteria are discarded. A typical boundary at this stage, shown in Figure 5.25a will contain gaps and multiply connected edge points. Small triangles are eliminated by deleting the longest side; small rectangles are replaced by their longest diagonal, as indicated in Figure 5.25b. Short spur lines are also deleted. At this stage, short gaps are bridged by straight-line connection. This form of edge linking can be used with a wide variety of edge detectors.

An alternative by Robinson [61] is a simple but effective edge linking algorithm in which edge points from an edge detector providing eight edge compass directions are examined in 3×3 blocks, as indicated in Figure 5.26. The edge point in the center of the block is declared a valid edge if it possesses directional

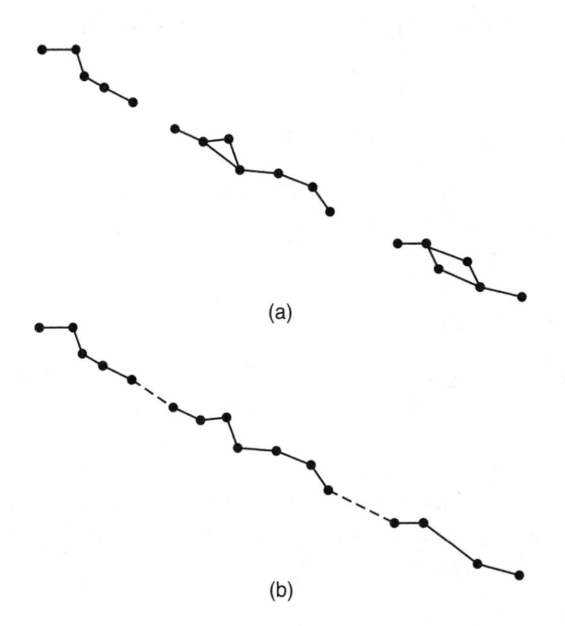

FIGURE 5.25 Example of Roberts edge linking. (a) Edge point linkages. (b) Elimination of multiple linkages and bridging.

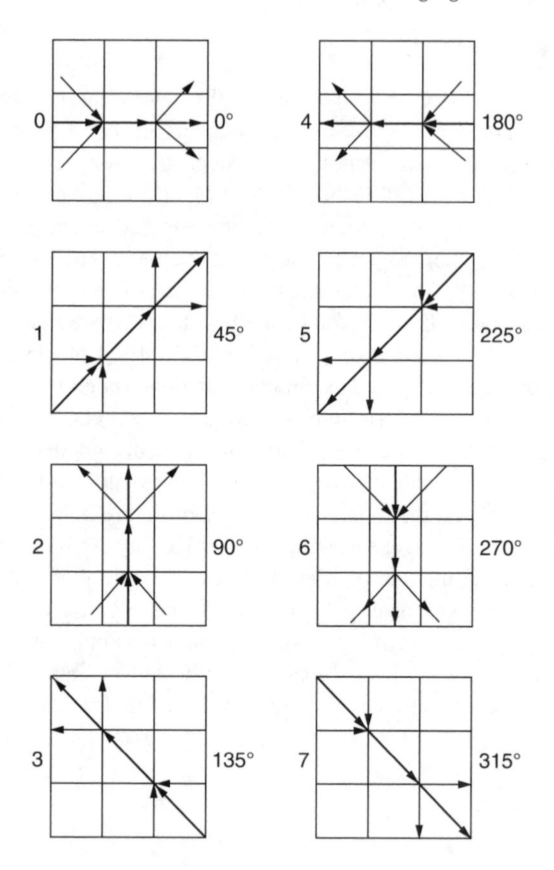

FIGURE 5.26 Robinson's edge linking rules.

neighbors in the proper orientation. Extensions to larger windows should be beneficial, but the number of potential valid edge connections will grow rapidly with window size.

5.4.5 Segment Labeling

A successful image segmentation eventually leads to the labeling of each pixel that lies within a specific segment. One way of labeling is to append to each pixel of an image the label number or index of its segment. A more succinct method is to specify the closed contour of each segment. If necessary, contour-filling techniques can be used to label each pixel within a contour. Further reading can be found in Gonzalez and Woods [26], Haralick and Shapiro [28], Marr [42], Pitas [52], and Rosenfeld and Kak [59].

The contour-following approach to image segment representation is conceptually illustrated as follows. In the binary image example of Figure 5.27, an imaginary person begins marching from the white background to the black pixel region indicated by the closed contour. When the person crosses into a black

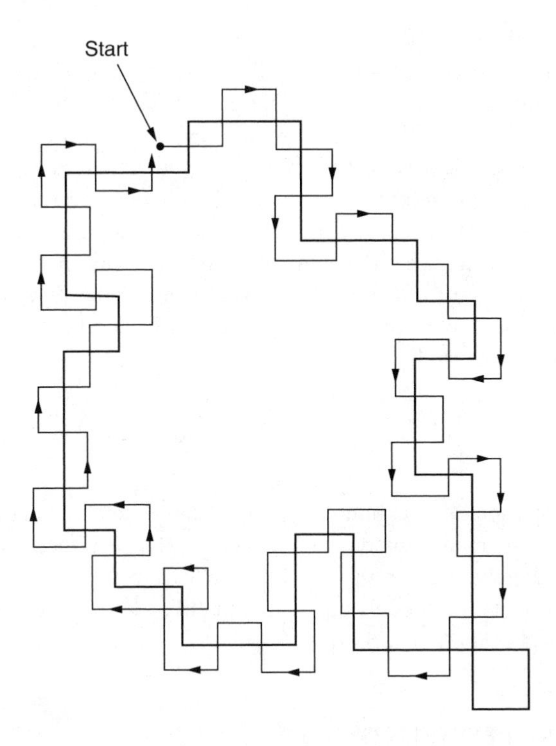

FIGURE 5.27 Example of contour following.

pixel, he makes a left turn and proceeds to the next pixel. If that pixel is black, he turns left again, but if the pixel is white, he turns right. The procedure continues until the person returns to the starting point. This follower may miss a spur pixel on a boundary. Figure 5.28a shows the boundary trace for such an example. This problem can be overcome by permitting the person to record his past steps and backtrack if his present course is erroneous.

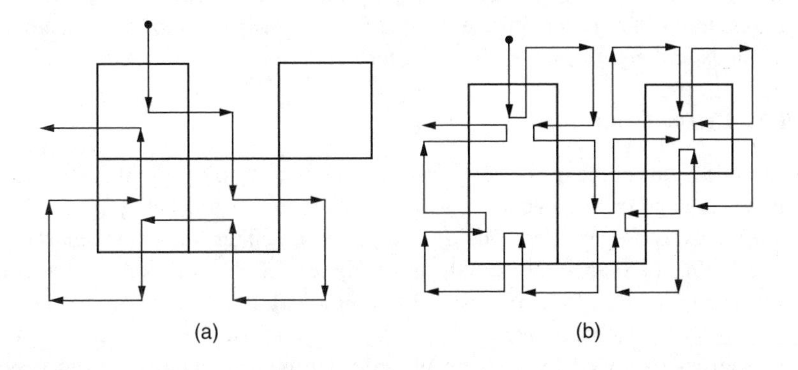

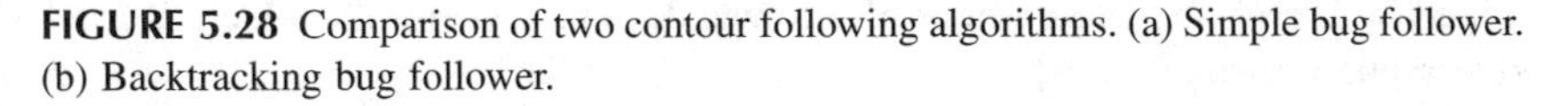
FIGURE 5.28 Comparison of two contour following algorithms. (a) Simple bug follower. (b) Backtracking bug follower.

Figure 5.28b illustrates the boundary trace for backtracking. In this algorithm, if the person makes a white-to-black pixel transition, he returns to the previous starting point and makes a right turn. The person makes a right turn whenever he makes a white-to-white transition. While the person is following a contour, he can create a list of the pixel coordinates of each boundary pixel. Alternatively, the coordinates of some reference pixels on the boundary can be recorded, and the boundary can be described by a relative movement code. One such simple code is the crack code, which is generated for each side p of a pixel on the boundary such that $C(p) = 0, 1, 2, 3$ for movement to the right, down, left, or up, respectively, as shown in Figure 5.28. The crack code for the object for this figure is listed below.

p:	1	2	3	4	5	6	7	8	9	10	11	12
$C(p)$:	0	1	0	3	0	1	2	1	2	2	3	3

Upon completion of the boundary trace, the value of the index p is the perimeter of the segment boundary. Freeman developed a method of boundary coding, called the chain code, in which the path from the centers of connected boundary pixels is represented by an eight-element code [17]. The Freeman chain code is discussed in the next section.

5.5 FEATURE REPRESENTATION

Image features are the primitive characteristics or attributes of an image. Some features are *natural*, in the sense that such features are defined by the visual appearance of an image, while other so-called *artificial* features result from specific manipulations of an image. Natural features include the luminance of a region of pixels and grayscale textural regions. Image amplitude histograms and spatial frequency spectra are examples of artificial features.

Feature representation and description become possible after an image has been segmented into regions. The next task for the computer is to describe the regions in terms of a chosen representation scheme. This section discusses a number of methods and algorithms for feature representation.

5.5.1 Chain Code

Chain code is one of the most commonly used algorithms to represent the boundary of an object in an image. In chain coding, the direction vectors between successive boundary pixels are encoded. The directions can be either four-directional, as in Figure 5.29a, or eight-directional, as in Figure 5.29b. For example, a commonly used chain code in Figure 5.29c employs eight directions, which can be coded by 3-bit code words. Typically, the chain code contains the start pixel address followed by a string of code words. Such codes can be generalized by increasing the number of allowed direction vectors between successive boundary pixels [16].

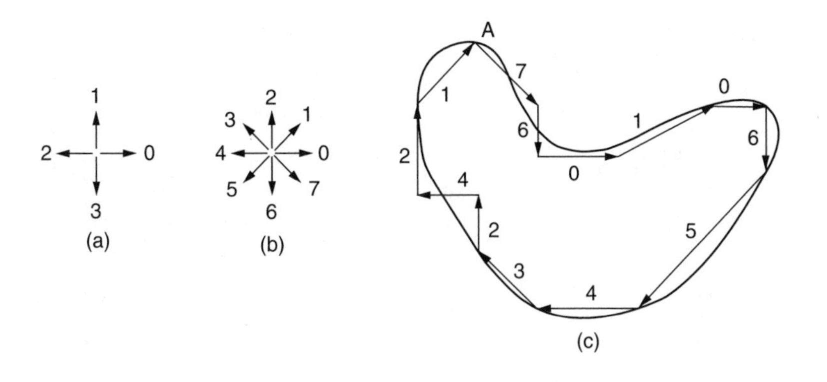

FIGURE 5.29 Chain code for boundary representation. (a) Four direction numbers for chain code elements. (b) Eight direction numbers for chain code elements. (c) Contour and chain code: (A), 76010655432421.

Figure 5.29c shows an object contour. Consider a starting point *A*. At point *A*, find the nearest edge pixel and code its orientation by the code words defined in Figure 5.29b. In case of a tie, choose the one with largest code value. In this case, the contour is formed clockwise. Continue until there are no more boundary pixels; that is, the contour is enclosed by returning to the starting point *A*.

The chain code of a boundary depends on the starting point. The code can be normalized with respect to the starting point by a straightforward procedure: we simply treat the chain code as a circular sequence of direction numbers and redefine the starting point so that the resulting sequence of numbers begins from an integer of minimum magnitude. These normalizations may help in matching.

Chain codes lend themselves to efficient calculation of certain parameters of the curves, such as area. The following algorithm computes the area enclosed by a four-neighbor chain code. The example is illustrated in Figure 5.30.

```
For a four-neighbor code (0: +x, 1: +y, 2: -x, 3:
-y) surrounding a region in a counterclockwise
sense, with starting point (x, y):

  begin Chain Code

1. area = 0

2. yposition := y;

3. for each element of chain code

       switch (case of element direction)

         0: area := area - yposition;

         1: yposition := yposition + 1;
```

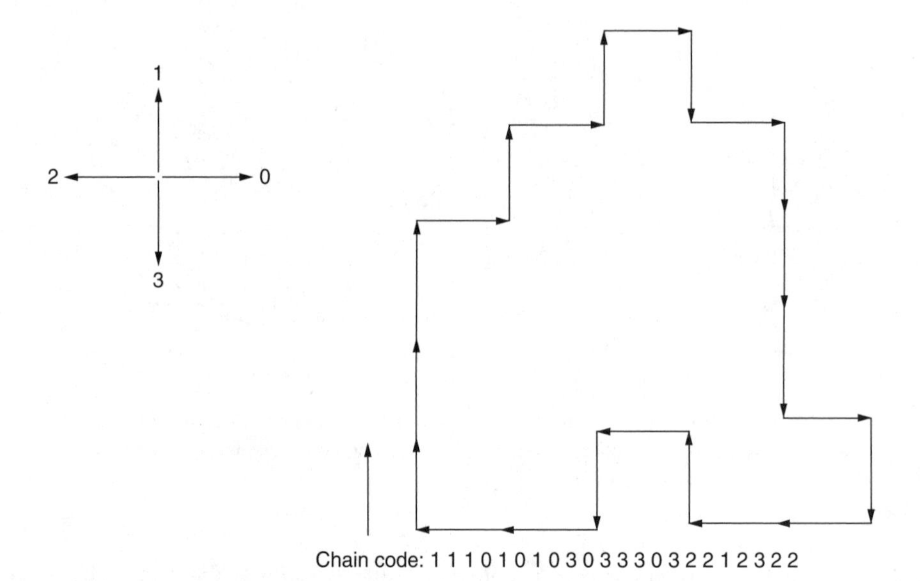

FIGURE 5.30 Example of chain code area.

```
        2: area := area + yposition
        3: yposition := yposition - 1;
    end of switch;
end Chain Code
```

5.5.2 Fourier Descriptor

After the boundary trace of an object is known, we can consider it as a pair of waveforms $x(t)$, $y(t)$. Any of the traditional one-dimensional signal representation techniques can be used. For any sampled boundary, we can define

$$u(n) = x(n) + jy(t), \quad n = 0, 1, ..., N-1 \tag{5.55}$$

where a closed contour would be periodic with period N. Its DFT representation is

$$u(n) = \frac{1}{N}\sum_{k=0}^{N-1} a(k)\exp\left(\frac{j2\pi kn}{N}\right), \quad 0 \le n \le N-1 \tag{5.56}$$

$$a(k) = \sum_{n=0}^{N-1} u(n)\exp\left(\frac{-j2\pi kn}{N}\right), \quad 0 \le k \le N-1 \tag{5.57}$$

The complex coefficients $a(k)$ are called the Fourier descriptors (FDs) of the boundary.

TABLE 5.2
Properties of Fourier Descriptors

Transformation	Boundary	Fourier Descriptors
Identity	$u(n)$	$a(k)$
Translation	$\tilde{u}(n) = u(n) + u_0$	$\tilde{a}(k) = a(k) + u_0\delta(k)$
Scaling or Zooming	$\tilde{u}(n) = \alpha \cdot u(n)$	$\tilde{a}(k) = \alpha \cdot a(k)$
Starting Point	$\tilde{u}(n) = u(n - n_0)$	$\tilde{a}(k) = a(k)e^{-j2n_0k/N}$
Rotation	$\tilde{u}(n) = u(n)e^{j\theta_0}$	$\tilde{a}(k) = a(k)e^{j\theta_0}$
Reflection	$\tilde{u}(n) = u*(n)e^{j2\theta} + 2\gamma$	$\tilde{a}(k) = a*(-k)e^{j2\theta} + 2\gamma\delta(k)$

Several geometrical transformations of a boundary or shape can be related to simple operations on FDs, as listed in Table 5.2. When the boundary is translated by

$$u_0 = x_0 + jy_0 \tag{5.58}$$

then the new FDs remain the same except at $k = 0$. The effect of scaling, that is, shrinking or expanding the boundary, results in scaling of $a(k)$. Changing the starting point in tracing the boundary results in a modulation of $a(k)$. Rotation of the boundary by an angle θ_0 causes a constant phase shift of θ_0 in the FDs. As shown in Figure 5.31, reflection of the boundary or shape about a straight line inclined at an angle θ,

$$Ax + By + C = 0 \tag{5.59}$$

gives the new boundary $\tilde{x}(n)$, $\tilde{y}(t)$ as

$$\tilde{u}(n) = u*(n)e^{j2\theta} + 2\gamma \tag{5.60}$$

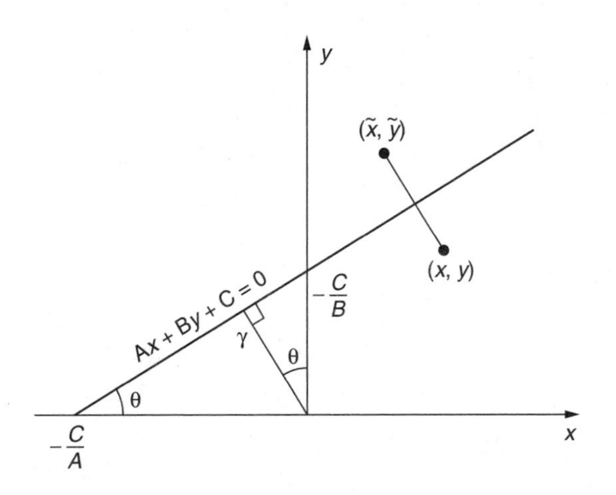

FIGURE 5.31 Reflection about a straight line.

where

$$\gamma = \frac{-(A + jB)C}{A^2 + B^2}, \quad e^{j2\theta} = \frac{-(A + jB)^2}{A^2 + B^2} \tag{5.61}$$

Fourier descriptors can also regenerate shape features through an inverse Fourier transformation. The number of descriptors needed for reconstruction depends on the shape and the desired accuracy. Figure 5.32 shows the effect of

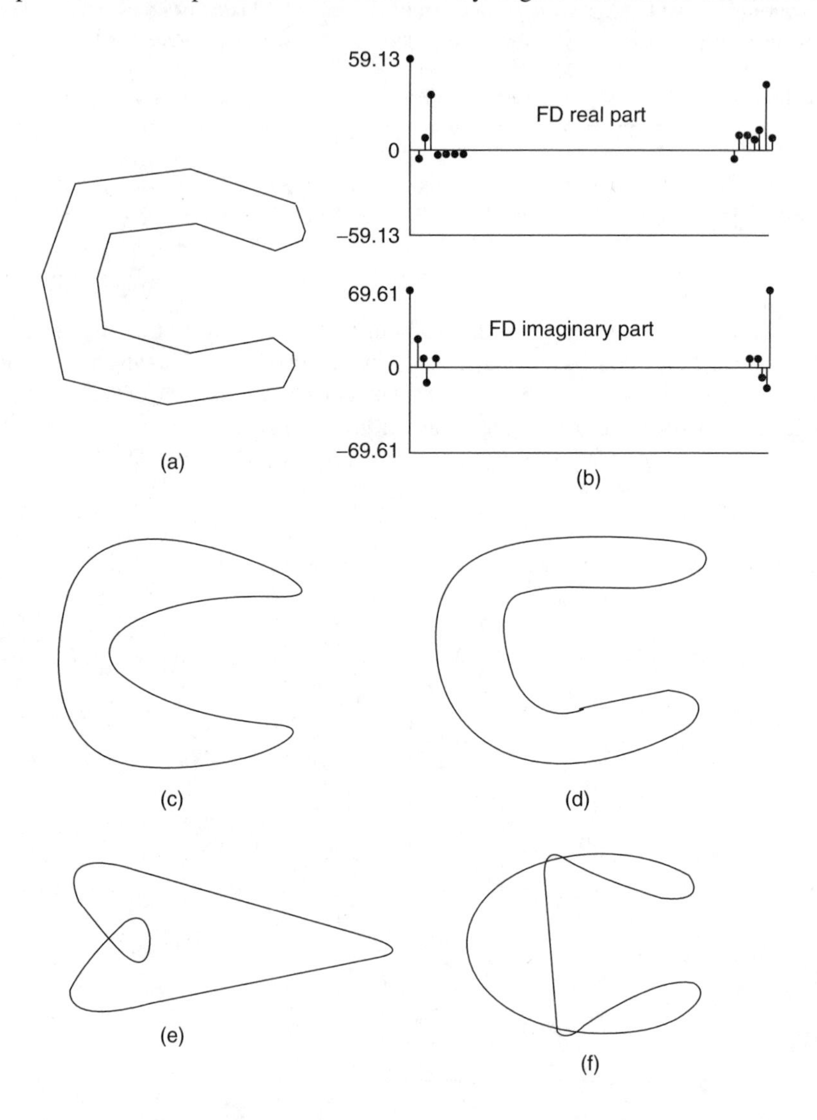

FIGURE 5.32 Fourier descriptions (FDs). (a) Given shape; (b) FDs, real and imaginary components; (c) shape derived from largest five FDs; (d) derived from all FDs quantized to 17 levels each; (e) amplitude reconstruction; and (f) phase reconstruction.

truncation and quantization of the FDs. From Table 2, it can be observed that the FD magnitudes have some invariant properties. For example, $|\tilde{a}(k)|$, k = 1, 2, …, $N-1$ are invariant to starting point, rotation, and reflection. The features $\tilde{a}(k)/|\tilde{a}(k)|$ are invariant to scaling. These properties can be used to detect shapes regardless of their size, orientation, and so on. However, the FDs' magnitude or phase alone is generally inadequate for reconstruction.

Fourier descriptors can correctly match similar shapes even if they have different size and orientation. If $a(k)$ and $b(k)$ are the FDs of two boundaries $u(k)$ and $v(k)$ respectively, then their shapes are similar if the distance

$$d(u_0, \alpha_0, \theta_0, n_0) = \min_{u_0, \alpha_0, \theta_0, n_0} \left\{ \sum_{n=0}^{N-1} \left| u(n) - \alpha v(n+n_0) e^{j\theta_0} - u_0 \right|^2 \right\} \tag{5.62}$$

is small. The parameters u_0, α_0, n_0, and θ_0 are chosen to minimize the effects of translation, scaling, starting points, and rotation, respectively. If $u(n)$ and $v(n)$ are normalized so that $\Sigma u(k) = \Sigma v(k) = 0$, then for a given shift n_0, the above distance is minimum when

$$\begin{aligned} u_0 &= 0 \\ \alpha &= \frac{\Sigma_k c(k)\cos(\varphi_k + k\phi + \theta_0)}{\Sigma_k |b(k)|^2} \\ \tan\theta_0 &= -\frac{\Sigma_k c(k)\sin(\varphi_k + k\phi)}{\Sigma_k c(k)\cos(\varphi_k + k\phi)} \end{aligned} \tag{5.63}$$

where $a(k)b*(k) = c(k)e^{j\phi_{k,\phi}} = -2\pi n_0/N$, and $c(k)$ is a real quantity. These equations give α and θ_0, from which the minimum distance d is given by

$$d = \min_{\phi}[d(\phi)] = \min_{\phi}\left\{ \sum_k |a(k) - \alpha b(k)\exp[j(k\phi + \theta_0)]|^2 \right\} \tag{5.64}$$

The distance $d(\phi)$ can be evaluated for each $\phi = \phi(n_0)$, $n_0 = 0, 1, \ldots, N-1$, and the minimum searched to obtain d. The quantity d is then a useful measure of difference between two shapes. The FDs can also be used for analysis of line patterns or open curves, skeletonization of patterns, computation of area of a surface, and so on.

Instead of using two functions $x(t)$ and $y(t)$, it is possible to use only one function when t represents the arc length along the boundary curve. Defining the arc tangent angle as shown in Figure 5.33,

$$\theta(t) = \tan^{-1}\left[\frac{dy(t)/dt}{dx(t)/dt}\right] \tag{5.65}$$

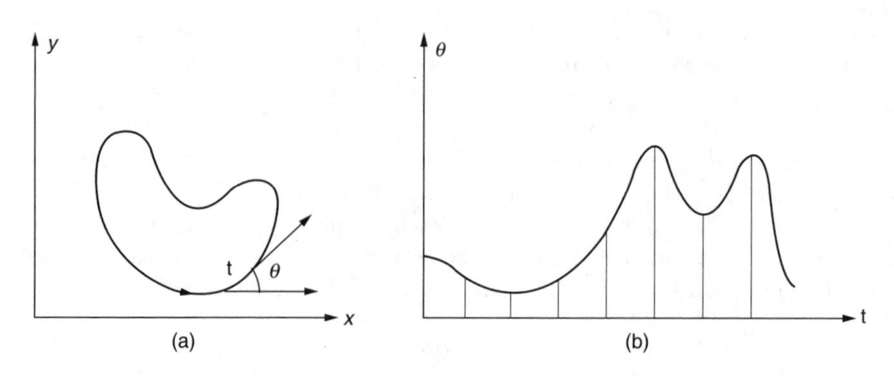

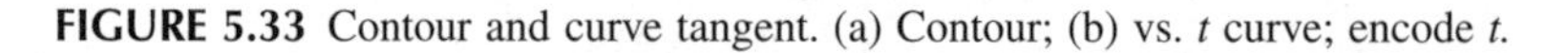
FIGURE 5.33 Contour and curve tangent. (a) Contour; (b) vs. t curve; encode t.

the curve can be traced if $x(t)$, $y(t)$, and $\theta(t)$ are known. Since t is the distance along the curve, it is true that

$$dt^2 = dx^2 + dy^2 \Rightarrow \left(\frac{dx}{dt}\right)^2 + \left(\frac{dy}{dt}\right)^2 = 1 \tag{5.66}$$

which gives $dx/dt = \cos\theta(t)$, $dy/dt = \sin\theta(t)$, or

$$\mathrm{x}(t) = \mathrm{x}(0) + \int_0^t \begin{bmatrix} \cos\theta(\tau) \\ \sin\theta(\tau) \end{bmatrix} d\tau \tag{5.67}$$

Sometimes, the FDs of the curvature of the boundary

$$\kappa(t) = \frac{d\theta(t)}{dt} \tag{5.68}$$

or those of the detrended function

$$\hat{\theta}(t) = \int_0^t \kappa(\tau)\, d\tau - \frac{2\pi t}{T} \tag{5.69}$$

are used. The latter has the advantage that $\hat{\theta}(t)$ does not have the singularities at corner points that are encountered in polygon shapes. Although we have now only a real scalar set of FDs, their rate of decay is found to be much slower than those of $u(t)$.

5.5.3 Moments

Moments are a useful alternative to series expansions for representing the shape of objects. Let $f(x, y) \geq 0$ be a real bounded function with support on a finite region R. We define its $(p + q)$-th order moment

$$m_{p,q} = \iint_R f(x, y) x^p y^q dx\, dy, \quad p, q = 0, 1, 2, \ldots \tag{5.70}$$

Note that setting $f(x, y) = 1$ gives the moments of the region R that could represent a shape. Thus the results presented here would be applicable to arbitrary objects as well as their shapes. Without loss of generality, we can assume that $f(x, y)$ is nonzero only in the region $= \{x \in (-1, 1), y \in (-1, 1)\}$. Then higher-order moments will in general have increasingly smaller magnitudes.

Certain functions of moments are invariant to geometric transformations such as translation, scaling, and rotation. Such features are useful in identifying objects with unique shapes regardless of their location, size, and orientation.

Translation—With the translation of coordinates $x' = x + \alpha$, $y' = y + \beta$, the central moments

$$\mu_{p,q} = \iint (x-\bar{x})^p (y-\bar{y})^q f(x,y) dx\, dy \tag{5.71}$$

are invariants, where $\bar{x} = m_{1,0}/m_{0,0}$, $\bar{y} = m_{0,1}/m_{0,0}$.

Scaling—With a scale change $x' = x + \alpha$, $y' = y + \beta$, the moments of $f(\alpha x, \alpha y)$ change to $\mu'_{p,q} = \mu_{p,q}/\alpha^{p+q+2}$. The normalized moments, defined as

$$\eta_{p,q} = \frac{\mu'_{p,q}}{(\mu'_{0,0})^{\gamma}}, \quad \gamma = (p+q+2)/2 \tag{5.72}$$

are then invariant to size change.

Rotation and Reflection—A linear coordinate transformation that is performed on $f(x, y)$

$$\begin{bmatrix} x' \\ y' \end{bmatrix} = \begin{bmatrix} \alpha & \beta \\ \gamma & \delta \end{bmatrix} \begin{bmatrix} x \\ y \end{bmatrix} \tag{5.73}$$

will lead to change in the moment-generating function. By the theory of algebraic invariants, it is possible to find certain polynomials of $\mu_{p,q}$ that remain unchanged under the transformation of Equation 5.73. For example, some moment invariants with respect to rotation (that is, for $\alpha = \delta = \cos\theta$, $\beta = -\gamma = \sin\theta$) and reflection ($\alpha = -\delta = \cos\theta$, $\beta = \gamma = \sin\theta$) are given in Table 5.3.

TABLE 5.3
Rotational Moment Invariants

First-order moments	$\mu_{0,1} = \mu_{1,0} = 0$
Second-order moments	$1 = \mu_{2,0} = \mu_{0,2}$
$(p + q = 2)$	$2 = (\mu_{2,0} - \mu_{0,2})^2 + 4\mu_{1,1}^2$
Third-order moments	$3 = (\mu_{3,0} - 3\mu_{1,2})^2 + (\mu_{0,3} - 3\mu_{2,1})^2$
$(p + q = 3)$	$4 = (\mu_{3,0} + \mu_{1,2})^2 + (\mu_{0,3} + \mu_{2,1})^2$
	$5 = (\mu_{3,0} - 3\mu_{1,2})(\mu_{3,0} + \mu_{1,2})[(\mu_{3,0} + \mu_{1,2})^2 - 3(\mu_{2,1} + \mu_{0,3})^2]$
	$+ (\mu_{0,3} - 3\mu_{2,1})(\mu_{0,3} + \mu_{2,1})[(\mu_{0,3} + \mu_{2,1})^2 - 3(\mu_{1,2} + \mu_{3,0})^2]$
	$6 = (\mu_{2,0} - \mu_{0,2})[(\mu_{3,0} + \mu_{1,2})^2 - (\mu_{2,1} + \mu_{0,3})^2] + 4\mu_{1,1}(\mu_{3,0} + \mu_{1,2})(\mu_{0,3} + \mu_{2,1})$

Applications of moment invariants can be found in many cases such as shape analysis, character recognition, and scene matching. Being invariant under coordinate transformations, the moment invariants are useful features in pattern-recognition problems. Using N moments, for instance, an image can be represented as a point in an N-dimensional vector space. This converts the pattern recognition problem into a standard decision theory problem, for which several approaches are available. For binary image we can set $f(x, y) = 1$, $(x, y) \in R$. Then the moment calculation reduces to the separable computation

$$m_{p,q} = \sum_{x} x^p \sum_{y} y^q. \tag{5.74}$$

5.5.4 Shape Features and Representation

The shape of an object refers to its profile and physical structure. These characteristics can be represented by the previously discussed boundary, region, moment, and structural representation. These representations can be used for matching shapes, recognizing objects, or making measurements of shapes; see Figure 5.34.

5.5.4.1 Geometry Features

Measurement of certain geometric attributes is one of primary image analysis and pattern recognition applications. The most important attributes and their definitions are listed below.

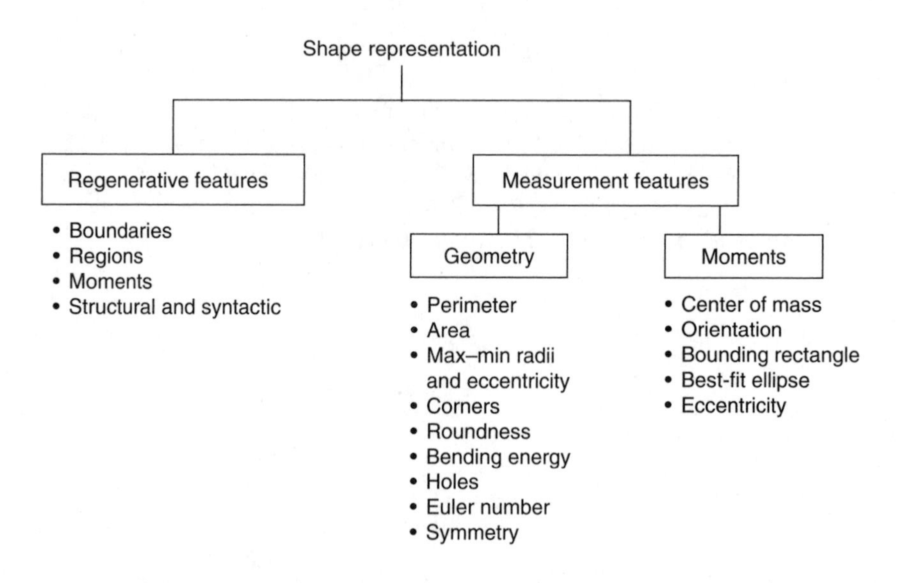

FIGURE 5.34 Shape features.

Perimeter—

$$T = \int \sqrt{x^2(t) + y^2(t)}dx \tag{5.75}$$

where t is the boundary parameter.

Area—

$$A = \iint_R dx\,dy = \int_{\partial R} y(t)\frac{dx(t)}{dt}dt - \int_{\partial R} x(t)\frac{dy(t)}{dt}dt \tag{5.76}$$

where R and ∂R denote the object region and its boundary, respectively.

Radii—R_{min} and R_{max} are the minimum and maximum distances, respectively, to boundary from the center of mass; see Figure 5.35a. Sometimes the ratio R_{max}/R_{min} is used as a measure of eccentricity or elongation of the object.

Corner—Corners are locations on the boundary where the curvature $\kappa(t)$ becomes unbounded. When t represents distance along the boundary,

$$|\kappa(t)|^2 = \left(\frac{d^2y}{dt^2}\right)^2 + \left(\frac{d^2x}{dt^2}\right)^2 \tag{5.77}$$

In practice, a corner is declared whenever $|\kappa(t)|$ assumes a large value; see Figure 5.35b.

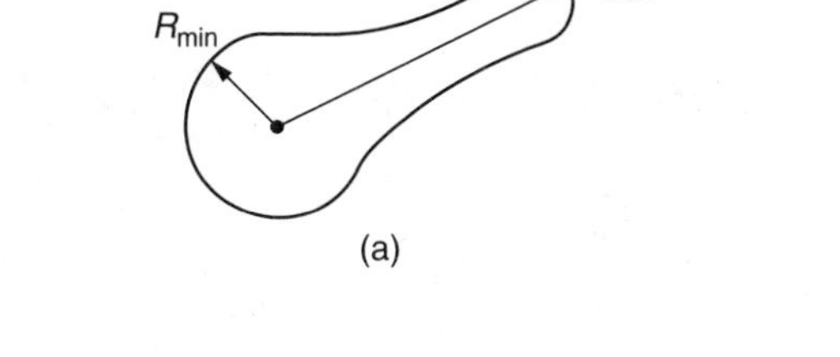

(a)

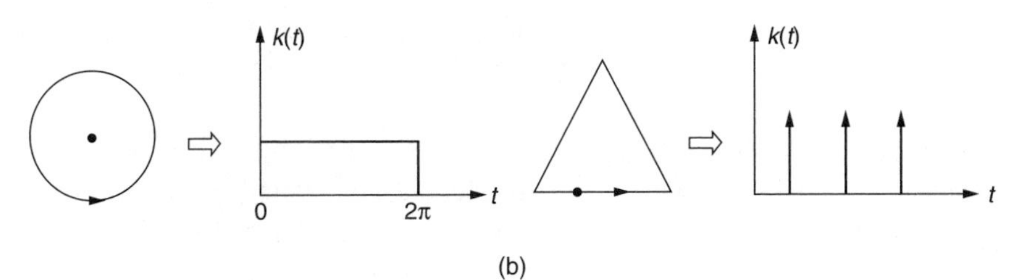

(b)

FIGURE 5.35 Geometry features. (a) Maximum and minimum radii; (b) curvature functions for corner detection.

Bending energy—This is an attribute of the curvature

$$E = \frac{1}{T}\int_0^T |\kappa(t)|^2\, dt \tag{5.78}$$

In terms of $\{a(k)\}$, the FDs of $u(t)$, this attribute is given by

$$E = \sum_{k=-\infty}^{\infty} |a(k)|^2 \left(\frac{2\pi k}{T}\right)^4 \tag{5.79}$$

Roundness—

$$\gamma = \frac{(T_{\text{parameter}})^2}{4\pi(A_{\text{area}})} \tag{5.80}$$

For a perfect disc, $\gamma = 1$, the minimum value.

5.5.4.2 Moment-Based Features

Many shape features can be conveniently represented in terms of moments. For a shape represented by a region R containing N pixels, we have the following moments:

Center of mass—

$$\bar{m} = \frac{1}{N}\sum\sum_{(m,n)\in R} m, \quad \bar{n} = \frac{1}{N}\sum\sum_{(m,n)\in R} n \tag{5.81}$$

The (p, q) order central moments become

$$\mu_{p,q} = \frac{1}{N}\sum\sum_{(m,n)\in R} (m-\bar{m})^p (n-\bar{n})^q \tag{5.82}$$

Orientation—Orientation is defined as the angle of axis of the least moment of inertia; see Figure 5.36a. It is obtained by minimizing with respect to θ the sum

$$I(\theta) = \sum\sum_{(m,n)\in R} D^2(m,n) = \sum\sum_{(m,n)\in R} [(n-\bar{n})\cos\theta - (m-\bar{m})\sin\theta]^2 \tag{5.83}$$

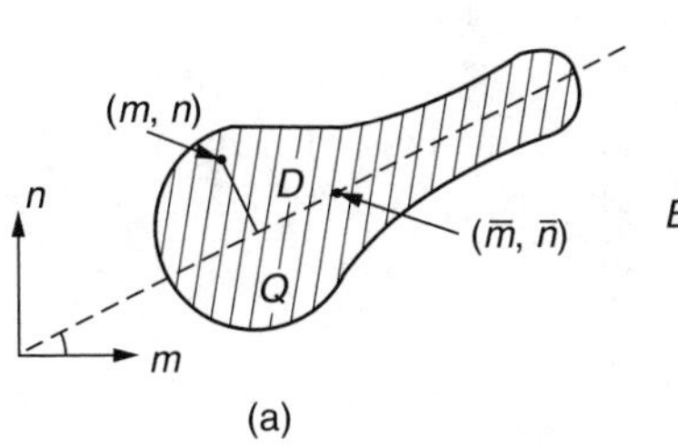

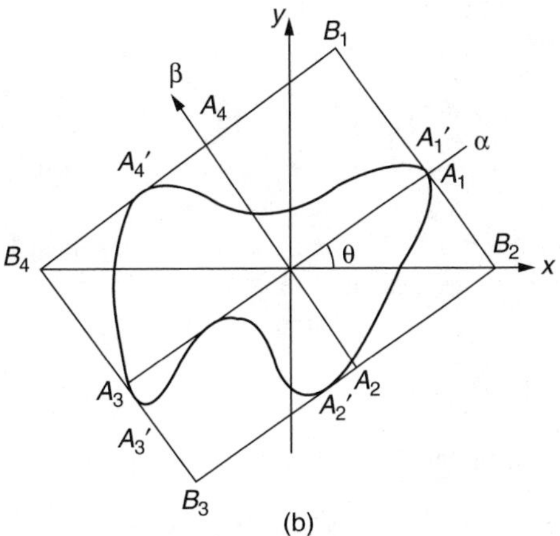

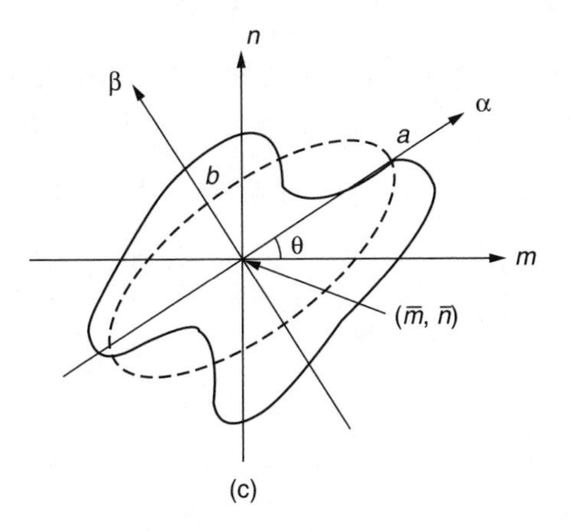

FIGURE 5.36 Moment-based features. (a) Orientation; (b) boundary rectangle; (c) best fit ellipse.

The result is

$$\theta = \frac{1}{2}\tan^{-1}\left[\frac{2\mu_{1,1}}{\mu_{2,0} - \mu_{0,2}}\right] \tag{5.84}$$

Bounding rectangle—The bounding rectangle is the smallest rectangle enclosing the object that is also aligned with its orientation; see Figure 5.36b. Once θ is known, we use the transformation

$$\alpha = x\cos\theta + y\sin\theta \tag{5.85a}$$

$$\beta = -x\sin\theta + y\cos\theta \tag{5.85b}$$

on the boundary points and search for α_{min}, α_{max}, and β_{max}. These give the locations of points A_3', A_1', A_2', and A_4', respectively; see Figure 5.36b. From these the bounding rectangle is known immediately with length $l_b = \alpha_{max} - \alpha_{min}$ and width $w_b = \beta_{max} - \beta_{min}$. The ratio $l_b w_b / A_{area}$ is also a useful shape feature.

Best-fit ellipse—The best-fit ellipse is the ellipse whose second moment equals that of the object. Let a and b denote the lengths of semi-major and semi-minor axes, respectively, of the best-fit ellipse; see Figure 5.36c. The last and the greatest moments of inertia for an ellipse are

$$I_{min} = \frac{\pi}{4} ab^3, \quad I_{max} = \frac{\pi}{4} a^3 b \tag{5.86}$$

For orientation θ, the above moments can be calculated as

$$I'_{min} = \sum_{(m,n)\in R}\sum [(n-\bar{n})\cos\theta - (m-\bar{m})\sin\theta]^2 \tag{5.87}$$

$$I'_{max} = \sum_{(m,n)\in R}\sum [(n-\bar{n})\sin\theta - (m-\bar{m})\cos\theta]^2 \tag{5.88}$$

For the best-fit ellipse we want $I_{min} = I'_{min}$ and $I_{max} = I'_{max}$, which gives

$$a = \left(\frac{4}{\pi}\right)^{1/4} \left[\frac{I'^3_{max}}{I'_{min}}\right]^{1/8}, \quad b = \left(\frac{4}{\pi}\right)^{1/4} \left[\frac{I'^3_{min}}{I'_{max}}\right]^{1/8} \tag{5.89}$$

Eccentricity—Measures of eccentricity include aspect ratios R_{max}/R_{min}, I'_{max}/I'_{min}, a/b, and the function

$$\varepsilon = \frac{(\mu_{2,0} - \mu_{0,2})^2 + 4\mu_{1,1}}{A_{area}} \tag{5.90}$$

For further reading, see Ballard and Brown [2], Freeman [18], Fu [20], Galbiati [23], Gonzalez and Woods [26], Jain [33], Paulus and Hornegger [49], Pratt [55], and Rosenfeld and Kak [59].

5.5.5 Skeletons

Skeletons record the information about shapes of objects and make available the least geometric representation. A skeleton reduces a structural shape to a graph. One may accomplish this reduction via a thinning, or skeletonizing, algorithm. Thinning procedures play a central role in a broad range of problems in image processing, ranging from automated inspection of circuit boards to counting cells in a complicated biological image. A powerful approach for obtaining skeletons is to employ morphological operations of combined opening and closing

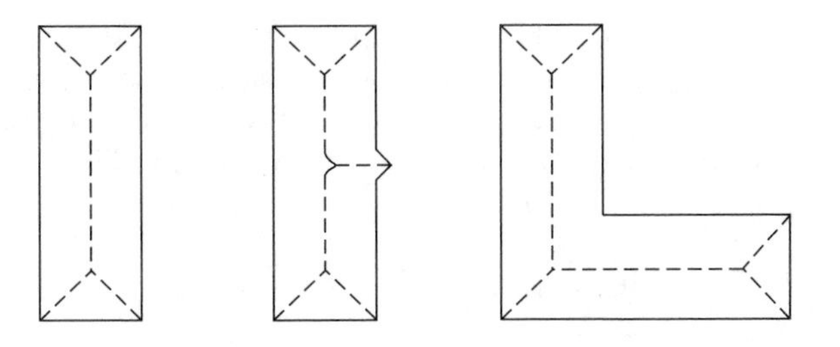

FIGURE 5.37 Medial axes (dashed lines) of simple regions.

operators, as discussed in the next section. Here the focus is on a transformation approach.

The skeleton of a region may be obtained via the medial axis transformation (MAT). The MAT of a region R with border B is as follows. For each point p in R, we find its closest neighbor in B. If p has more than one such neighbor, it is said to belong to the medial axis (skeleton) of R. The concept of *closest* depends on the definition of a distance. Figure 5.37 shows some examples using the Euclidean distance. The same results would be obtained with the maximum disk.

The medial axis transformation of a region has an intuitive definition based on the so-called "prairie fire concept." Consider an image region as a prairie of uniform, dry grass, and suppose that a fire is lit along its border. All fire fronts will advance into the region at the same speed. The MAT of the region is the set of points reached by more than one fire front at the same time. The MAT of a region yields an intuitively pleasing skeleton, and the advent of computer technologies has made the comprehensive requirement for computing an easy task. Implementation potentially involves calculating the distance from every interior point to every point on the boundary of a region. Numerous algorithms have been proposed for improving computational efficiency. Typically these are thinning algorithms that iteratively delete edge points of a region subject to the constraints that deletion of these points (*a*) does not remove the end point, (*b*) does not break connectivity, and (*c*) does not cause excessive erosion of the region.

A straightforward approach to extract medial axes from a binary region is as follows. Let B be the set of boundary points. For each point P in R, find its closest neighbors on the region boundary. If more than one boundary point is the minimum distance from point x, then x is on the skeleton of R. The skeleton is the set of pairs $\{x, d_s(x, B)\}$when $d_s(x, B)$ is the distance from x to the boundary, as defined above. Since each x in the skeleton retains the information on the minimum distance to the boundary, the original region may be recovered conceptually as the union of disks centered on the skeleton points.

Some common shapes have simply structured medial axis transform skeletons. In the Euclidian metric, a circle has a skeleton consisting of only its central point. A convex polygon has a skeleton consisting of linear segments; if the polygon is nonconvex, the segments may be parabolic or linear. Skeletonization is sensitive to noise at the boundary. Reducing this sensitivity may be accomplished by smoothing the boundary, using a polygonal boundary approximation, or including only those points in the skeleton that are greater than some distance from the boundary. The latter scheme can lead to disconnected skeletons. See Levialdi [36], Maragos and Schafer [41], and Peleg and Rosenfeld [51] for more information.

5.6 MORPHOLOGICAL IMAGE PROCESSING AND ANALYSIS

Mathematical morphology has many advantages over conventional statistical processing techniques in performing image shape analysis and recognition. In recent years, applications of mathematical morphology in the machine vision area have solved many problems associated with pattern recognition and industrial inspection.

The concepts of mathematical morphology provide some very powerful tools with which low-level image analysis can be performed. Low-level analysis, by its very nature, involves repeated computations over large, regular data structures. Parallelism appears to be a necessary attribute of any hardware system which is to perform such image-analysis tasks efficiently. This parallelism can take a variety of forms. The tools of morphology can be used to realize inherent advantages of speed and flexibility in many image processing applications.

5.6.1 Morphological Transformations of Binary Images

Morphological transformations are transformations of signals based on the theory of mathematical morphology. In this theory, each signal is viewed as a set in a Euclidean space. Morphological transformations are also referred to as morphological filtering. In this section, only consider 2D binary (set) images are considered; grayscale images (functions) are considered in Section 5.6.2. Additional information on morphology can be found in Dougherty [11], Heijmans [30], Maragos and Schafer [39, 40, 41], Serra [64, 65], and Zhuang and Haralick [79].

The four basic set operations are *erosion* and *dilation* (which are special cases of the hit-or-miss set transformation), and *opening* and *closing*. The erosion and dilation of sets correspond to set shrinking and expanding, respectively. For binary images (viewed as sets), the erosion, dilation, opening, and closing operations are the simplest morphological filters. These filters, originally defined for sets or binary signals, were expanded to functions, such as multilevel signals or grayscale images.

5.6.1.1 Morphological Set Transformations

If the topological space is restricted to the Euclidean space, $E = \mathbf{R}^d$, the new operations are defined on the space P(E). The subsets of $E = \mathbf{R}^d$ are denoted by capital English letters, A, B,.... The space E can be either the continuous space $\mathbf{R}^2$ or the discrete space $\mathbf{Z}^2$. For example, a single binary sampled image, which can be mathematically represented by a binary function of two independent discrete variables, is viewed as a subset of $\mathbf{Z}^2$ (i.e., as a set of integer pairs). These integer pairs can be viewed as a subset of the coordinates with respect to two basis unit vectors whose lengths equal the sampling periods in each direction. The representation of sampled binary images as subsets of $\mathbf{Z}^2$ is suitable for rectangular-sampled digital images.

5.6.1.2 Minkowski's Addition

Minkowski's addition $A \oplus B$ of two sets A and B consists of all points c that can be expressed as an algebraic vector addition $c = a + b$, where the vectors a and b belong to the sets A and B, respectively.

$$A \oplus B = \{a + b : a \in A, b \in B\} \tag{5.91}$$

For $x \in \mathbf{R}^d$, $A \oplus x$ is a *translate* of A by the translation x, denoted as A_x. Then *Minkowski's addition* is expressed as

$$A \oplus B = \bigcup_{b \in B} A_b \tag{5.92}$$

In other words, this operation can be viewed as the *union* of the translated sets of A_b, for all $b \in B$.

5.6.1.3 Minkowski's Subtraction

$$A \ominus B = (A^c \ominus B)^c = \{x : x + b \in A \text{ for every } b \in B\} \tag{5.93}$$

That is, *Minkowski's subtraction* $A \ominus B$ of two sets A and B consists of only points a that are in A and in A_b for all $b \in B$. By this definition, *Minkowski's subtraction* can be expressed as

$$A \ominus B = \bigcap_{b \in B} A_b \tag{5.94}$$

This means that *Minkowski's subtraction* can also be the intersection of all translated sets A_b, for $b \in B$.

5.6.1.4 Dilation

Let the set $B^s = \{-b : b \in B\}$, that is, B^s is the *symmetrical set* of B with respect to the origin. Dilation of A by B is defined as the set $\{z : A \cap B_z\} \neq \phi$, where ϕ is an empty set, of the points z such that A meets the translate B_z, denoted by $A \oplus B^s$:

$$A \oplus B^s = \left\{z : B_Z^s \cap A \neq \phi\right\} = \bigcup_{b \in B} A_{-b} \tag{5.95}$$

5.6.1.5 Erosion

The *erosion* of A by B is defined as the set $\{z : B_z \subset A\}$ of the points z such that the translate B_z is included in A and, by Equation 5.93, is equal to $A \ominus B^s$:

$$A \ominus B^s = \{z : B_Z^s \subset A\} = \bigcap_{b \in B} A_{-b} \tag{5.96}$$

Figure 5.38 shows the set A dilated by B and compares the dilation with the Minkowski's addition. Figure 5.39 shows the set A eroded by B and compares the erosion with the Minkowski's subtraction. Both dilation and erosion are nonlinear operations, which are generally noninvertible. Only Minkowski's addition is commutative and associative, that is:

$$A \oplus B = B \oplus A \tag{5.97}$$

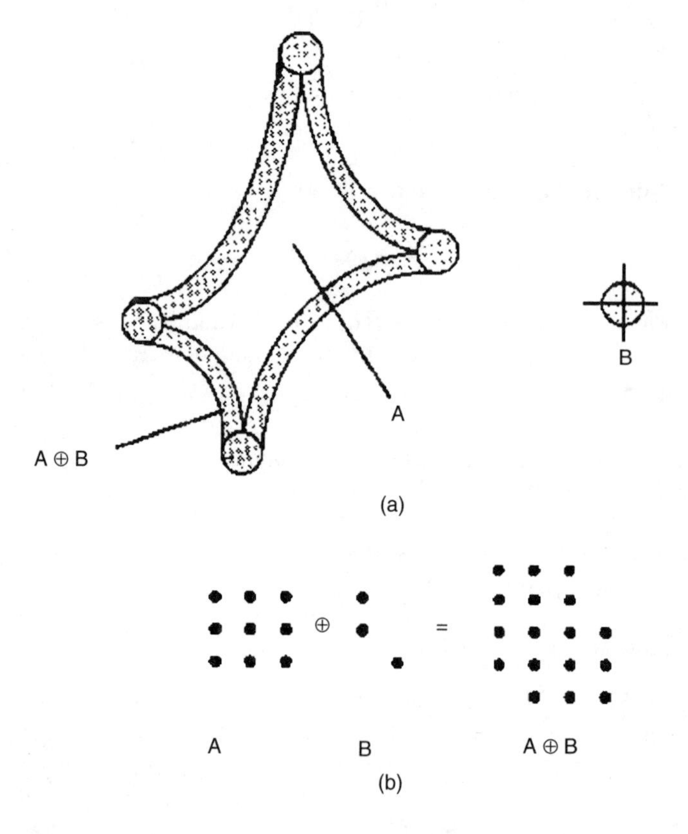

FIGURE 5.38 Illustration of dilation as an expansion process. (a) A is the object, and B is the structuring element. (b) Dilation of A by B in discrete domain.

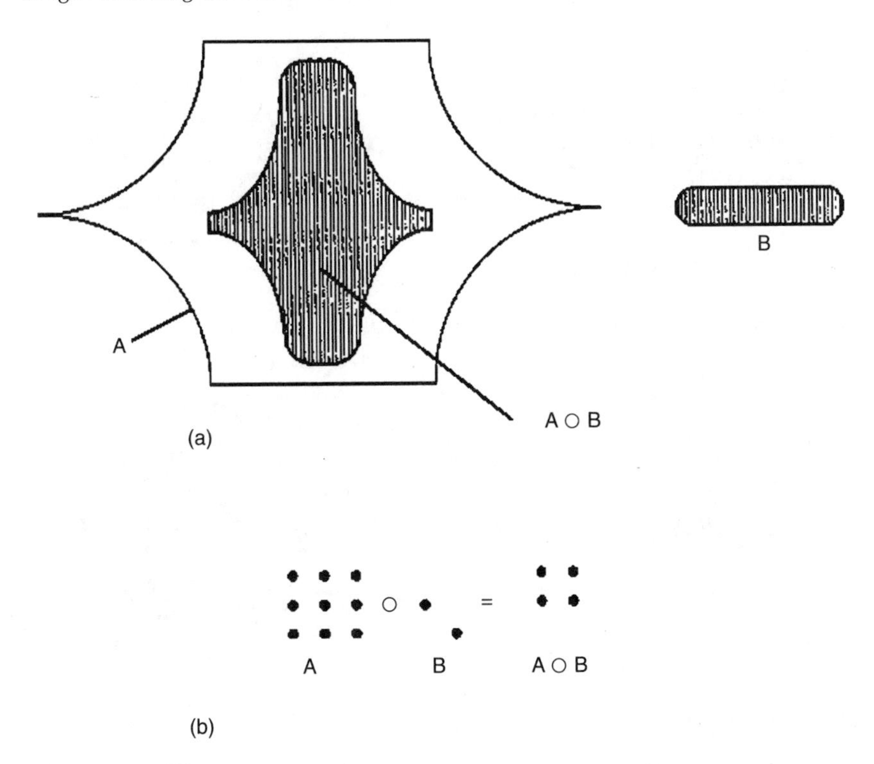

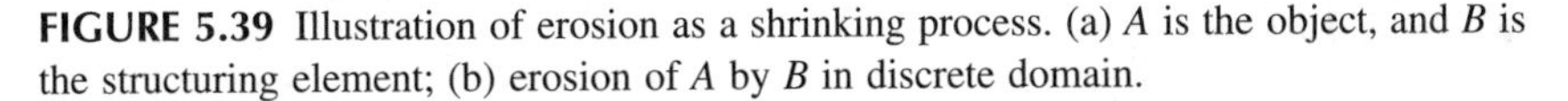

FIGURE 5.39 Illustration of erosion as a shrinking process. (a) A is the object, and B is the structuring element; (b) erosion of A by B in discrete domain.

and

$$(A \oplus B) \oplus C = A \oplus (B \oplus C) \tag{5.98}$$

The dilation and the erosion are analogous to the Minkowski's addition and subtraction, respectively. If B is a symmetrical set (i.e., $B = B^s$), then there is no difference between dilation and Minkowski's addition, and between erosion and Minkowski's subtraction.

In some literature, the dilation is denoted by $\oplus$ and the erosion by $\ominus$. We notice that there are differences between the Minkowski's addition (respectively subtraction) and the dilation (respectively erosion). Figure 5.40 shows that when the structuring element is not symmetric, the results can be quite different.

Dilation and erosion have the following properties:

Translation invariance—The set from A translated by x then dilated (or eroded) by B is equivalent to the set that results from dilating (or eroding) of A by B then translating by x. That is, the dilation and erosion are invariant under translation

$$(A)_x \oplus B^s = (A \oplus B^s)_x \tag{5.99}$$

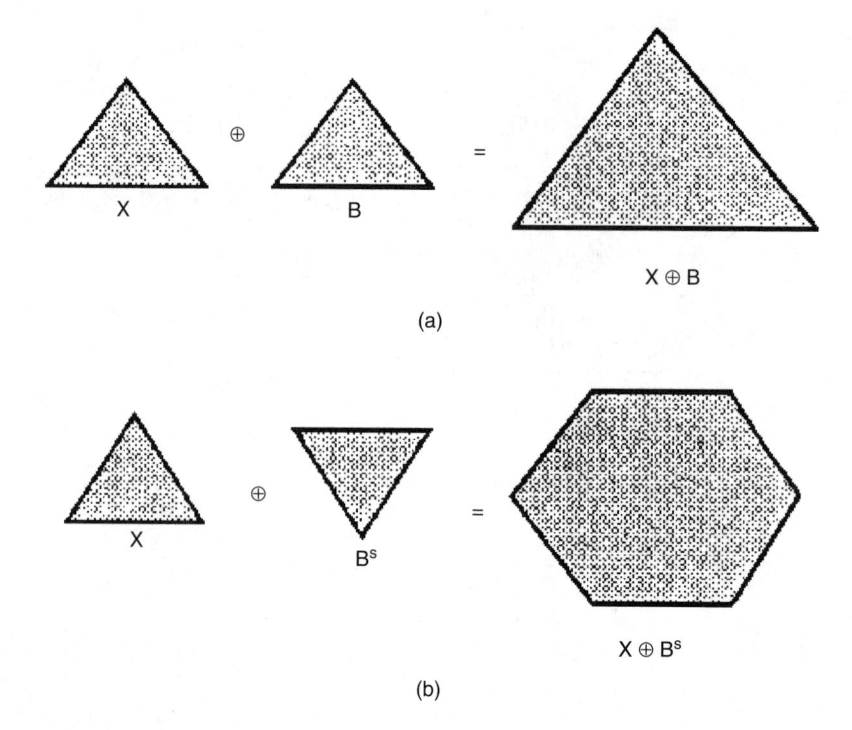

FIGURE 5.40 Difference between (a) Minkowski's addition defined as $X \oplus B$ and (b) the dilation defined by $X \oplus B^s$.

and

$$(A)_x \ominus B^s = (A \ominus B^s)_x \tag{5.100}$$

Increasing and inclusion properties—The dilation and the erosion are *increasing* transformations. That is, for $A_1 \subseteq A_2$,

$$A_1 \oplus B^s \subseteq A_2 \oplus B^s \tag{5.101}$$

and

$$A_1 \ominus B^s \subseteq A_2 \ominus B^s \tag{5.102}$$

The erosion is an *inclusion* with respect to a *structuring element*. That is, for $B_1 \subseteq B_2$,

$$A \ominus B_1^s \subseteq A \ominus B_2^s \tag{5.103}$$

Parallel composition—The operation of dilation distributes (commutes) with set union, whereas erosion distributes with set intersection. In addition, erosion by a union of two elements is equal to the intersection of the individual erosions:

$$(A \cup B) \oplus C^s = (A \oplus C^s) \cup (B \oplus C^s) \tag{5.104a}$$
$$(A \cap B) \ominus C^s = (A \ominus C^s) \cup (B \ominus C^s) \tag{5.104b}$$
$$A \ominus (B \cup C)^s = (A \ominus C^s) \cup (B \ominus C^s) \tag{5.104c}$$

Serial composition—Successively dilating (respectively eroding) a set A by B and then C is equivalent to dilating (respectively eroding) A by the dilation of B and C:

$$A \oplus (B \oplus C)^s = (A \oplus B^s) \oplus C^s \tag{5.105a}$$
$$A \ominus (B \oplus C)^s = (A \ominus B^s) \ominus C^s \tag{5.105b}$$

For the bounded set A and the origin o, it is true that $A \ominus A^s = o$. From the definitions of the Minkowski's addition and subtraction, the inclusion can be extended to the set union and intersection in the form of dilation and erosion:

$$A \oplus (B \cap C) \subset (A \oplus B) \cap (A \oplus C) \tag{5.106a}$$
$$A \ominus (B \cap C) \supset (A \ominus B) \cup (A \ominus C) \tag{5.106b}$$
$$(B \cup C) \ominus A \supset (B \ominus A) \cup (C \ominus A) \tag{5.106c}$$

Two additional morphological operations are opening and closing, which are the combined operations of erosion and dilation. The *opening* A_B of A by B is defined as

$$A_B = (A \ominus B^s) \oplus B \tag{5.107}$$

Note that A_B is the result of A eroded by B then dilated by B^s (or Minkowski's adding B to the erosion of A by B). The *closing* A^B of A by B is defined as:

$$A^B = (A \oplus B^s) \ominus B \tag{5.108}$$

Figure 5.41 illustrates that opening suppresses the sharp cusps and cuts the narrow isthmuses of the object, whereas closing fills in the thin gulfs and small holes. Thus, if the structuring element B has a regular shape, both opening and closing can be thought of as nonlinear filters which smooth the contours of the object. Both operations are generally noninvertible.

The opening and the closing are increasing operations. That is, $A \subset A'$ implies $A_B \subset A'_B$ and $A^B \subset A'^B$. They are *idempotent*, i.e., $(A_B)_B = A_B$ and $(A^B)^B = A^B$. They are dual to each other, i.e., $(A^c)_B = (A^B)^c$ and $(A^c)^B = (A_B)^c$. Moreover, the closing

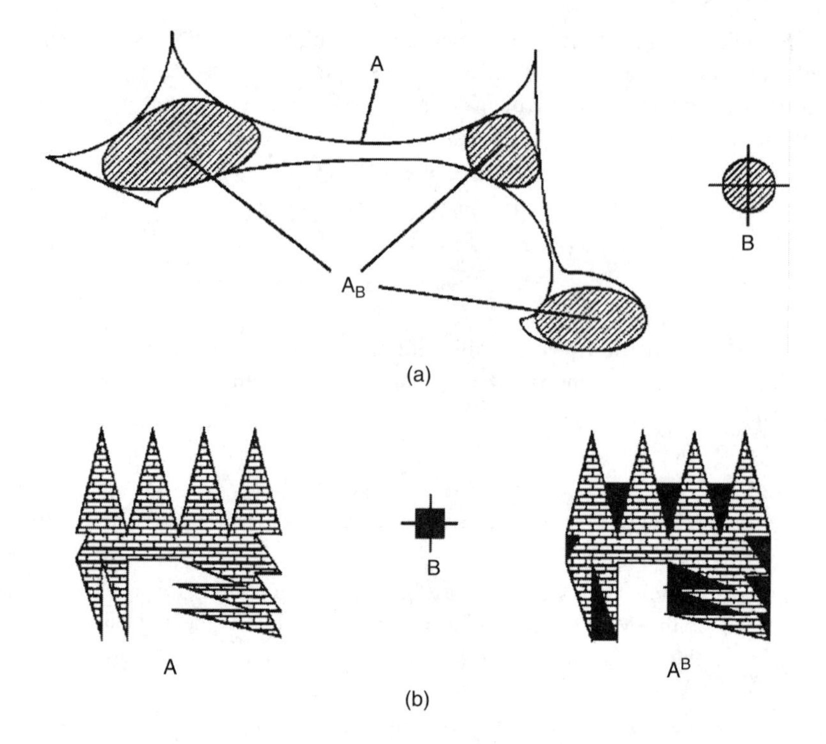

FIGURE 5.41 Illustration of opening and closing operations. (a) Opening of the object A by the structuring element B; the shaded area inside A is the resulting set of opening; (b) closing of the object A by the structuring element B, and the resulting set with valleys being filled.

is *extensive* and opening is *antiextensive*, that is, $A_B \subset A \subset A^B$. Figure 5.42 shows an example of the closing operation where a group of small clusters is merged into a large cluster.

A set A is *open with respect to* B if $A_B = A$, and *closed with respect to* B if $A^B = A$. The set A is open (respectively closed) with respect to B if and only if there exists a set C such that $A = C \oplus B$ (respectively $A = C \ominus B$). Thus, we have that $C \oplus B = (C \oplus B)_B$, $\forall B, \forall C$.

As a particular case, choose $A = o$; then B is always open with respect to itself, since $B = o \oplus B$; i.e., $B = (B \ominus B^s) \oplus B$. This characterization of opening leads to further properties:

1. If C is open with respect to B, then the openings A_B and A_C and the closing A^B and A^C of any set A satisfy the following inclusions: $A_C \subset A_B \subset A \subset A^B \subset A^C$.
2. If C is open with respect to B, then for every set A: $(A_B)_C = A_C$.
3. If B is open with respect to C, then, on the contrary: $(A_B)_C = A_B$.

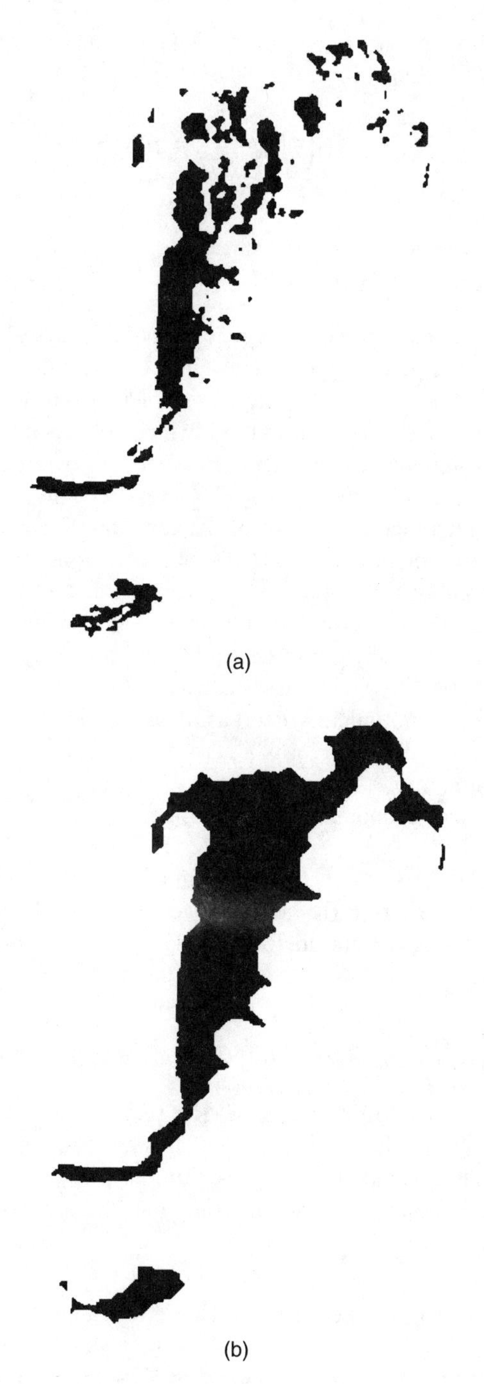

FIGURE 5.42 An example of a closing operation to merge neighboring clusters. (a) Object image; (b) the resulting image after closing.

From hit-or-miss topology point of view, the opening of A by B is the union of the translates B_y included in A, that is,

$$A_B = \bigcup \{B_y : y \in E, B_y \in A\} \tag{5.109}$$

5.6.2 Morphological Transformations of Grayscale Images

Process imaging applications most frequently produce grayscale or color images instead of binary images. Therefore, it is natural to extend the set transformations described in the previous section to the corresponding function transformations. These transformations are restricted to those which correspond to set transformations. Thus, the main issue here is to represent functions by sets. Functions can be represented by an ensemble of sets obtained as cross-sections of the function or by defining a single set called the *umbra* of the function. Figure 5.43 shows a 1D function f, one of its cross-sections, and its umbra. The function f is defined in $\mathbf{R}^2$, and the *cross-section* $X_t(f)$ at level t is the set x such that $X_t(f) = \{x : f(x) \geq t\}$.

Let $\mathbf{D}$ denote the space representing the domain of function f. Then the Euclidean space is denoted by $E = \mathbf{D} \times \mathbf{R}$, or $\mathbf{D} \times \mathbf{Z}$. Obviously, $X_t \subseteq \mathrm{Dom(f)} \subseteq \mathbf{D}$. By thresholding f at $t \in \mathbf{R}$, we can associate a family of sets with the function f. This is a family of monotonically decreasing sets as t monotonically increasing. This leads to that the family f_i is a class of *upper semicontinuous function (u.s.c.)*, on $\mathbf{D}$. Furthermore, if t_i is chosen to be monotonically decreasing then the family f_i becomes monotonically increasing, that is *lower semicontinuous function (l.s.c.)*.

Sternberg [70] suggested the notion of umbra. This is a link between functions $f \in F_u(\mathbf{R}^2)$ and the sets $Y \in F_u(\mathbf{R}^3)$ of the space. Points of $\mathbf{R}^3$ are parameterized by their projection x on $\mathbf{R}^2$ and their altitude t on an axis perpendicular to $\mathbf{R}^2$. The umbra U^y, $y \in F_u(\mathbf{R}^3)$ is the dilate of Y by the positive axis $[0, +\infty)$ of the t's:

$$U^y = Y \oplus [0, -\infty) = \{(x, t') : (x, t) \in Y;\ t' \leq t\} \tag{5.110}$$

The class of U of the umbrae of $F_u(\mathbf{R}^3)$ plays a fundamental role in the morphological study of the functions F_u. It is stable for intersections and for finite unions, and is compact in $F(\mathbf{R}^2)$. Each umbra U induces a unique function $f(U)$, whose value at point x is the sum of the t's such that $(x, t) \in U$. For a given function f, the set

$$U(f) = \{(x, t) : f(x) \geq t\}, \quad x \in \mathrm{R}^2, \quad t \in \mathrm{R} \tag{5.111}$$

is by construction an umbra, and it follows

$$U(f) = \{(x, t) : x \in X_t(f)\}, \quad X_t(f) = \{x : (x, t) \in U(f)\} \tag{5.112}$$

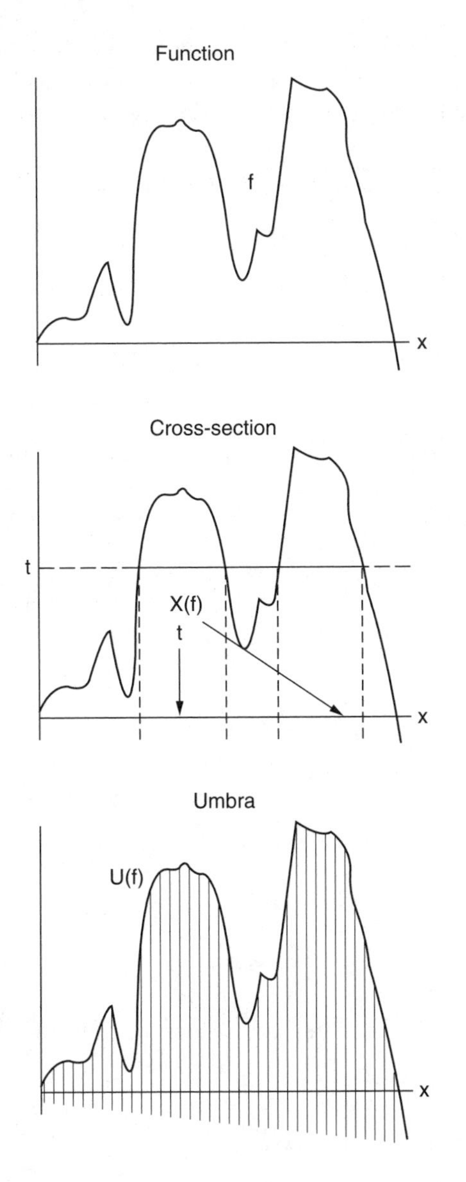

FIGURE 5.43 A function, f, its cross-section $X_t(f)$ at level t, and its umbra $U(f)$ represented as the shaded region.

Hence, there is a unique umbra U and a unique u.s.c. function f associated with each closed set X of $\mathbf{R}^3$; conversely, each u.s.c. function f corresponds to a unique umbra $U(f)$.

A function f can be morphologically transformed by another function g, the *structuring function*. Transforming a function by a set B is a special case of transforming it by a function g, i.e., g is a cylinder.

For this reason, we define the Minkowski sum $f \oplus B$ of the function f by the structuring element B via the umbrae according to $U(f \oplus B) = U(f) \oplus B$.

We extend B to functions by taking for $B = U(g)$ the umbra of function g; then

$$f \oplus g \Rightarrow U(f \oplus g) = U(f) \oplus B = U(f) \oplus U(B) = U(f) \oplus U(g) \quad (5.113)$$

$$f \ominus g \Rightarrow U(f \ominus g) = U(f) \ominus B^r = U(f) \ominus U^r(B) = U(f) \ominus U^r(g) \quad (5.114)$$

with

$$-(f \ominus g) = (-f) \oplus g \quad (5.115)$$

Now the opening of function f by function g is defined from their umbrae as follows:

$$U(f_g) = [U(f) \ominus U^s(g)] \oplus U(g) \quad (5.116)$$

and, noting that $U^s(g) = U^s(g^s)$, the result is

$$f_g = (f \ominus g^s) \oplus g \quad (5.117)$$

Likewise, the closing of function f by function g using their umbrae is expressed as follows:

$$U(f^g) = [U(f) \oplus U(g^s)] \ominus U^s(g^s) \quad (5.118)$$

$$f^g = (f \oplus g^s) \ominus g \quad (5.119)$$

The morphological function transformations by functions are illustrated in Figure 5.44.

5.6.3 Hit-or-Miss Transformation

A practical and pioneer approach to mathematically defined hit-or-miss transformation was devised to use morphological tools to realize automated shape recognition [10]. For 2D images, the problem of shape recognition is equated to that of detecting occurrences of shapes within an image. In the case of binary images, the problem of shape recognition is simpler since the function $f(x, y)$ describing the image on a 2D domain takes on only binary values (0 or 1).

The method by Crimmins and Brown [10] can detect the occurrence of a specified object within a given 2D image. The method employs a hit-or-miss morphological operator and is illustrated by Figure 5.45. Applying an erosion

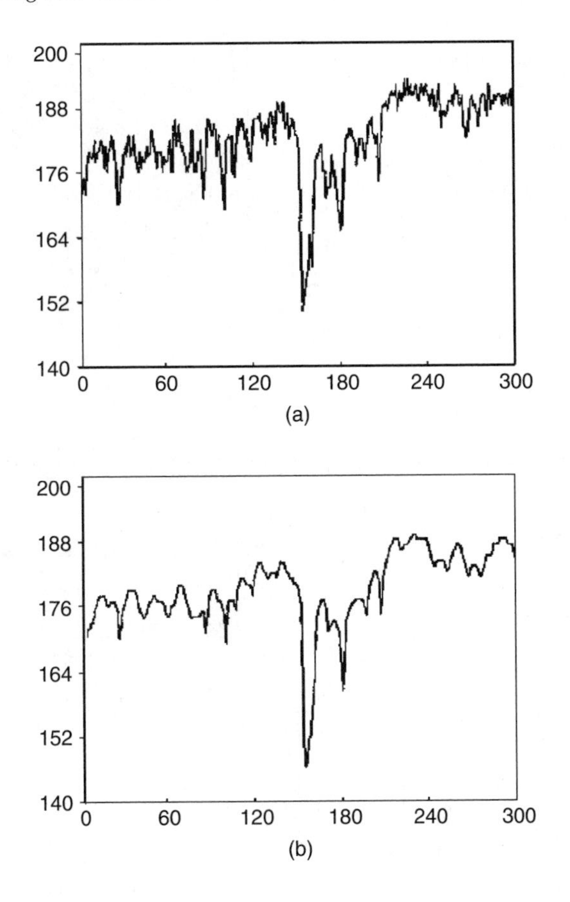

FIGURE 5.44 An example of function operations. (a) A function with detailed variation; (b) the resulting function of f^g; f is the function in (a), and $g = (1, 2, 3, 3, 3, 2, 1)$.

operation to the image using the medium-sized disk as the structuring element will eliminate the small disk, shrink the corresponding medium disk in the image into one and only one point, and preserve that part of the large disk in which the medium disk can fit.

Then a window, W, is introduced which contains the medium-sized disk A and does not contain any parts of other disks. Taking the complement of A inside W yields the set $W \cap A^c$. Applying an erosion operation using a structuring element $W \cap A^c$ to the complement of the original image, I^c, will leave one and only one point located at the same point in image I as would result when eroding I by A. After erosion, large disks are completely eliminated and a set of points is present at the places where smaller disks are located. The entire process is summarized in a theorem according to which the shape recognition problem is essentially that of detecting the occurrences of the shapes to be recognized within an image. The theorem is as follows:

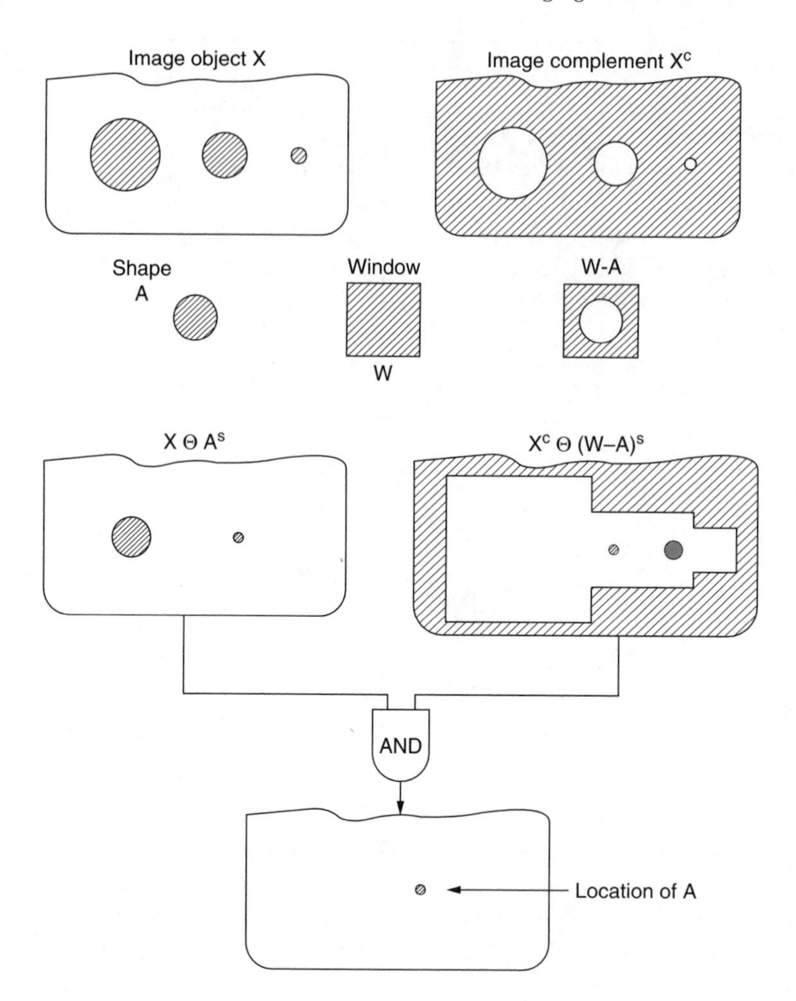

FIGURE 5.45 Hit-or-miss operator for basic shape recognition. (From Crimmins, T., and Brown, W., *IEEE Trans. Aerospace and Electron. Syst. Vol. AES-21*:60–69, Jan. 1985. With permission.)

Theorem 5.1

The shape in window W occurs in the image I at, and only at, the following set of locations:

$$[I \ominus A] \cap [I^c \ominus (W \cap A^c)] \tag{5.120}$$

The erosion $I \ominus A$ produces the set of all points in which the shape A can fit, while I^c $(W \cap A^c)$ presents the set of all locations where the windowed background of A fits into the complement of I. Shapes having points within $I \ominus A$ do not have any points that fall into $I^c \ominus (W \cap A^c)$ and vice versa, except of course for shape

A. Therefore, the intersection of the results of the above two erosions will leave those points corresponding to shape A only.

In many situations the images obtained are not perfect, i.e., the shapes of objects in the acquired field of view may not be represented exactly the same as the ideal shapes used for the structuring elements. Generally, only imperfect images are available for subsequent automated processing due to a variety of realistic noise sources interfering with the actual scenes. In the case of imperfect images, a pattern is taken as a family of sets A, $\gamma \in \Gamma$ along with the convention that the pattern $\{A(\gamma)\}$ occurs at z, if and only if $A_Z(\gamma)$ occurs in the image for some $\gamma \in \Gamma$. This concept is formalized as another theorem:

Theorem 5.2

The pattern $\{A(\gamma)\}$ occurs in the image I at the set of locations

$$\cap_{\gamma \in \Gamma}[I \ominus A(\gamma)] \cap [I^c \ominus (W - A(\gamma))] \tag{5.121}$$

The hit-or-miss operator is exclusively used in the matching processing for automatic recognition of perfect and imperfect shapes. Hierarchical shape generation is used for adaptive thresholding to find locations of local maxima and local minima and provides precise size measurements with a property of rotation-invariance.

5.6.4 Morphological Skeletons

Morphological skeletons are similar to those obtained by MAT; see Figure 5.46 for examples of objects and their skeletons. They are one type of geometric representation. In general, the term *skeleton* has been used to describe a line-thinned caricature of the binary image which summarizes its shape and conveys information about its size, orientation, and connectivity. Let $S_\gamma(X)$, $r > 0$ denote the rth *skeleton subset*, i.e., the set of the centers of the maximal disks

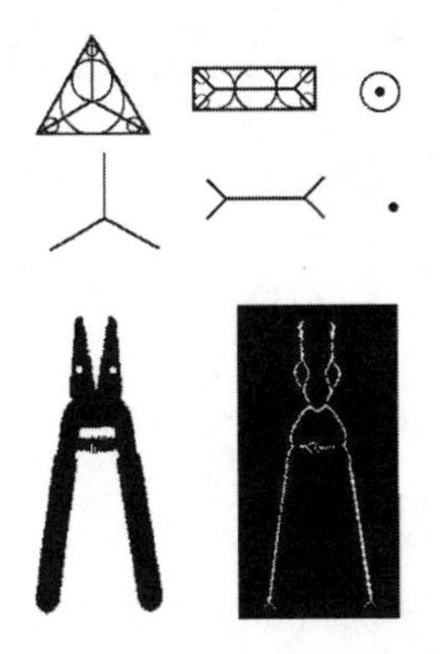

FIGURE 5.46 Examples of objects and their skeletons.

whose radius is equal to r. These skeleton subsets can be obtained using morphological erosions and openings. A morphological skeleton can be represented by the following operations:

$$SK(X) = \bigcup_{r>0} S_r(X) = \bigcup_{r>0} [(X \ominus rB) - (X \ominus rB)_{drB}] \tag{5.122}$$

where rB denotes the open disk of radius r and drB is a closed disk of infinitesimally small radius dr. The boundaries of the eroded sets $(X \ominus rB)$ can be viewed as the propagating wavefronts where the propagation time coincides with the radius r. Subtracting from these eroded versions of X their opening by drB retains only the angular points, points belonging to the skeleton.

Although the analysis of skeletonization is based on Euclidean space, we can still define the morphological skeleton of binary images sampled on a rectangular or hexagonal grid. In addition, digitization gives us a practical advantage because we can program the shape and size of the fundamental structuring element. Let the subset X of Z^2 represent a discrete binary image. It makes no difference whether X is open or closed because all subsets of Z^2 are both open and closed in the context of Euclidean topology. Morphological skeletons $SK(X)$ can be determined according to

$$S_n(X) = (X \ominus nB^s) - (X \ominus nB^s)_B, \quad n = 0, 1, 2, \ldots, N \tag{5.123}$$

and

$$SK(X) = \bigcup_{n=0}^{N} S_n(X) \tag{5.124}$$

where $S_n(X)$ denotes the nth skeleton subset of X. Equation 5.124 implies that the skeleton $SK(X)$ of X is obtained as the finite union of these $(N + 1)$ skeleton subsets [41].

A very important property of the morphological skeleton transformation is that it has an inverse. That is, the discrete binary image X can be exactly reconstructed as the finite union of its $(N + 1)$ skeleton subsets dilated by the hexagonal structuring element of proper size; that is,

$$X = \bigcup_{n=0}^{N} [S_n(X) \oplus nB] \tag{5.125}$$

5.6.5 Watersheds

Watersheds are one of the classics in the field of topography. The Great Divide, which separates the United States into two regions, constitutes a typical example of a watershed line. A drop of water falling on one side of this line flows down

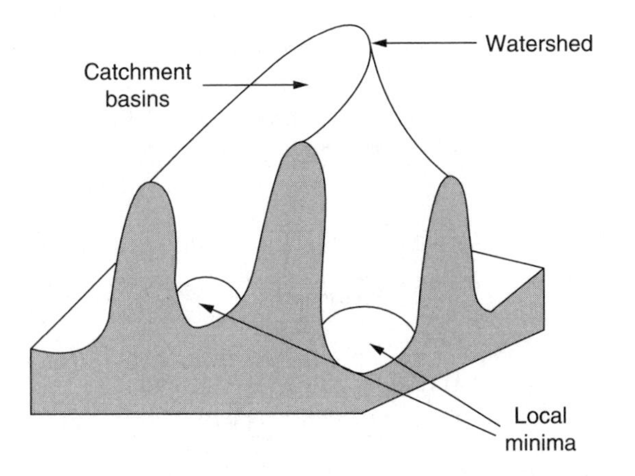

FIGURE 5.47 Illustration of watershed.

until it reaches the Atlantic Ocean, whereas a drop falling on the other side flows down to the Pacific Ocean. The two regions the Great Divide separates are called the catchment basins of the Atlantic and the Pacific Oceans, respectively. The two oceans are the minima associated with these catchment basins [30, 64, 65].

In the field of image processing and more particularly in mathematical morphology, grayscale images are often considered as topographic reliefs. This representation is very useful because notions such as minima, catchment basins and watersheds can be well defined for grayscale images under this representation, as shown in Figure 5.47.

Figure 5.48 shows the watershed algorithm flowchart. The morphological operations are coupled with some algorithmic operations and linear operations such as gradient. In Figure 5.49, the watershed algorithm is applied to segmenting the holes in an engine head [78]. The image of the engine head is a range image, where the image function values represent the relative distance from sensor to object. The fiducial points are selected and used as inner-markers. The inner-markers image is dilated with a 3×3 mask. Outer-markers are obtained by applying a morphological skeletonization algorithm to the background of the inner-markers image. The gradient image is generated without any filtering of the original image. Even if the gradient image from the operator is not as desired, the segmentation result using a watershed algorithm appears to be reasonable. In Figure 5.49, (a) is the range image of an engine head, (b) shows the gradient image, (c) illustrates the markers image consisting of inner-markers and outer-markers, and (d) and (e) are two presentations of watershed, where (d) shows the encircled area and (e) displays the segmented objects.

5.6.6 Applications of Morphological Processing

Image morphology has found many applications in image analysis and processing, machine vision, and pattern recognition. Typically, they are:

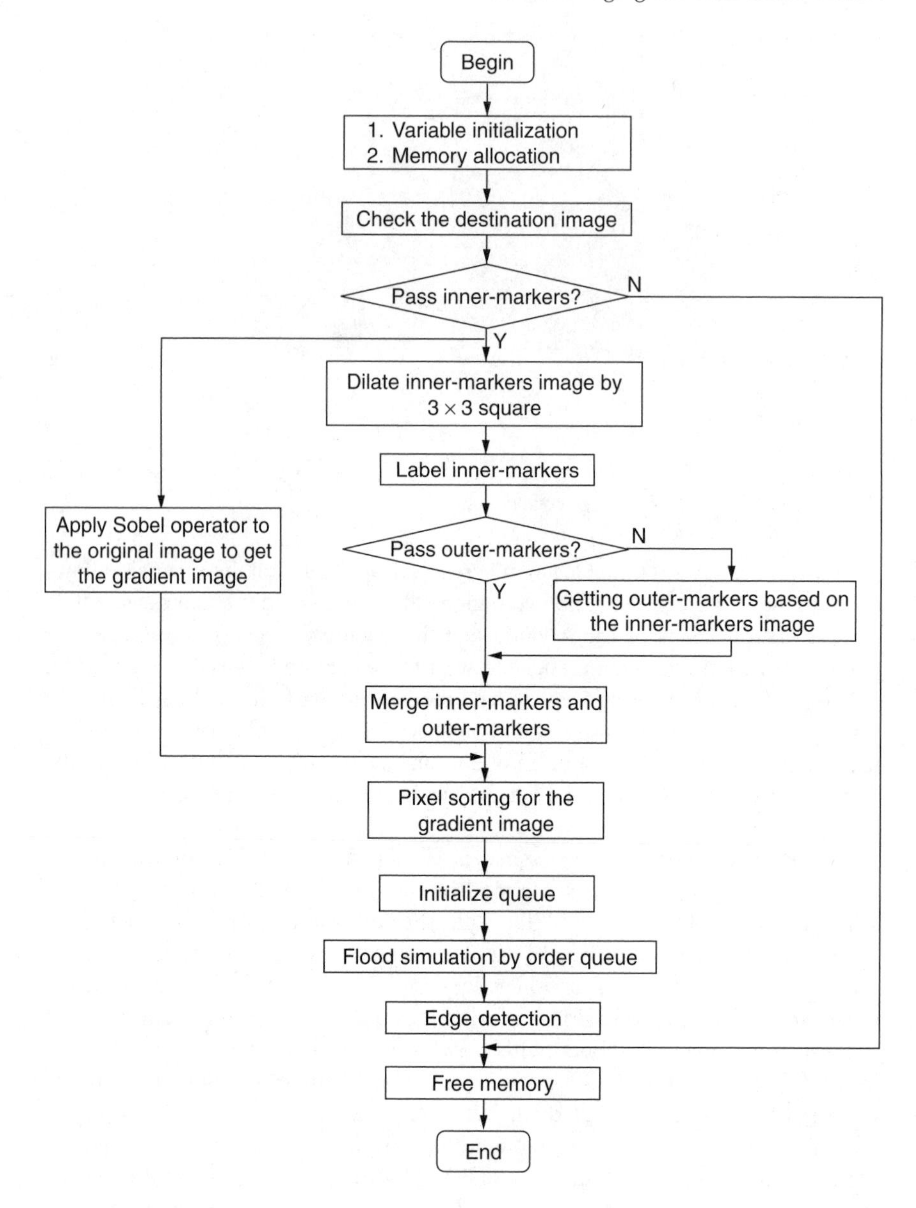

FIGURE 5.48 Flowchart of watershed algorithm.

5.6.6.1 Image Processing and Analysis

Morphological filters are a class of nonlinear image processing filters. These filters have the property of nonlinear translation-invariance and locally modify the geometric features of signals such as their peaks and valleys. They involve the interaction, through set operations, of the graph (and its interior) of signals

(a)

(b)

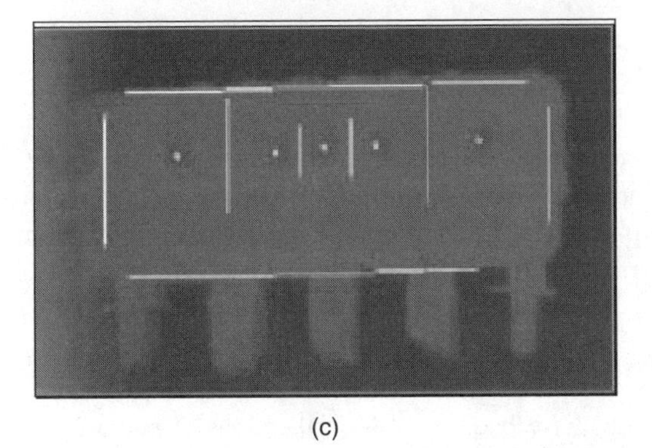

(c)

FIGURE 5.49 An example of applying watershed algorithm for segmentation. (a) Original range image; (b) gradient image by Sobel operator; (c) markers image (inner-markers and outer-markers); (d) resultant watersheds as encircled; and (e) another representation of watersheds for specifying segmented objects.

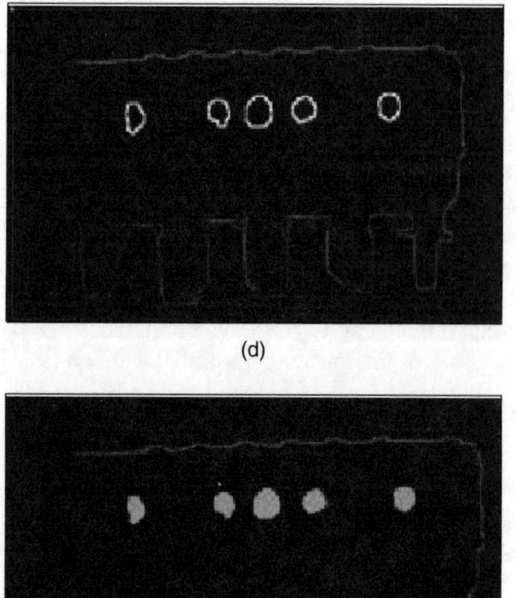

(d)

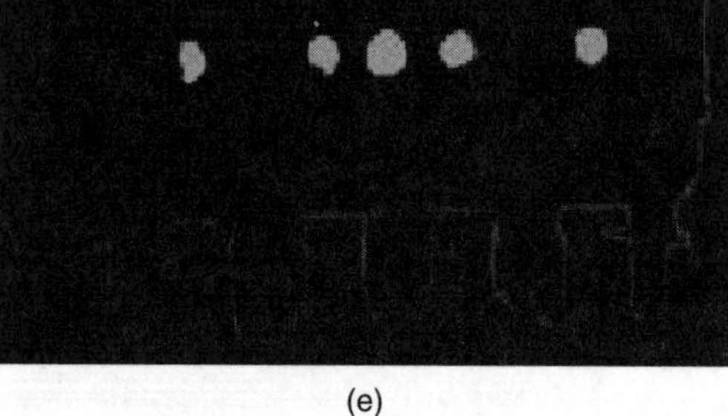

(e)

FIGURE 5.49 (Continued).

and certain compact sets, called the structuring elements. Data processing often employs a combination of a variety of structuring elements and operators. The results of morphological processing usually present more geometrical features than those of traditional global image processing [70, 71, 75].

5.6.6.2 Pattern Analysis

Research activities have been notable in the areas of pattern analysis and recognition. The *pattern spectrum* in particular provides a means of quantitative analysis for shape size distribution and shape representation. The shape representation through skeletonization links the shape size and its geometric property. The optimal geometric representation of shapes by morphologic skeletons reflects a new approach to the problem of pattern recognition. Because morphological processing utilizes the geometric properties of the structuring elements, some pattern recognition problems that used to employ relatively high-level algorithms can be solved by morphological transformation [10, 29, 42].

5.6.6.3 Biomedical and Medical Imaging

The application of morphological filtering in biomedical and medical imaging has been under investigation since the advent of mathematical morphology. Methods that have been developed for cytology in automatic screening of cervical smears include autofocusing, segmentation of nuclei, selection of alarms, elimination of artifacts, and the calibration of cells.

Application to the analysis of 2D electrophoretic gels of biological materials has been reported. The study of the shape and contour of live and moving cells is of fundamental importance to biology. Recent developments also include image analysis for computed tomography (CT), which includes 3D mathematical morphology transforms corresponding to high-pass, bandpass, and lowpass filtering with remarkable properties of sharp cutoffs without phase shifts [45, 69].

5.6.6.4 Texture Analysis

It has been found that morphological granulometry is a proper method to investigate the local texture of an image. Refinement can be accomplished by employing a number of optimally formed structuring elements to generate variant granulometries, each revealing different texture qualities. The application of morphological filtering also produces the measurement of geometrical surface size distributions and evaluates the surface smoothness.

Morphological operations provide tools for structural approaches to texture analysis using grammatical models, such as shape grammars [13, 65].

5.6.6.5 Machine Vision and Inspection

Morphological filtering has been used in machine vision and industrial inspection since the beginning of the development of morphological filtering. A typical application is that of automatic inspection of printed circuits, which uses the operations of dilation and erosion alternatively to find irregularities. Many algorithms have been developed for the specific needs of industrial vision tasks. Automatic shape recognition is another example of applications in machine vision [67, 75].

5.6.7 Conclusion

In general, mathematical morphology is a new tool for image analysis. Its application involves tasks in the machine vision area. The research effort is gradually being placed on more general image processing problems, such as labeling, grouping, extracting, and matching. Thus, from low to intermediate to high-level vision, morphological processing plays an important role.

REFERENCES

1. H. Andrews and B. Hunt. *Digital Image Restoration,* New York: Prentice Hall, 1977.
2. D. Ballard and C. Brown. *Computer Vision,* Englewood Cliffs, NJ: Prentice Hall, 1982.
3. G. Baxes. *Digital Image Processing: Principles and Applications,* New York: John Wiley, 1994.
4. A. Bovik, T. S. Huang, and D. Munson. A generalization of median filtering using linear combinations of order statistics. *IEEE Trans. Acoust., Speech, Signal Process.,* ASSP-21:1342–1349, Dec. 1983.
5. K. Bubna and C. Stewart. Model selection techniques and merging rules for range data segmentation algorithms. *Computer Vision and Image Understanding: CVIU,* 80(2), pp. 215–245, Nov. 2000.
6. R. Campbell and P. Flynn. A survey of free-form object representation and recognition techniques. *Computer Vision and Image Understanding: CVIU,* 81(2), pp. 166–210, Feb. 2001.
7. J. Canny. A computational approach to edge detection. *IEEE Trans. Pattern Anal. Machine Intelligence,* 8(6):679–698, Nov. 1986.
8. K. Castleman. *Digital Image Processing,* New York: Prentice Hall, 1996.
9. R. Chellappa and A. Jain (editors). *Markov Random Fields: Theory and Applications,* New York: Academic Press, 1993.
10. T. Crimmins and W. Brown. Image algebra and automatic shape recognition. *IEEE Trans. Aerospace and Electron. Syst.,* Vol. AES-21:60–69, Jan. 1985.
11. E. Dougherty. *An Introduction to Morphological Image Processing,* Bellingham, WA: SPIE Press, 1992.
12. E. Dougherty. *Digital Image Processing Methods,* New York: Marcel Dekker, 1994.
13. E. Dougherty and J. Astola. *Mathematical Nonlinear Image Processing,* Dordrecht: Kluwer, 1992.
14. R. Duda and P. Hart. Use of the Hough transformation to detect lines and curves in pictures. *Commun. ACM,* 15-1:11–15, Jan. 1972.
15. D. Dudgeon and R. Mersereau. *Multidimensional Digital Signal Processing,* Englewood Cliffs, NJ: Prentice Hall, 1984.
16. M. Duff and T. Fountain. *Cellular Logic Image Processing,* London: Academic Press, 1986.
17. H. Freeman. On encoding arbitrary geometric configurations. *IRE Trans. Electronic Comput.,* 10:260–268, 1961.
18. H. Freeman. *Machine Vision for Inspection and Measurement,* San Diego: Academic Press, 1989.
19. K. S. Fu. *Digital Pattern Recognition,* Heidelberg: Springer-Verlag, 1976.
20. K. S. Fu. *Applications of Pattern Recognition,* Boca Raton, FL: CRC Press, 1982.
21. K. S. Fu. *Syntactic Pattern Recognition and Applications,* Englewood Cliffs, NJ: Prentice Hall, 1982.
22. K. Fukunaga. *Introduction to Statistical Pattern Recognition* (2nd ed.), New York: Academic Press, 1990.
23. L. Galbiati, Jr. *Machine Vision and Digital Image Processing Fundamentals,* Englewood Cliffs, NJ: Prentice Hall, 1990.

24. C. Giardina and E. Dougherty. *Morphological Methods in Image and Signal Processing,* Englewood Cliffs, NJ: Prentice Hall, 1987.
25. A. Glassner. *Principles of Digital Image Synthesis,* San Francisco: Morgan Kaufmann, 1995.
26. R. Gonzalez and R. Woods. *Digital Imaging Processing,* Reading, MA: Addison-Wesley, 1992.
27. W. Green. *Digital Image Processing: A Systems Approach,* New York: Van Nostrand Reinhold, 1989.
28. R. Haralick and L. Shapiro. *Computer and Robot Vision,* Vol. I, Reading, MA: Addison-Wesley, 1992.
29. R. Haralick and L. G. Shapiro. *Computer and Robot Vision,* Vol. II, Reading, MA: Addison-Wesley, 1993.
30. H. Heijmans. *Morphological Image Operators,* Boston: Academic Press, 1994.
31. T. S. Huang, G. Yang, and Q. Tang. A fast two-dimensional median filtering algorithm. *IEEE Trans. Acoust., Speech Signal Proc.,* 27(1):13–18, 1979.
32. B. Horn. *Robot Vision,* Cambridge: MIT Press, 1987.
33. A. Jain. *Fundamentals of Digital Image Processing,* Englewood Cliffs, NJ: Prentice Hall, 1989.
34. R. Jain, R. Kasturi, and B. Schunk. *Machine Vision,* New York: McGraw-Hill, 1995.
35. H. Kabir. *High Performance Computer Imaging,* New York: Prentice Hall, 1996.
36. S. Levialdi. On shrinking binary picture patterns. *Comm. ACM* 15: 7–10, Jan. 1972.
37. H. Lim. *Two-Dimensional Signal and Image Processing,* New York: Prentice Hall, 1990.
38. T. Lindeberg. *Scale-Space Theory in Computer Vision,* Dordrecht: Kluwer Academic Publishers, 1994.
39. P. Maragos and R. Schafer. Morphological filters: Part I: Their set-theoretic analysis and relations to linear shift invariant filters. *IEEE Trans. Acoust., Speech, Signal Processing,* ASSP-35:1153–1169, Aug. 1987.
40. P. Maragos and R. Schafer. Morphological filters: Part II: Their relations to median, order-statistic, and stack filters. *IEEE Trans. Acoust., Speech, Signal Processing,* ASSP-35:1170–1184, Aug. 1987.
41. P. Maragos and R. Schafer. Morphological skeleton representative and coding of binary images. *IEEE Trans. Acoust., Speech, Signal Processing,* ASSP-34: 1228–1244, Oct. 1986.
42. D. Marr. *Vision,* San Francisco: W.H. Freeman and Company, 1982.
43. D. Marr and E. Hildreth. Theory of edge detection. *Proc. Royal Society of London B,* 207:187–217, 1980.
44. G. J. McLachlan. *Discriminant Analysis and Statistical Pattern Recognition,* New York: John Wiley, 1992.
45. F. Meyer. Automatic screening of cytological specimens. *Comput. Vision Graph. Image Process.* 35:356–369, 1986.
46. A. Oppenheim and R. Schafer. *Digital Signal Processing,* New York: Prentice Hall, 1975.
47. A. Oppenheim and R. Schafer. *Discrete-Time Signal Processing,* New York: Prentice Hall, 1989.
48. A. Oppenheim, A. Willsky, and S. Navab. *Discrete Time Signal Processing,* Englewood Cliffs, NJ: Prentice Hall, 1989.
49. D. Paulus and J. Hornegger. *Pattern Recognition and Image Processing in C++,* Braunschweig: Vieweg, 1995.

50. T. Pavlidis. *Algorithms for Graphics and Image Processing,* Rockville, MD: Computer Science Press, 1982.
51. S. Peleg and A. Rosenfeld. A min-max medial axis transformation. *IEEE Trans. Pattern Anal. Machine Intell.,* PAMI-3:208–210, March 1981.
52. I. Pitas. *Digital Image Processing Algorithms,* Englewood Cliffs, NJ: Prentice Hall, 1993.
53. I. Pitas. *Digital Image Processing Algorithms and Applications.* New York: John Wiley & Sons, 2000.
54. J. Prewitt. Object enhancement and extraction. In: *Picture Processing and Pschopictorics*, B. Lipkin and A. Rosenfeld, New York: Academic Press, 1970, pp. 75–150.
55. W. Pratt. *Digital Image Processing* (2nd ed.), New York: John Wiley & Sons, 1991.
56. H. Preston, Jr., M. Duff, S. Levialdi, P. Norgren, and J. Toriwaki. Basics of cellular logic with some applications in medical image processing. *Proc. IEEE* 67: 826–856, 1979.
57. A. Rosenfeld. *Picture Processing by Computer,* New York: Academic Press, 1969.
58. A. Rosenfeld and A. Kak. *Digital Picture Processing,* Vol. 1, Orlando: Academic Press, 1982.
59. A. Rosenfeld and A. Kak. *Digital Picture Processing,* Vol. 2, Orlando: Academic Press, 1982.
60. L. Roberts. Machine perception of three-dimensional solids. In: *Optical and Electro-optical Information Processing,* J. Tippet (ed.), Cambridge, MA: MIT Press, 1965.
61. G. Robinson. Detection and coding of edges using directional masks. *Proc. SPIE Conf. on Advances in Image Transmission Techniques,* San Diego, CA, Aug. 1976.
62. J. Russ. *The Image Processing Handbook,* Boca Raton, FL: CRC Press, 1992.
63. J. Sanz. *Image Technology: Advances in Image Processing, Multimedia, and Machine Vision.* Heidelberg: Springer-Verlag, 1995.
64. J. Serra. *Image Analysis and Mathematical Morphology,* London: Academic Press, 1982.
65. J. Serra, ed. *Image Analysis and Mathematical Morphology,* Vol. 2, New York: Academic Press, 1988.
66. R. Schalkoff. *Digital Image Processing and Computer Vision,* New York: John Wiley & Sons, 1989.
67. L. Schmitt and S. Wilson. The AIS-5000 parallel processor. *IEEE Trans. Pattern Anal. Machine Intell.*, PAMI-10: 320–330, May 1988.
68. J. Shen and S. Castan. An optimal linear operator for step edge detection. *Computer Vision, Graphics and Image Processing: Graphical Models and Image Processing,* 54(2):112–133, 1992.
69. S. Sternberg. Biomedical image processing. *IEEE Computer.,* Jan. 1983:22–24.
70. S. Sternberg. Grayscale morphology. *Comput. Vision Graph. Image Processing,* 35:333–355, 1986.
71. R. Stevenson and G. Arce. Morphological filters: Statistics and further syntactic properties. *IEEE Trans. Circuits and Systems,* CAS-34:1292–1305, Nov. 1987.
72. S. Umbaugh. *Computer Vision and Image Processing,* New York: Prentice Hall, 1998.
73. D. Vernon. *Machine Vision: Automated Visual Inspection and Robot Vision,* New York: Prentice Hall, 1991.
74. J. Wilder. Finding and evaluating defects in glass. In: *Machine Vision for Inspection and Measurement,* H. Freeman, Ed. San Diego: Academic Press, 1989.

75. S. Wilson. Morphological networks. *Visual Communications and Image Processing IV (1989),* Nov. 1989:483–493.
76. G. Winkler. *Image Analysis: Markov Fields and Dynamic Monte Carlo Methods,* New York: Springer-Verlag, 1995.
77. T. Young and K. S. Fu. *Handbook of Pattern Recognition and Image Processing,* Orlando: Academic Press, 1986.
78. D. Zhao and X. Zhang. Range-data-based object surface segmentation via edges and critical points. *IEEE Trans. Image Processing,* 6:826–830, 1997.
79. X. Zhuang and R. Haralick. Morphological structuring element decomposition. *Comput. Vision Graph. Image Process.,* 35:370–382, 1986.

6 State Estimation

Jari P. Kaipio, Stephen Duncan, Aku Seppänen, Erkki Somersalo, and Arto Voutilainen

CONTENTS

6.1 INTRODUCTION

Modern optimal control systems are based on state space representations of the processes that are to be controlled. State estimation is a topic in probability and statistics. More precisely, it is a subfield of estimation theory, in which questions about unknown quantities are posed based on observations and other information. The observations are always treated as random variables, whereas statistics is divided into two paradigms depending on whether the unknowns are interpreted as "unknown but deterministic" or as "random variables with some probability distribution." These paradigms are usually referred to as the frequentist and Bayesian paradigms, respectively. We will not delve into the difference between the two interpretations other than giving a general comment on literature in which the treatment may be confusing due to the differences in paradigm. In what follows, we take the Bayesian point of view.

We wish to obtain estimates that are optimal in some sense. Denote the available observations with y. If we take the unknown vector x to be a random variable, we would like to compute an estimate $\hat{x}$ for x based on the observations y so that some criterion $\chi(x, \hat{x})$ is minimized. The most obvious criterion is the two-norm criterion $\chi(x, \hat{x}) = \|x - \hat{x}\|_2$. The estimator can be thought as a mapping $\hat{x} : y \mapsto \hat{x}$, so that the optimization is in fact over mappings of y. If the idea is to obtain an optimal formula with best (with respect to the two-norm criterion) mean performance over the distribution of x, we would arrive at the mean square estimate problem $\min_{\hat{x}} E\|x - \hat{x}\|_2$, whose solution is known to be the conditional expectation of x given y, denoted by $E(x\,|\,y)$. The distribution of x is called the prior distribution, that is, the distribution of the unknown prior to obtaining observations. The prior distribution can usually be considered as a model for the unknown that can be based on rigorous physical modeling, more or less ad hoc assumptions, or something in between.

The state estimation problem is a special version of the more general mean square estimation problem in which the prior distribution for the unknown state variable x_t at time t is given by a temporal evolution model. In this chapter we discuss the state space representation in its most typical form as well as the most common statements of the problem.

A particular subfield of state estimation problems that is relevant to process imaging is the estimation and control of distributed parameters. This means that the variables of interest are inherently infinite dimensional, such as concentration of a chemical substance, the void fraction, or the velocity field in a three-dimensional (3D) vessel. We will not delve into infinite dimensional estimation problems but rather consider the approximation of the variables in finite dimensional subspaces.

There are numerous excellent books on state space models and state estimation theory. Almost all the material in Section 6.1 through Section 6.5 can be found for example in Anderson and Moore [1], Chui and Chen [2], and Durbin and Koopman [3], which we recommend as general references. Good texts on more general estimation theory in finite dimensional problems (including state estimation theory) are Melsa and Cohn [4] and Sorenson [5].

6.1.1 State Space Representation of Dynamical Systems

The state space representation of a dynamical system consists of the specification of the evolution model for the state variables and the observation model that links the observations/measurements to the state variables. In state space theory, the state variables and observations are taken to be stochastic, or random, processes. For most of this chapter we consider linear evolution and observation models. Furthermore, the standard state space theory relies on the evolution model exhibiting first order Markov property. This and the additional assumptions are considered in more detail in Section 6.2. For the moment, we will simply state the rudiments of the most conventional state estimation problem.

Let the state variable $x_t \in \Re^N$ satisfy the following linear first order (stochastic) difference equation:

$$x_{t+1} = F_t x_t + s_t + B_t u_t + w_t \tag{6.1}$$

where $F_t \in \Re^{N \times N}$ is the state (evolution) matrix, s_t is the deterministic (known but uncontrolled) input of the system, $u_t \in \Re^p$ is the (possible) control input, and $B_t \in \Re^{N \times p}$ determines how the control affects the state. Finally, $w_t \in \Re^N$ is the state noise process, which is usually assumed to obey Gaussian statistics with known mean and covariance. Also, the matrices F_t and B_t are assumed to be known for all t.

The observations $y_t \in \Re^M$ are linked to the state via the model

$$y_t = G_t x_t + v_t \tag{6.2}$$

where $G_t \in \Re^{M \times N}$ and $v_t \in \Re^M$ is the (Gaussian) measurement noise process with known mean and covariance.

Equation 6.1 and Equation 6.2, together with the specification of the associated matrices, specify the state space representation of the dynamical system. The actual state estimation problem can be posed in a variety of ways of which the most common is the following: Given a distribution for the initial state x_0, compute the best estimate $\hat{x}_t$ for the current state x_t based on the complete measurement history $\{y_1, \ldots, y_t\}$ each time a new measurement y_t is obtained. There are different possibilities for the optimality criteria, but the almost exclusively chosen one is the mean square criterion, which amounts to the minimization of the expectation of the squared estimate error norm, that is, $\min E\|x_t - \hat{x}_t\|_2^2$, where E denotes expectation.

Because the estimates are optimal with respect to the whole history of observations, it is clear that the dimension of the estimation problem increases linearly with time. The solution of the problem would thus become infeasible without a recursive estimation scheme. It turns out that such a scheme exists. A recursive solution to the estimation problem was solved in the early 1960s independently by Rudolf Kalman [6] and Peter Swerling [7] in somewhat different forms. The solution, however, is known by the name Kalman filter only.

The algorithm was a remarkable feat and has been termed as one of the mathematical algorithms that have had most impact in technology and science—another one is the fast Fourier transform (according to the *SIAM News* series "Mathematics

That Counts" in 1993). The applications are almost countless, ranging from spacecraft guidance to the analysis of brain-related electrical signals. A quick search on World Wide Web with the terms "Kalman filter" gives about 250,000 results. The reason is obvious: most physical variables of interest depend on time and cannot be measured directly. For such problems, the state space representation has turned out to be the most feasible.

There are two main categories of state space models: the discrete-time and the continuous-time models. The latter is relevant if the observations are continuous functions of time. In most engineering systems, the state variables represent some physical quantity that is inherently continuous in time. However, the observations are nowadays usually obtained by sampling at discrete times. Thus the most usual scheme is to consider discrete-time models in which the continuous-time state is (formally) integrated over the sampling intervals. In the sequel, we concentrate on discrete-time state space models and discuss briefly how to treat continuous-time state evolution models. For more information on continuous-time models and respective filtering, see the original paper by Kalman and Bucy [8] and two by Balakrishnan [9, 10].

An important concept of observability is related to whether the state can be estimated from a set of observations. This notion is considered in Chapter 7 of this volume.

6.1.2 Applications to Imaging and Control

The exact specification of the state estimation problem depends on the end-use of the state estimates. Let us fix the optimality criterion as the mean square criterion. In terms of statistics, the mean square estimate based on a set observations $\mathcal{P}$ is the conditional expectation of x_t given $\mathcal{P}$ and is denoted $E(x_t|\mathcal{P})$. Let, however, the available set of observations be variable, so that we denote the optimal estimate for x_t as

$$x_{t|k} = E(x_t \mid y_1, \ldots, y_k) \tag{6.3}$$

Now, depending on whether $t < k$, $t = k$, or $t > k$, we have different relevant applications. It is customary to refer to an estimated variable with a "hat," such as $\hat{x}_t$. This notation is not needed in this chapter since the estimates can always be recognized by the conditioned subscript, as in $x_{t|k}$.

Assume that we are currently at time t and wish to estimate the state at time $t + q$, $q > 0$. This would be a relevant estimation problem if we wish to obtain a warning of an unwanted event. This is called *state prediction*: compute the estimate $x_{t+q|t}$ at each time t when a new observation is obtained. At the other end of the scale, assume that we are interested in a transient effect that occurs in the time interval $t = 1, \ldots, T$. It is then reasonable to use all available data to compute the state estimates at each time, that is, to compute $x_{t|T}$, $t = 1, \ldots, T$ once all observations are obtained. This is called the fixed-interval smoothing problem and obviously cannot be implemented with a single recursion. Luckily, the fixed-interval smoother can be implemented as a succession of a forward recursion and a backward (with respect to time) recursion, thus saving us from the dimensionality problem.

It is intuitively obvious and mathematically provable that estimates based on a set of measurements cannot be worse than those based on a subset of those measurements. Especially, the estimates $x_{t|t+q}$, $q > 0$, must be better than or equal to $x_{t|t}$. In contrast to the prediction problem, however, in the fixed-lag smoothing problem, we compute the estimates $x_{t-q|t}$, $q > 0$, at time t. In other words, at each time, we obtain estimates for the state that prevailed q instants earlier. This is usually feasible in imaging applications since the estimation error for $x_{t-5|t}$ might be significantly smaller than what it is for $x_{t-5|t-5}$ but might correspond to a lag of only a few milliseconds. Thus, for quality control type problems, the fixed-lag smoothers might be the best bet. However, for automatic control problems, only the estimates $x_{t|t}$ and $x_{t|t-1}$, that is, the Kalman filter estimate and the one-step prediction estimate, play a relevant role. These provide the best state estimates when the control u_t is adjusted after obtaining the measurement y_t.

Although Kalman filtering is extremely useful in its own right as a method for estimating the underlying states of a system from a series of measurements, the technique is particularly powerful when combined with state feedback control, where the estimates of the states are used to generate the inputs to actuators in order to achieve a desired response. The aim of the feedback control law is to determine the inputs that minimize (in some sense) the difference between the actual states of the system and some desired, reference states. Initially, the control strategy was based on linear quadratic Gaussian controllers [11], but in the 1980s the limitations of this approach were identified, particularly when the controller was derived from an uncertain model of the process. To address these shortcomings, the $\mathcal{H}_\infty$ methods were developed [12, 13], where the controller was designed to remain stable and to perform well, even when there was a known degree of uncertainty in the underlying process model. Together with model predictive control methods [14], $\mathcal{H}_\infty$ methods have become the most commonly used approaches to designing controllers for systems with multiple inputs and multiple outputs and with many states. The systems that are typically monitored by process imaging fall into this category, and the control design procedures for these systems will be described in Chapter 7.

6.2 REAL-TIME RECURSIVE ESTIMATION: KALMAN PREDICTORS AND FILTERS

This section deals with the classical recursive discrete time estimation problem. We restate the state space model

$$x_{t+1} = F_t x_t + s_t + B_t u_t + w_t \tag{6.4}$$

$$y_t = G_t x_t + v_t \tag{6.5}$$

We make the following assumptions:

The initial state obeys the Gaussian distribution $x_0 \sim \mathcal{N}(\bar{x}_0, \Gamma_0)$.
The matrices F_t, G_t, and B_t, the vector s_t, and the control u_t are known at respective times.
The state and observation noise processes are Gaussian with zero mean and (known) covariances Γ_{w_t} and Γ_{v_t}, respectively. Furthermore, both are vector valued white noise processes so that $\mathrm{cov}(w_t, w_k) = \delta_{tk}\Gamma_{w_t}$ and $\mathrm{cov}(v_t, v_k) = \delta_{tk}\Gamma_{v_t}$, where δ_{tk} is the Kronecker delta.
We make the additional assumption of $\mathrm{cov}(w_t, v_k) = 0$ for all t, k. This assumption could be avoided but is usually valid.

Since x_0, w_t, and v_t are Gaussian and the mappings are linear, y_t and x_t are also Gaussian for all t. Thus it follows that the conditional probability distributions $\pi(x_t \mid y_1, \ldots, y_k)$ are also Gaussian, and it is sufficient to compute only the means and covariances for these distributions. The estimate covariances are denoted by $\mathrm{cov}(x_{t|k}) = \Gamma_{t|k}$. The standard innovation form Kalman filter and (one-step) predictor recursions take the form

$$x_{t|t-1} = F_{t-1}x_{t-1|t-1} + s_{t-1} + B_{t-1}u_{t-1} \tag{6.6}$$

$$\Gamma_{t|t-1} = F_{t-1}\Gamma_{t-1|t-1}F_{t-1}^T + \Gamma_{w_{t-1}} \tag{6.7}$$

$$K_t = \Gamma_{t|t-1}G_t^T\left(G_t\Gamma_{t|t-1}G_t^T + \Gamma_{v_t}\right)^{-1} \tag{6.8}$$

$$\Gamma_{t|t} = (I - K_tG_t)\Gamma_{t|t-1} \tag{6.9}$$

$$x_{t|t} = x_{t|t-1} + K_t(y_t - G_tx_{t|t-1}) \tag{6.10}$$

where $x_{t|t}$ and $x_{t|t-1}$ are the conditional means of the filtering and prediction densities, respectively, and K_t is the so-called Kalman gain.

It must be noted that the notations and the form of the recursions are highly variable, and each source should be carefully investigated to find the associated assumptions. Indeed, there are multiple forms for identical problems. The differences lie in analytical tractability and computational stability and efficiency. However, Equation 6.6 and Equation 6.7, and Equation 6.8 through Equation 6.10 are called the time-update and measurement-update equations, respectively. We shall return to some of these issues in Section 6.6.

The recursions (Equation 6.6 through Equation 6.10) allow for an important interpretation. Once an estimate $x_{t|t}$ is available, the evolution equation is propagated to time $t+1$ to obtain the predictor estimate $x_{t+1|t}$. When the observation at time $t+1$ is obtained, the predicted state is corrected based on the predicted measurement $y_{t+1|t} = G_{t+1}x_{t+1|t}$ and the resulting prediction error $y_{t+1} - y_{t+1|t}$.

The practical success of the Kalman filter is naturally related to the accuracy—or at least feasibility—of the state evolution and observation models. Apart from this, some problems are inherently easier than others. If either of the models is known to be inadequate, this should be taken into account. For example, if the observation model is not accurate, the observation error covariance should be made larger than the measurement system analysis suggests in order to accommodate the errors due to the computational model approximation. Correspondingly, if the evolution model is poor, the state noise covariance has to be set large, which in turn can result in noisy state estimates. However, once the state space representation is derived, the estimate error analysis can in principle be performed before any measurements are conducted. This is due to the fact that (with linear state space models) the covariances do not depend on the data y_t. However, the covariances are the actual covariances only in the case in which the actual system is governed by the same equations as the model. Thus, the theoretical error covariances should be considered only if the state space model is known to be relatively accurate.

The version of Kalman filter that does not assume independence of the state and observation noise is somewhat more complex and can be found for example in Anderson and Moore [1].

In cases where the system matrices are independent of time, the recursions for the state covariance assume the form of a Riccati equation, which can be solved off-line. If the system is stable, the Riccati equation has an asymptotic solution, which means that the Kalman gain has an asymptotic solution. Note that the time-independence of system matrices means that the observation's model is time-independent. Thus Equation 6.7 through Equation 6.9 need not be recomputed during the iteration. This is likely to reduce the computational complexity of the problem considerably. In some cases the dimension of the observation vector is larger than the dimension of the state vector. In such cases another form of the Kalman filter, called the information filter, will be more stable and efficient.

Regardless whether the assumptions on Gaussianity hold or not, the Kalman filter is the optimal linear (affine) filter with respect to the mean square criterion. For a treatment that is based on the Hilbert space formalism of random variables, see Brockwell and Davis [15]. However, to obtain this particular mode of optimality, one should compute the exact covariances and modify the recursions to accommodate nonvanishing means. This is a straightforward task but can be computationally tedious.

6.3 ON-LINE AND TRANSIENT ESTIMATION: SMOOTHERS

As noted above, on-line imaging for example for quality assurance purposes but without automatic control is feasible even when the estimates $\hat{x}_t$ are not obtained immediately after the observation y_t. Furthermore, in transient type situations the estimates can possibly be computed completely off-line. The two relevant schemes for these two cases are the fixed-lag smoother and the fixed-interval smoother. We treat the fixed-lag smoother first.

It is most instructive to begin with the fixed-lag smoother since it gives an excellent insight into the versatility of the state space representation. The problem is to obtain an estimate for $x_{t-q|t}$, $q > 0$. The Kalman filter provides the optimal estimate for the state X_t at time t after observing y_t. We may write in principle whatever we want as the state variable X_t as long as we can write the corresponding state and observation models. Now, assume that Equation 6.1 and Equation 6.2 form a valid model, and define an augmented state vector $X_t^T = (x_t^T, x_{t-1}^T, \ldots, x_{t-q}^T)$. By adding the identities $x_{t-k} = x_{t-k}$, $k = 1, \ldots, q$, to the state evolution model, we arrive at an augmented state space model:

$$X_{t+1} = \begin{pmatrix} F_t & & & \\ I & & & \\ & \ddots & & \\ & & I & 0 \end{pmatrix} X_t + \begin{pmatrix} s_t \\ 0 \\ \vdots \\ 0 \end{pmatrix} + \begin{pmatrix} B_t u_t \\ 0 \\ \vdots \\ 0 \end{pmatrix} + \begin{pmatrix} w_t \\ 0 \\ \vdots \\ 0 \end{pmatrix} \tag{6.11}$$

$$y_t = (G_t, 0, \ldots, 0) X_t + v_t \tag{6.12}$$

It is clear that this representation is exactly equal to Equation 6.1 and Equation 6.2, in the sense that the dependence between x_t (the first block of X_t) and y_t is the same as before, with straightforward associations and modifications to the variables of the state space representation, including the initial state mean and covariance and state noise covariance. Let us apply the Kalman filter to the system in Equation 6.11 and Equation 6.12. Then, the last block of the Kalman filtered estimate $X_{t|t}$ is actually $x_{t-q|t}$, as desired.

While the above form could be used to obtain the fixed-lag estimates, we can exploit the sparse structure of the matrices. There are again a number of variations that differ in computational cost and stability. We use the following one, which is actually an addition to the standard Kalman filtering recursion. At each time t compute [1]:

$$x_{t-i|t} = x_{t-i|t-1} + K_{t-i}(y_t - G_t x_{t|t-1}), \quad i = 0, \ldots, q \tag{6.13}$$

$$K_{t-i} = \Gamma_{t|t-1}^{(i,0)} G_t^T \left(G_t \Gamma_{t|t-1} G_t^T + \Gamma_{v_t} \right)^{-1}, \quad i = 0, \ldots, q \tag{6.14}$$

$$\Gamma_{t+1|t}^{(i+1,0)} = \Gamma_{t|t-1}^{(i,0)} (I - K_t G_t)^T F_t^T, \quad i = 0, \ldots, q \tag{6.15}$$

where the covariance matrices $\Gamma_{t|t-1}^{(i,0)}$ are defined as $\Gamma_{t|t-1}^{(i,0)} = E\{(x_{t-i} - x_{t-i|t-1})(x_t - x_{t|t-1})^T\}$, and thus, $\Gamma_{t|t-1}^{(0,0)} = \Gamma_{t|t-1}$ is obtained from the filter in Equation 6.7. Note that, depending on q, we may have to store a very large number of matrices and vectors.

The fixed-interval smoother consists of the conventional computation of the Kalman filter and predictor estimates and storing these estimates as well as the backward gain matrices:

$$A_t = \Gamma_{t|t} F_t \Gamma_{t+1|t}^{-1}, \quad t = 1, \ldots, T-1 \tag{6.16}$$

Then the following backward iteration is carried out:

$$x_{t-1|T} = x_{t-1|t-1} + A_{t-1}(x_{t|T} - x_{t|t-1}), \quad t = T, T-1, \ldots, 2 \tag{6.17}$$

$$\Gamma_{t-1|T} = \Gamma_{t-1|t-1} + A_{t-1}(\Gamma_{t|T} - \Gamma_{t|t-1})A_{t-1}^T \tag{6.18}$$

Note again that the storage requirement can be significant. However, the backward gain matrices are not needed simultaneously, and so the requirement is not on the core memory.

How much better the smoothed estimates are when compared to the real-time Kalman filtered estimates depends on the overall state space model in a nontrivial way. In some cases, the filtered estimate errors possess a delay type structure that is largely absent in the smoothed estimates. This is typical behavior, especially for cases in which the observation model is not exceptionally informative (the maximum likelihood problem max $\pi(y_t|x_t)$ is not stable) and the state evolution model is not very accurate. As an example, if we use the random walk model as the state evolution model and the actual evolution exhibits convection, we would obtain estimates that would be delayed and would show "tails." In some other cases, the decrease of estimation error can be practically negligible, especially when compared to the increased computational complexity.

6.4 NONLINEAR AND NON-GAUSSIAN STATE ESTIMATION

6.4.1 Extended Kalman Filters

We have previously assumed that the evolution and observation models are linear. In many interesting applications, this is not the case. Especially, the observation models for capacitance and impedance tomography are nonlinear. Furthermore, the feasible state evolution models in complex flows are easily nonlinear. In addition, the noise term in the observation model may not be additive. An example is when the observations are counts due to radioactive processes such as in emission tomography. In this case, the observations would be Poisson distributed, and the additive noise model is not appropriate. Finally, several induction type observation errors are multiplicative rather than additive.

In this section we treat only the additive noise case. The nonlinear state space model that we discuss is

$$x_{t+1} = F_t(x_t) + B_t(u_t) + w_t \tag{6.19}$$

$$y_t = G_t(x_t) + v_t \tag{6.20}$$

where $F_t : \Re^N \to \Re^N$ is the state (evolution) operator, $u_t \in \Re^p$ is the (possible) control input, and $B_t : \Re^p \to \Re^N$ and $G_t : \Re^N \to \Re^M$. More complex models exist.

The idea in extended Kalman filters is straightforward: the nonlinear mappings are approximated with the affine mappings given by the first two terms of the Taylor expansion. We note that the notion of extended Kalman filter is not completely fixed, and this term could refer to several levels of refinement. The state in which the linearization is computed will be denoted with x_t^*.

In the global linearization approach, the mappings F_t and G_t are linearized in some time invariant point $x_t^* \equiv x^*$ so that we have $F_t(x_t) \approx F_t(x^*) + J_{F_t}|_{x^*} (x_t - x*) = b_t + J_{F_t}|_{x^*} x_t$ and similarly for G_t. Here J_{F_t} is the Jacobian mapping of F_t. The rationale behind global linearization is that the Jacobian does not have to be recomputed during the iteration. It is clear that a good guess of the "mean state" x^* is a prerequisite for this approximation to be successful.

The version of extended Kalman filter that is most commonly used is the local linearization version in which the mappings are linearized in the (at the time) best available state estimates, either the predicted or the filtered state. This necessitates the recomputation of the Jacobians at each time instant. Note that the linearization is not necessarily needed in the time update when the predictor $x_{t|t-1}$ is computed. This applies also for the measurement update. Furthermore, one does not have to approximate the control term when the state estimates are computed. The recursions take the form

$$x_{t|t-1} = F_{t-1}(x_{t-1|t-1}) + B_{t-1}(u_{t-1}) \tag{6.21}$$

$$\Gamma_{t|t-1} = J_{F_{t-1}} \Gamma_{t-1|t-1} J_{F_{t-1}}^T + \Gamma_{w_{t-1}} \tag{6.22}$$

$$K_t = \Gamma_{t|t-1} J_{G_t}^T \left(J_{G_t} \Gamma_{t|t-1} J_{G_t}^T + \Gamma_{v_t} \right)^{-1} \tag{6.23}$$

$$\Gamma_{t|t} = \left(I - K_t J_{G_t} \right) \Gamma_{t|t-1} \tag{6.24}$$

$$x_{t|t} = x_{t|t-1} + K_t (y_t - G_t(x_{t|t-1})) \tag{6.25}$$

The linearizations are needed only in the computation of the covariances and the Kalman gain. However, if the computation of the Jacobians is faster than for example $G_t(x_{t|t-1})$, then the affine approximations could be used in Equation 6.21 and Equation 6.25.

We shall discuss the final version, the iterated extended Kalman filter for the case in which the state evolution equation is linear and the observation model is nonlinear. The treatment of the nonlinear state is equivalent, but the restricted problem is the one studied later in this chapter and occurs more frequently. The time-varying parameter estimation problem is an exception and is discussed in Section 6.5.2.

The idea is most easily explained based on the Bayesian interpretation of the Kalman filter. To keep the treatment short, we note that the measurement update is equivalent to the computation of the conditional mean estimate for the state x_t given the measurement history $(y_1, \ldots, y_t)$, which is $E(x_t | y_1, \ldots, y_t)$. In other words, the computation is of the mean of the posterior density $\pi(x_t \mid y_1, \ldots, y_t)$, which in the Gaussian case coincides with the maximum of the density. It can be shown that we have

$$\pi(x_t \mid y_1, \ldots, y_t) \propto \pi(y_t \mid x_t)\pi(x_t \mid y_1, \ldots, y_{t-1}) \tag{6.26}$$

where the first density on the right hand side is called the likelihood of the observation and the latter is called the prediction density. Assume that $\Gamma_{t|t-1}$ and Γ_{v_t} are positive definite for all t so that the Cholesky factorizations $\Gamma_{t|t-1}^{-1} = L_2^T L_2$ and $\Gamma_{v_t}^{-1} = L_1^T L_1$ exist. In our case, the maximization of the posterior density is equivalent with the minimization of the following quadratic functional, so that we can write

$$x_{t|t} = \text{solution of } \min_x \left\{ \| L_1(y_t - G_t(x)) \|_2^2 + \| L_2(x - x_{t|t-1}) \|_2^2 \right\} \tag{6.27}$$

which can be solved for example with the Gauss–Newton algorithm. Thus, in the iterated extended Kalman filter, we would compute Equation 6.21 and Equation 6.22 as before, but Equation 6.23 through Equation 6.25 would be replaced by first computing the estimate $x_{t|t}$ by minimizing Equation 6.27 and then computing $\Gamma_{t|t}$ from Equation 6.23 and Equation 6.24, so that the Jacobians J_{G_t} are recomputed at $x_{t|t}$. The case of a nonlinear state evolution equation is an isolated problem of solving a nonlinear difference equation. The evaluation of the predictor covariance, however, may call for a Taylor series approximation.

In all cases, it must be noted that the recursions in this section are not optimal with respect to any simple criteria. While the computational load is sure to increase from versions 1 to 3, the increase of the quality of the estimates is not guaranteed. Trying all versions out in a particular application and choosing the simplest adequate one is therefore always recommended.

Note also that even in the case of evolution models that are naturally linear with respect to a physical parameter, say temperature, an unconventional choice for state variables may render the state evolution model highly nonlinear. See Kolehmainen et al. [16] for an example in which the state variables comprise a parametric representation for the boundary of a temporally varying inclusion in a vessel.

Optimal filters for the general nonlinear non-Gaussian problems are based on Monte Carlo–based approaches and are discussed in the following section.

6.4.2 Sampling and Particle Filters

Assume that $\{x_t | t = 0, 1, \ldots\}$ is a Markov process with initial density $\pi(x_0)$ and that the observation y_t depends only on the current state x_t of the system. Thus, a general stochastic state space model can be expressed using probability densities such as

$$\pi(x_{t+1} \mid x_t) \tag{6.28}$$

$$\pi(y_t \mid x_t) \tag{6.29}$$

Then the posterior (filtering) density is given by Equation 6.26, where the prior (prediction) density is of the form

$$\pi(x_t \mid y_1, y_2, \ldots, y_{t-1}) = \int \pi(x_t \mid x_{t-1}) \pi(x_{t-1} \mid y_1, y_2, \ldots, y_{t-1}) \tag{6.30}$$

In principle, this recursion can be used to determine the posterior densities and compute their expectations. Unfortunately, the required integrals can only be evaluated in closed form for a few special cases. For example, linear state space models with additive Gaussian noises lead to Gaussian posterior densities that can be obtained by using the recursion in Equation 6.6 through Equation 6.10.

Recently, sequential Monte Carlo (SMC) methods, also known as particle filters, have become increasingly important tools for obtaining approximations for posterior densities and posterior means in nonlinear and non-Gaussian problems. In SMC methods, each probability density is approximated with a discrete density, and the support and frequency function evolves with time according to the state space model.

The basic particle filter is called the sampling importance resampling (SIR) filter. It is discussed by Doucet, de Freitas, and Gordon [17]; Liu and Chen [18]; Doucet, Godsill, and Andrieu [19]; and Carpenter et al. [20], but a brief description of the SIR filter is now given. Let $\{x_{t-1}^{(i)} | i = 1, \ldots, N\}$ be a sample from posterior density $\pi(x_{t-1} \mid y_1, \ldots, y_{t-1})$. This set can be understood as a discrete approximation for $\pi(x_{t-1} \mid y_1, \ldots, y_{t-1})$ when each sample (particle) has equal probability mass (weight). A set of particles with equal weights representing the prior density of next time instant is obtained by drawing $\tilde{x}_t^{(i)}$ from $\pi(x_t \mid x_{t-1}^{(i)})$, $i = 1, \ldots, N$. Next, an approximation for the expectation of the posterior density is determined by using the importance sampling method. The posterior density consists of likelihood density and prior density; see Equation 6.26. The prior density is represented by the set $\{\tilde{x}_t^{(i)} \mid i = 1, \ldots, N\}$, and thus $\pi(x_t | y_1, \ldots, y_{t-1})$ is selected as the proposal density. Then the normalized importance weights are of the form $w_t^{(i)} = c_t^{-1} \pi(y_t \mid x_t^{(i)})$, where $c_t = \sum_{j=1}^{N} \pi(y_t \mid x_t^{(j)})$. Now the support of the posterior density is represented by the set $\{\tilde{x}_t^{(i)} | i = 1, \ldots, N\}$, and the associated probability masses are $\{w_t^{(i)} \mid i = 1, \ldots, N\}$. An approximation for the posterior mean is $E(x_t | y_1, \ldots, y_t) \approx \sum_{i=1}^{N} w_t^{(i)} \tilde{x}_t^{(i)}$. In order to proceed to the next time step, an equally weighted set of particles representing the posterior density must be generated. One can do this simply by drawing from the nonweighted set according to the weights (probability masses).

The SIR filter can be summarized as follows:

`Step 1. Initialization:` Set $t=0$. Draw a sample $\{x_0^{(i)} \mid i=1,\ldots,N\}$ from $\pi(x_0)$.

`Step 2. Prediction:` Set $t \leftarrow t+1$. Draw $\{\tilde{x}_t^{(i)} \mid i=1,\ldots,N\}$ from $\pi(x_t \mid x_{t-1}^{(i)})$.

`Step 3. Weights:` Calculate

$$w_t^{(i)} = \frac{\pi\left(y_t \mid \tilde{x}_t^{(i)}\right)}{\sum_{j=1}^{N} \pi\left(y_t \mid \tilde{x}_t^{(j)}\right)} \tag{6.31}$$

`Step 4. Expectation:` $E(x_t \mid y_1,\ldots,y_t) \approx \sum_{i=1}^{N} w_t^{(i)} \tilde{x}_t^{(i)}$.

`Step 5. Resampling:` Draw N samples (with replacement) from $\{\tilde{x}_t^{(i)} \mid i=1,\ldots,N\}$ so that $\tilde{x}_t^{(i)}$ is selected with probability $w_t^{(i)}$. Denote the new set $\{x_t^{(i)} \mid i=1,\ldots,N\}$. Go to Step 2.

The most important feature of the SIR filter is that it is not restricted to any specific state space model but is applicable with virtually any model, including nondifferentiable and discontinuous observation models, multiplicative noise, and noise distributions with point masses. The only requirements are that one must be able to sample $\pi(x_{t+1} \mid x_t)$ and evaluate $\pi(y_t \mid x_t)$ at given points. The implementation of SIR filters is straightforward, and different types of spatial constraints can be easily realized. The estimates obtained with SIR filter converge to correct values as $N \to \infty$, but the convergence rate decreases as the state dimension increases. Thus, the SIR filter is applicable in relatively low-dimensional problems; otherwise, the computational burden may become unreasonable. In addition to the basic SIR filter, a number of different particle filter modifications are designed to improve the performance in various situations; see Doucet, de Freitas, and Gordon [21]; van der Merwe et al. [22]; Chen and Liu [23]; Andrieu and Doucet [24]; and Pitt and Shephard [25].

6.5 PARTIALLY UNKNOWN MODELS: PARAMETER ESTIMATION

In practical problems, the case is usually that one or more of the state space terms are at least partially unknown. While careful analysis of the measurement system usually pins the observation model down relatively accurately, the state evolution model is always inaccurate to some extent. In particular, the state noise covariance is very likely to be a parameter that is not well known. In such cases, there are basically three alternatives.

The first is the subjective "eyeball norm" minimization approach, in which the parameters are tuned manually based on previous experience. The two other approaches are more objective. In the batch type optimization approach, the task is to estimate the best time-invariant parameters, for example by the maximum likelihood method. However, if the unknown parameters are assumed to be time-varying, it is possible to augment the state variable to include the unknown parameters.

6.5.1 Batch-Type Optimization: Time-Invariant Parameters

The estimation of unknown parameters in a state space model is often called *state space identification*. Denote the unknown parameters by $\psi \in \Re$. Several parameters of the state space representation might depend on the parameters ψ. For example, we might be able to postulate that $\Gamma_{w_t} \equiv \sigma_w^2 I$ but be unable to specify σ_w^2.

The idea is to compute the likelihood L of the observations $Y_T = (y_1, \ldots, y_T)$ given the parameters ψ and maximize the likelihood with respect to ψ, that is,

$$\max_{\psi} L(Y_T \mid \psi) \tag{6.32}$$

The likelihood depends on the unknown parameters. For example, consider the case in which the initial state distribution is known. Then the likelihood $L(Y_T)$ can be written in the form [3]

$$L(Y_T \mid \psi) = -\frac{MT}{2}\log 2\pi - \frac{1}{2}\sum_{t=1}^{T}\left(\log\left|\Gamma_{t|t-1}\right| + e_t^T \Gamma_{t|t-1} e_t \mid \psi\right) \tag{6.33}$$

where $|\cdot|$ denotes determinant, $e_t = y_t - G_t x_{t|t-1}$ is the prediction error, and the notation $|\psi$ on the left-hand side refers to all variables being calculated with parameters ψ.

There are certain other analytical forms for the likelihood with various types of unknown parameters. However, most often it is necessary to use Newton or quasi-Newton type methods with numerically approximated gradients to compute the maximum of Equation 6.33. A particularly suitable algorithm is the BFGS (Broyden–Fletcher–Goldfarb–Shanno) quasi-Newton algorithm. Some numerical considerations are given in the general state space references, and more details can be found for example in Nocedal and Wright [26]. See also Shumway [27] for the application of the expectation–maximization (EM) algorithm in this problem.

6.5.2 Augmented State Models: Time-Varying Parameters

It is most likely that the unknown parameters ψ occur in the state evolution model. If the parameters were known, we could employ the state evolution model $x_{t+1} = F_t(x_t; \psi_t) + s_t + B_t u_t + w_t$, where we have stated the explicit time dependence of the parameter ψ_t.

The straightforward approach is to augment the state to include the unknown parameters so that $X_t = (x_t, \psi_t)$. The usual choice would be to postulate a random walk model for ψ_t. Although straightforward, this approach can turn out to be tedious and can easily lead to an unobservable state space model.

Typically, we would then aim for using an affine approximation for the original state evolution model to arrive at

$$X_{t+1} = \begin{pmatrix} x_{t+1} \\ \psi_{t+1} \end{pmatrix} = \begin{pmatrix} b_t + s_t \\ 0 \end{pmatrix} + \bar{F}_t X_t + \begin{pmatrix} B_t \\ 0 \end{pmatrix} u_t + \begin{pmatrix} w_t \\ w'_t \end{pmatrix} \tag{6.34}$$

where typically

$$\bar{F}_t = \begin{pmatrix} \frac{\partial F_t}{\partial x_t}\Big|_{X_t^*} & \frac{\partial F_t}{\partial \psi_t}\Big|_{X_t^*} \\ 0 & I \end{pmatrix}, \quad \text{cov}\begin{pmatrix} w_t \\ w'_t \end{pmatrix} = \begin{pmatrix} \Gamma_{w_t} & 0 \\ 0 & \sigma_{w'}^2 I \end{pmatrix} \tag{6.35}$$

$$b_t = F_t(X_t^*) - \begin{pmatrix} \frac{\partial F_t}{\partial x_t}\Big|_{X_t^*} & \frac{\partial F_t}{\partial \psi_t}\Big|_{X_t^*} \end{pmatrix} X_t^* \tag{6.36}$$

and where $\sigma_{w'}^2$ is still to be fixed to adjust the tracking properties of the algorithm, and $X_t^* = X_{t|t}$ or some time-invariant state. This is the approach taken in Section 6.8.4, where we estimate the total flux assuming that the flow profile in a pipeline is known. Note again that the linearization is actually only needed for the computation of the covariances and the Kalman gain.

6.6 FURTHER TOPICS

6.6.1 General Modeling Issues and Inverse Problems

Much of the success of the state estimation is naturally based on the quality and accuracy of the models. While in most cases it can be argued that one should be able to determine the observation model based on engineering considerations, the model can be too complex to realize completely. Nevertheless, there is seldom a significant amount of inherent uncertainty in the observation model. The state evolution model, however, is usually subject to much more uncertainty.

A straightforward ad hoc choice for the evolution model would be to employ the discrete random walk model [28] with orthogonal state noise covariance structure, that is,

$$x_{t+1} = x_t + w_t \quad \text{cov}(w_t) = \sigma_w^2 I \tag{6.37}$$

The random walk is essentially a model for very slow state evolution. In cases where the use of a random walk model is more or less ad hoc, but where the actual state evolution is not slow, it has turned out that the state estimates given by the Kalman filter are often useless. This is due to the fact that in order to allow for adequately fast tracking, the state noise covariance has to be tuned relatively large, which means that the state estimate (error) covariance will also be large. The smoothed estimates are usually better, but the problem prevails. It should also be noted that when the covariance of the state noise tends to infinity, the state estimates tend essentially to (weighted) least squares estimates of the state based on the measurements at the time only (the maximum likelihood estimate). That is,

$$x_{t|t} = \text{solution of } \min_x \| L_v(y_t - G_t x) \|, \quad L_v^T L_v = \Gamma_{v_t}^{-1} \tag{6.38}$$

This result is usually meaningless if G_t does not have full row rank, and often $N >> M$ so that this is not the case. In process tomography applications, the dimension of the state is often much larger than the dimension of the observations. For example, for a typical 3D EIT problem in a tank geometry, we may have N between 2000 and 5000, while $M = 80$. Even when there are fewer states, estimating states from tomographic measurements is usually classified as an ill-posed inverse problem; such problems have the property that the computation of the unknowns (the state) based on the observations is an unstable problem, even when the observations are noiseless [29].

In stationary inverse problems, a conventional strategy is to regularize the problem by introducing a regularizing functional. This is commonly called Tikhonov regularization. For a moment, assume that there is no evolution model and x is just a vector-valued random variable. Then, under Gaussian assumptions, the Tikhonov regularized problem takes the form (in our case)

$$\min\{\| L_v(y_t - G_t x) \|^2 + \alpha \| L_x(x - x_t^*) \|^2\} \tag{6.39}$$

where α is called a regularization parameter.

This approach, which often bears the interpretation of spatial regularization, can readily be incorporated into the state estimation scheme by introducing an augmented (fictitious) observation model,

$$Y_t = \begin{pmatrix} y_t \\ \sqrt{\alpha} L_x x_t^* \end{pmatrix} = \begin{pmatrix} G_t \\ \sqrt{\alpha} L_x \end{pmatrix} x_t + \begin{pmatrix} v_t \\ v_t' \end{pmatrix} \tag{6.40}$$

where $\text{cov}(v_t, v_t') = \text{diag}(\Gamma_{v_t}, I)$. Note that here the dimension of the augmented observation vector is approximately $M + N >> M$, so that the information filter would be a feasible choice.

The idea behind the regularizing functional is that the regularization operator L_x should be chosen so that the norm as $\|L_x(x_t - x_t^*)\|$ is small for all expected values x_t and large for all unexpected values x_t. Typically, in our case the state represents a spatial variable such as conductivity or void fraction. In such cases, it is customary that L_x is based on a discretization of a differential operator such as the Laplacian. For examples on how to construct spatial regularization functionals, see Vauhkonen et al. [30, 31] and Kaipio et al. [32, 33]. For more details and examples on the spatial regularization of state space models, see Kaipio and Somersalo [34], and Baroudi et al. [35].

6.6.2 Construction of the Finite Dimensional Discrete-Time Evolution Model

Most often, the relevant state evolution models are based on differential equations, systems of differential equations, or partial differential equations. Moreover, the task is to obtain computational models for the stochastic counterparts of these models. How to obtain the discrete-time state evolution model from a continuous-time stochastic (differential equation) model? Fortunately, in the case of linear differential models this turns out to be a straightforward task since the determination of the state evolution operator F_t and the state noise covariance Γ_{w_t} can be separated.

The stochastic differential equations can often be expressed in the form

$$dx = f(x,t)\,dt + \sigma(x,t)\,dB(t) \tag{6.41}$$

where $B(t)$ is the continuous time Brownian motion which has zero mean. Stochastic differential equations of this type are straightforward to handle. The state transition mapping $F_t(x_t)$ can be obtained by integrating (possibly numerically) Equation 6.41 from the physical time τ_t to τ_{t+1}. In the case of a linear differential equation $f(x;t) \equiv \mathcal{F}x$, this can in principle be done analytically:

$$x_{t+1} \doteq x(\tau_{t+1}) = e^{\mathcal{F}(\tau_{t+1}-\tau_t)}x_t = F_t x_t \tag{6.42}$$

although numerical integration methods may be more feasible in this case as well. Note that the observation interval $\tau_{t+1} - \tau_t$ does not have to be a constant.

The determination of the structure of the state noise covariance is in principle a difficult analytical problem [36]. However, if $\sigma(x,t) \equiv \sigma$ is a constant, we have

$$E\int_{\tau_t}^{\tau_{t+1}} \sigma dB = 0 \qquad \operatorname{cov}\int_{\tau_t}^{\tau_{t+1}} \sigma dB = (\tau_{t+1} - \tau_t)\sigma\sigma^T = \Gamma_{w_t} \tag{6.43}$$

The standard choice is to set $\Gamma_{w_t} = (\tau_{t+1} - \tau_t)\bar{\sigma}_w^2 I$. The variance $\bar{\sigma}_w^2$ is a quantity that has either to be tuned manually or estimated as described above. However, the construction of the state evolution model may easily result in more involved models for Γ_{w_t}; see Section 6.8.2. In the case of a more general stochastic differential equation model for the state evolution equation, the integration of Equation 6.41 has to be carried out explicitly. Especially if $\sigma = \sigma(x_t)$, the computation of expectation and covariance of $\int \sigma dB$ is not a trivial task. See Kloeden and Platen [37] for details in the general case. For the mathematical formulation of the stochastic convection–diffusion equation, see Pikkarainen [38].

However, some of the most feasible state evolution models are partial differential equations, such as the convection–diffusion, thermal conductivity, fluid dynamical models such as Navier–Stokes; free boundary models such as the Stefan problem; and several more general transport equations. The most feasible scheme to approach these models is the so-called semidiscrete scheme, which is also known as the method of lines. The following is a very formal simplified presentation of the overall idea for models such as thermal diffusion equation, which we denote by $\mathcal{L}x = f$ where $\mathcal{L}$ is the (partial) differential operator.

In the semidiscrete scheme related to the finite element method, the unknown variable, say temperature, $x(\xi, t)$, $\xi \in \Re^{2,3}$, $t \in [0, T]$, is approximated in the form $x(\xi, t) \approx x^h(\xi, t) = \sum_k \beta_k(t)\varphi_k(\xi)$. Since $\mathcal{L}x - f = 0$, we should of course have $\int(\mathcal{L}x - f)\varphi = 0$ for all nonnegative (test) functions φ where the integration is carried out over the domain. In the Galerkin approach, we supplant $x(\xi, t)$ by $x^h(\xi, t)$ and choose the basis functions $\varphi_k(\xi)$ as test functions $\varphi(\xi)$. Completing the problem by assigning initial and boundary values and using Green's formula in the integration will then yield a system of ordinary differential equations of the form

$$\bar{M}\frac{\partial \beta}{\partial t} = \bar{D}\beta + \bar{d} \tag{6.44}$$

Rewriting this in the form of Equation 6.41, equipping with the Brownian motion, and then integrating over the observation intervals yields

$$\beta_{t+1} = D\beta_t + d_t + w_t \tag{6.45}$$

where β_t would be identified as the state x_t, $F_t = D$, d_t is a term involving boundary conditions of the original initial-boundary value problem, and w_t is the state noise process (with specified covariance).

In addition to state evolution models, several observation models are also governed by partial differential equations. The most important class is diffuse tomographic imaging; this includes diffuse optical tomography, capacitance tomography, and electrical impedance tomography, which are governed by elliptic PDEs. Also, certain modalities using sound probing lead to either hyperbolic or potential problems. For general treatments of partial differential equations and finite element methods, see Ciarlet [39], Renardy and Rogers [40], and Brenner and Scott [41]. We refer to Curtain and Zwart [42] for infinite-dimensional system analysis.

One area in the process industry in which large dimensional (nonlinear) finite difference and finite element models are routinely employed is bio-reactor control. However, these treatments usually favor non-optimal asymptotic output tracking control instead of (stochastic) state estimation and optimal control; see Dochain [43] and Tervo et al. [44].

It must be noted that the boundary and initial values may not be completely specified, as is the case in the example in Section 6.8.2. In such a case it may be feasible to model for example the boundary data as random variables or stochastic processes too. These could then be absorbed into the state noise process. Equivalently, the modelling errors in observation equation can in principle be embedded into observation noise.

6.6.3 Higher Order Evolution Models

If the physical model is of the second order (in time), the conventional approach is to consider the corresponding system of coupled differential equations, semi-discretize, and modify the observation model accordingly. This is the approach for example in the case of wave equation.

Let us assume that we have constructed the state space representation of the system and applied it to some data. After computing the state estimates and further the estimates for the state noise, we check for the correlation of w_t. It would not be surprising to find out that the state noise is correlated.

The straightforward approach is to carry out multivariable time series analysis for the state noise process and to find an approximate whitening filter $\Psi(z) = \sum_\ell \psi_\ell z^{-\ell}$ which is a delay operator such as

$$e_t \doteq \Psi(z) w_t = \psi_0 w_t + \psi_1 w_{t-1} + \cdots \tag{6.46}$$

so that e_t would be approximately uncorrelated.

As an example, assume that a simple difference $e_t = w_t - w_{t-1}$ of the state noise process would be serially uncorrelated. We could then subtract the state equations at times $t+1$ and t to yield

$$x_{t+1} = F_t x_t + x_t - F_{t-1} x_{t-1} + w_t - w_{t-1} \tag{6.47}$$

which can be further written in the form

$$\begin{pmatrix} x_{t+1} \\ x_t \end{pmatrix} = \begin{pmatrix} F_t + I & -F_{t-1} \\ I & 0 \end{pmatrix} \begin{pmatrix} x_t \\ x_{t-1} \end{pmatrix} + \begin{pmatrix} e_t \\ 0 \end{pmatrix} \tag{6.48}$$

in which the state noise process is now uncorrelated. The observation model naturally has to be modified accordingly.

It is clear that when the state dimension is hundreds or thousands, this approach may turn out to be impractical if simple subtraction is not adequate. Furthermore, it is recommended that the (indirectly or directly) estimated noise processes w_t and v_t be checked both for lack of serial correlation and for covariance structure; see especially Durbin and Koopman [3].

6.6.4 Stability and Computational Issues

Two of the main problems in applying state estimation techniques to process imaging, as noted earlier, are that the state dimension may be very large and that, in practice, the state covariance matrices are never sparse. It is thus essential to use all available means to carry out the computations in a stable and efficient manner. It is also likely that a practical real-time or on-line state estimation problem necessitates the adoption of parallel computing.

Consider the innovation form Kalman filter given in Equation 6.21 through Equation 6.25. A particular problem arises when the measurements are very accurate so that Γ is very small. It can turn out that the covariances lose positive definiteness due to numerical reasons. This may induce several stability issues.

In such cases, it is possible to revert to the so-called square root filters. In these versions of Kalman filters and smoothers, the Cholesky factors of the covariances are updated instead of the covariances themselves. This will ensure positive definiteness and thus stability. The computational cost, however, will be greater. A brute force approach to enforce positive definiteness of the estimated covariances is to use larger observation and state noise covariances than the measurement and evolution models and system analysis would suggest.

6.7 OBSERVATION AND EVOLUTION MODELS IN THE PROCESS INDUSTRY

6.7.1 Observation Modalities

The observation models are treated comprehensively in other chapters of this book. We, however, stress here that the state estimation scheme does not preclude in any way that the observation model remains the same from time to time. In other words, we can in principle combine several different modalities either simultaneously or sequentially as long as we are able to write the observation model in terms of the same state variable.

6.7.2 Evolution Models

In our example cases, we only consider single-phase flow models. We model the evolution of a chemical substance by using convection–diffusion equation and the Navier–Stokes equations. In such a case, the state variable is the concentration of the chemical. The Navier–Stokes model is suitable only for the case of laminar flow. However, in the case of turbulent flow, we can replace the Navier–Stokes model with a turbulent flow model; see for example Durbin and Koopman [3]. The process models for multiphase systems have been discussed by Hewitt [45], Dyakowski [46], and Levy [47]. In the case of multiphase flow, the state variables would be different from those in the single-phase flow. However, tomography systems have limited resolution; especially in EIT, the spatial resolution is not very high. Thus, in many cases it is not possible to locate small volumes of different phases based on tomographic measurements. For this reason, homogenization models are often needed in the case of multiphase systems. For example, in the case of gas–liquid flow, the void fraction could be considered as the state variable.

6.8 EXAMPLE: CONVECTION–DIFFUSION MODELS

Here is an example that demonstrates the different tasks and considerations needed to create a state space representation for the dynamical estimation (imaging) of the concentration distribution in a pipeline. The problem is governed by three partial differential equations. These equations are discretized to yield the state evolution and observation models.

The overall situation is modelled as follows. We are interested in the distribution of a chemical substance in a flow in a pipeline. The substance is assumed to be conductive, so that the concentration differences in the pipeline will also induce conductivity differences. Thus, impedance tomography is a feasible measurement modality. The following is a condensed account of an application, the details of which can be found in Seppänen et al. [48–55]. In this section, we consider only numerical examples. Note, however, that we have recently evaluated the approach also with experimental results; see Seppänen et al. [55]. See also the experimental papers Vauhkonen et al. [56] and Kim et al. [57, 58, 59], in which the random walk model has been used as the evolution model in the state space representation.

The state is assumed to obey a convection–diffusion equation, which is a parabolic partial differential equation,

$$\frac{\partial c}{\partial t} = -\overline{v} \cdot \nabla c + \nabla \cdot \kappa \nabla c \tag{6.49}$$

where $c = c(\xi, t)$ is concentration, $\overline{v}$ is the velocity field of the fluid, and κ is the diffusion coefficient. We take the (discretized) concentration as the state variable.

The velocity field is usually difficult to measure, and computational fluid dynamics results are usually considered inaccurate. If an accurate velocity field is available, then this is used. However, we assume here that the (measured) velocity field is not available but that it can be approximated with the (mean) Navier–Stokes flow, which is a model for laminar incompressible single-phase flow. We emphasize that in practical situations, this is unlikely to be a good model for the flow, as the flows are often turbulent and possibly multiphase. We are, however, in fact constructing a feasible model for the concentration — not the flow. It is an old trick to model turbulent flows with approximate velocity fields and then to increase the diffusion coefficient.

The numerical solution of the Navier–Stokes equations is a challenging problem, and it is subject to intensive research [60]. The difficulties in the Navier–Stokes problem are associated with the high nonlinearity of the problem and the incompressibility constraint. Furthermore, the resulting differential equations in the FEM scheme are usually stiff. However, for a moment we take the velocity field as predetermined, and hence it acts as a known distributed parameter in the convection–diffusion equation.

The observations are conventional impedance tomography measurements with a sequence of predetermined current patterns, and all voltage measurements related to a current pattern are assumed to be carried out simultaneously. The most accurate known model for EIT measurements is the so-called complete electrode model, which is an elliptic partial differential equation with mixed boundary conditions.

For different electrolytes, there are different models for the mapping from concentration to conductivity. For example, in the case of strong electrolytes, the conductivity σ is related to concentration c of solution by [61]:

$$\sigma(c) = \Lambda_0 c - \beta c^{\frac{3}{2}} \tag{6.50}$$

where Λ_0 and β are known coefficients. However, we take the mapping to be linear in this example. This does not induce any particular relief since the observation model is nonlinear in any case.

6.8.1 Impedance Tomography and the Measurement Model

Impedance tomography has been treated more comprehensively elsewhere in this book. We restate the complete electrode model [62]:

$$\nabla \cdot (\sigma \nabla u) = 0, \quad x \in \Omega \tag{6.51}$$

$$u + z_\ell \sigma \frac{\partial u}{\partial n} = U_\ell, \quad x \in e_\ell, \quad \ell = 1, 2, \ldots, L \tag{6.52}$$

$$\int_{e_\ell} \sigma \frac{\partial u}{\partial n} dS = I_\ell, \quad x \in e_\ell, \quad \ell = 1,2,\ldots, L \tag{6.53}$$

$$\sigma \frac{\partial u}{\partial n} = 0, \quad x \in \partial\Omega \backslash \cup_{\ell=1}^{L} e_\ell \tag{6.54}$$

where $u = u(x)$ is the electric potential, e_ℓ is the ℓth electrode, z_ℓ is contact impedance between the ℓth electrode and contact material, U_l is the potential on ℓth electrode, I_ℓ is the injected current (on that electrode), and n is outward unit normal. In addition, to fix the zero potential and to fulfill the charge conservation, we set

$$\sum_{\ell=1}^{L} U_\ell = 0 \quad \sum_{\ell=1}^{L} I_\ell = 0 \tag{6.55}$$

Equation 6.51 through Equation 6.55 are discretized with the finite element approximation. This leads to the equation

$$U = R(\sigma) I \tag{6.56}$$

for each current pattern I. Here $U \in \Re^M$ denotes the potential differences between electrode pairs, and $\sigma \in \Re^N$ is a vector that gives the conductivities at each FEM node.

For the computation of the covariances, we need to compute the Jacobian mappings and the Taylor approximations

$$U(\sigma) \approx U(\sigma^*) + J_R(\sigma^*, I)(\sigma - \sigma^*) \tag{6.57}$$

We complete the specification of the observation model by assuming that the observation noise is purely additive and that all errors are independent and have the same variance. Thus we have

$$U_t = R(\sigma_t, I_t) + v_t \tag{6.58}$$

where I_t is the current pattern injected at time t, and we have $\Gamma_{v_t} = \sigma_v^2 I$ where σ_v^2 is determined by the measurement system properties, the demodulation time, and external electromagnetic fields. We will identify $y_t = U_t$ and $G_t(\sigma_t) = R(\sigma_t, I_t)$. We note that if the external electromagnetic fields play a significant role, the diagonal structure of the covariance is most likely not a good approximation.

Next we write the observation model in terms of the state variable, which is the concentration c in this example. First we write the model between the conductivity and the concentration in the form $\sigma = \sigma(c)$. For example, in the case of strong electrolytes we can use the model in Equation 6.50. We substitute this model in the observation model (Equation 6.58) and obtain

$$U_t = R(\sigma(c_t), I_t) + v_t = R^*(c_t, I_t) + v_t \tag{6.59}$$

The Jacobian of the mapping R^* is simply

$$J_{R^*}(c_t) = J_R(\sigma(c_t)) \cdot J_\sigma(c_t) \tag{6.60}$$

For details on the computational model for impedance tomography used in this computational example, see Vauhkonen [63] and Kaipio et al. [33].

6.8.2 Convection–Diffusion Model

In order to complete the specification of the convection–diffusion equation (6.49), we set the following initial and boundary conditions

$$c(x, 0) = c_0(x) \tag{6.61}$$

$$c(x, t) = c_{\text{in}}(t), \quad x \in \partial\Omega_{\text{in}} \tag{6.62}$$

$$\frac{\partial c}{\partial n} = 0, \quad x \in (\partial\Omega \backslash \partial\Omega_{\text{in}}) \tag{6.63}$$

where $c_{\text{in}}(t)$ is the time-varying concentration on the input boundary $\partial\Omega_{\text{in}}$ and n is the outward unit normal. Since the input concentration c_{in} is not always known, we consider c_{in} as a stochastic function. Thus, we write

$$c_{\text{in}}(t) = \bar{c}_{\text{in}}(t) + \eta(t) \tag{6.64}$$

where $\bar{c}_{\text{in}}(t)$ is the deterministic part of the input and η is a stochastic process.

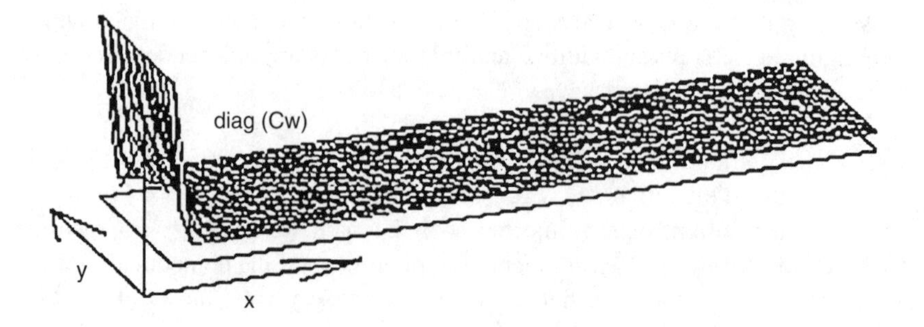

FIGURE 6.1 The variances of the state noise w_t (the diagonal of the state noise covariance matrix Γ_{w_t}).

Next we apply the FEM to the convection–diffusion equation (Equation 6.49). The FEM solution of the convection–diffusion equation used here is described in detail by Seppänen et al. [49]. In the FEM solution, we write the variational form for the weak solution of the PDE and construct a system of differential equations based on the variational form. The system of differential equations is solved numerically using the backward Euler method to give a concentration profile that is discretized in time:

$$c_{t+1} = Fc_t + s_{t+1} + w_{t+1} \tag{6.65}$$

where the vector $c_t \in \Re^N$ includes the discretized values of the concentration at time t, $F \in \Re^{N\times N}$ is the evolution matrix, $s_t \in \Re^N$ is deterministic input of the system, and $w_t \in \Re^N$ is a stochastic process.

It turns out that the success of the overall state estimation scheme in this application rests on the accurate modelling of the state noise process. This task involves the accumulation of the state noise process from the convection–diffusion equation and the process due to the unknown boundary data. The surface plot in Figure 6.1 illustrates the diagonal of the (accumulated) state noise covariance matrix Γ_{w_t}, that is, the variance of w_t, in a 2D pipe-flow case. Since the input concentration c_{in} is the main source of uncertainty in the evolution model, the variance of the state noise w_t is highest on the input layer of the pipe.

6.8.3 Example 1: Known Velocity Field

In the first example, we assume a time-independent (stationary) Navier–Stokes flow in a straight 2D pipe. In such a case, the velocity profile is known to be parabolic [64]. The evolution of the chemical substance was determined by using the convection–diffusion model with this flow field. The EIT observations were computed by approximating numerically the complete electrode model introduced above. We also added noise to the observations. The noise consisted of two Gaussian components: one with standard deviation of 0.1% of the maximum difference of the noiseless observations, and the other with standard deviation of 1%.

The state estimates were computed by using the fixed lag Kalman smoother. When the evolution model was constructed for the state estimation scheme, the velocity field and the diffusion coefficient were assumed to be known. Figure 6.2 represents the estimated states together with the true concentration distributions at certain times.

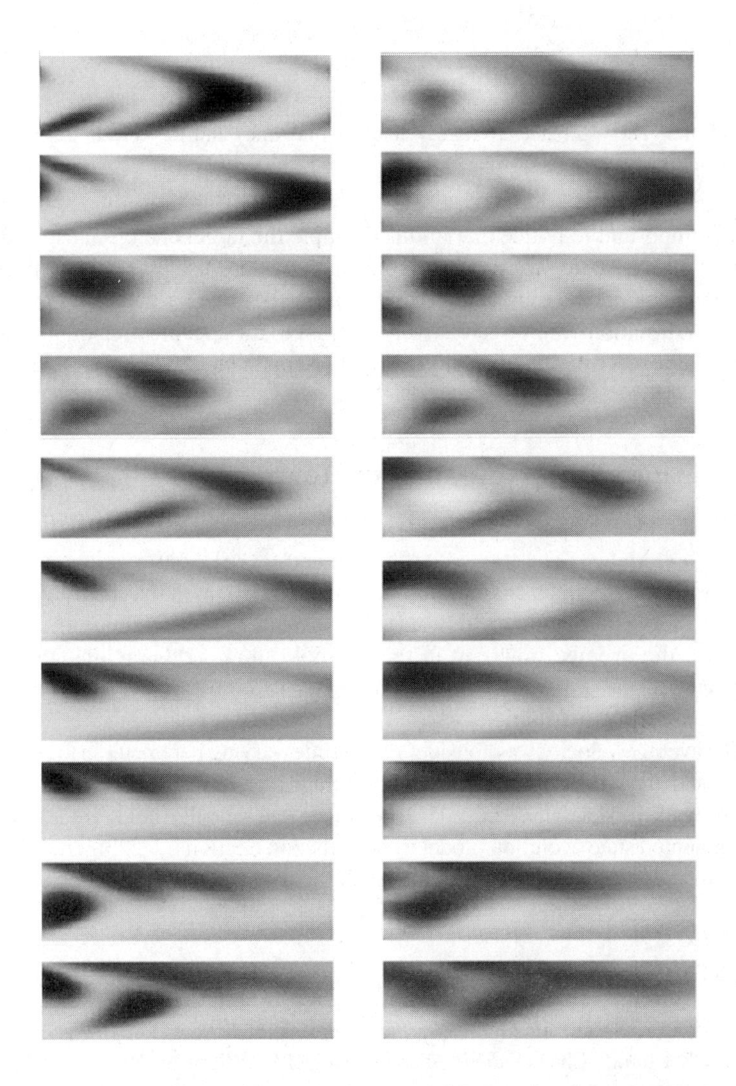

FIGURE 6.2 The true concentration distribution (left column) and the reconstructed distribution (right column).

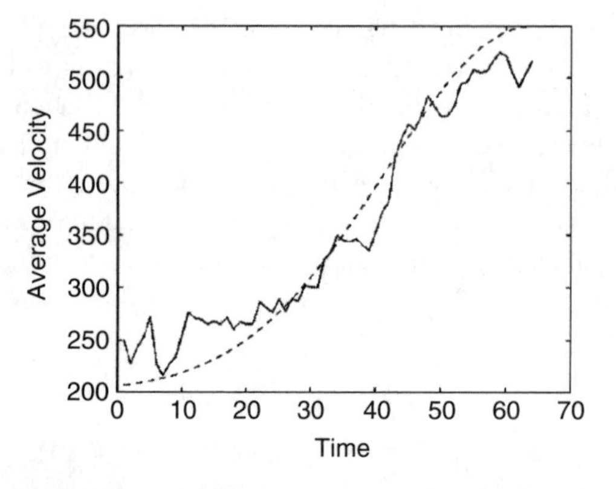

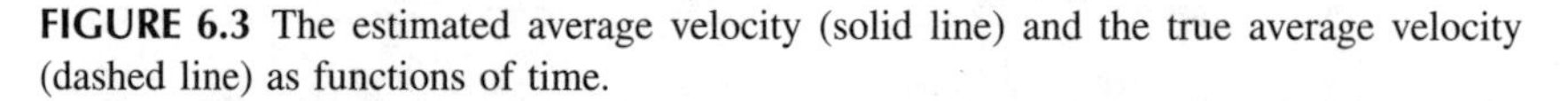

FIGURE 6.3 The estimated average velocity (solid line) and the true average velocity (dashed line) as functions of time.

It is important to notice that in this example the target changes at such a high rate that all the stationary reconstruction methods are useless. Furthermore, the state estimates obtained by using the random walk model are not feasible. However, the estimation scheme is relatively tolerant to misspecification of the velocity fields in the convection–diffusion model. Seppänen et al. have studied this feature [51].

6.8.4 Example 2: Unknown Rate of Flow

In the second example, we consider the case in which the velocity field is (partially) unknown. The target distribution and the EIT observations were simulated as in the previous example. The only difference was that here the average velocity in the parabolic field was time varying. When computing the state estimates, the average flow rate in the parabolic flow field was unknown. That is, there was one unknown velocity parameter to be estimated — together with the concentration distribution — based on the EIT observations.

The average velocity was considered as a time-varying parameter in the evolution model, and the time-dependence of the average velocity was described with the random walk model. Thus, in order to estimate the velocity field, we constructed the augmented state evolution model as described in Section 6.5.2. Figure 6.3 shows the estimated average velocity together with the true average velocity.

REFERENCES

1. BDO Anderson, JB Moore. *Optimal Filtering.* New York: Prentice Hall, 1979.
2. CK Chui, G Chen. *Kalman Filtering.* Heidelberg: Springer-Verlag, 1987.
3. J Durbin, SJ Koopman. *Time Series Analysis by State Space Methods.* Oxford: Oxford University Press, 2001.

4. JL Melsa, DL Cohn. *Decision and Estimation Theory.* New York: McGraw-Hill, 1978.
5. HW Sorenson. *Parameter Estimation: Principles and Problems.* New York: Marcel Dekker, 1980.
6. RE Kalman. A new approach to linear filtering and prediction problems. *J. Basic Eng.* 82D:35–45, 1960.
7. P Swerling. A proposed stagewise differential correction procedure for satellite tracking and prediction. *J. Astronaut Sci.* 6, 1959.
8. RE Kalman, RS Bucy. New results in linear filtering and prediction theory. *J. Basic Eng.* 83D:35–45, 1961.
9. AV Balakrishnan. *Elements of State Space Theory.* New York: Optimization Software, 1983.
10. AV Balakrishnan. *Kalman Filtering Theory.* New York: Optimization Software, 1984.
11. T Kailath. *Linear Systems.* Englewood Cliffs, NJ: Prentice Hall, 1980.
12. M Green, DJN Limebeer. *Linear Robust Control.* Englewood Cliffs, NJ: Prentice Hall, 1995.
13. S Skogestad, I Postlethwaite. *Multivariable Feedback Control: Analysis and Design.* Chichester, U.K.: Wiley, 1996.
14. DE Clarke, ed. *Advances in Model-based Predictive Control.* Oxford: Oxford University Press, 1994.
15. PJ Brockwell, RA Davis. *Time Series: Theory and Methods.* Heidelberg: Springer-Verlag, 1991.
16. V Kolehmainen, A Voutilainen, JP Kaipio. Estimation of non-stationary region boundaries in EIT: State estimation approach. *Inv. Probl.* 17:1937–1956, 2001.
17. A Doucet, N de Freitas, N Gordon. An introduction to sequential Monte Carlo methods. In: A Doucet, N de Freitas, N Gordon, eds. *Sequential Monte Carlo Methods in Practice.* Heidelberg: Springer-Verlag, 2001, pp. 3–14.
18. JS Liu, R Chen. Sequential Monte Carlo methods for dynamic systems. *J. Am. Stat. Assoc.* 93:1032–1044, 1998.
19. A Doucet, S Godsill, C Andrieu. On sequential Monte Carlo sampling methods for Bayesian filtering. *Stat. Comput.* 10:197–208, 2000.
20. J Carpenter, P Clifford, P Fearnhead. Improved particle filter for nonlinear problems. *IEE P-Radar Son. Nav.* 146:2–7, 1999.
21. A Doucet, N de Freitas, N Gordon, eds. *Sequential Monte Carlo Methods in Practice.* Heidelberg: Springer-Verlag, 2001.
22. R van der Merwe, A Doucet, N de Freitas, E Wan. The unscented particle filter. Technical Report CUED/F-INFENG/TR380, Cambridge University, Engineering Department, 2000.
23. R Chen, JS Liu. Mixture Kalman filters. *J. Roy. Stat. Soc. B* 62:493–508, 2000.
24. C Andrieu, A Doucet. Particle filtering for partially observed Gaussian state space models. *J. Roy. Stat. Soc. B* 64:827–836, 2002.
25. MK Pitt, N Shephard. Filtering via simulation: auxiliary particle filters. *J. Am. Stat. Assoc.* 94:590–599, 1999.
26. J Nocedal, SJ Wright. *Numerical Optimization.* Heidelberg: Springer-Verlag, 1999.
27. RH Shumway. *Applied Statistical Time Series Analysis.* New York: Prentice Hall, 1988.
28. M Niedzwiecki. *Identification of Time-Varying Processes.* New York: John Wiley & Sons, 2000.

29. CW Groetsch. *Inverse Problems in the Mathematical Sciences.* Braunschweig: Vieweg, 1993.
30. M Vauhkonen, JP Kaipio, E Somersalo, PA Karjalainen. Electrical impedance tomography with basis constraints. *Inv. Probl.* 13:523–530, 1997.
31. M Vauhkonen, D Vadász, PA Karjalainen, E Somersalo, JP Kaipio. Tikhonov regularization and prior information in electrical impedance tomography. *IEEE Trans. Med. Imaging* 17:285–293, 1998.
32. JP Kaipio, V Kolehmainen, E Somersalo, M Vauhkonen. Statistical inversion and Monte Carlo sampling methods in electrical impedance tomography. *Inv. Probl.* 16:1487–1522, 2000.
33. JP Kaipio, V Kolehmainen, M Vauhkonen, E Somersalo. Inverse problems with structural prior information. *Inv. Probl.* 15:713–729, 1999.
34. JP Kaipio, E Somersalo. Nonstationary inverse problems and state estimation. *J. Inv. Ill-Posed Problems* 7:273–282, 1999.
35. D Baroudi, JP Kaipio, E Somersalo. Dynamical electric wire tomography: Time series approach. *Inv. Probl.* 14:799–813, 1998.
36. B Øksendal. *Stochastic Differential Equations.* Heidelberg: Springer-Verlag, 1998.
37. P Kloeden, E Platen. *Numerical Solution of Stochastic Differential Equations.* Heidelberg: Springer-Verlag, 1992.
38. H Pikkarainen. The mathematical formulation of the state evolution equation in electrical process tomography. In: *3rd World Congress on Industrial Process Tomography,* Banff, Canada, 2003, pp. 660–663.
39. PG Ciarlet. *The Finite Element Method for Elliptic Problems.* Amsterdam: North-Holland, 1978.
40. M Renardy, RC Rogers. *An Introduction to Partial Differential Equations.* Heidelberg: Springer-Verlag, 1992.
41. SC Brenner, LR Scott. *The Mathematical Theory of Finite Element Methods.* Heidelberg: Springer-Verlag, 1994.
42. RF Curtain, H Zwart. *An Introduction to Infinite-Dimensional Linear Systems Theory.* Heidelberg: Springer-Verlag, 1995.
43. D Dochain. Contribution to the analysis and control of distributed parameter systems with application to (bio)chemical processes and robotics. Docent thesis, Université Catholique de Louvain, Belgium, 1994.
44. J Tervo, M Vauhkonen, PJ Ronkanen, JP Kaipio. A three-dimensional finite element model for the control of certain nonlinear bioreactors. *Math. Methods Appl. Sci.* 23:357–377, 2000.
45. GF Hewitt. *Measurement of Two Phase Flow Parameters.* London: Academic Press, 1978.
46. T Dyakowski. Tomography in a process system. In: RA Williams, MS Beck, eds., *Process Tomography.* Oxford: Butterworth-Heinemann, 1995.
47. S Levy. *Two-Phase Flow in Complex Systems.* New York: John Wiley & Sons, 1999.
48. A Seppänen, M Vauhkonen, PJ Vauhkonen, E Somersalo, JP Kaipio. State estimation with fluid dynamical evolution models in process tomography: An application to impedance tomography. *Inv. Probl.* 17:467–484, 2001.
49. A Seppänen, M Vauhkonen, PJ Vauhkonen, E Somersalo, JP Kaipio. State estimation with fluid dynamical evolution models in process tomography: EIT application. Technical Report 6/2000, University of Kuopio, Department of Applied Physics, 2000. http://venda.uku.fi/research/IP/publications/deptrep/fluiddyneit.pdf.

50. A Seppänen, M Vauhkonen, E Somersalo, JP Kaipio. State space models in process tomography: Approximation of state noise covariance. *Inv. Prob. Eng.* 9:561–585, 2001.
51. A Seppänen, M Vauhkonen, PJ Vauhkonen, E Somersalo, JP Kaipio. Fluid dynamical models and state estimation in process tomography: Effect due to inaccuracies in flow fields. *J. Electr. Imaging* 10(3):630–640, 2001.
52. A Seppänen, M Vauhkonen, E Somersalo, JP Kaipio. Effects of inaccuracies in fluid dynamical models in state estimation of process tomography. In: H McCann, DM Scott, eds. *Process Imaging for Automatic Control, Proc. SPIE, Vol.* 4188, 2000, pp. 69–80.
53. A Seppänen, M Vauhkonen, PJ Vauhkonen, E Somersalo, JP Kaipio. State estimation in three dimensional impedance imaging: Use of fluid dynamical evolution models. *Proc. 2nd World Congress on Industrial Process Tomography,* Hannover, Germany, 2001, pp. 198–206.
54. A Seppänen, M Vauhkonen, E Somersalo, JP Kaipio. Inference of velocity fields based on tomographic measurements in process industry. *Proc. 4th International Conference on Inverse Problems in Engineering: Theory and Practice,* Angra dos Reis, Rio de Janeiro, Brazil, 2002.
55. A Seppänen, L Heikkinen, T Savolainen, E Somersalo, JP Kaipio. An experimental evaluation of state estimation with fluid dynamical models in process tomography. *Proc. 3rd World Congress on Industrial Process Tomography,* Banff, Canada, 2003, pp. 541–546.
56. PJ Vauhkonen, M Vauhkonen, T Mäkinen, PA Karjalainen, JP Kaipio. Dynamic electrical impedance tomography: Phantom studies. *Inv. Prob. Eng.* 8:495–510, 2000.
57. KY Kim, BS Kim, MC Kim, YJ Lee, M Vauhkonen. Image reconstruction in time-varying electrical impedance tomography based on the extended Kalman filter. *Measur. Sci. Technol.* 12:1032–1039, 2001.
58. KY Kim, SI Kang, MC Kim, S Kim, YJ Lee, M Vauhkonen. Dynamic image reconstruction in electrical impedance tomography with known internal structures. *IEEE Trans. Magn.* 38:1301–1304, 2002.
59. KY Kim, SI Kang, MC Kim, S Kim, YJ Lee, M Vauhkonen. Dynamic electrical impedance tomography with known internal structures. *Inv. Prob. Eng.* 11:1–19, 2003.
60. S Turek. *Efficient Solvers for Incompressible Flow Problems.* Berlin: Springer, 1999.
61. R Chang. *Physical Chemistry with Applications to Biological Systems.* New York: Collier-Macmillan, 1981.
62. E Somersalo, M Cheney, D Isaacson. Existence and uniqueness for electrode models for electric current computed tomography. *SIAM J. Appl. Math.* 52:1023–1040, 1992.
63. M Vauhkonen. Electrical impedance tomography and prior information. Ph.D. thesis, University of Kuopio, Kuopio, Finland, 1997.
64. R Bird, E Stewart, N Lightfoot. *Transport Phenomena.* New York: John Wiley & Sons, 1960.

7 Control Systems

Stephen Duncan, Jari Kaipio, Anna R. Ruuskanen, Matti Malinen, and Aku Seppänen

CONTENTS

7.1 INTRODUCTION

Most control systems attempt to regulate the values of one or more discrete properties, for example, a voltage or a temperature. Sometimes these properties may be "bulk" measurements, such as the average temperature of an object or the mean flow rate of a fluid. These control systems are referred to as lumped parameter systems, and the individual variables that are to be controlled can be represented by a vector that is changing over time. In distributed parameter control systems [1], the property being controlled varies continuously in both space and time, and the aim is to regulate the profile of the property over a particular region of space. The control variable is described as a function of space and time.

It is important to distinguish between a distributed parameter control system, which controls a variable that is distributed in space, and a distributed control system (DCS), which controls a number of discrete variables by using a set of separate controllers that are distributed in space and connected by a communications network. Although the names of the two controllers are similar, they are very different: in the former, it is the control variable that is distributed in space, while in the latter, it is the controllers.

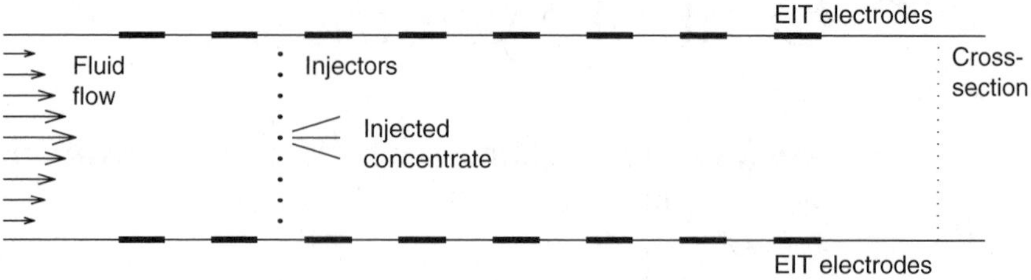

FIGURE 7.1 A cross-section through a system for regulating the concentration of a substance in a fluid that is flowing along a cylindrical pipe.

An example of a distributed parameter control system is the regulation of the concentration of a substance in the fluid flowing along a pipe [2, 3], which was introduced in Chapter 6. Referring to Figure 7.1, the concentration of the substance at point, $\mathbf{r} \in \Omega$, and time, t, can be denoted by $c(\mathbf{r}, t)$, where Ω denotes the region of space inside the pipe. A distributed parameter control system could be used to regulate $c(\mathbf{r}, t)$, so that the concentration profile over a cross-section taken at a given position near the end of the pipe is uniform, even though the concentration entering the pipe is varying as a result of fluctuations in upstream processes. A second example would be minimizing the concentration over the upper half of the pipe at a given point along the pipe, by moving the substance to the lower half, in preparation for a downstream separation process.

Distributed parameter systems also arise in many other industrial applications, for example, any process that involves the control of *flow* of heat or of mass. Processes that involve the control of the flow of heat include the thawing and defrosting of foodstuffs [4], control of surface temperature during welding and metal spraying [5, 6], and thermal camber in metal rolling [7]. Industrial 2D processes that regulate the flow of mass include papermaking [8], the extrusion of plastic film [9], coating and converting processes [10], galvanizing [11], and printing. Distributed parameter control also arises in many chemical processes, such as batch reactors, where it is used in controlling the concentration of a particular chemical through a process such as a catalytic cracker. Here it is necessary to control the flows of both heat and matter (i.e., the chemical) with the additional complication that there is usually more than one species to regulate and the concentration of species changes as the chemicals react [12, 13]. A further layer of complexity is added if there is a flow mechanism that is moving the species through the reactor. The resulting control problem is highly complex, and sophisticated techniques are required to design and implement systems for these processes. Another important class of process is the control of flexible structures [14, 15, 16], such as robotic arms, space structures, slice lips and die lips in paper making, and the active damping of structures during earthquakes.

Four important requirements must be met before a controller can be designed for any of these processes:

Control specification — The aim of the controller has to be specified. For the example in Figure 7.1, where the concentration profile in the fluid flow along a pipe is regulated, one common control specification is to ensure that $c(\mathbf{r}, t)$ matches a desired profile, $d(\mathbf{r}, t)$. It is usually not possible to make $c(\mathbf{r}, t)$ match $d(\mathbf{r}, t)$ throughout Ω, the whole region of the pipe. Instead, the control specification is defined over a region of the pipe, $\Omega^c \subset \Omega$, which for the regulation of the concentration is often the cross-section at a given point along the pipe. For example, in Figure 7.1, Ω^c corresponds to the dotted line at the right-hand end of the pipe. In practice, it is not possible to make $c(\mathbf{r}, t)$ exactly equal to $d(\mathbf{r}, t)$ over Ω^c. Instead, the best that can be done is to minimize some criterion penalizing the discrepancy between the actual and desired variables, such as:

$$S^2(t) = \int_{\Omega^c} | d(\mathbf{r}, t) - c(\mathbf{r}, t) |^2 \, dV \tag{7.1}$$

which is the squared deviation between $c(\mathbf{r}, t)$ and $d(\mathbf{r}, t)$ over Ω^c. For most applications in the process industry, it is usual for the desired profile, $d(\mathbf{r}, t)$, to remain constant or to change relatively infrequently, for example following a set point change. As a result, the aim of the controller is to bring $c(\mathbf{r}, t)$ close to this desired value as quickly as possible following a start-up and then to minimize the squared deviation, $S^2(t)$, in the presence of disturbances. The control specification will therefore include a requirement on the speed of response of the system following a start-up or a set point change, and one way of doing this is to keep $S^2(t)$ small throughout the change.

Actuation mechanism — In order to minimize the cost function, and hence minimize the difference between $c(\mathbf{r}, t)$ and $d(\mathbf{r}, t)$, it is necessary to have an actuation mechanism that will adjust the concentration $c(\mathbf{r}, t)$ along the pipe. For regulating the concentration profile along the pipe, one possible set of actuators would be a series of injectors that could inject extra substance into the flow, which has the effect of locally increasing the concentration. The combined effect of the fluid flow and the diffusion would then redistribute the extra substance, and the aim of the controller is to adjust the injection rate in order to satisfy the control specification. For this reason, the injection rate is referred to as the control (or input) variable. A number of practical difficulties accompany the use of injectors as the actuation mechanism. The main problem is that injectors can put extra substance into the flow to increase the concentration, but they cannot remove it in order to decrease the concentration. This makes the actuation mechanism nonlinear; methods for handling this constraint will be discussed. A second problem is that it may not be possible to adjust the injection rate sufficiently quickly to achieve the desired speed of response. In practice, actuators are usually designed so that their dynamic response (i.e., their dynamic bandwidth) is faster than dynamic response of process. For the process considered here, the dynamic response of the process is determined by the convection rate of the flow and the diffusion of the substance within that flow. It is likely that a flow valve will control the injection rate, and it needs to be possible to adjust the valve faster than the process dynamics. If this is not the case, then the response of the actuator should be included in the controller design.

Process model—To determine how the injection rate should be adjusted to achieve the desired concentration profile, it is necessary to be able to predict the effect of a change of injection rate on concentration. This is usually done by using a model of the process, and because the concentration profile $c(\mathbf{r}, t)$ evolves with time, the model must be dynamic. Most modern control designs are based on a mathematical model. When a mathematical model is used as the starting point in the design, we refer to model-based control design to stress the difference to ad hoc controllers such as PID controllers. For relatively simple processes, such as the concentration profile in a steady flow along a straight cylindrical pipe, the model can be derived from a physical understanding of the process. For more complex processes, such as flow of pulp from a headbox on a paper machine [8], it is not always possible to derive an adequate physical model, and the dynamic response has to be identified from a series of identification experiments. The controller uses a model of the process to determine the input variable that should be applied to the actuators in order to achieve the control specification.

Sensing system—Although a process model is required to predict the effect that adjusting the flow rate to the injectors has on the concentration profile, in practice, any model will be incomplete in the sense that it cannot model all of the disturbances entering the process. In addition, there are likely to be inaccuracies in the model, so it is necessary to include a method for sensing the actual concentration profile within the pipe. The estimation of the concentration within (part of the pipeline) using all available information on the process, including the uncertainties, is referred to as *state estimation*, which was reviewed in Chapter 6. Industrial process tomography provides an efficient, noninvasive method of sensing the concentration profile within pipeline flow, and its potential as a sensor within a feedback loop has been recognized for some time [17]. However, it is only recently that controllers have been proposed that include tomographic sensors [2, 3].

This chapter describes how these four components can be combined to derive a controller for regulating the concentration profile within the flow.

7.2 MODELLING THE PROCESS

Because distributed parameter systems describe properties that vary in both space and time, they are usually modelled as partial differential equations (PDEs). The evolution of $c(\mathbf{r}, t)$, the concentration of the substance in a fluid flow, can be described by the convection–diffusion equation,

$$\frac{\partial c(\mathbf{r},t)}{\partial t} + \mathbf{v}(\mathbf{r})\cdot\nabla c(\mathbf{r},t) = \nabla\cdot(\kappa\nabla c(\mathbf{r},t)) + f(\mathbf{r},t) \qquad (7.2)$$

where $\mathbf{v}(\mathbf{r})$ is the (steady) velocity profile of the flow, κ is the diffusion coefficient, and $f(\mathbf{r}, t)$ is due to the extra substance being injected into the flow by the actuators. Associated with this PDE are a series of boundary conditions:

$$\begin{aligned} c(\mathbf{r}, 0) &= c_0(\mathbf{r}) \\ c(\mathbf{r}, t) &= c_{\text{inlet}}(t), \quad \mathbf{r} \in \partial\Omega_{\text{inlet}} \\ \hat{\mathbf{n}} \cdot \nabla c(\mathbf{r}, t) &= 0, \quad \mathbf{r} \in (\partial\Omega_{\text{inlet}} \not\subset \partial\Omega) \end{aligned} \tag{7.3}$$

where $c_{\text{inlet}}(t)$ describes the time-varying concentration profile at the inlet of the pipe (denoted $\partial\Omega_{\text{inlet}}$), and $\hat{\mathbf{n}}$ is the outward normal unit vector at the walls of the pipe.

Although it is possible to design a controller directly from the PDE description of the process [1], one common approach is to base the controller design on $c^h(\mathbf{r}, t)$, a finite dimensional approximation to the full concentration profile,

$$c(\mathbf{r}, t) \approx c^h(\mathbf{r}, t) = \sum_{n=1}^{N} q_n(t)\phi_n(\mathbf{r}) \tag{7.4}$$

where the concentration variable has been expanded in terms of a finite number of spatial basis functions, $\phi_n(\mathbf{r})$, with time-varying coefficients, $q_n(t)$. There are a number of possibilities for the basis functions, but the most natural choice for the convection–diffusion case considered here is piecewise linear finite elements, primarily because this basis is used in the reconstruction of the concentration profiles from the tomographic sensor. Using finite elements also has the useful property that the overall error in the finite dimensional approximation

$$\int_{\Omega} | c(\mathbf{r}, t) - c^h(\mathbf{r}, t) |^2 \, dV \tag{7.5}$$

can be bounded *a priori* [18] and that if $\mathbf{r}_1, \mathbf{r}_2, \ldots, \mathbf{r}_N$ denote the locations of the nodes of N finite elements, then $c^h(\mathbf{r}_n, t) = q_n(t)$.

Following the approach described in Chapter 6, it is possible to substitute the approximation in Equation 7.4 into the PDE in Equation 7.3 to obtain a finite dimensional state space model of the form

$$\dot{\mathbf{q}}(t) = \mathbf{F}^c\mathbf{q}(t) + \mathbf{B}^c\mathbf{u}(t) + \mathbf{s}(t) + \mathbf{w}^s(t) \tag{7.6}$$

where $\mathbf{q}(t) \in \Re^N$ is a state vector containing $q_n(t)$, the values of the concentrations at each of the N nodes; $\mathbf{u}(t) \in \Re^M$ is a vector of the control inputs to the system, which correspond to the amount of extra substance being injected at each of the M injectors; $\mathbf{s}(t) \in \Re^N$ denotes the concentration profile entering at the start of the pipe; and $\mathbf{w}^s(t) \in \Re^N$ represents the state noise. $\mathbf{F}^c \in \Re^{N\times N}$ and $\mathbf{B}^c \in \Re^{N\times M}$ are the state evolution and input matrices, respectively. The PDE in Equation 7.3 has been reduced to a set of first-order ordinary differential equations by expressing the concentration profile, $c(\mathbf{r}, t)$, in terms of the finite dimensional approximation in Equation 7.4. In practice, control systems are usually operated in "sample and hold" mode, where measurements are taken at discrete time intervals and these measurements are then used to determine the inputs, $\mathbf{u}(t)$. These values of the inputs are then applied to the actuators, and the inputs are held fixed until the next measurements are received. Under these circumstances, the ordinary differential equations in Equation 7.6 can be expressed as a set of difference equations describing the evolution of the state between successive sample times:

$$\mathbf{q}_{t+1} = \mathbf{F}\mathbf{q}_t + \mathbf{B}\mathbf{u}_t + \mathbf{s}_t + \mathbf{w}_t^s \tag{7.7}$$

where $\mathbf{q}_t$, $\mathbf{u}_t$, $\mathbf{s}_t$, and $\mathbf{w}_t^s$ denote the values of the state, the input to the actuators, the entry concentration, and the state noise respectively, sampled at time t with the time between samples being taken as unity. In the sequel, the subscript t refers to time index (with respect to the sampling) instead of actual physical time (in seconds). The matrices $\mathbf{F}$ and $\mathbf{B}$ are obtained by integrating Equation 7.6 over the period t to $t+1$.

The model is illustrated in Figure 7.2, which shows the evolution of the concentration profile along the pipe. The figure consists of a series of "snapshots" taken at successive time intervals, where the flow is moving from left to right. The concentration profile at the start of the pipe, $\mathbf{s}_t$ (i.e., at the left-hand end), varies over time, and the plots show how the concentration profile varies as the flow moves along the pipe. It can be seen that the effect of the diffusion "smoothes out" the initial concentration profile. The aim of the control system is to inject extra substance into the stream in order to achieve the desired concentration profile over the region Ω^c.

7.3 FEEDBACK CONTROL

The aim of the control system is to ensure that the actual concentration profile, $c(\mathbf{r}, t)$, matches as closely as possible the desired profile, $d(\mathbf{r}, t)$, over the region Ω^c, as stated in Equation 7.1. If $d(\mathbf{r}, t)$ is expressed in terms of the same finite dimensional basis function expansion as $c(\mathbf{r}, t)$, as given in Equation 7.4, so that

$$d(\mathbf{r}, t) \approx \sum_{n=1}^{N} q_n^d(t)\phi_n(\mathbf{r}) \tag{7.8}$$

then the minimization criterion in Equation 7.1 becomes

$$\begin{aligned} S^2(t) &= \int_{\Omega^c} |d(\mathbf{r}, t) - c(\mathbf{r}, t)|^2 \, dV \\ &= \int_{\Omega^c} \left[\sum_{n=1}^{N} \left(q_n^d(t) - q_n(t) \right) \phi_n(\mathbf{r}) \right] \left[\sum_{n'=1}^{N} \left(q_{n'}^d(t) - q_{n'}(t) \right) \phi_{n'}(\mathbf{r}) \right] dV \\ &= (\mathbf{q}^d(t) - \mathbf{q}(t))' \mathbf{Q} (\mathbf{q}^d(t) - \mathbf{q}(t)) \end{aligned} \tag{7.9}$$

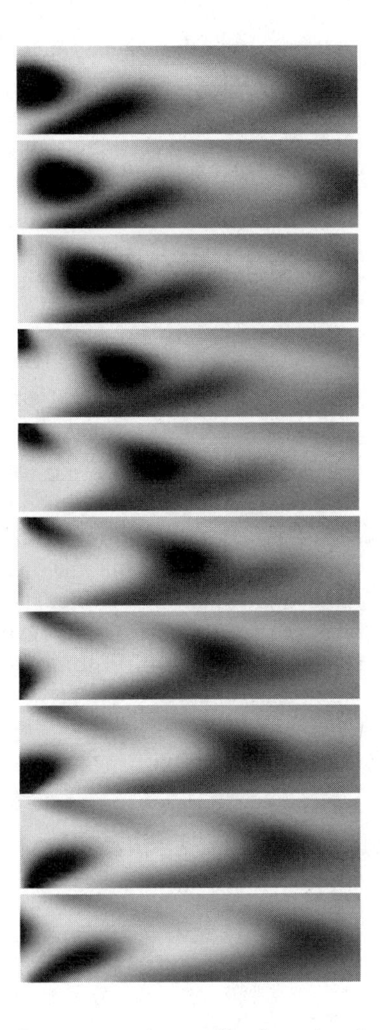

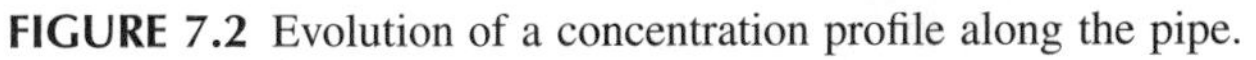

FIGURE 7.2 Evolution of a concentration profile along the pipe.

Since the aim is to make $c(\mathbf{r}, t)$ match $d(\mathbf{r}, t)$ over a subregion of the pipe, $\Omega^c \subset \Omega$ (typically the cross-section at a particular point along the pipe), then $\mathbf{q}^d(t)$ will only be defined at the $N_1 \leq N$ nodes associated with the finite elements lying within Ω^c. As a result, $\mathbf{Q} \in \Re^{N \times N}$ is a positive semidefinite, sparse matrix with N_1 nonvanishing rows and columns associated with these nodes, given by

$$[\mathbf{Q}]_{n,n'} = \int_{\Omega^c} \phi_n(\mathbf{r})\phi_{n'}(\mathbf{r})\,dV \tag{7.10}$$

since only those integrals in which the supports of both basis functions intersect Ω^c are nonvanishing.

For the sample and hold, discrete time model in Equation 7.7, the minimization criterion becomes

$$S_t^2 = \left(\mathbf{q}_t^d - \mathbf{q}_t\right)' \mathbf{Q}\left(\mathbf{q}_t^d - \mathbf{q}_t\right) \tag{7.11}$$

In practice, the design of a control system has to be a trade-off between minimizing the difference between the actual and desired concentration profiles and avoiding excessive control inputs. For this reason, S_t^2 is usually augmented to

$$S_t^2 = \left(\mathbf{q}_t^d - \mathbf{q}_t\right)' \mathbf{Q}\left(\mathbf{q}_t^d - \mathbf{q}_t\right) + \mathbf{u}_t^T \mathbf{R} \mathbf{u}_t \tag{7.12}$$

where the positive definite matrix, $\mathbf{R} \in \Re^{M \times M}$, defines the weighting that is applied to the control inputs. The magnitude of the elements of this matrix relative to the elements of $\mathbf{Q}$ is a parameter of the control design, and making the elements of $\mathbf{R}$ small relative to the elements of $\mathbf{Q}$ means that the control system attempts to make the actual concentration profile match the desired profile closely — although there is the possibility of large control inputs, which may exceed the operating range of the actuators. By contrast, increasing the size of $\mathbf{R}$ reduces the magnitude of the control signals at the expense of larger errors between the actual and desired profiles.

One approach to formalizing the design problem is to define a variable $\mathbf{z}_t$,

$$\mathbf{z}_t = \begin{bmatrix} \mathbf{Q}^{1/2}\left(\mathbf{q}_t^d - \mathbf{q}_t\right) \\ \mathbf{R}^{1/2}\mathbf{u}_t \end{bmatrix} = \begin{bmatrix} \mathbf{Q}^{1/2} \\ \mathbf{0} \end{bmatrix}\left(\mathbf{q}_t^d - \mathbf{q}_t\right) + \begin{bmatrix} \mathbf{0} \\ \mathbf{R}^{1/2} \end{bmatrix}\mathbf{u}_t \tag{7.13}$$

so that

$$S_t^2 = \mathbf{z}_t' \mathbf{z}_t \tag{7.14}$$

The signal $\mathbf{z}_t$ is referred to as the objective signal, and the aim of the controller is to determine the control inputs, $\mathbf{u}_t$, that will minimize S_t^2 in the presence of exogenous inputs, such as unknown disturbances and the demand signal. If $\mathbf{s}_t$, the concentration profile at the inlet to the pipe, were known at every time step, then in the absence of any state noise, the state at all future times, $t > t_1$, can be predicted by using Equation 7.7, provided that the parameters of the model are known accurately and the state $\mathbf{q}_{t_1}$ is known at any time, t_1. Under these circumstances, it would be possible to determine suitable future control inputs, $\mathbf{u}_t$, over the interval $t > t_1$ that would minimize S_t^2. However, in practice, neither the model nor the concentration profile is known accurately, and there are significant disturbances within the process. For this reason, it is necessary to take measurements on the process in order to determine the actual (rather than the predicted) state of the process. As described in preceding chapters, tomographic sensing systems provide a noninvasive method of obtaining (partial) information on the concentration profile within the flow. As shown in Chapter 6, with EIT, the measurements are the voltages between the electrodes when a current is applied to each pair in turn. If the voltages are stacked into a vector, $\mathbf{V}_t \in \Re$, where P is the number of electrode pairs, then

$$\mathbf{V}_t = \mathbf{U}_{0,t} + \mathbf{J}_t[\mathbf{q}_t - \mathbf{q}_0] + \mathbf{w}_t^m \tag{7.15}$$

which is derived from the observation model by linearizing about the concentration profile $c_0(\mathbf{r})$, with $\mathbf{q}_0$ being the states associated with the basis function expansion of $c_0(\mathbf{r})$, $\mathbf{U}_{0,t}$ being the result of the mapping from concentration to voltage associated with $c_0(\mathbf{r})$, and $\mathbf{J}_t$ being the Jacobian of the mapping. The noise associated with the measurement is represented by $\mathbf{w}_t^m \in \Re^P$. Both $\mathbf{U}_{0,t}$ and $\mathbf{q}_0$ are known at time t, so Equation 7.15 can be rearranged to

$$\mathbf{y}_t = \mathbf{C}_t\mathbf{q}_t + \mathbf{w}_t^m \tag{7.16}$$

where $\mathbf{y}_t = \mathbf{V}_t - \mathbf{U}_{0,t} + \mathbf{J}_t\mathbf{q}_0$ and $\mathbf{C}_t = \mathbf{J}_t$. The elements of $\mathbf{y}_t$ can be regarded as the "measurements" obtained by the EIT sensor, and Equation 7.16 reflects the way that these measurements are related to the underlying states, $\mathbf{q}_t$. The overall aim of the control system is to extract information about the current states from these measurements and then to determine the input signals, $\mathbf{u}_t$, that will satisfy the performance specification in Equation 7.14.

Bringing together the state evolution equation in Equation 7.7, the expression for the objective signal in Equation 7.13, and the measurement equation in Equation 7.16,

$$\begin{aligned}
\mathbf{q}_{t+1} &= \mathbf{F}\mathbf{q}_t + \mathbf{B}\mathbf{u}_t + \mathbf{s}_t + \mathbf{w}_t^s \\
\mathbf{z}_t &= \begin{bmatrix} \mathbf{Q}^{1/2} \\ \mathbf{0} \end{bmatrix}\left(\mathbf{q}_t^d - \mathbf{q}_t\right) + \begin{bmatrix} \mathbf{0} \\ \mathbf{R}^{1/2} \end{bmatrix}\mathbf{u}_t \\
\mathbf{y}_t &= \mathbf{C}_t\mathbf{q}_t + \mathbf{w}_t^m
\end{aligned} \tag{7.17}$$

These expressions can be written more compactly by combining the exogenous signals into a single vector, $\mathbf{w}_t$:

$$\mathbf{w}_t = \begin{bmatrix} \mathbf{q}_t^d \\ \mathbf{s}_t \\ \mathbf{w}_t^s \\ \mathbf{w}_t^m \end{bmatrix} \tag{7.18}$$

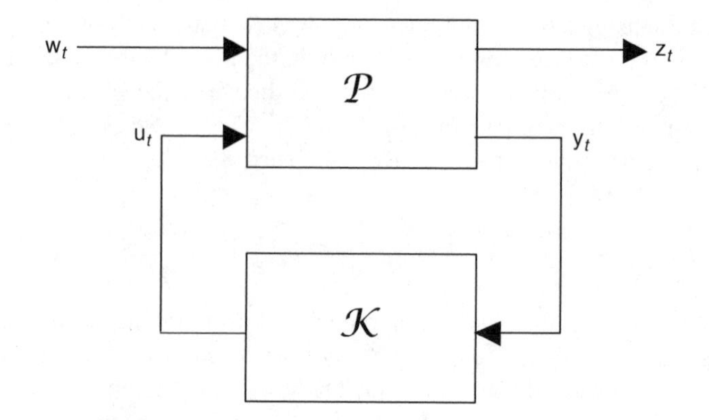

FIGURE 7.3 The structure of the generalized plant.

leading to a generalized plant model of the process [19, 20, 21]:

$$\begin{bmatrix} \mathbf{q}_{t+1} \\ \mathbf{z}_t \\ \mathbf{y}_t \end{bmatrix} = \begin{bmatrix} \mathbf{A}^p & \mathbf{B}_1^p & \mathbf{B}_2^p \\ \mathbf{C}_1^p & \mathbf{D}_{11}^p & \mathbf{D}_{12}^p \\ \mathbf{C}_2^p & \mathbf{D}_{21}^p & \mathbf{D}_{22}^p \end{bmatrix} \begin{bmatrix} \mathbf{q}_t \\ \mathbf{w}_t \\ \mathbf{u}_t \end{bmatrix} \tag{7.19}$$

where

$$\begin{aligned} &\mathbf{A}^p = \mathbf{F} && \mathbf{B}_1^p = \begin{bmatrix} \mathbf{0} & \mathbf{I}_N & \mathbf{I}_N & \mathbf{0} \end{bmatrix} && \mathbf{B}_2^p = \mathbf{B} \\ &\mathbf{C}_1^p = \begin{bmatrix} -\mathbf{Q}^{1/2} \\ \mathbf{0} \end{bmatrix} && \mathbf{D}_{11}^p = \begin{bmatrix} \mathbf{Q}^{1/2} & \mathbf{0} & \mathbf{0} & \mathbf{0} \\ \mathbf{0} & \mathbf{0} & \mathbf{0} & \mathbf{0} \end{bmatrix} && \mathbf{D}_{12}^p = \begin{bmatrix} \mathbf{0} \\ \mathbf{R}^{1/2} \end{bmatrix} \\ &\mathbf{C}_2^p = \mathbf{C}_t && \mathbf{D}_{21}^p = \begin{bmatrix} \mathbf{0} & \mathbf{0} & \mathbf{0} & \mathbf{I}_P \end{bmatrix} && \mathbf{D}_{22}^p = \mathbf{0} \end{aligned} \tag{7.20}$$

The structure of the generalized plant is shown in Figure 7.3, and the optimal control problem is to design a controller that uses the measurements, $\mathbf{y}_t$, to determine the control inputs, $\mathbf{u}_t$, that minimize $\mathbf{z}_t$, while the plant is subjected to the exogenous variables, $\mathbf{w}_t$. In essence, this is minimizing the closed loop mapping from $\mathbf{w}_t$ to $\mathbf{z}_t$. The next section describes two approaches to solving this problem.

7.4 CONTROL DESIGN

7.4.1 CONTROLLABILITY AND OBSERVABILITY

Before starting the design of a control system, it is necessary to consider the controllability and the observability of the process. Within the context of the pipeline flow process being considered here, controllability can be thought of as the ability of the actuators (i.e., the injectors) to change the (discretized) state, that is, the concentration at all nodes in the finite element node mesh. If the rate of diffusion of the substance within the flow is slow compared to the convection of the process, then the concentration at any point upstream of injectors cannot be affected by the control system because any adjustments to the concentration by the injectors will be moved downstream by the flow. In addition, the level of control over the concentration is limited by both the spatial and dynamic bandwidths of the actuators, which in turn are determined by the spatial "footprint" of the injectors and the speed at which the flow from the injectors can be adjusted. By contrast, it is possible for the sensing system to observe the concentration at any state associated with a node that is either upstream or downstream of the sensor. Although the concentration profile can be measured as the flow passes the sensor, and this measurement can be used to predict the concentration profile at a point further downstream, there may be disturbances or modelling errors that affect the flow downstream of the sensor which cannot be directly measured. Hence, only the states associated with nodes that lie downstream of the first injection point and upstream of the last sensing location are both controllable and observable. For the flow system, an additional practical constraint is that the injectors can only add substance to the flow; they cannot remove it. The effect of this constraint will be discussed in more detail later. For the moment, the controllability of an unconstrained process will be considered.

Within control systems theory, the concept of controllability is defined more formally as the existence of a sequence of control inputs, $\mathbf{u}_t$, that will take the system from an initial state, $\mathbf{q}_0$, to the origin, $\mathbf{q}_t = \mathbf{0}$, within a finite time [22]. If there is a fixed amount of substance in the flow at the start of the pipe, then the concentration at a point upstream of the injectors cannot be reduced to zero because the injectors have no effect on concentration at this point, and, hence, states associated with nodes upstream of the injectors are uncontrollable. If $\mathbf{s}_t = \mathbf{0}$ throughout, and a sequence of N control inputs is applied to a system with N states, then [23]

$$\begin{aligned}\mathbf{q}_N &= \mathbf{F}^N\mathbf{q}_0 + \mathbf{F}^{N-1}\mathbf{B}\mathbf{u}_0 + \mathbf{F}^{N-2}\mathbf{B}\mathbf{u}_1 + \cdots + \mathbf{B}\mathbf{u}_{N-1} \\ &= \mathbf{F}^N\mathbf{q}_0 + \Gamma^c\mathbf{U}\end{aligned} \tag{7.21}$$

where

$$\Gamma^c = [\mathbf{B} \quad \mathbf{FB} \quad \cdots \quad \mathbf{F}^{N-2}\mathbf{B} \quad \mathbf{F}^{N-1}\mathbf{B}] \quad \mathbf{U} = \begin{bmatrix} \mathbf{u}_{N-1} \\ \mathbf{u}_{N-2} \\ \vdots \\ \mathbf{u}_1 \\ \mathbf{u}_0 \end{bmatrix} \tag{7.22}$$

Provided that Γ^c, the controllability matrix, has rank N, it is possible to find N control inputs, $\mathbf{u}_0, \mathbf{u}_1, \ldots, \mathbf{u}_{N-2}, \mathbf{u}_{N-1}$, that will take the states of the system from $\mathbf{q}_0$ to $\mathbf{q}_N = \mathbf{0}$. This means that one can test the controllability of a system by examining the rank of its controllability matrix. By a similar argument, a time invariant system will be observable if its observability matrix has full rank [22]. However, for the process being considered here, the matrix $\mathbf{C}_t$ in the measurement equation (Equation 7.16) is time varying, so the test for observability is defined in terms of a requirement on the observability grammian having rank N [22].

Even though all the states may be controllable (in the sense defined), this does not necessarily mean that all controllable states can be regulated at every time instant. The problem is that controllability requires that the states of the system can be brought from one given state to another state (which is taken to be zero) within N steps. Although the desired state is reached at the end of N steps, the value of the state at time steps $t = 1, 2, \ldots, N-1$ may be unacceptable. Since it is usual to require that the states be controlled at every time step, a further restriction is required on the number of states that can be regulated. One way of seeing this is to consider the case where the aim is to apply control inputs that will hold the state at a given steady value, $\mathbf{q}^d$, so that

$$\mathbf{z}^* = \mathbf{Q}^{1/2}(\mathbf{q}^d - \mathbf{q}^*) \tag{7.23}$$

where $\mathbf{q}^*$ and $\mathbf{z}^*$ are the steady values of the state and the objective. In order to make $\mathbf{z}^* = \mathbf{0}$ for all times, then in the absence of any disturbances, the steady control inputs, $\mathbf{u}^*$, must satisfy

$$\mathbf{q}^* = \mathbf{F}\mathbf{q}^* + \mathbf{B}\mathbf{u}^* + \mathbf{s}^* \tag{7.24}$$

or

$$(\mathbf{I} - \mathbf{F})\mathbf{q}^* = \mathbf{B}\mathbf{u}^* + \mathbf{s}^* \tag{7.25}$$

Assuming that the matrix $\mathbf{B}$ has full rank, it will (generally) only be possible to find a solution, $\mathbf{u}^*$, that makes $\mathbf{z}^* = \mathbf{0}$ when the number of actuators, M, is greater than or equal to the number of states, N. In practice, the number of controllable and observable states in the finite element model far exceeds the number of injectors, so it is only possible to control either a subset of M states, or alternatively, M linear combinations of states. For example, it is possible to regulate the average concentration over M regions of the pipe, where the averages are obtained by taking the weighted sum of the states within each region. To achieve this, the $\mathbf{Q}$ matrix in the objective function, $\mathbf{z}_t$, is restricted to have rank M. If the rank of $\mathbf{Q}$ exceeds M, then in general it will not be possible to make the objective, $\mathbf{z}_t$, equal to zero, even in steady state. Under these circumstances, the best that can be done is to minimize the magnitude of the objective at each time step, where the magnitude is usually defined in terms of the norm, $\mathbf{z}_t'\mathbf{z}_t$. Using a similar approach, if $P < N$ measurements are taken by the EIT sensor, it will only be possible to determine the value of P states exactly. In practice, as described in Chapter 6, the states are estimated by using an observer, where the aim is not to reconstruct P states exactly, but instead to estimate all N states by minimizing the error between actual states and the estimates at all nodes.

7.4.2 Linear Quadratic Gaussian Control

Although the aim of the controller is to determine the control inputs, $\mathbf{u}_t$, that will ensure that $\mathbf{q}_t$ remains close to $\mathbf{q}_t^d$ (and hence, that $c(\mathbf{r}, t)$ is close to $d(\mathbf{r}, t)$), in practice, the target profile remains fixed, or, at most, changes infrequently. As a result, the controller can be considered as a regulator, where the aim is to maintain the $\mathbf{q}_t$ close to $\mathbf{q}_t^d$ in the presence of disturbances, such as $\mathbf{s}_t$, the term associated with the unknown concentration profile at the start of the pipe, and $\mathbf{w}_t^s$, the state noise. One possible approach to designing a control law is to choose the control inputs, $\mathbf{u}_t$, to minimize the expected value of $\mathbf{z}_t'\mathbf{z}_t$:

$$E[\mathbf{z}_t'\mathbf{z}_t] = \lim_{T\to\infty}\frac{1}{T}\sum_{t=1}^{T}\mathbf{z}_t'\mathbf{z}_t \tag{7.26}$$

Given a set of measurements, $\mathbf{y}_t$, the linear quadratic Gaussian (LQG) controller generates the inputs $\mathbf{u}_t$ that will minimize $E[\mathbf{z}_t'\mathbf{z}_t]$, for the case where ($a$) the disturbances, $\mathbf{w}_t^s$ and $\mathbf{w}_t^m$, are Gaussian, white noise sequences and (b) the reference state, $\mathbf{q}_t^d$, is zero. It is often reasonable to assume that the underlying disturbances are Gaussian random signals, which are not correlated in time, so that

$$E\left[\mathbf{w}_t^s\left(\mathbf{w}_{t+\tau}^s\right)'\right] = \begin{cases}\mathbf{0} & \text{for} \quad \tau \neq 0\\ \mathbf{C}^s & \text{for} \quad \tau = 0\end{cases} \tag{7.27}$$

$$E\left[\mathbf{w}_t^m\left(\mathbf{w}_{t+\tau}^m\right)'\right] = \begin{cases}\mathbf{0} & \text{for} \quad \tau \neq 0\\ \mathbf{C}^m & \text{for} \quad \tau = 0\end{cases} \tag{7.28}$$

In the following, we also assume that the underlying state and measurement noises are uncorrelated, so that $E[\mathbf{w}_t^s(\mathbf{w}_{t+\tau}^m)'] = 0$ for all τ.

It is less straightforward to accommodate the requirement that the reference state, $\mathbf{q}_t^d$, is zero. One approach, which is discussed below but not entirely satisfactory, is to redefine the state, $\mathbf{q}_t$ [24, 2]. Assume that a steady state control input, $\mathbf{u}^*$, is applied to the actuators and that this generates a steady state response, $\mathbf{q}^*$:

$$\begin{aligned}\mathbf{q}^* &= \mathbf{F}\mathbf{q}^* + \mathbf{B}\mathbf{u}^* + \mathbf{s}^*\\ &= (\mathbf{I}_N - \mathbf{F})^{-1}(\mathbf{B}\mathbf{u}^* + \mathbf{s}^*)\end{aligned} \tag{7.29}$$

where $\mathbf{s}^*$ is taken to be the known steady term associated with the concentration profile at the start of the pipe. In practice, this concentration profile will be neither steady nor known, but by considering $\mathbf{s}_t$ as a Gaussian random signal whose distribution is known, one can set $\mathbf{s}^*$ to $E[\mathbf{s}_t]$ and incorporate the random variations about this expected level into the state noise, $\mathbf{w}_t^s$. The corresponding steady state value of $\mathbf{z}_t$ is

$$\begin{aligned}\mathbf{z}^* &= \mathbf{Q}^{1/2}(\mathbf{q}^d - \mathbf{q}^*) \\ &= \mathbf{Q}^{1/2}\left(\mathbf{q}^d - (\mathbf{I}_N - \mathbf{F})^{-1}(\mathbf{B}\mathbf{u}^* + \mathbf{s}^*)\right)\end{aligned} \tag{7.30}$$

The optimal steady state input is the value of $\mathbf{u}^*$ that minimizes $\mathbf{z}^*$, and given this value of $\mathbf{u}^*$, together with the corresponding value of $\mathbf{q}^*$ obtained from Equation 7.29, the time-varying input and the state can be redefined in terms of these steady state values:

$$\begin{aligned}\mathbf{u}_t &= \mathbf{u}^* + \bar{\mathbf{u}}_t \\ \mathbf{q}_t &= \mathbf{q}^* + \bar{\mathbf{q}}_t\end{aligned} \tag{7.31}$$

The aim of the controller is now to drive $\bar{\mathbf{q}}_t$ to zero, so that $\mathbf{q}_t$ approaches the optimal state, $\mathbf{q}^*$. The steady state (infinite time horizon) LQG state feedback law [20] that achieves this aim is

$$\bar{\mathbf{u}}_t = -\mathbf{K}\bar{\mathbf{q}}_t \tag{7.32}$$

where $\mathbf{K}$ is the state feedback gain matrix

$$\mathbf{K} = (\mathbf{R} + \mathbf{B}'\mathbf{S}\mathbf{B})^{-1}\mathbf{B}'\mathbf{S}\mathbf{F} \tag{7.33}$$

with $\mathbf{S}$ being the solution of the discrete time, algebraic Riccati equation

$$\mathbf{S} = \mathbf{F}'\mathbf{S}\mathbf{F} + \mathbf{Q} - \mathbf{F}'\mathbf{S}\mathbf{B}(\mathbf{R} + \mathbf{B}'\mathbf{S}\mathbf{B})^{-1}\mathbf{B}'\mathbf{S}\mathbf{F} \tag{7.34}$$

By substituting for $\bar{\mathbf{u}}_t$ and $\bar{\mathbf{q}}_t$ from Equation 7.31 into Equation 7.32, the state feedback law becomes

$$\mathbf{u}_t = \mathbf{u}^* - \mathbf{K}(\mathbf{q}_t - \mathbf{q}^*) \tag{7.35}$$

In order to implement this control law, it is necessary to estimate the states, $\mathbf{q}_t$, by using the measurements from the EIT sensor. If $\hat{\mathbf{q}}_{t|t}$ denotes the expected value of $\mathbf{q}_t$, based on the voltage measurements taken up to time t, so that

$$\hat{\mathbf{q}}_{t|t} = E[\mathbf{q}_t \mid \mathbf{V}_k\,; k = 1, 2, \ldots, t] \tag{7.36}$$

then following the approach described in the previous chapter, $\hat{\mathbf{q}}_{t|t}$ can be computed recursively by using a Kalman filter:

$$\begin{aligned}
\hat{\mathbf{q}}_{t|t-1} &= \mathbf{F}\hat{\mathbf{q}}_{t-1|t-1} + \mathbf{B}\mathbf{u}_t + \mathbf{s}^* \\
\mathbf{P}_{t|t-1} &= \mathbf{F}\mathbf{P}_{t-1|t-1}\mathbf{F}' + \mathbf{C}^s \\
\mathbf{L}_t &= \mathbf{P}_{t|t-1}\mathbf{J}_t'\left(\mathbf{J}_t\mathbf{P}_{t|t-1}\mathbf{J}_t' + \mathbf{C}^m\right)^{-1} \\
\mathbf{P}_{t|t} &= (\mathbf{I} - \mathbf{L}_t\mathbf{J}_t)\mathbf{P}_{t|t-1} \\
\hat{\mathbf{q}}_{t|t} &= \hat{\mathbf{q}}_{t|t-1} + \mathbf{L}_t(\mathbf{y}_t - \hat{\mathbf{y}}_t)
\end{aligned} \tag{7.37}$$

where $\mathbf{P}_{t|t}$ denotes the covariance of $\hat{\mathbf{q}}_{t|t}$. The term $\mathbf{y}_t - \hat{\mathbf{y}}_t$ in the final equation represents the difference between the measured output, $\mathbf{y}_t$, and the estimated output, $\hat{\mathbf{y}}_t = \mathbf{C}_t\hat{\mathbf{q}}_{t|t-1}$. By using the definition of $\mathbf{y}_t$ in Equation 7.15 and Equation 7.16, one can express this in terms of the voltage measurements taken by the EIT sensor as

$$\mathbf{y}_t - \hat{\mathbf{y}}_t = \mathbf{V}_t - \mathbf{U}_{0,t} - \mathbf{J}_t(\hat{\mathbf{q}}_{t|t-1} - \mathbf{q}_0) \tag{7.38}$$

The state estimate, $\hat{\mathbf{q}}_{t|t}$, can be used in the state feedback law to give

$$\mathbf{u}_t = \mathbf{u}^* - \mathbf{K}(\hat{\mathbf{q}}_{t|t} - \mathbf{q}^*) \tag{7.39}$$

We note that there is another (observer) Riccati type matrix difference equation for which the recursions (Equation 7.37) yield a recursive solution. If $\mathbf{J}_t$ were time-independent (corresponding to using a single current pattern), a steady state solution for $\mathbf{P}_{t|t}$ would exist so that the Kalman gain, $\mathbf{L}_t$, could be computed off-line. However, when the steady state does not exist, the Kalman recursions in Equation 7.37 have to be computed on-line.

The results of applying this control law are shown in Figure 7.4, where three sequences of plots show the evolution of concentration profile along the pipe. Figure 7.4a shows the evolution of the substance along the pipe in the absence of any controller. As a result of the varying concentration entering the pipe (left-hand end), the concentration at the end of the pipe (right-hand end) is not constant. Figure 7.4b shows the concentration injected by the control system, and Figure 7.4c shows the combined effect of the original evolution of concentration plus the concentration due to the injectors. The profile at the right-hand end of the pipe is significantly more uniform than the uncontrolled profile. This is confirmed by Figure 7.5 and Figure 7.6, which show respectively the evolution of the maximum and minimum concentrations at the right-hand end of the pipe and the mean squared deviation from the desired concentration profile. In both cases, the results with and without control are shown; the effect of the control reduces the variation in concentration and at the same time maintains the mean concentration close to the desired value.

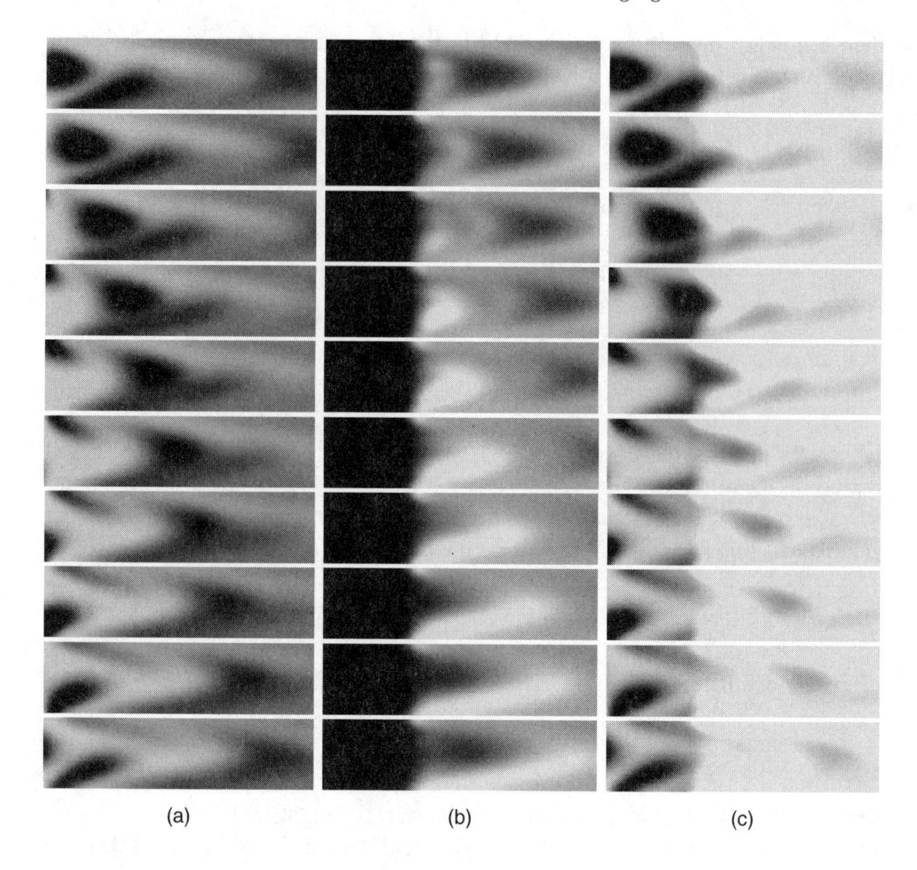

FIGURE 7.4 (a) The evolution of concentration without control, (b) the evolution of controlled injections into the flow, and (c) the controlled concentration evolution.

7.4.3 $\mathcal{H}_\infty$ Control

Some of the assumptions behind LQG control design are restrictive, in the sense that it is assumed that the underlying disturbances have a Gaussian distribution and that the state is controlled to zero. A nonzero reference state was incorporated by calculating the steady state input, $\mathbf{u}^*$, which generates the required state, $\mathbf{q}^d$, and then using the controller to drive the state of the system to the corresponding fixed state, $\mathbf{q}^*$. This approach essentially includes a feed-forward element in the controller, and it relies on the state evolution model of the process in Equation 7.7 being accurate. In practice, there will be errors in this model, which can mean that the controller will drive the state to an incorrect value. A more fundamental problem is that it is not possible to guarantee that the controller will remain stable in the presence of errors between the actual process and the model [19]. An alternative approach that overcomes all of these issues is to use an $\mathcal{H}_\infty$ control law, which allows for nonzero reference states and does not place restrictions on

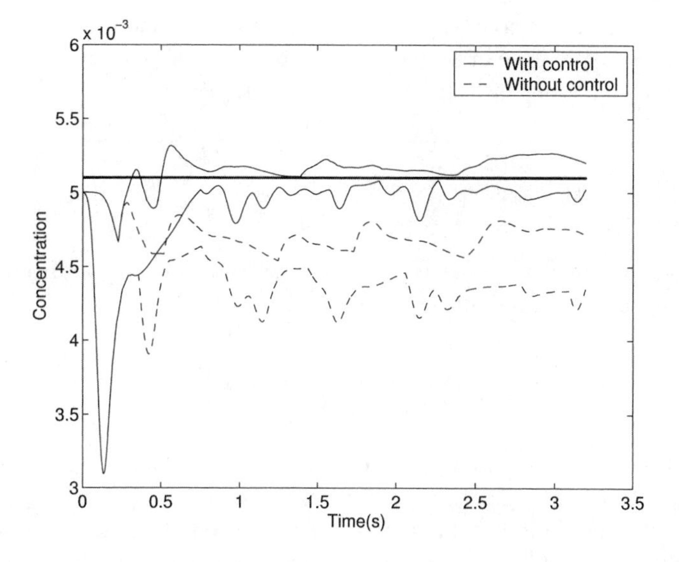

FIGURE 7.5 The concentration over the cross-section without and with control. Only the minimum and maximum values of the concentration on the nodes corresponding to the cross-section are plotted at each time. The thick line denotes the desired value.

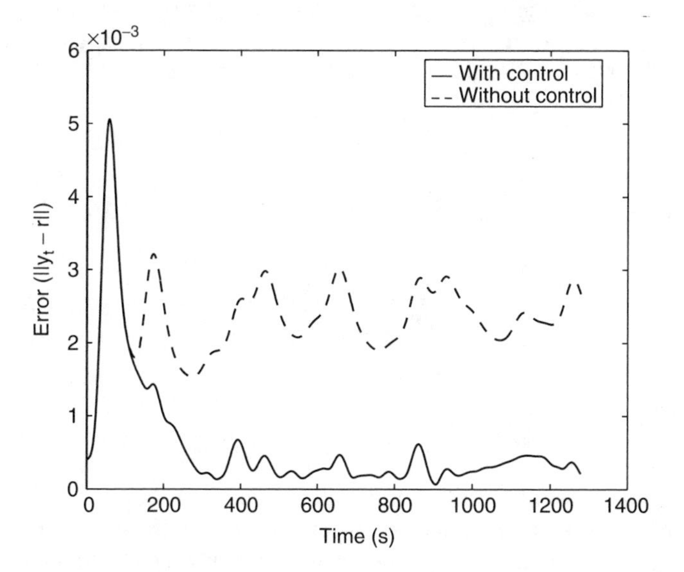

FIGURE 7.6 The deviations of the concentration from the desired value over the cross-section without and with control.

the form of the exogenous variables. It can also guarantee the stability of the controller in the presence of a range of errors between the process and the model.

If the $L_{2,[0,T]}$ norms of $\mathbf{z}_t$ and $\mathbf{w}_t$ are defined as [19]

$$\|\mathbf{z}_t\|_{2,[0,T]} = \sqrt{\sum_{t=0}^{T} \mathbf{z}_t' \, \mathbf{z}_t} \quad \|\mathbf{w}_t\|_{2,[0,T]} = \sqrt{\sum_{t=0}^{T} \mathbf{w}_t' \, \mathbf{w}_t} \tag{7.40}$$

then the aim of the controller is to use the measurements, $\mathbf{y}_t$, to determine the control inputs, $\mathbf{u}_t$, that minimize

$$\sup_{\|\mathbf{w}_t\|_{2,[0,T]} \neq 0} \frac{\|\mathbf{z}_t\|_{2,[0,T]}}{\|\mathbf{w}_t\|_{2,[0,T]}} \tag{7.41}$$

In essence, the controller determines $\mathbf{u}_t$ that will minimize the norm of $\mathbf{z}_t$ in the presence of the "worst case" (in the $L_{2,[0,T]}$ sense) exogenous input. Rather than find the control inputs that optimize the criterion in Equation 7.41 directly, a suboptimal control law is found that satisfies a specific performance bound, γ:

$$\sup_{\|\mathbf{w}_t\|_{2,[0,T]} \neq 0} \frac{\|\mathbf{z}_t\|_{2,[0,T]}}{\|\mathbf{w}_t\|_{2,[0,T]}} < \gamma \tag{7.42}$$

The optimal control law can be recovered by finding the lowest value of γ for which a solution to the optimization problem in Equation 7.42 exists.

The control input, $\mathbf{u}_t$, which minimizes Equation 7.42 must also minimize

$$\sup_{\|\mathbf{w}_t\|_{2,[0,T]} \neq 0} \left\{ \frac{\|\mathbf{z}_t\|^2_{2,[0,T]}}{\|\mathbf{w}_t\|^2_{2,[0,T]}} \right\} < \gamma^2 \tag{7.43}$$

so that there exists an ε, such that [19]

$$\frac{\|\mathbf{z}_t\|^2_{2,[0,T]}}{\|\mathbf{w}_t\|^2_{2,[0,T]}} \leq \gamma^2 - \varepsilon^2 \tag{7.44}$$

or

$$\|\mathbf{z}_t\|^2_{2,[0,T]} - \gamma^2 \|\mathbf{w}_t\|^2_{2,[0,T]} \leq -\varepsilon^2 \|\mathbf{w}_t\|^2_{2,[0,T]} \tag{7.45}$$

If this inequality is satisfied for all possible disturbances and for some value of ε, then the criterion in Equation 7.42 is satisfied. Hence by making the left-hand side of Equation 7.45 an objective function and finding the $\mathbf{u}_t$ that minimizes this objective function, then by checking that the minimum value satisfies the criterion in Equation 7.45, one can find a $\mathbf{u}_t$ that will also minimize Equation 7.42. In what follows, it is assumed that $\mathbf{D}_{12}^{P\prime} \mathbf{C}_1^P = 0$ and $\mathbf{B}_1^P \mathbf{D}_{21}^{P\prime} = 0$, which will be the case for the model given in Equation 7.21.

In terms of general model given in Equation 7.20, one solution to this optimization problem can be obtained from the solution of two Riccati equations:

$$\mathbf{S}_{t-1} = \mathbf{A}^{p\prime}\mathbf{S}_t\mathbf{A}^p - \mathbf{A}^{p\prime}\mathbf{S}_t\begin{bmatrix}\mathbf{B}_1^p & \mathbf{B}_2^p\end{bmatrix}\begin{bmatrix}-\gamma^2\mathbf{I}+\mathbf{B}_1^{p\prime}\mathbf{S}_t\mathbf{B}_1^p & \mathbf{B}_1^{p\prime}\mathbf{S}_t\mathbf{B}_2^p \\ \mathbf{B}_2^{p\prime}\mathbf{S}_t\mathbf{B}_1^p & \mathbf{R}+\mathbf{B}_2^{p\prime}\mathbf{S}_t\mathbf{B}_2^p\end{bmatrix}^{-1}$$

$$\times\begin{bmatrix}\mathbf{B}_1^{p\prime} \\ \mathbf{B}_2^{p\prime}\end{bmatrix}\mathbf{S}_t\mathbf{A}^p + \mathbf{C}_1^{p\prime}\mathbf{C}_1^p \tag{7.46}$$

which is solved backward from a given $\mathbf{S}(T) \geq \mathbf{0}$, and

$$\mathbf{P}_{t+1} = \mathbf{A}^p\mathbf{P}_t\mathbf{A}^{p\prime} - \mathbf{A}^p\mathbf{P}_t\begin{bmatrix}\mathbf{C}_1^{p\prime} & \mathbf{C}_2^{p\prime}\end{bmatrix}\begin{bmatrix}-\gamma^2\mathbf{I}+\mathbf{C}_1^p\mathbf{P}_t\mathbf{C}_1^{p\prime} & \mathbf{C}_1^p\mathbf{P}_t\mathbf{C}_2^{p\prime} \\ \mathbf{C}_2^p\mathbf{P}_t\mathbf{C}_1^{p\prime} & \mathbf{I}+\mathbf{C}_2^p\mathbf{P}_t\mathbf{C}_2^{p\prime}\end{bmatrix}^{-1}$$

$$\times\begin{bmatrix}\mathbf{C}_1^p \\ \mathbf{C}_2^p\end{bmatrix}\mathbf{P}_t\mathbf{A}^{p\prime} + \mathbf{B}_1^p\mathbf{B}_1^{p\prime} \tag{7.47}$$

which is solved forward from $\mathbf{P}(0) \geq \mathbf{0}$. The performance bound in Equation 7.42 will only be satisfied provided that the spectral radius $\rho(\mathbf{S}_t\mathbf{P}_t) < \gamma^2$ [19].

Given $\mathbf{S}_t$ and $\mathbf{P}_t$, a discrete time $\mathcal{H}_\infty$ controller is given by [25]

$$\begin{aligned}
\hat{\mathbf{q}}_{t+1} &= \mathbf{A}^p\hat{\mathbf{q}}_t + \mathbf{B}_1^p\hat{\mathbf{w}}_t^* + \mathbf{B}_2^p\mathbf{u}_t + \bar{\mathbf{L}}_y(\mathbf{y}_t - \hat{\mathbf{y}}_t) \\
\hat{\mathbf{w}}_t^* &= \mathbf{K}_w\hat{\mathbf{q}}_t \\
\hat{\mathbf{y}}_t &= \mathbf{C}_2^p\hat{\mathbf{q}}_t + \mathbf{D}_{21}^p\hat{\mathbf{w}}_t^* \\
\mathbf{u}_t &= -\mathbf{K}_u\hat{\mathbf{q}}_t - \mathbf{K}_{uw}\hat{\mathbf{w}}_t^* - \bar{\mathbf{L}}_{uy}(\mathbf{y}_t - \hat{\mathbf{y}}_t)
\end{aligned} \tag{7.48}$$

where

$$\begin{aligned}
\mathbf{K}_u &= \left(\mathbf{R}+\mathbf{B}_2^{p\prime}\mathbf{S}_t\mathbf{B}_2^p\right)^{-1}\mathbf{B}_2^{p\prime}\mathbf{S}_t\mathbf{A}^p \\
\mathbf{K}_{uw} &= \left(\mathbf{R}+\mathbf{B}_2^{p\prime}\mathbf{S}_t\mathbf{B}_2^p\right)^{-1}\mathbf{B}_2^{p\prime}\mathbf{S}_t\mathbf{B}_1^p \\
\mathbf{K}_w &= -\left(\mathbf{B}_1^{p\prime}\mathbf{S}_t\mathbf{B}_2^p\left(\mathbf{R}+\mathbf{B}_2^{p\prime}\mathbf{S}_t\mathbf{B}_2^p\right)^{-1}\mathbf{B}_2^{p\prime}\mathbf{S}_t\mathbf{B}_1^p - \left(-\gamma^{-2}\mathbf{I}-\mathbf{B}_1^{p\prime}\mathbf{S}_t\mathbf{B}_1^p\right)\right)^{-1} \\
&\quad\times\left(\mathbf{B}_1^{p\prime}\mathbf{S}_t\mathbf{A}^p - \mathbf{B}_1^{p\prime}\mathbf{S}_t\mathbf{B}_2^p\left(\mathbf{R}+\mathbf{B}_2^{p\prime}\mathbf{S}_t\mathbf{B}_2^p\right)^{-1}\mathbf{B}_2^{p\prime}\mathbf{S}_t\mathbf{A}^p\right)
\end{aligned} \tag{7.49}$$

$$\bar{\mathbf{L}}_y = \bar{\mathbf{A}}^p \bar{\mathbf{P}}_t \bar{\mathbf{C}}_2^{p\prime} \left(\mathbf{I} + \bar{\mathbf{C}}_2^p \bar{\mathbf{P}}_t \bar{\mathbf{C}}_2^{p\prime}\right)^{-1}$$
$$\bar{\mathbf{L}}_{uy} = \mathbf{R}^{-1/2} \bar{\mathbf{C}}_1^p \bar{\mathbf{P}}_t \bar{\mathbf{C}}_2^{p\prime} \left(\mathbf{I} + \bar{\mathbf{C}}_2^p \bar{\mathbf{P}}_t \bar{\mathbf{C}}_2^{p\prime}\right)^{-1} \tag{7.50}$$

The term

$$\bar{\mathbf{P}}_t = \mathbf{P}_t \left(\mathbf{I} - \gamma^{-2} \mathbf{S}_t \mathbf{P}_t\right)^{-1} \tag{7.51}$$

can be calculated directly from the solutions of the Riccati equations in Equation 7.46 and Equation 7.47, and

$$\bar{\mathbf{A}} = \mathbf{A}^p + \mathbf{B}_1^p \mathbf{K}_w$$
$$\bar{\mathbf{C}}_1^p = \mathbf{R}^{1/2} (\mathbf{K}_u + \mathbf{K}_{uw} \mathbf{K}_w) \tag{7.52}$$
$$\bar{\mathbf{C}}_2^p = \mathbf{C}_2^p + \mathbf{D}_{21}^p \mathbf{K}_w$$

As with the LQG control law, the expression in Equation 7.38 can be used in place of $\mathbf{y}_t - \hat{\mathbf{y}}_t$. The controller is also simplified from the general case because $\mathbf{D}_{22}^p = 0$ [25].

In the limit $T \to \infty$, the Riccati equation for $\mathbf{S}_t$ in Equation 7.46 converges to a steady state solution,

$$\mathbf{S} = \mathbf{A}^{p\prime} \mathbf{S} \mathbf{A}^p - \mathbf{A}^{p\prime} \mathbf{S} \begin{bmatrix} \mathbf{B}_1^p & \mathbf{B}_2^p \end{bmatrix} \begin{bmatrix} -\gamma^2 \mathbf{I} + \mathbf{B}_1^{p\prime} \mathbf{S} \mathbf{B}_1^p & \mathbf{B}_1^{p\prime} \mathbf{S} \mathbf{B}_2^p \\ \mathbf{B}_2^{p\prime} \mathbf{S} \mathbf{B}_1^p & \mathbf{R} + \mathbf{B}_2^{p\prime} \mathbf{S} \mathbf{B}_2^p \end{bmatrix}^{-1}$$
$$\times \begin{bmatrix} \mathbf{B}_1^{p\prime} \\ \mathbf{B}_2^{p\prime} \end{bmatrix} \mathbf{S} \mathbf{A}^p + \mathbf{C}_1^{p\prime} \mathbf{C}_1^p \tag{7.53}$$

which can be precalculated off-line. However, the Riccati equation for $\mathbf{P}_t$ in Equation 7.66 does not converge to a steady state solution because, as defined in Equation 7.38, the $\mathbf{C}_2^p$ matrix is time varying, so that $\mathbf{P}_t$ has to be calculated recursively on line.

The control law obtained from the $\mathcal{H}_\infty$ design is suboptimal in the sense that it determines the optimal controller for a given value of γ, rather than the lowest possible value of γ. One can find the optimal controller by using the γ-iteration, where the control design is repeated for decreasing values of γ, until the condition on the spectral radius, $\rho(\mathbf{S}_t \mathbf{P}_t) < \gamma^2$, no longer holds. In practice, the key step in designing both the LQG and the $\mathcal{H}_\infty$ controllers is the choice of the weighting matrices in the objective function. In the control designs presented here, the weightings have been assumed to be static, but the weightings are often chosen to have dynamic responses, so that different weights can be applied over the range of operating frequencies. For example, it is often much more important for the control system to regulate slowly varying changes in the concentration profile rather than rapid, short-term fluctuations. Under these circumstances, the weighting in the objective function would penalize low frequency deviations. One can implement frequency dependent weightings by augmenting the state vector, $\mathbf{q}_t$, with states describing the dynamics of the weightings [21].

7.5 PRACTICALITIES OF IMPLEMENTING CONTROLLERS

7.5.1 ROBUSTNESS

One of the main difficulties associated with designing any control system is that the control law is based on a model of the process. Invariably this model is inaccurate, and the performance of the controller will be degraded because the response of the actual process is different from the response predicted by the model. This is particularly true in the flow process being considered here, as many of the parameters of the model are uncertain, especially the velocity profile of the flow. Although the controller may often perform acceptably in the presence of these uncertainties, in some circumstances, the performance may be degraded to the extent that the system becomes unstable. The amount of degradation that results from the mismatch between the model and the actual response depends on the robustness of the controller, and one of the key criticisms of LQG control is that it is not possible to guarantee a specific level of performance in the presence of model mismatch. By contrast, the performance of an $\mathcal{H}_\infty$ controller can be guaranteed in the presence of a specified amount of mismatch. This performance guarantee can be incorporated within the control design by expressing the effect of a mismatch as an additional input signal $\mathbf{w}_t^\Delta$ that is generated by passing a signal from the process, $\mathbf{z}_t^\Delta$, through an element, Δ, that represents the uncertainty, as shown in Figure 7.7.

The uncertainty, Δ, is usually considered to be a dynamic element, which reflects the fact that the uncertainty may not be the same at different frequencies. The effect of the uncertainty can be incorporated into the model given in Equation 7.19 by augmenting the exogenous signal, $\mathbf{w}_t$, to include $\mathbf{w}_t^\Delta$, and augmenting $\mathbf{z}_t$ with $\mathbf{z}_t^\Delta$. If an $\mathcal{H}_\infty$ controller is designed with a performance bound γ, then the closed loop system will remain stable for any model mismatch, Δ, for which [19]

$$\sup_{\|\mathbf{z}_t^d\|_{2,[0,T]} \neq 0} \frac{\left\|\mathbf{w}_t^d\right\|_{2,[0,T]}}{\left\|\mathbf{z}_t^d\right\|_{2,[0,T]}} < \frac{1}{\gamma^2} \tag{7.54}$$

This shows that when γ is made small, the closed loop system can accommodate a large exogenous input signal due to the modelling errors without the

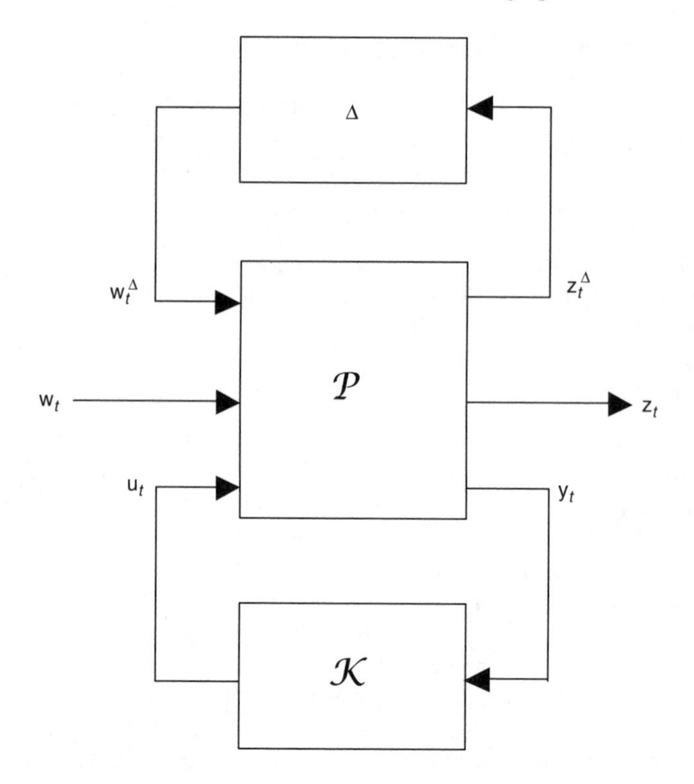

FIGURE 7.7 Structure of generalized plant including modeling uncertainty.

system becoming unstable. In principle, it is possible to adjust the performance weights to make γ small and hence to make the system very robust to modelling uncertainties. Unfortunately, improved robustness tends to be achieved at the expense of performance in terms of factors such as the bandwidth of disturbance rejection and the speed of response to any changes in the desired state, $\mathbf{q}_t^d$. In practice, it is usual to include user-specified, frequency-dependent weightings in the system to allow the designer to balance the requirement for a robust system with the need to maintain performance [21, 26, 20]. A further complication is that the robust stability criterion applies to any possible model mismatch that satisfies the criterion in Equation 7.54. This tends to make the controller conservative, in the sense that it accommodates uncertainties that are very unlikely to occur in practice. For this reason, it is better to design a controller that is robust to more specific, structured uncertainties. These could include uncertainties in specific parameters in the model, such as the diffusion constant, or in specific terms, such as the velocity profile. By using the μ-synthesis methodology [19], one can design controllers that are robust to specific uncertainties and, as a result, are less conservative.

7.5.2 Actuator Constraints

Up to this point, it has been assumed that there is no limitation on the range of control inputs, $\mathbf{u}_t$, which can be applied to the actuators, but in practice, the size of the inputs will be constrained. One particular feature of using injectors as the actuators is that they can only add extra substance to the flow; they cannot remove excess substance. This effectively constrains the inputs so that $\mathbf{u}_t \geq \mathbf{0}$, making the controller nonlinear. In principle, it would be possible to handle this constraint (along with other constraints, such as the maximum rate at which substance can be injected into the flow) by designing a model predictive controller (MPC) [27, 28]. The problem with this controller is that it performs a constrained optimization at every time step, which is computationally demanding. Given that the finite element model on which the control design is based tends to result in models with large numbers of states, it is unlikely that an MPC could be implemented, because it would not be possible to complete the necessary computations within the time between samples, even if fast algorithms, such as interior point methods [29], are used.

A current topic of research is examining whether a suboptimal solution to the constrained MPC problem could be obtained that would allow control inputs to be generated in real time. Although the LQG and $\mathcal{H}_\infty$ controllers do not handle constraints explicitly, they have the advantage that the control law is defined in terms of matrices, most of which can be precomputed off-line. The exception in both cases is that the Riccati equation for the observer has to be computed recursively, because the measurement equation is time varying. This means that the amount of on-line computation is relatively low. It is possible to handle the actuator constraints in a suboptimal manner, by increasing $\mathbf{R}$, the weighting on the control action in the objective function, until the magnitudes of the control inputs lie within their limits for the expected range of variations. For the specific case of LQG control, it is also possible to use nonnegative least squares [30] to choose the values of the steady state inputs $\mathbf{u}^*$ that minimize $\mathbf{z}^*$ in Equation 7.30, subject to the constraint that $\mathbf{u}^* \geq \mathbf{0}$ [2].

7.5.3 Computer-Based Implementation

Although most of the computation associated with the LQG and $\mathcal{H}_\infty$ controllers presented here can be precalculated off-line, a significant amount of computation needs to be implemented in real time. The major computational load is associated with the recursion of the observer Riccati equation in Equation 7.37 or Equation 7.47 and the updating of the state from the voltage measurements obtained from the EIT sensor. These computations only require matrix multiplication, but because the number of states generated by the finite element model tends to be high, the matrices have large dimensions. At the same time, the time between successive samples needs to be sufficiently short to be able to respond to the variations within the process [23]. The bandwidth of the process itself is usually dominated by the flow velocity along the pipe, and typically at least six to ten samples must be taken in the time the flow takes to move from the actuator to the point where the desired concentration profile is specified. Even for relatively slow flow rates, this means that the sample rate is typically greater than 1 Hz. Although, as shown in preceding chapters in this book, the EIT sensor can take measurements of the voltages between each pair of electrodes at a much higher rate than this, the limitation is the time required to process the measurements and to calculate the inputs to the injectors. For this reason, it is necessary that controller be implemented in a real-time control system, where it can be guaranteed that the calculations can be completed within the time between control samples [31].

7.6 CONCLUSION

This chapter has shown how a process imaging system (in this case, a tomographic sensor) can be used as a noninvasive measurement device within a feedback loop for the control of a distributed parameter system. Although the control of concentration profile within the flow of fluid along a pipe has been used as an example, the approach is general and can be applied to other imaging applications [32]. The key component in the design of the control system is the dynamic model of the process, which, as in the example presented here, is usually a finite element representation of the underlying PDE. This dynamic model can be used within an observer to reconstruct the state of the process, which in this case is the value of the concentration profile at each of the nodes in the finite element model. Both the LQG and $\mathcal{H}_\infty$ approaches to the design of state feedback controllers have been presented, and the robustness of the controllers to uncertainties in the underlying model has been discussed. One of the main shortcomings of these approaches is the lack of an optimal method of handling the constraints on the actuator inputs; although satisfactory suboptimal solutions have been developed, improving on these methods is an area of current research. The performance of the control approach has been illustrated through simulations. Because the control laws are based on a finite element model, which tends to result in a high number of states, the implementation of the controller requires that large matrix manipulations be carried out on-line. This means that a computer-based controller must be implemented using a real-time system. Nevertheless, the combination of imaging technology with advanced control design provides a unique and useful set of tools for implementation of automatic control systems.

REFERENCES

1. A El Jai, AJ Pritchard. *Sensors and Controls in the Analysis of Distributed Systems,* Chichester, U.K.: Halstead Press, 1988.
2. AR Ruuskanen, A Seppänen, SR Duncan, E Somersalo, JP Kaipio. Optimal control in process tomography. *Proceedings of 3rd World Congress on Industrial Process Tomography,* Banff, Canada, 2003, pp. 245–251.
3. SR Duncan. Using process tomography as a sensor in a system for controlling concentration in fluid flow. *Proceedings of 2nd World Congress on Industrial Process Tomography,* Hannover, Germany, 2001, pp. 378–386.

4. CJ Cottee, SR Duncan. Design of matching circuit controllers for radio-frequency heating. *IEEE Transactions on Control Systems Technology,* 11(1):91–100, 2003.
5. O Vayega, MA Demetriou, H Doumanidis. An LQR-based optimal actuator guidance scheme in thermal processing of coatings. *Proceedings of American Control Conf.,* Chicago, IL, 2000.
6. PDA Jones, SR Duncan, T Rayment, PS Grant. Control of temperature for a spray deposition process. *IEEE Transactions on Control Systems Technology,* 11(5):656–667, 2003.
7. JM Allwood. Model-based evaluation of the effect of horizontal roll offset on cross-directional control performance in cold-strip rolling. *IEE Proc. Control Theory Appl.,* 149(5):463–470, 2002.
8. EM Heaven, IM Jonsson, TM Kean, MA Maness, RN Vyse. Recent advances in cross machine profile control. *IEEE Control Syst. Magazine,*14(5):35–46, 1994.
9. SR Duncan. Cross-directional control of web forming processes. Ph.D. thesis, University of London, London, U.K., 1989.
10. AP Featherstone, JG van Antwerp, RD Braatz. *Identification and Control of Sheet and Film Processes.* Berlin, Germany: Springer-Verlag, 2000.
11. OLR Jacobs. Designing feedback controllers to regulate deposited mass in hot-dip galvanizing. *Control Eng. Pract.,* 3(11):1529–1542, 1995.
12. BW Bequette. Non-linear control of chemical processes: A review. *Ind. Eng. Chem. Res.,* 30:1391–1413, 1991.
13. BA Ogunnaike, WH Ray. *Process Dynamics, Modelling and Control.* New York: Oxford, 1994.
14. AR Fraser, RW Daniel. *Perturbation Techniques for Flexible Manipulators.* London: Kluwer Academic, 1991.
15. NW Hagood, A von Flotow. Damping of structural vibrations with piezoelectric materials and passive electrical networks. *J. Sound Vibration,* 146(2):243–268, 1991.
16. D Halim, SOR Moheimani. Spatial resonant control of flexible structures: Application to a piezoelectric laminate beam. *IEEE Trans. Control Syst. Technol.,* 9(1):37–53, 2001.
17. SP Luke, RA Williams. Industrial applications of electrical tomography to solids conveying. *Measurement & Control,* 30(7):201–205,1997.
18. C Johnson. *Numerical Solution of Partial Differential Equations by the Finite Element Method.* Cambridge, U.K.: Cambridge University Press, 1992.
19. M Green, DJN Limebeer. *Linear Robust Control.* Englewood Cliffs, NJ: Prentice Hall, 1995.
20. JB Burl. *Linear Optimal Control.* Menlo Park, CA: Addison Wesley Longman, 1998.
21. S Skogestad, I Postlethwaite. *Multivariable Feedback Control: Analysis and Design.* Chichester, U.K.: Wiley, 1996.
22. T Kailath. *Linear Systems.* Englewood Cliffs, NJ: Prentice Hall, 1980.
23. KJ Astrom, B Wittenmark. *Computer-Controlled Systems: Theory and Design* (2nd ed.). Englewood Cliffs, NJ: Prentice Hall, 1990.
24. GF Franklin, JD Powell, A Emami-Naeini. *Feedback Control of Dynamic Systems* (3rd ed.). New York: Addison Wesley, 1994.
25. A-K Christiansson. A general framework for hybrid $\mathcal{H}_\infty$-control. Licentiate thesis, Chalmers University of Technology, Goteborg, Sweden, 2000.
26. JM Maciejowski. *Multivariable Feedback Design.* New York: Addison Wesley, 1989.

27. BW Bequette. *Chemical Process Control Using MATLAB.* Upper Saddle River, NJ: Prentice Hall, 2003.
28. JA Rossiter. *Model Based Predictive Control: A Practical Approach.* Boca Raton, FL: CRC Press, 2003.
29. YE Nesterov, AS Nemirovskii. *Interior-Point Polynomial Algorithms in Convex Programming.* Philadelphia, PA: SIAM Press, 1994.
30. GH Golub, CF van Loan. *Matrix Computations* (3rd ed.). Baltimore, MD: John Hopkins University Press, 1996.
31. S Bennett. *Real-Time Computer Control.* New York: Prentice Hall, 1988.
32. M Malinen, T Huttunen, JP Kaipio. An optimal control approach for ultrasound induced heating. *Int. J. Control,* 76(13):1323–1336, 2004.

8 Imaging Diagnostics for Combustion Control

Volker Sick and Hugh McCann

CONTENTS

8.1 INTRODUCTION

It is estimated that about 90% of the energy that is used worldwide is made available through combustion processes [1]. Mainly fossil fuels are used for this purpose, and fears loom of running out of natural resources over time. Any improvement in combustion performance, even to a minute extent, can therefore bring about a substantial absolute saving in fuel resources. Furthermore, combustion typically leads to a number of side effects, such as waste heat and pollutants that can be harmful to nature in general and to humans and their immediate environment directly. Globally, massive efforts are being undertaken in research and development programs to improve combustion processes in order to increase efficiency, to reduce pollutant formation, and to improve reliability.

While combustion usage and combustion related technologies have developed gradually over the last million years [2], the last 200 years have brought an explosive development of new applications: automobiles and airplanes that are propelled with combustion engines of various kinds, and electricity generation via combustion-driven power plants. Improvements of these devices have been substantial since their introduction. Legislation has often driven improvements by setting standards for efficiencies and pollutant release, e.g., for internal combustion engines [3]. Similar achievements have occurred for other combustion devices, such as electrical power plants and gas turbines. Much of today's effort is now being devoted to controlling precisely the operation of combustion processes to improve stability and adapt to changing boundary conditions (e.g., cold start or acceleration of automobiles).

The complex interactions of physical and chemical processes control the release of useful energy through combustion in these devices. Hence, understanding the related fundamental principles of combustion is the key to breakthrough technology and improvement of existing devices. The insights of Lavoisier [4], which helped to overcome the phlogiston theory for combustion, and Michael Faraday's *Chemical History of a Candle* [5] are examples of early combustion science that led to focused efforts to make combustion devices more useful. In recent years, the arsenal of new diagnostic techniques has led to an understanding of the complexity of combustion processes that now can be explored for the optimization of combustion technology. Detailed understanding now provides means to develop strategies that allow control of the combustion process to ensure that in spite of undesirable changes in operating conditions, the process is kept at (or returned to) its optimal performance.

For a control strategy to work, parameters that impact the combustion process (e.g., the fuel/air ratio) must be measured and used to counteract detrimental processes. For some technical applications, control strategies have already been implemented and are crucial for proper operation. A prominent example is the sensor that monitors the oxygen concentration in the exhaust pipe of automobiles for precise control of the fuel/air ratio to ensure optimized performance of the catalyst that cleans the exhaust gases [3]. A second example is the removal of nitric oxide from flue gases in power plants or waste incinerators through the addition of urea or ammonia. The amount added must be carefully controlled to achieve optimized reduction of nitric oxide and to avoid the risk that excessive amounts of added urea or ammonia themselves become pollutants [6, 7]. Additional examples and details can be found in Kohse-Höinghaus and Jeffries [8].

Most control strategies, including those mentioned above, measure a global parameter and act on a global variable. However, many technical processes are carried out in such a way that there are substantial spatial inhomogeneities, e.g., of fuel distribution or temperature, and it is possible for a control strategy to miss an optimum operating condition because the process is sampled in nonrepresentative areas. This issue will eventually lead us into a discussion of imaging and tomography as potential means to generate control parameters that include information about spatial distribution. Measurements that provide information about processes directly from the combustion process itself are very demanding because technical combustion processes are usually operated in a hostile

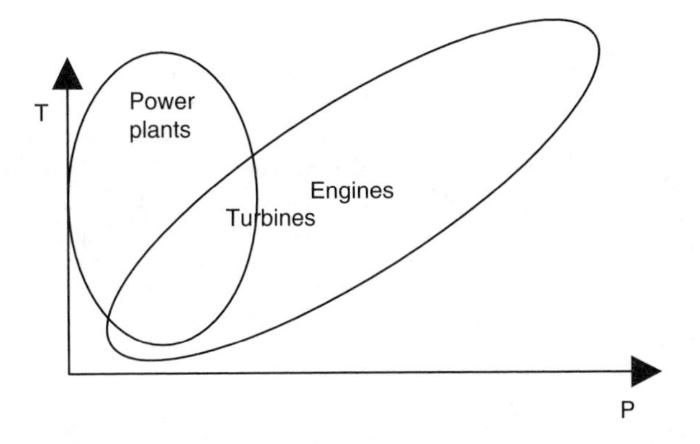

FIGURE 8.1 Schematic temperature T and pressure P operating range of technical combustion devices.

environment for diagnostics. For example, in automotive engines the combustion chamber pressure can be of the order of 100 bar, and gas temperature can exceed 2500 K, with steady-state wall temperature typically 450 K. The combustion takes place in an electrically hostile environment (due to spark ignition) and repeating at typically 30 Hz in each cylinder! Even simple access to the measurement subject can become a major hurdle for any type of measurement in devices such as large-scale power plants, internal combustion engines, and gas turbines (see Figure 8.1).

Spatially resolving techniques that are in principle able to provide data necessary for control strategies are introduced in this chapter, and their basic principles are discussed together with examples of applications. Furthermore, the difficulties of implementing such measurements for continuous control strategies are described, and an outlook of prospects is given. For better understanding, a brief introduction to the main aspects of some combustion processes is given to illustrate the importance of controlling, for example, the local fuel/air ratio or temperature. As will be seen, many of the techniques and approaches described will provide information that could be used for control purposes. Few of these techniques are used in everyday practice because the instrumentation is prohibitively expensive and often not robust enough to be implemented in combustion devices on a large scale. Nevertheless, many of the techniques discussed here allow the identification of crucial parameters in technical combustion devices and thus help to identify efficient control strategies.

8.2 COMBUSTOR TYPES

The intent of this section is to give a very simple description of the distinguishing features of the combustion process for a range of technically important combustion devices and to identify the crucial features where image-based controls can help to improve the process. Internal combustion engines will be treated with

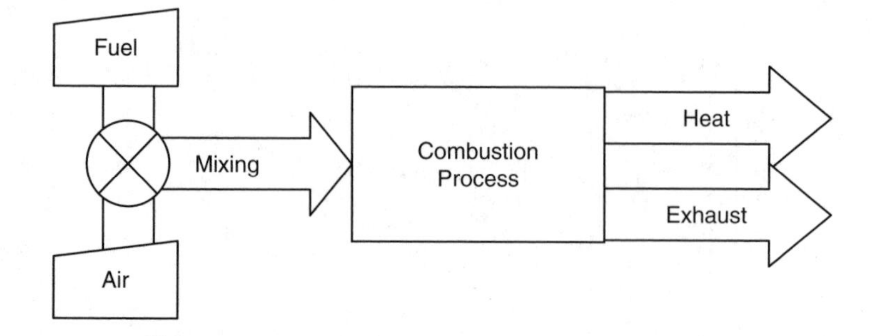

FIGURE 8.2 Schematic depiction of a generic combustion process. Independent of the actual device, the fuel/air mixing and handling of the generated heat and exhaust products are among the most important issues of the overall process.

more detail than other devices, but the principles and issues for diagnostics and control are similar for all of them.

A simplistic description of combustion is that the addition of air to fuel yields both heat and exhaust (see Figure 8.2). In reality, the conversion of fuel and air into heat and combustion products is a complex scheme of many coupled chemical reactions [1]. It is this multiplicity of reactions that explains why combustion processes are so dependent on initial conditions such as the ratio of fuel and air, the type of fuel, the temperature, the pressure, and the flow conditions. Each of the chemical reactions that are part of the combustion process has its own rate, which will depend on temperature, pressure, and the concentration of the participating species. If any of those parameters change, the rate of the reaction can change, and if this particular reaction has a significant impact on the overall combustion rate, then the entire combustion process will change.

Furthermore, transport processes can affect combustion significantly, since transport of species to and from the flame front can be a rate-determining step that in turn influences the combustion progress. This is most obvious in nonpremixed (diffusion) flames, where fuel and air are partially mixed immediately before the reaction occurs. Often this effect leads to subsequent problems, as will be illustrated below. Still, nonpremixed flames are very important flames in technical combustion processes since they allow the fuel and air to be kept apart as long as possible, which is a big advantage in terms of safety when no combustible mixtures have to be handled. Transport processes also affect premixed flames (where air and fuel are carefully mixed prior to ignition). For such flames (for example, in many current automobile engines), the overall rate at which the fuel is consumed depends significantly on the level of turbulence in the flow. In general, the consumption rate increases as turbulence increases; see Figure 8.3. It is a fortunate coincidence that the level of turbulence increases as the speed of an engine increases; otherwise, the fuel could not be burnt in time during an engine cycle at higher speeds. The increase of the fuel consumption rate with turbulence is also beneficially exploited in other combustion processes where

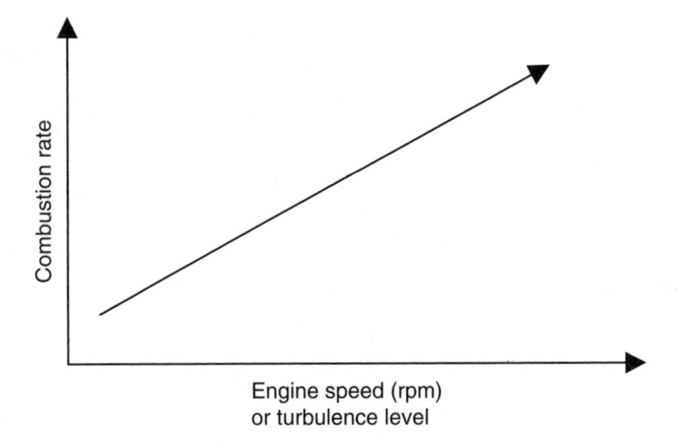

FIGURE 8.3 The rate of combustion increases as the level of turbulence increases (up to a certain point, beyond which the flame will extinguish). This is an important factor that allows it to operate automobile engines at variable speeds.

increasing turbulence leads to more heat released per time per given volume of, for example, a combustor.

The following sections discuss some selected combustion devices to explain why controlling certain parameters can help to improve combustion performance and pollutant formation. It will become clear that while controlling a combustion process based on information obtained from imaging might be desirable, the design of the combustion device might not allow this, or the cost of an imaging system might be prohibitive for continuous operation. These are cases where the use of imaging techniques can help the developer to understand and improve performance and ensure stable and optimized operation.

8.2.1 Internal Combustion Engines

The single most important parameter for internal combustion engine operation is the proper preparation of the fuel/air mixture. The equivalence ratio (or fuel/air ratio or stoichiometry) influences the amount of power generated, the propagation speed of the flame, the formation of nitric oxide, the amount of unburnt hydrocarbons that are potentially released from the engine, and the formation and release of carbon monoxide. For port-fuel-injected engines, still today the most widespread type of spark-ignited (SI) engine, the fuel/air mixture is initially prepared outside of the cylinders and then introduced through the intake valves into the cylinder, where more dilution with air occurs (see Figure 8.4). By the time the spark is fired—at a timing optimized for best performance—the incoming mixture and the gases inside the cylinder have mixed well enough that, usually, the resulting mixture composition (and temperature) is spatially uniform. Flow control can then be used to manipulate the combustion progress via the turbulence dependence of the combustion rate.

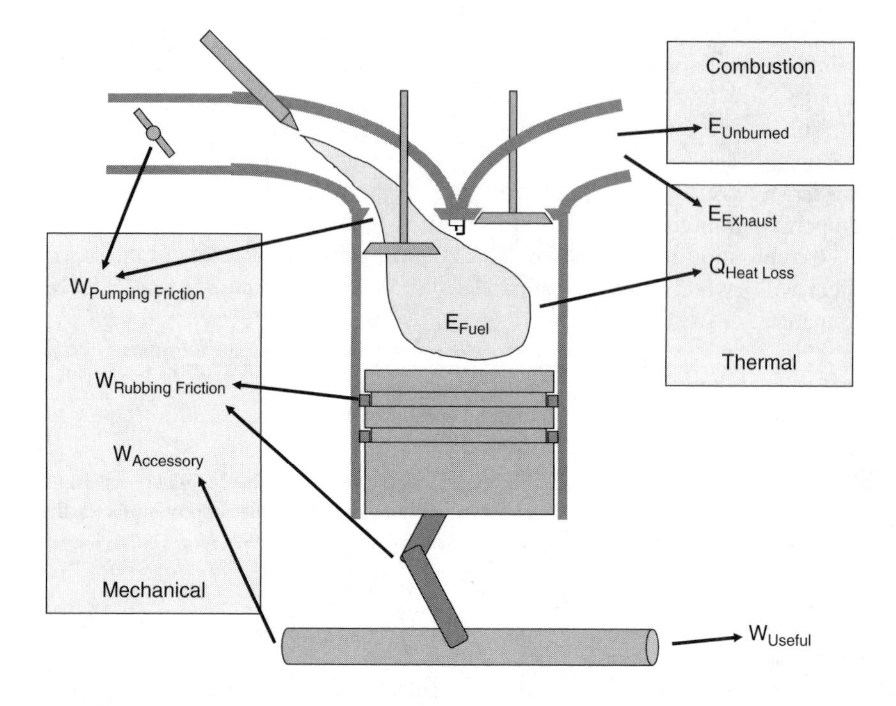

FIGURE 8.4 A number of factors affect the overall performance and efficiency (η) of an internal combustion engine.

More recently, and in part because better control mechanisms are now available, direct-injected engines have gained substantial attention. In this type of engine, regardless of whether spark- or compression-ignited (CI), the fuel is directly injected into the cylinder, and the amount of fuel injected controls the overall stoichiometry, which now can be chosen to be very lean to help lower the formation of nitric oxide.

Fuel direct-injection has a number of significant advantages [9], but the fact that liquid fuel is introduced into the cylinder means that there is less time for proper mixture preparation. In a time-critical manner, the liquid fuel must be atomized, evaporated, and mixed with the surrounding in-cylinder gas. Furthermore, for some concepts, it is required that the fuel/air mixture be stratified in the combustion chamber to allow reliable ignition and ensure that no flame extinction occurs because of mixtures that are too lean. While an overall lean fuel/air ratio helps to minimize pollutant formation, it poses a problem for exhaust gas after-treatment. The standard three-way catalyst used for most SI engines will only work well if the fuel/air ratio is set to stoichiometric values. Thus, new catalysts for lean operation are needed to remove nitric oxide from the exhaust gases. Those catalysts have been improved over the last few years but still do not come near the efficiency of the three-way catalyst. Other strategies use a trap to store the nitric oxide temporarily; when the trap is full, an automated recycling takes place that requires the engine to switch instantaneously into stoichiometric operating conditions to ensure that the nitric oxide can be removed properly through catalytic converters.

The stratified mixture preparation in the cylinder can produce locally rich regions that can lead to soot formation in SI engines, a problem that previously was largely associated with Diesel engines. Also, the potentially large spatial and temporal variations in the fuel/air ratio will lead to variations in temperature, since combustion at different fuel/air ratios will yield different temperatures. This effect will have a significant impact on the formation of nitric oxide, which is dramatically enhanced as temperature increases, and thus even small regions in the cylinder where the temperature is higher than the average temperature will lead to a substantial increase in the formation of this pollutant. This illustrates why the formation of a well-controlled fuel/air mixture is crucial for a stable and clean operation of direct-injected SI engines.

The same arguments hold for Diesel engines, where the main difference is that the fuel/air mixture is not ignited via a spark but through auto-ignition. The compression ratio of Diesel engines is so high that the compression of the intake air leads to temperatures that are high enough to ignite the fuel. Here the time for mixing of fuel and air is too short to evaporate all of the fuel before ignition takes place, and combustion starts even in the presence of liquid fuel. The formation of soot is inevitable under such conditions, and research aims at finding flow conditions that will promote mixing so that the soot is burnt as much as possible inside the cylinder.

Although not new as a combustion concept for internal combustion engines, homogeneous-charge compression ignition (HCCI), also known by various other names (e.g., controlled auto-ignition), is currently receiving a lot of attention in research and development [10]. This concept, although it can work with Diesel fuel, is best described as a hybrid of a regular gasoline SI engine and a Diesel engine. The fuel and air are homogeneously mixed (more or less) as in a standard gasoline engine, and combustion initiation takes place through auto-ignition just as in a Diesel engine. Traditionally, auto-ignition is exactly what developers of SI engines have always feared and striven to avoid. Unintentional auto-ignition in a gasoline engine is known as "engine knock," which can lead to the destruction of the engine; the onset of knock is a function of the chemical and thermal inhomogeneity of the unburned gas while a flame is propagating through it [11]. As a first measure to avoid this, the overall fuel/air ratio (ϕ) is chosen to be very lean ($\phi=0.5$ or less, compared to $\phi=1.0$ at stoichiometry for regular SI engines), but this means that the power output will be lower for this type of engine. Furthermore, stratifying the fuel/air mixture or diluting with exhaust gas can also help to decrease the rate of heat release and thus minimize the risk of destroying the engine. The low fuel/air ratio leads to reduced peak temperatures, which is a big advantage in terms of nitric oxide formation. HCCI engines have demonstrated operation with nitric oxide levels orders of magnitude lower than in standard engines [10]. Operation at a fixed load set point and engine speed can be achieved quite easily and made to continue in a very stable manner. This is an operating condition that can be used in a stationary power generator, for example, but not in an engine that needs to propel an automobile. Acceleration, deceleration, and load changes can require adjustments to the engine that need to be implemented during operation on a very short time scale. Performance of such an engine critically depends on many factors, and careful selection and control of operating conditions are crucial.

Every one of the engine concepts discussed above has its merits and weaknesses. As direct-injection engines mature, the potential arises to operate these engines in different modes depending on the power and speed requirements. This, of course, will be possible only with sophisticated and robust control schemes.

8.2.2 Turbines

Turbines are primarily important in two areas: power generation and engines for aircraft. Turbines are highly complex mechanical devices that must function over many years of sustained operation. Jet engines have to operate under dramatically changing boundary conditions (temperature, pressure, and humidity change substantially during a typical flight), and combustion stability becomes a serious problem. For readers who are interested in details on turbines, a recent review by Correa gives a good overview of the status and remaining problems with turbines [12].

8.2.3 Solid Fuel Burners

As was illustrated with the examples above, the mixture composition of fuel and air has a significant impact on performance and pollutant formation in combustion processes. For gaseous or liquid fuels, mixing can be managed in a relatively well-controlled manner. Mixing becomes a real issue with the use of solid fuels, as found in waste incinerators. The quality of the fuel (waste) changes over a large range of heating values, humidity, and compactness. All of these factors will affect the amount of heat produced and the local temperature. As was the case for nitric oxide formation in internal combustion engines, temperature inhomogeneities in waste incinerators can lead to the formation of highly dangerous toxic substances such as dioxins. A means to control the temperature distribution in the furnace of a waste incinerator is thus a critical element in its safe and clean operation.

Another aspect of solid fuel combustors (and other combustion processes) is the need for treatment of the exhaust gas to remove oxides of nitrogen and sulfur dioxide. Nitric oxide removal is achieved by injecting a chemical (urea or ammonia) into the exhaust stack to react with the nitric oxide, thereby converting it to water and nitrogen [13]. Given that smokestacks of large power plants have diameters of several meters, the injection of the reagents can lead to spatial inhomogeneity. Controlling the amount of locally available reagent is a necessity to guarantee best possible removal of nitric oxide.

8.3 IMAGING IN COMBUSTORS

8.3.1 Concepts

In order to extract useful control signals that are based on spatially resolved information, measurements must be performed that provide at least two-dimensional (2D) data with adequate temporal resolution and at a suitable repetition rate (i.e., framing rate). This section describes different means to acquire image information from combustion-related processes. Techniques to measure parameters such as species concentration, temperature, and velocity are discussed, along with the extraction of information that can be used for control purposes.

8.3.1.1 Sensing Principles

An important factor in imaging diagnostics is the spatial resolution that is required. For a direct imaging detector (e.g., CCD camera), the spatial resolution is determined by the number of pixels and the overall area that needs to be imaged (see Chapter 3). Direct imaging techniques obtain 2D information directly without further processing, even if subsequent enhancement or processing of the image is required to obtain data for control purposes. Indirect imaging techniques, on the other hand, generate images through a tomographic reconstruction process, as discussed in Chapter 4. The spatial resolution of tomographic systems is a complex function of the number of measurement channels, their geometrical configuration and noise properties, and the image reconstruction algorithm. Both imaging approaches can use either passive detection, such as recording the luminosity of flames or positron emission tracking (PET), or active detection, such as fluorescence measurements or electrical capacitance tomography (ECT).

There are major differences between passive and active imaging in terms of the information that can be obtained as well as the ability to locate the origin of the signal. Passive 2D direct imaging usually means that the recorded signal in each pixel is a line-of-sight integral over part of the third dimension. This integration could be a problem if significant gradients in the measured parameter in the third dimension could affect the control signals being extracted [14]. On the positive side of passive direct imaging is its relative ease of implementation, at least on research systems, and the need for only one (usually large) optical access port. For active direct imaging, it is often necessary to provide optical access for a laser light sheet and additional optical access for the image detector perpendicular to the laser-illuminated plane. Requirements for tomographic systems are somewhat "in between" in the sense that they require mostly access only in two dimensions but for best performance need to have access points all around the measurement object; these can, however, be relatively small windows or lenses.

Passive imaging usually records electromagnetic emissions, often in the visible or near-visible spectral range. These emissions can either be thermal blackbody radiation (Planck distribution) or specific emissions from molecules with a distinct band structure. Blackbody radiation can provide access to temperature and can also be used for measurements of soot concentration. Molecular infrared

TABLE 8.1
Features of Direct and Indirect Imaging Methods

	Direct	Indirect
Access required	Large-scale, often 3D	Small, distributed, 2D
Localization of pixels	Integrating in third dimension	Accurately defined in 3D
Spatial resolution	Large number of pixels/unit area; high resolution	Relatively low
Temporal resolution	Excellent	Good
Framing rate	Limited by excitation source or by detection camera; usually poor, typically 1 fps	Usually excellent, >1000 fps, continuously; limited by detector noise

emissions are strong for prominent combustion-related molecules such as water and carbon dioxide; these have been used in tomographic imaging of laboratory flames [15].

Any form of imaging, whether direct or indirect, active or passive, will be subject to some limits on temporal resolution and framing rate. Some broad indications of these parameters can be given for each general type of imaging, but they do not necessarily apply to every example within the given class. In Table 8.1 and Table 8.2, we attempt to give broad-brush indicators of the key issues and performance status for each class of imaging.

During combustion, molecules with excess energy content can radiate in the visible or ultraviolet spectral range via electronic transitions. Passive detection by arrays of optical fibers was explored by Spicher et al. [16] and Ault and Witze [17]. This has been extended to tomographic detection and reconstruction of the chemiluminescence emissions from flames in engines using passive imaging

TABLE 8.2
Features of Active and Passive Imaging Methods

	Passive	Active
Access required	Detector only	Excitation and detection
Sensing	Electromagnetic emission	Wide range of techniques
Spatial resolution	Not determined by whether the technique is passive or active.	
Temporal resolution	Determined by detector sensitivity, for fixed source strength	In principle, under full control of system designer
Framing rate	Can be excellent, >1000 fps, continuously; limited by image detector noise and source strength	Wide range of capabilities, depending on technique used; e.g., can be limited by excitation source or by detector

mode [18]. Prominent sources for this chemiluminescence are the ultraviolet emissions from OH radicals, blue emission from CH radicals, green from C_2, and red from H_2.

In contrast to passive imaging, active imaging provides greater flexibility in terms of species selectivity. With a broad range of excitation sources available, light wavelength from optical sources can be specifically matched with absorption lines of the molecules of interest. If the absorption leads to subsequent fluorescence that can be isolated from signals of other molecules, then a sensitive imaging strategy emerges to allow the detection of even trace species. This approach is called planar laser induced fluorescence (PLIF) imaging. Scattering techniques will provide information about droplets or solid particles. Examples later in this chapter will illustrate the versatility of active imaging. The following sections describe techniques that are used to image combustion processes.

8.3.1.2 Sensing Techniques

Absorption is a spatially integrating technique that typically yields one data point for each absorption path. The optical path can be defined by collimation of the launch, receive optics, or both. In shadowgraphy setups, it is typical to use a direct imaging detector, imaging in the plane perpendicular to the line of sight. The attenuation of bulk illuminating signals is then measured to provide some information about the spatial extent of an object, e.g., a spray, which is placed between the light source and the detector. This method yields a 2D projection of a typically three-dimensional (3D) object. Although a one-dimensional (1D) projection can be obtained by using multiple parallel absorption paths [19], it is more useful to arrange collimated beams to cross each other within a plane; then, spatially resolved information in the plane of the beams is obtained via tomographic reconstruction algorithms [20, 21].

The absorption technique can be used in the collimated line-of-sight approach to provide accurate measurements of the path integral of concentration for various species, such as hydrocarbon fuel [21], O_2 [22, 23], H_2O [24–26], CO [26, 27], CO_2 [27, 28], and NO [29]. It can also yield temperature measurement along a line of sight [26, 30]. All of this recent work is based on the use of diode lasers, with varying degrees of tunability.

Electric impedance measurements are similar in nature to line-of-sight absorption measurements in the sense that a spatially integrated signal is obtained. However, the measurements obtained between any two electrodes are sensitive to the distribution of material over the whole cross-section. Sophisticated reconstruction is needed to evaluate signals. This technique is discussed in Chapter 4. However, it should be noted that when the subject to be imaged is a flame, the electrical properties are highly unusual compared with other subjects of electrical tomography [31].

Laser induced fluorescence imaging is a combination of absorption (usually weak, to avoid noticeable attenuation of the laser light) and fluorescence emission from excited states [32]. This is a sensitive technique that provides opportunities

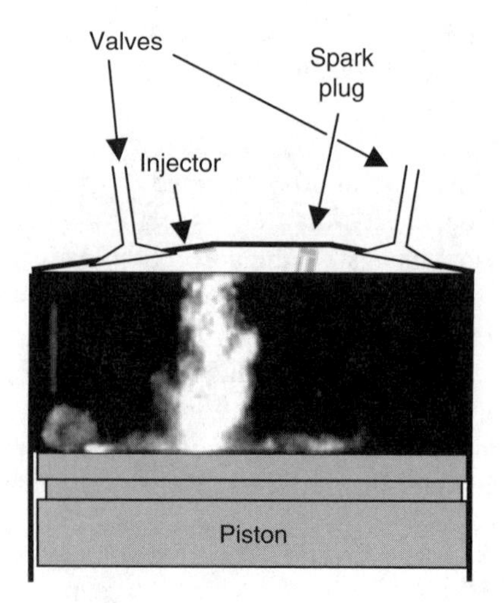

FIGURE 8.5 Planar laser induced fluorescence imaging visualizes the fuel spray inside a direct-injection engine. The spray hits the piston surface and then spreads out horizontally.

for measurement of many different species and quantities. Besides high sensitivity, it offers high selectivity through the combination of molecule-specific absorption and emission transitions, much like fingerprinting. This technique is schematically illustrated in Figure 8.5 and Figure 8.6.

For combustion-related measurements, imaging of fuel distributions is one of the most important applications of PLIF. Figure 8.7 shows the schematic of a PLIF experiment for engine studies. The figure also indicates some of the spectroscopic processes that are involved in the measurement.

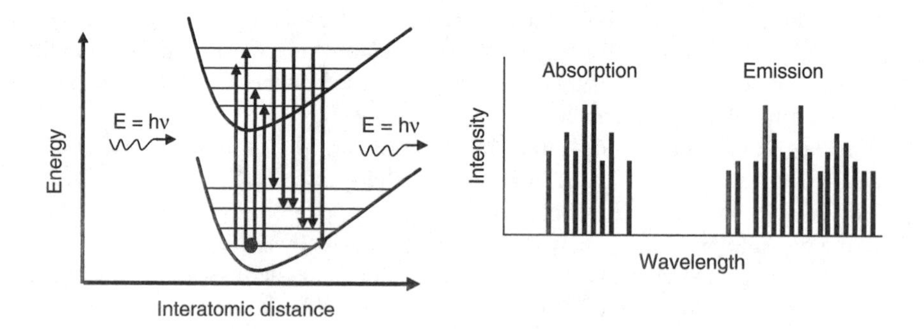

FIGURE 8.6 "Fingerprinting" principle of laser induced fluorescence. Matching the wavelength of the laser light to molecular absorption allows species-selective excitation, and, subsequently, species-specific emissions allow an overall highly selective detection of trace species in the presence of large amounts of other species.

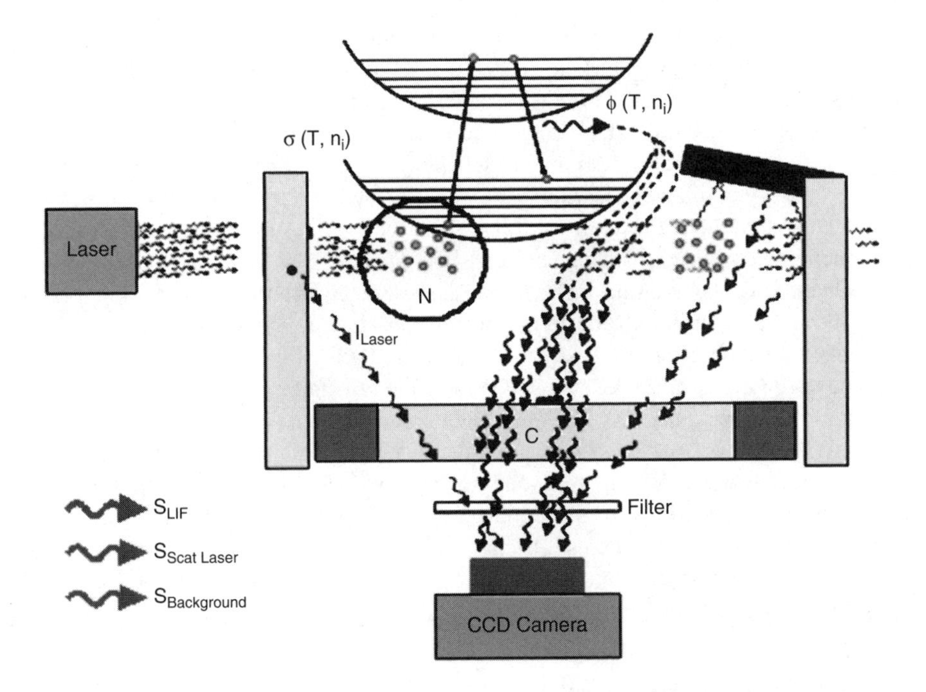

FIGURE 8.7 Quantitative PLIF imaging to determine the number *N* of molecules (shown here for measurements in engines) requires the isolation of the fluorescence signals (S_{LIF}) from background signals ($S_{background}$) and scattered laser light ($S_{scat\ laser}$). Temperature and number density dependent absorption cross-sections and fluorescence quantum yields need to be known for quantification of the measured PLIF signals.

While some constituents of regular fuels will produce PLIF signals, it is very difficult to quantify these signals, and usually such approaches are only qualitative in nature. For quantitative measurements, tracers are added to nonfluorescent model fuels, e.g., iso-octane, n-heptane, or blends of a few hydrocarbons [20]. Tracers for different fuels have been introduced. Some tracers, such as ketones, track the fuel concentration [33, 34], while others (e.g., toluene) directly measure the equivalence ratio (fuel/air ratio) without a separate oxygen measurement [35, 36].

Tracer-based LIF imaging can be used for both liquid and gas phase studies; there are even tracer combinations that allow the distinction between vapor and liquid phase [37]. Hydroxyl radicals are important intermediates in combustion processes that can be conveniently imaged via PLIF [38, 39]. The distribution of hydroxyl radicals provides information about reactive zones. Like hydroxyl radicals, nitric oxide is a minority species in combustion processes of high importance for technical devices. Several PLIF schemes have been developed to image nitric oxide directly in various combustion devices [40–46]. It is also feasible to image temperature distributions by PLIF, as discussed in Section 8.4.1.

Particle image velocimetry (PIV) is a technique that tracks the motion of either droplets or solid particles to deduce 2D or 3D velocity fields. The signals that are recorded are based on Mie scattering from the particles, and correlating techniques are employed to determine vector fields from the data [47]. This technique is more fully described in Chapter 3.

Rayleigh scattering has very limited use in technical combustion devices since the signal has the same wavelength as the exciting (laser) light, making it susceptible to reflections from surfaces. In contrast to PIV, where signals are also resonant with the exciting light, these measurements are based on the intensity of the signal; therefore, reflected light will introduce a systematic error into the measurement. For details, see M. B. Long in Taylor [48]. Rayleigh scattering depends on the number density, and thus, if the effective scattering cross-section is known, temperature distributions can be inferred from the measurements [49].

With *Raman scattering* it is possible to measure majority species quantitatively even under harsh environments such as engines. However, due to the very low Raman scattering cross-sections, 2D measurements are possible only in unusually dense environments [50] or with high laser powers that pose additional technical problems [51]. One-dimensional measurements are much more easily implemented [52].

8.3.2 Enabling Technology

Given the wide range of demanding applications in the combustion arena, it is no surprise that the field is in a position to benefit from a wide range of recent technology innovations. A similar situation arose in the 1980s with the rapid adoption of (then) advanced microprocessor technology for engine management systems.

Access is an important issue in combustion imaging. For example, in a highly sooting combustion system, electrical impedance tomography may have advantages over optical techniques for flame imaging, motivating electrical access for suitably large electrodes [31]. On the other hand, chemically specific images require optical access; if PLIF is adopted, then large-scale optical access is necessary. Hence, access is normally a limiting feature in all combustion imaging systems. It is worth noting at this point that optical fibers have added a new dimension to optical combustion imaging systems since their applications to engines were pioneered by Spicher et al. [16] and Ault and Witze [17]. The same is true of diode lasers coupled to optical fiber delivery systems, although only recently has their promise been demonstrated in subsystems relevant to combustors; see Allen [22] and Hindle et al. [53].

Robustness of the technology used is paramount given the hostile environment of the combustion system and the demanding targets for combustion measurements over a wide range of operating conditions. The practitioners of each measurement technique have developed particular solutions for the multifaceted problems they face. However, many of these solutions only apply to research engines or pilot combustors. It is highly desirable (especially for control) to operate imaging systems on plant that is as near as possible to "normal" operating conditions. Much of the future of measurement engineering in this field will be devoted to such technological development.

Instrumentation technology is a crucial enabling discipline for combustion imaging. Combustion imaging techniques have benefited greatly from advances in electronics and data acquisition systems. For example, the CCD camera is a highly sophisticated device that is now available with many optimized properties for this application, e.g., good temporal resolution (~10 ns) and background rejection, extended spectral sensitivity to record UV and NIR signals, programmable readout, and high spatial resolution with more than 10^6 pixels. Technological uses of the CCD have been greatly facilitated by its pervasive use in consumer electronics, e.g., leisure photography and security systems. The remaining Achilles' heel of the CCD camera is its relatively slow readout rate, which is about 100 frames per second. Specialized systems are available that combine multiple CCD sensors in one unit which then can be read out at much higher rates (on the order of thousands of frames per second).

Data analysis of image data streams is demanding in at least two respects:

1. The rate of raw data accumulation when imaging even at modest rates.
2. The need to interpret an image and extract features that provide a control signal, and to complete the analysis on a sufficiently short time scale.

For the long-term exploitation of combustion imaging for control, it is essential to overcome or circumvent these problems. Hence the techniques discussed in Chapter 3 through Chapter 7 are of central concern for this goal. Computing power is already heavily exploited in the automotive industry for engine management. It is anticipated that the combustion-based industries will be among the early adopters of image-based control as soon as the basic utility and robustness of such systems are established.

8.4 RESULTS FROM COMBUSTOR IMAGING

8.4.1 ENGINES

Most of the applications to date have used modified engines where parts of the cylinder head, the cylinder liner, or the piston have been replaced with optically transparent material [36]. Figure 8.8 shows an optical engine. Once optical access is provided, a number of imaging techniques can be applied to study flows, fuel/air mixing, and combustion.

A crucial parameter that controls the fuel/air mixing and combustion progress is the flow and its associated turbulence. PIV can measure instantaneous velocity fields from which turbulence quantities such as turbulent kinetic energy, dissipation rates, and length scales can be determined [54]. PIV investigations have shown how sensitively the flow reacts to changes of external control mechanisms

FIGURE 8.8 Example of an optically accessible direct-injection engine (see Frieden and Sick [36] for details). Reprinted with permission from SAE Paper No. 2003-01-0068 © SAE International.

in an engine [55]. Forcing the flow into a strongly swirling mode not only has a strong impact on cycle-to-cycle variations in the flow pattern [56] but also substantially changes turbulence levels [54]. Using a flow control device in the intake manifold of an engine substantially changes the flow structure, as Figure 8.9 shows for three different levels of in-cylinder swirl flow.

As discussed in Section 8.2.1, more engines are being designed based on direct-injection concepts where the fuel is directly injected into the combustion chamber. The placement of fuel at the right location at the right time is essential for optimum performance and lowest emissions. Planar direct imaging of Mie scattering from the fuel droplets in laboratory spray rigs provides strong signals that visualize the extent of the fuel spray [57, 58]. These data can be used to tune injector location and injection timing for best operation. The direct visualization helps developers to understand why some operating conditions are better than others. Figure 8.10 shows an example of such a direct image, following a technique described by Hargrave et al. [59].

Alternatively, fuel sprays can also be visualized via PLIF imaging with the advantage that the PLIF signals are relatively immune to scattered light [36, 60, 61]. However, unless exciplex techniques [62] are used, the recorded signal will be the sum of signals from fuel in the liquid and gas phases. Combined with Mie scattering imaging, PLIF signals can be qualitatively separated into gas and liquid phase signals [63] and also can be used to obtain some information about the mean droplet diameter distribution [60, 64, 65]. Spray visualization of that nature was crucial during the development of direct-injection gasoline engines to

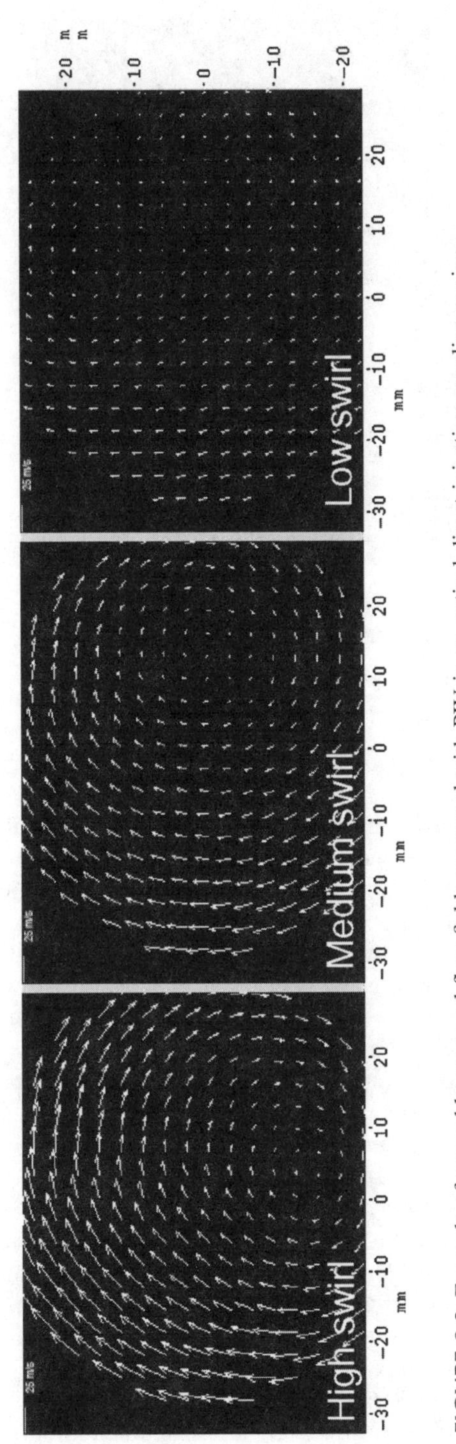

FIGURE 8.9 Example of ensemble-averaged flow fields measured with PIV in an optical direct-injection gasoline engine.

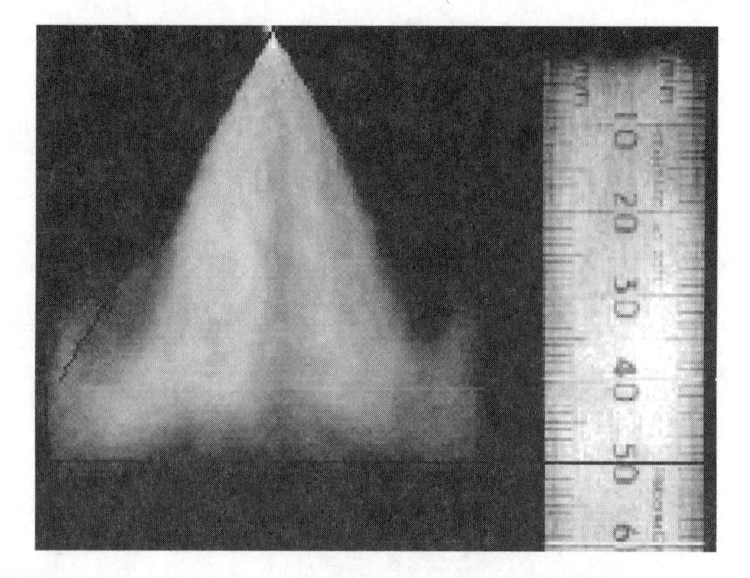

FIGURE 8.10 Mie scattering image of a fuel spray from a pressure swirl injector as commonly used for automotive applications.

improve the guidance of the spray toward the spark plug to ensure reliable ignition [66–68]. Integrating shadowgraphy spray investigations are useful as well in this context [40]. Recently, the application of x-ray absorption for spray studies has been reported [69].

Tomographic spray measurements to date have used the apparent attenuation of collimated light beams due to Mie scattering. These techniques have the potential to enable very rapid imaging of the large-scale properties of the spray. Figure 8.11 [53] shows cross-sectional images of the same gasoline direct-injection (GDI) spray of iso-octane as shown in the photographic side-view of Figure 8.10; images are shown at 2 mm above the vessel base (bottom row), 10 mm and 18 mm, and at different times after the nominal start of injection.

In this type of case, it may be valid under many circumstances to use a model-based reconstruction (as discussed in Chapter 4) and hence improve the definition of the required spray parameters. Although NIR penetration of GDI sprays reaches as low as about 1 to 5%, the optical tomographic reconstruction techniques developed for dense systems in medicine [70, 71] also promise to be of utility for combustion spray systems.

An ideal application of PLIF is the visualization of gaseous fuel distribution and equivalence ratio maps. Given the critical dependence of combustion performance on the equivalence ratio, this is a valuable tool for determining the optimal operating conditions of an engine. Two-dimensional fuel or equivalence ratio distributions are reported in the literature for many different engine configurations (see SAE Technical Paper Series). Scanning into the third dimension, as shown in Figure 8.12 through Figure 8.14, provides additional information about the

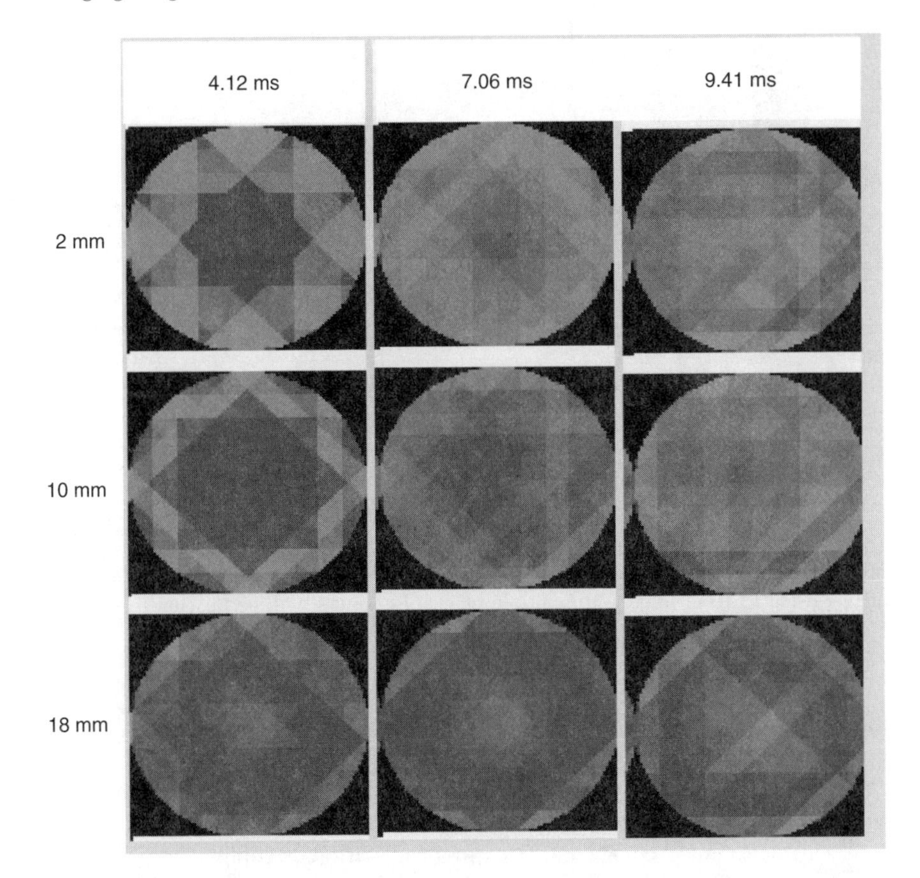

FIGURE 8.11 Tomographic reconstruction of the fuel distribution in a fuel spray (same spray as shown in Figure 8.10).

combustion process [36, 66]. Figure 8.12 shows details of a PLIF setup where the laser light sheet is scanned to illuminate horizontal planes at various heights in the engine. As a result, averaged image sequences can be used to compose a 3D representation of the fuel spray development inside the engine. Figure 8.13 illustrates how much the fuel spray structure changes when the in-cylinder swirl flow is increased. For a selected plane that intersects the position where the spark plug ignites the fuel, image sequences shown in Figure 8.14 demonstrate that the flow substantially affects the rate of combustion as well. For the higher swirl conditions, the flame propagates much faster through the fuel/air mixture. A multilaser system has also been devised to allow single-pulse 3D measurements [72], but the complexity and cost of such a system limit the applicability to selected situations.

Many of the fuel imaging studies have used 3-pentanone as fluorescence tracer in iso-octane as a base fuel. This technique enables quantitative measurements of the fuel concentration distribution to be obtained [73–75]. Toluene is

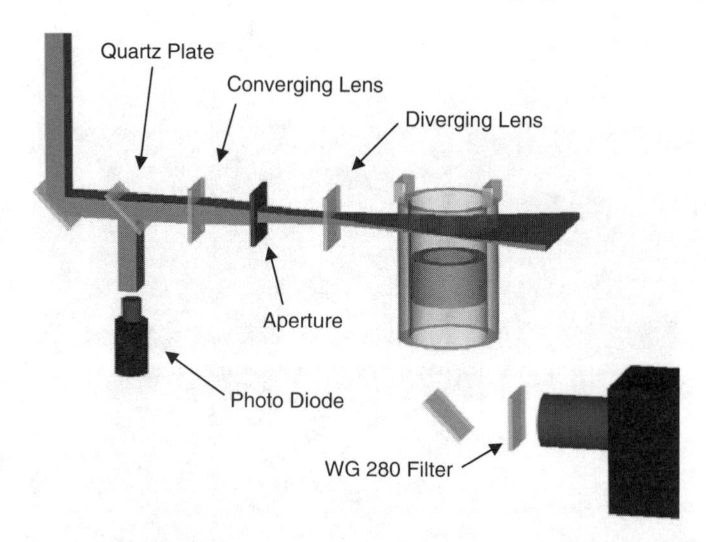

FIGURE 8.12 Setup for quasi-3D measurements of the equivalence ratio in an optical engine using PLIF of toluene in multiple planes. Reprinted with permission from SAE Paper No. 2003-01-0068 © SAE International.

also used as tracer and, because of its susceptibility to oxygen quenching, PLIF imaging of toluene gives a reading of the equivalence ratio [35, 36, 76]. The combination of both tracers allows the quantitative determination of oxygen concentrations in an engine and hence also of the amount of residual gas [77, 78].

Real fuels, however, are mixtures of many different components, each of which has individual evaporation characteristics. Improved engine studies therefore should include effects of multicomponent fuels. In a combination of PLIF and IR-absorption tomography, mixing effects as a function of fuel component in a multicomponent fuel were investigated. The higher spatial resolution of the PLIF imaging diagnostics and the quantitative nature of the absorption-based measurements formed a complementary measurement strategy [20]. PLIF imaging of real gasoline was used as well, and the results were compared to tracer-based PLIF imaging [66]. PLIF was calibrated with a premixed homogeneous evaporation of gasoline blended with 3-pentanone (to de-emphasize the LIF signals of gasoline that stem mostly from higher boiling point components); subsequently, stratified charge operation points were studied [79].

For control purposes, it is useful to image the fuel distribution quasi-continuously throughout large portions of the engine cycle with good temporal resolution, e.g., every few degrees CA over a period of 80° CA at engine speeds up to 6000 rpm. These speeds are beyond the current capabilities of PLIF but are well within the range of NIR absorption tomography (NIRAT) in an engine simulator, using a diode-laser– and fiber-based 32-channel system operating at 1700 nm [21]. Imaging rates of 3500 frames per second have been achieved [53], and model-based

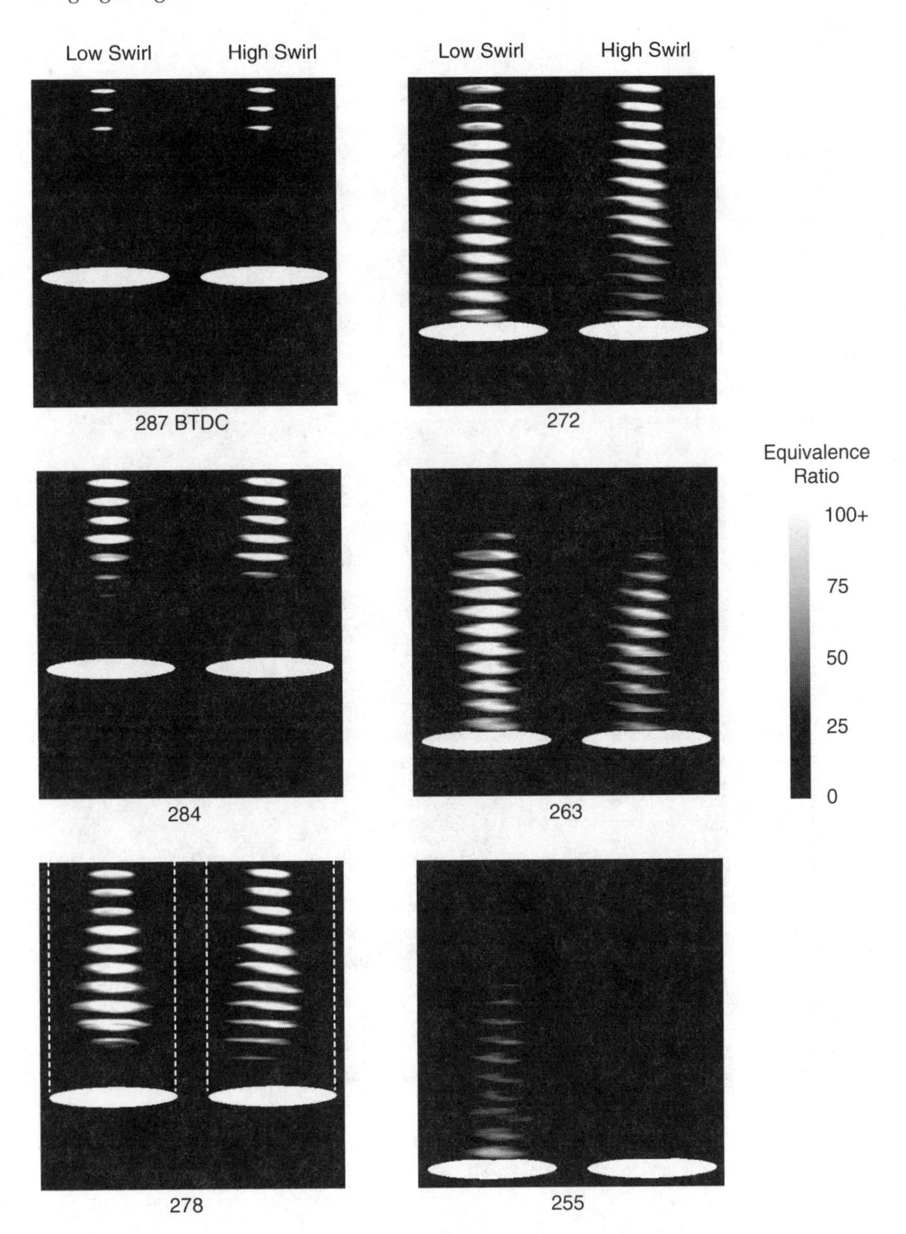

FIGURE 8.13 2000-rpm fuel spray images for low- and high-swirl flow conditions in an optical direct-injection gasoline engine. Crank position varies as indicated in °BTDC (degrees Before Top Dead Center), i.e., the rotational position of the crank shaft before the piston reaches its highest position in the engine (closest to the spark plug). Piston location is shown in white. Elevations (from top) = 1, 4, 8, 12, 16 mm, etc. Reprinted with permission from SAE Paper No. 2003-01-0068 © SAE International.

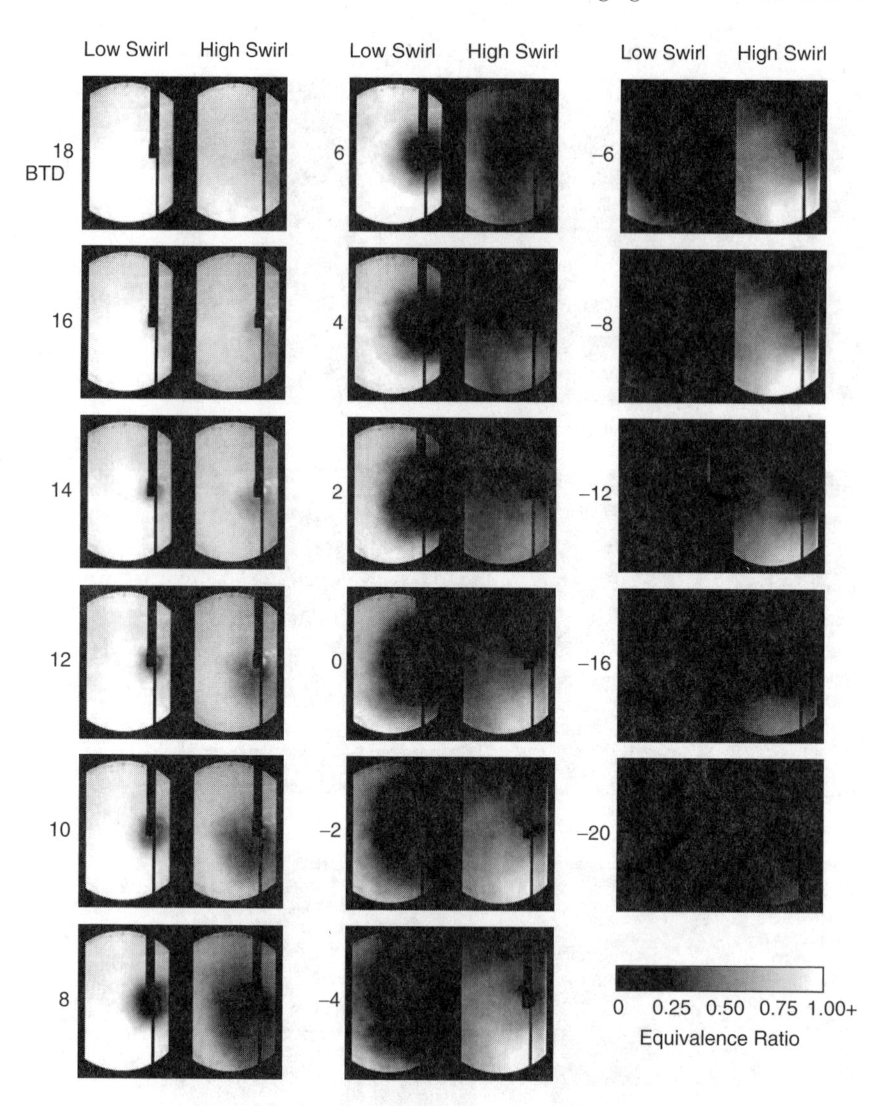

FIGURE 8.14 Combustion images for 2000-rpm low- and high-swirl conditions. The images were taken at an elevation of $z = -1$, close to the spark plug gap. All times are in °BTDC. (See Frieden and Sick [36] for details.) Reprinted with permission from SAE Paper No. 2003-01-0068 © SAE International.

image reconstruction has been implemented to produce physically reasonable images [80]. Figure 8.15 shows an illustrative sequence of NIRAT images of two propane gas injections into an engine simulator. The two injections were separated in space by about $^1/_3$ of the chamber diameter, and in time by 100 ms.

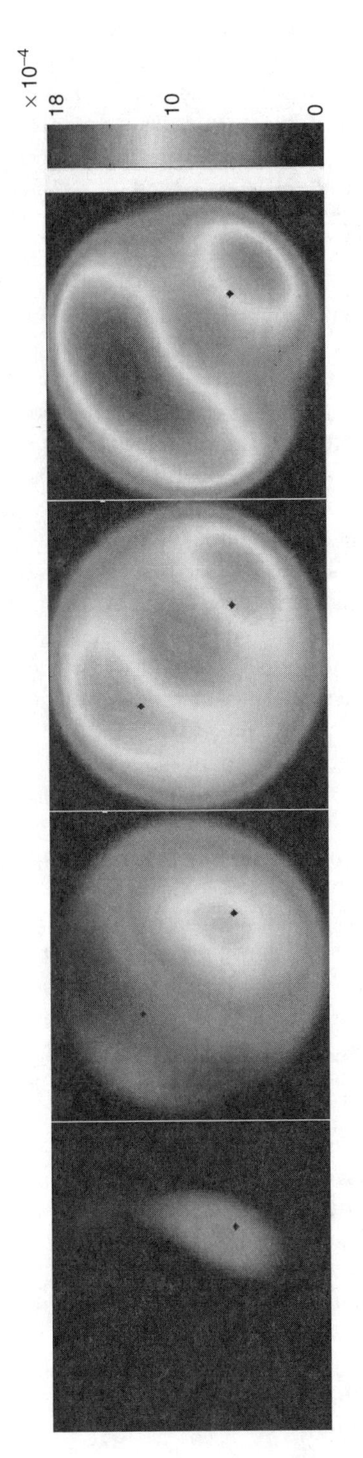

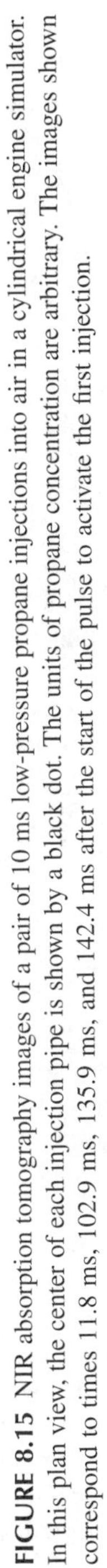

FIGURE 8.15 NIR absorption tomography images of a pair of 10 ms low-pressure propane injections into air in a cylindrical engine simulator. In this plan view, the center of each injection pipe is shown by a black dot. The units of propane concentration are arbitrary. The images shown correspond to times 11.8 ms, 102.9 ms, 135.9 ms, and 142.4 ms after the start of the pulse to activate the first injection.

Once a fuel/air mixture is ignited, the location of combustion is indicated in the fuel imaging experiment by areas where the PLIF fuel signal disappears. The rate of combustion can be deduced from images taken at different times after ignition [36, 81]. PLIF imaging of hydroxyl radicals can also be used for this purpose. Such images provide complementary information to the images of "missing" fuel, i.e., where the fuel has already been consumed [39, 82]. Hydroxyl radicals (OH) persist until late in the exhaust zone, albeit at lower concentration at the thermal equilibrium level. See Figure 8.16.

OH also exhibits chemiluminescence that can be recorded with imaging detectors to monitor combustion activity and at least a 2D representation of a 3D emission field. This technique is especially interesting because it can be recorded with a high repetition rate in contrast to PLIF imaging. Through chemiluminescence, information is obtained about the spatial and temporal occurrence of combustion, and its intensity is often also correlated with the release of heat [83].

Early work on passive fiber-based systems exploited broadband chemiluminescence. Large arrays of fibers (up to about 400) embedded in the cylinder wall, piston, and cylinder head were used to monitor flame-front velocity and the transition to knock [16]. Radially averaged flame velocity measurements, resolved in angle around the cylinder axis, were obtained [17] by using a small fiber array mounted in the spark plug in conjunction with electrical ionization detectors mounted in the head gasket. By using collimated optical collection of chemiluminescence via fibers mounted in a head gasket, tomographic images of the flame were obtained with excellent temporal resolution [18].

Like OH, CH radicals produce chemiluminescence that is a good indicator of combustion activity. CH emissions originate only in the flame front and not in the burnt gas region [84, 85]. The thermal radiation that is produced by soot can also be detected and is used to extract information about the soot temperature and soot volume fraction [86, 87]. See Figure 8.17.

All of the above techniques for combustion monitoring rely on optical access. However, the chemi-ionization of a flame region is detectable by electronic techniques. A series of experiments with electrical impedance tomography [31] have shown the potential for flame imaging without optical access, even with very crude image reconstruction. Due to the need for large electrodes and stable field patterns, this technique seems mostly of interest for turbine or furnace flame monitoring.

Burnout during the postcombustion part of the engine cycle can be visualized by following hydroxyl radicals that are present in their equilibrium concentration after the flame front has passed. After that time, the main products of combustion, water and carbon dioxide, can be used to image the mixing of exhaust with fresh charge; this process is known in the automobile industry as exhaust gas recirculation (EGR) and is used to reduce the formation of nitric oxide. PLIF imaging of water uses a weak two-photon process that is hard to quantify and as such is not used routinely in technical processes [88]. Carbon monoxide and carbon dioxide can be detected with infrared LIF [89], and recently UV LIF detection of CO_2 was investigated as well [90].

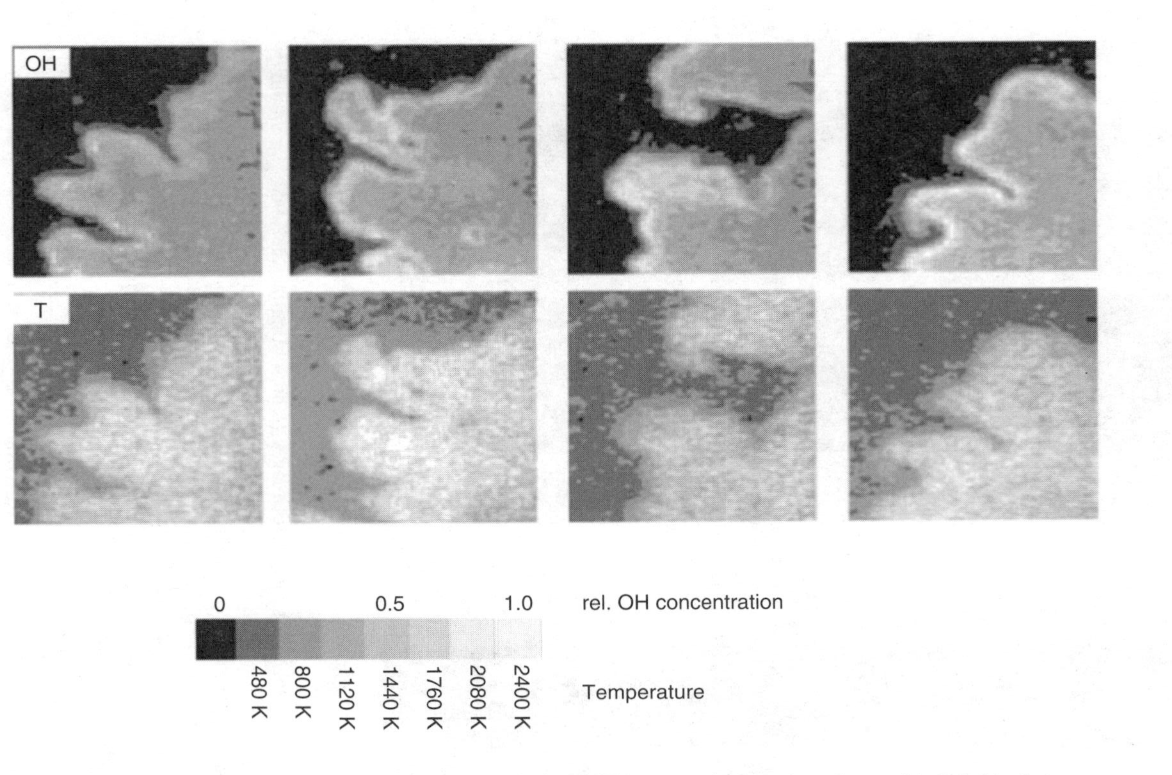

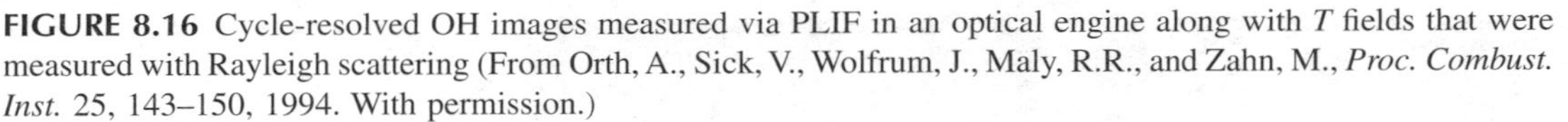

FIGURE 8.16 Cycle-resolved OH images measured via PLIF in an optical engine along with *T* fields that were measured with Rayleigh scattering (From Orth, A., Sick, V., Wolfrum, J., Maly, R.R., and Zahn, M., *Proc. Combust. Inst.* 25, 143–150, 1994. With permission.)

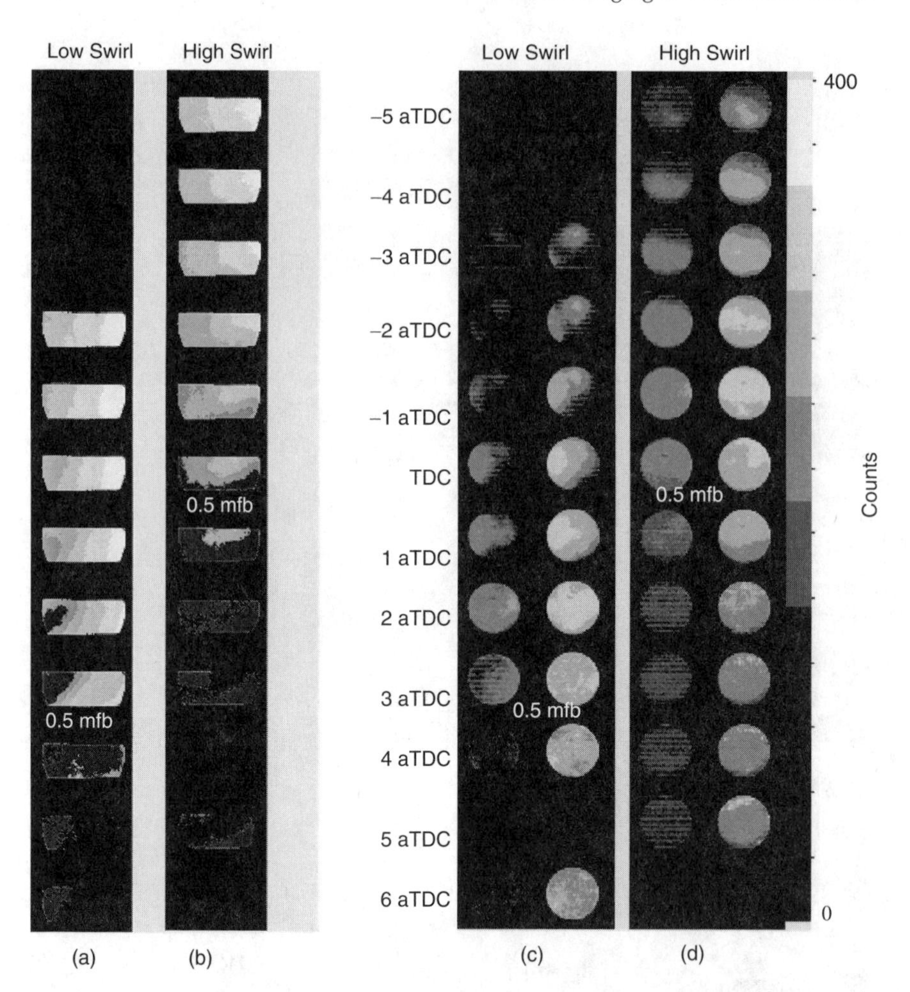

FIGURE 8.17 Averaged toluene PLIF images (representing the fuel) taken in an optical HCCI engine for low (a) and high (b) swirl flow conditions. Compare the averaged chemiluminescence images acquired under the same conditions (c and d). The images reveal how different in-cylinder swirl affects the combustion phasing and where and when the fuel is consumed inside the cylinder. The left-hand images in columns c and d show OH^* chemiluminescence, while the right-hand side shows chemiluminescence from CH, and potentially other species [83].

As with PLIF imaging of fuel via the use of fluorescing tracers, the exhaust gas can be tracked with a tracer. SO_2 was proposed as a suitable tracer since it is being formed during combustion and represents burned gases [91]. Quantification of SO_2 PLIF has not yet been demonstrated, however.

The importance of temperature distribution in the gas-phase region in combustion systems has motivated many sophisticated attempts to access it. Point (or "voxel") measurements of temperature by coherent anti-Stokes Raman spectroscopy

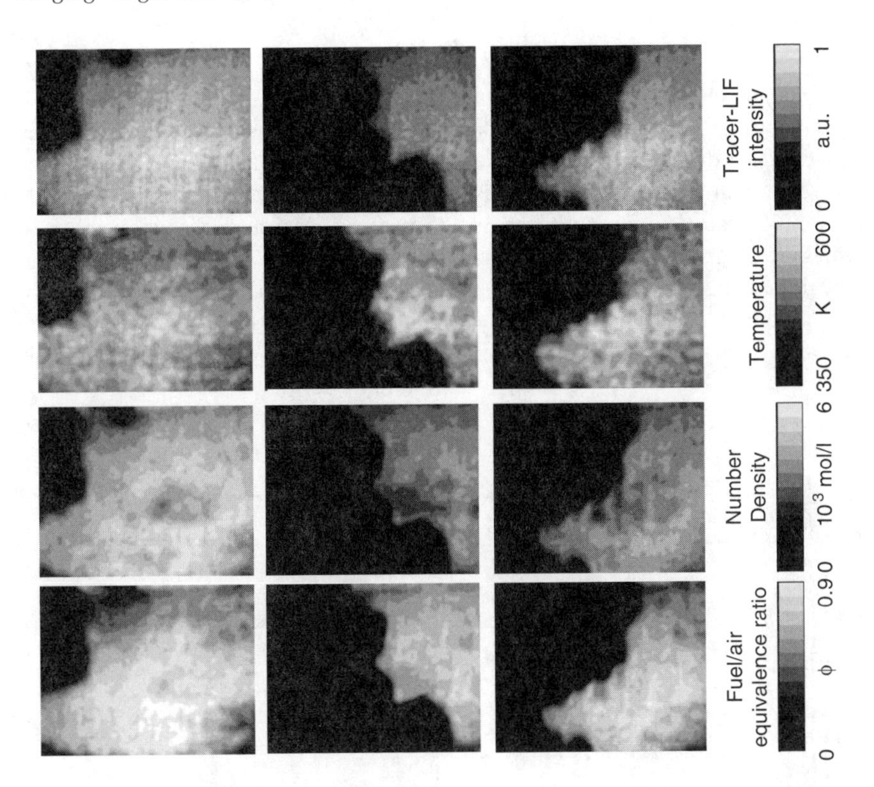

FIGURE 8.18 Measurements of the tracer-LIF intensity (first row), temperature (second row), fuel number density (third row), and fuel/air equivalence ratio (fourth row) in a two-stroke engine. The columns show three randomly chosen data sets for instantaneous measurements. (From Einecke, S., Schulz, C., and Sick, V., *Appl. Phys. B* 71, 717–723, 2000.) ©Springer-Verlag.

(CARS) have been used (see for example Kalghatgi et al. [92] and Lucht et al. [93]) for engine studies, and CARS has also been demonstrated in measurements along a line [94]. Rayleigh imaging in technical combustion devices is difficult because of potential interference from scattered laser light and is successful only in special cases. Temperature distributions in an engine model were reported along with hydroxyl images [49] and in combination with imaging of nitric oxide distributions [46]. Temperatures have been measured with LIF of indium atoms that were seeded as a salt into the fuel [95]. Two-line excitation of 3-pentanone, a tracer that is also used for fuel distribution measurements, was employed to measure temperature fields in a two-stroke engine to investigate the role of low temperature chemistry in ignition processes [33, 96] (see Figure 8.18). Recent work [30] has demonstrated the potential of diode-laser–based systems to unravel the spatial distribution of temperature by use of tunable diode laser absorption spectroscopy (TDLAS).

8.4.2 Turbines

Despite the difficult conditions for imaging diagnostics, atomization and fuel/air mixing [63], soot formation via laser-induced incandescence [97], and even temperature distribution measurements are reported in the literature [98]. Often these measurements are accomplished by using subsystem models to study principle features of importance in the process. Although the application of imaging diagnostics to gas turbine control applications presents a difficult technical challenge, significant progress has been made. Chemiluminescence techniques are frequently used for these applications [99–101]. A good overview of current issues related to gas turbines was given by Correa [12]. Details on

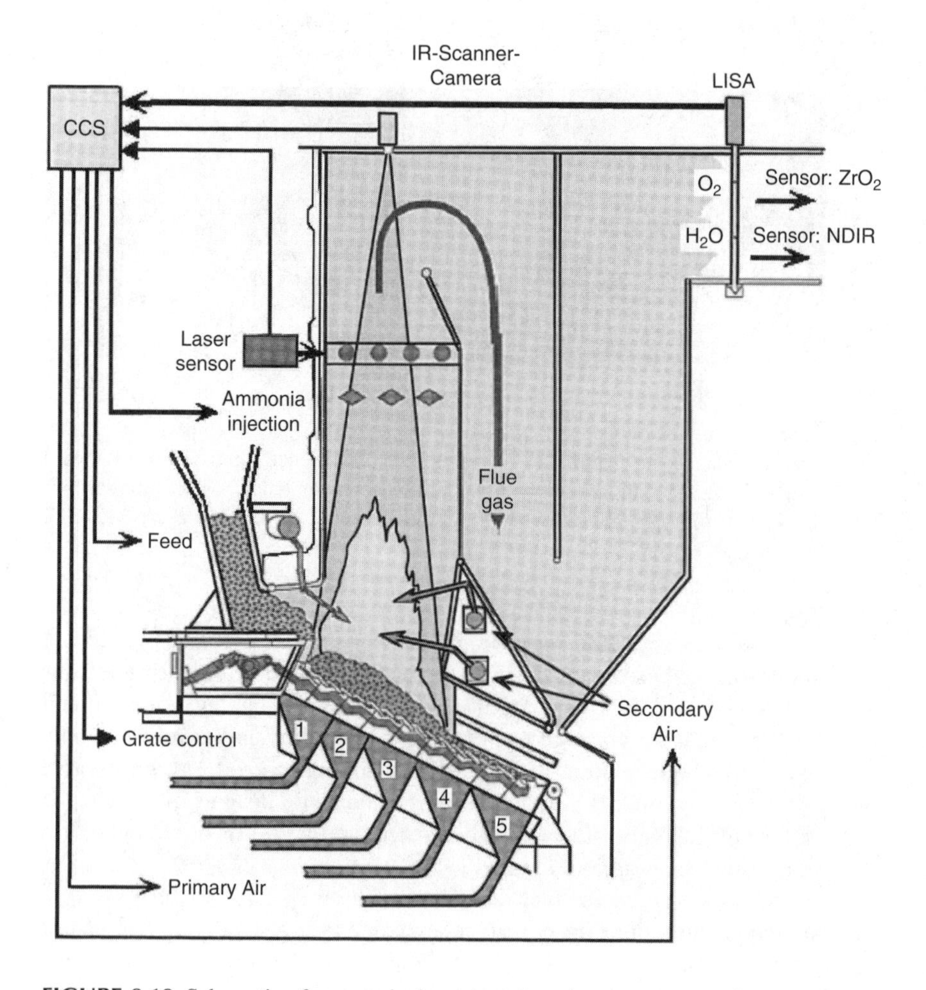

FIGURE 8.19 Schematic of a waste incineration plant showing a range of sensors that is used for active control of combustion performance. (From Wolfrum, J., *Proc. Combust. Inst.* 27, 1–41, 1998. With permission.)

diagnostics applications for turbines can be found in several chapters of Kohse-Höinghaus and Jeffries [8].

8.4.3 Waste Incinerators

Passive detection of thermal radiation from burning waste by an infrared camera was used to extract spatially resolved temperature information of the entire grate area in a waste incinerator plant [102]. Spatially equalizing temperature is important to suppress the formation of unwanted toxins. The temperature images are used to obtain control signals that are fed into the control of the air inlet ducts. This feedback enables the targeted adjustment of local fuel/air ratios and thus adjustment of the local temperature. See Figure 8.19.

In addition to controlling the temperature distribution via the primary air intake, control signals can also be obtained via measurements in the gas phase through infrared diode laser absorption spectroscopy [6]. This technique allows measurements of different species (e.g., oxygen vs. water) and temperature. Finally, ammonia (or urea) can be added to the exhaust gases as part of chemical processes that remove nitric oxide. Careful control of the amount of ammonia is needed for proper operation. A control system based on absorption using a CO_2-waveguide laser [103, 104] has been applied in a tomographic arrangement [105].

8.5 CONCLUSIONS

The discussion in this chapter has illustrated the application of imaging diagnostics to the design and development of combustor systems. In some cases, the insights gained from these applications have been applied to the generation of control maps for routine operation, such as in the case of direct-injection automotive engines. It is anticipated that this mode of combustor system development will grow as the imaging technology matures.

An example of how imaging diagnostics can be a crucial step on the path of designing and testing a control strategy is the current development of HCCI engines. Robust and reliable operation of these engines in automobiles requires continuous control to account for changes in engine speed, load demands, and other factors. Electronic engine controllers will have to use models that predict necessary action. Imaging diagnostics can greatly assist the development of such models. Models will then be implemented for engine control based on global parameters that are measured with nonimaging sensors.

The economic and environmental efficiency of most combustion schemes depends on spatially distributed parameters. The study of combustion over many decades has generated a wide variety of imaging diagnostic systems that provide access to a broad range of parameters. The range and capability of these diagnostic systems are increasing rapidly, exploiting advances in many fields, such as computing and opto-electronics. These imaging systems have played a crucial role in the development of modern combustors, including the development of control schemes, and are now indispensable tools in the field. We feel confident that, in the wider chemical processing industry, there are many instances where the combustion engineer's toolkit of imaging diagnostics could play a similar role in process and product development. Considerable gains may be had from implementing selected imaging techniques to control combustors in routine service. The technological challenges are considerable, but progress in the diagnostic techniques themselves and in associated computing and control techniques offers the prospect that these challenges can be overcome.

REFERENCES

1. J Warnatz, U Maas, RW Dibble. *Combustion,* Berlin: Springer-Verlag, 1999.
2. RM Fristrom. *Flame Structure and Processes,* New York: Oxford University Press, 1995.
3. JB Heywood. *Internal Combustion Engine Fundamentals,* New York: McGraw-Hill, 1988.
4. A Lavoisier. *Essays Physical and Chemical,* London: F. Cass Ltd., 1970.
5. M Faraday. *Faraday's Chemical History of a Candle: Twenty-Two Experiments and Six Classic Lectures,* Chicago: Chicago Review Press, 1988.
6. J Wolfrum. Lasers in combustion: From basic theory to practical devices. *Proc. Combust. Inst.* 27:1–41, 1998.
7. R Hemberger, S Muris, K-U Pleban, J Wolfrum. An experimental and modeling study of the selective noncatalytic reduction of NO by ammonia in the presence of hydrocarbons. *Combustion and Flame* 99:660–68, 1994.
8. K Kohse-Höinghaus, JB Jeffries. *Applied Combustion Diagnostics,* New York: Taylor and Francis, 2002.
9. F Zhao, M-C Lai, DL Harrington. Automotive spark-ignited direct-injection gasoline engines. *Progress in Energy and Combustion Science* 25:437–562, 1999.
10. F Zhao, TW Asmus, DN Assanis, JE Dec, JA Eng, PM Najt (ed.), *Homogeneous Charge Compression Ignition (HCCI) Engines: Key Research and Development Issues.* Pittsburgh, PA: Society of Automotive Engineers, 2003.
11. J Pan, CGW Sheppard, A Tindall, M Berzins, SV Pennington, JM Ware. End gas inhomogeneity, autoignition and knock. SAE Technical Paper 982616, 1998.
12. SM Correa. Power generation and aeropropulsion gas turbines: From combustion science to combustion technology. *Proc. Combust. Inst.* 27:1793–807, 1998.
13. A Arnold, H Becker, R Hemberger, W Hentschel, W Ketterle, M Köllner, W Meienburg, P Monkhouse, H Neckel, M Schäfer, K-P Schindler, V Sick, R Suntz, J Wolfrum. Laser *in-situ* monitoring of combustion processes. *Applied Optics* 29: 4860–72, 1990.
14. BD Stojkovic, TD Fansler, MC Drake, V Sick. High-speed imaging of OH* and soot temperature and concentration in a stratified-charge direct-injection gasoline engine. *Proc. Combust. Inst.* 30: 2657–2665, 2004.
15. H Burkhardt, E Stoll, in RA Williams, MS Beck (eds.), *Process Tomography: Principles, Techniques and Applications.* Oxford: Butterworth-Heinemann, 1995.
16. U Spicher, G Schmitz, H-P Kollmeier. Application of a new optical fibre technique for flame propagation diagnostics in IC engines. SAE Technical Paper 881637, 1988.
17. JR Ault, PO Witze. Evaluation and optimization of measurements of flame kernel growth and motion using a fibre-optic spark plug probe. SAE Technical Paper 981427, 1998.

18. H Philip, A Plimon, A Hirsch, G Fraidl, E Winklhofer. A tomographic camera system for combustion diagnostics in SI engines. SAE Technical Paper 95068, 1995.
19. E Winklhofer, A Plimon. Monitoring of hydrocarbon fuel/air mixtures by means of a light extinction technique in optically accessed research engines. *Optical Engineering* 30:1262–68, 1991.
20. H Krämer, S Einecke, C Schulz, V Sick, SR Nattras, JS Kitching. Simultaneous mapping of the distribution of different volatility classes using tracer LIF and NIR tomography in an IC engine. *SAE Transactions* 107:1048–59, 1998.
21. FP Hindle, SJ Carey, KB Ozanyan, DE Winterbone, H McCann. Measurement of gaseous hydrocarbon distribution by a near infra-red absorption tomography system. *J. Electronic Imaging* 10:593–600, 2001.
22. MG Allen. Diode laser absorption sensors for gas-dynamic and combustion flows. *Meas. Sci. Technol.* 9:545–62, 1998.
23. J Wang, ST Sanders, JB Jeffries, RK Hanson. Oxygen measurements at high pressures with vertical cavity surface-emitting lasers. *Applied Physics B: Lasers and Optics* 72:865–72, 2001.
24. MP Arroyo, S Langlois, RK Hanson. Diode-laser absorption technique for simultaneous measurements of multiple gasdynamic parameters in high-speed flows containing water vapour. *Appl. Opt.* 33:3296–307, 1994.
25. K Salem, D Mewes, E Tsotsas. Near infra-red tomography for concentration measurements of exhausting water vapour of a packed bed adsorber. *3rd World Congress on Industrial Process Tomography,* Banff, Canada, 2003.
26. H Teichert, T Fernholz, V Ebert. Simultaneous *in situ* measurement of CO, H_2O, and gas temperatures in a full-sized coal-fired power plant by near-infrared diode lasers. *Applied Optics* 42:2043–51, 2003.
27. DM Sonnenfroh, MG Allen. Observation of CO and CO_2 absorption near 1.57 μm with an external-cavity diode laser. *Applied Optics* 36:3298–300, 1997.
28. M Webber, S Kim, ST Sanders, DS Baer, RK Hanson, Y Ikeda. *In situ* combustion measurements of CO_2 by use of a distributed-feedback diode-laser sensor near 2.0 μm. *Applied Optics* 40:821–28, 2001.
29. DM Sonnenfroh, MG Allen. Absorption measurements of the second overtone band of NO in ambient and combustion gases with a 1.8 μm room-temperature diode laser. *Applied Optics* 36:7970–77, 1997.
30. ST Sanders, J Wang, JB Jeffries, RK Hanson. Diode-laser absorption sensor for line-of-sight gas temperature distributions. *Applied Optics* 40:4404–15, 2001.
31. RC Waterfall, R He, P Wolanski, Z Gut. Monitoring flame position and stability in combustion cans using ECT. *Proceedings 1st World Congress on Industrial Process Tomography,* Buxton, U.K., 1999, pp. 35–38.
32. AC Eckbreth. *Laser Diagnostics for Combustion Temperature and Species,* Amsterdam, The Netherlands: Gordon and Breach, 1996.
33. S Einecke, C Schulz, V Sick. Measurement of temperature, fuel concentration and equivalence ratio fields using tracer LIF in IC engine combustion. *Appl. Phys. B* 71:717–23, 2000.
34. M Thurber, B Kirby, R Hanson. Instantaneous imaging of temperature and mixture fraction with dual-wavelength acetone PLIF. *36th AIAA Aerospace Sciences Meeting and Exhibit,* Reno, NV, 1998.
35. J Reboux, D Puechberty. A new approach of PLIF applied to fuel/air ratio measurement in the compressive stroke of an optical SI engine. SAE Technical Paper 941988, 1994.

36. D Frieden, V Sick. Investigation of the fuel injection, mixing and combustion processes in an SIDI engine using quasi-3D LIF imaging. SAE Technical Paper 2003-01-0068, 2003.
37. PG Felton, FV Bracco, MEA Bardsley. On the quantitative application of exciplex fluorescence to engine sprays. SAE Technical Paper 930870, 1993.
38. A Arnold, H Becker, R Suntz, P Monkhouse, J Wolfrum, R Maly, W Pfister. Flame front imaging in an internal-combustion engine simulator by laser-induced fluorescence of acetaldehyde. *Opt. Lett.* 15:831–33, 1990.
39. A Arnold, A Buschmann, B Cousyn, M Decker, V Sick, F Vannobel, J Wolfrum. Simultaneous imaging of fuel and hydroxyl radicals in an in-line four cylinder SI engine. *SAE Transactions* 102:1–9, 1993.
40. A Arnold, F Dinkelacker, T Heitzmann, P Monkhouse, M Schäfer, V Sick, J Wolfrum, W Hentschel, K-P Schindler. DI Diesel engine combustion visualized by combined laser techniques. *Proc. Combust. Inst.* 24:1605–12, 1992.
41. A Bräumer, M Decker, S Ro, V Sick, J Wolfrum. 2D-laser temperature and NO concentration measurements in a domestic natural gas burner. *Third International Conference on Combustion Technologies for a Clean Environment,* Lisbon, Portugal, 1995.
42. L Eigenmann, J Meisl, R Koch, S Wittig, H Krämer, V Sick, J Wolfrum. *Validierung von pdf Modellansätzen mittels Messungen mit laserinduzierter Fluoreszenz, 19. Deutscher Flammentag.* VDI Verlag, Delft, Netherlands, 1997, pp. 561–66.
43. F Hildenbrand, C Schulz, V Sick, H Jander, HG Wagner. *Applicability of KrF Excimer Laser Induced Fluorescence in Sooting High-pressure Flames,* VDI Flammentag Dresden. VDI Verlag, Delft, Netherlands, 1999, pp. 269–74.
44. G Josefsson, I Magnusson, F Hildenbrand, C Schulz, V Sick. Multidimensional laser diagnostic and numerical analysis of NO formation in a gasoline engine. *Proc. Combust. Inst.* 27:2085–92, 1998.
45. C Schulz, V Sick, J Heinze, W Stricker. *Laser Applications to Chemical and Environmental Analysis,* OSA Technical Digest Series. Optical Society of America, Washington DC, 1996, pp. 133–35.
46. C Schulz, V Sick, J Wolfrum, V Drewes, M Zahn, R Maly. Quantitative 2D single-shot imaging of NO concentrations and temperatures in a transparent SI engine. *Proc. Combust. Inst.* 26:2597–604, 1996.
47. M Raffel, CE Willert, J Kompenhans. *Particle Image Velocimetry: A Practical Guide,* Berlin: Springer-Verlag, 1998.
48. AMKP Taylor, ed. *Instrumentation for Flows with Combustion.* London: Academic Press, 1993.
49. A Orth, V Sick, J Wolfrum, RR Maly, M Zahn. Simultaneous 2D-single shot imaging of OH concentrations and temperature fields in a SI engine simulator. *Proc. Combust. Inst.* 25:143–50, 1994.
50. M Decker, A Schik, UE Meier, W Stricker. Quantitative Raman imaging investigations of mixing phenomena in high pressure cryogenic jets. *Appl. Opt.* 37:5620–27, 1998.
51. DC Kyritsis, PG Felton, FV Bracco. On the feasibility of quantitative, single-shot, spontaneous Raman imaging in an optically accessible engine cylinder. SAE Technical Paper 1999-01-3537, 1999.
52. G Grünefeld, V Beushausen, P Andresen, W Hentschel. Spatially resolved Raman scattering for multi-species and temperature analysis in technically applied combustion systems: Spray flame and four-cylinder in-line engine. *Appl. Phys. B* 58:333–42, 1994.

53. FP Hindle, SJ Carey, KB Ozanyan, DE Winterbone, E Clough, H McCann. Near infra-red chemical species tomography of sprays of volatile hydrocarbons. *Technisches Messen* 69: 2002.
54. CO Funk, V Sick, DL Reuss, WJA Dahm. Turbulence properties of high and low swirl in-cylinder flows. SAE Technical Paper 2002-01-2841, 2002.
55. DL Reuss, T Kuo, B Khalighi, D Haworth, M Rosalik. Particle image velocimetry measurements in a high-swirl engine used for evaluation of computational fluid dynamics calculations. SAE Technical Paper 952381, 1995.
56. DL Reuss. Cyclic variability of large-scale turbulent structures in directed and undirected IC engine flows. SAE Technical Paper 2000-01-0246, 2000.
57. V Sick, KD Driscoll. Experimental investigation of droplet size distributions in a fan spray. *ILASS Americas, 14th Annual Conference on Liquid Atomization and Spray Systems,* Dearborn, MI, 2001.
58. G Wigley. *Optical Diagnostics for Flow Processes,* New York: Plenum Press, 1994.
59. GK Hargrave, G Wigley, J Allen, A Bacon. Optical diagnostics and direct injection of liquid fuel sprays. *Proceedings of International Conference on Optical Methods and Data Processing in Heat and Fluid Flow,* London, 1998, pp. 121–134.
60. BD Stojkovic, V Sick. Evolution and impingement of an automotive fuel spray investigated with simultaneous Mie/LIF techniques. *Applied Physics B: Lasers and Optics* 73:75–83, 2001.
61. R Steeper, E Stevens. Characteristics of combustion, piston temperatures, fuel sprays and fuel-air mixing in a DISI optical engine. SAE Technical Paper 2000-01-2900, 2000.
62. LA Melton. Planar liquid and gas visualization. *Ber. Bunsenges. Phys. Chem.* 97: 1560–67, 1993.
63. M Rachner, M Brandt, H Eickhoff, C Hassa, A Braümer, H Krämer, M Ridder, V Sick. A numerical and experimental study of fuel evaporation and mixing for lean premixed combustion at high pressure. *Proc. Combust. Inst.* 26:2741–48, 1996.
64. C-N Yeh, H Kosaka, T Kamimoto. Measurement of drop sizes in unsteady dense sprays. In: Kuo, KK (ed.) *Recent Advances in Spray Combustion, Volume I: Spray Atomization and Drop Burning Phenomena;* Progress in Astronautics and Aeronautics, Vol. 166, Reston, VA: American Institute of Aeronautics and Astronautics (AIAA), 1996, pp. 297–308.
65. MC Jermy, D Greenhalgh. Planar dropsizing by elastic and fluorescence scattering in sprays too dense for phase Doppler measurement. *Appl. Phys. B* 71:703–10, 2000.
66. T Fansler, D French, MC Drake. Fuel distributions in a firing direct-injection spark-ignition engine using laser-induced fluorescence imaging. SAE Technical Paper 950110, 1995.
67. W Hentschel, B Block, T Hovestadt, H Meyer, G Ohmstede, V Richter, B Stiebels, A Winkler. Optical diagnostics and CFD-simulations to support the combustion process development of the Volkswagen FSI direct injection gasoline engine. SAE Technical Paper 2001-01-3648, 2001.
68. K Kuwahara, K Ueda, H Ando. Mixing control strategy for engine performance improvement in a gasoline direct injection engine. SAE Technical Paper 980158, 1998.
69. Y Yue, CF Powell, R Poola, J Wang, JK Schaller. Quantitative measurements of Diesel fuel spray characteristics in the near-nozzle region using x-ray absorption. *Atomization and Sprays* 11, 471–490, 2001.

70. JC Hebden, A Gibson, RM Yusof, N Everdell, EMC Hillman, DT Delpy, SR Arridge, T Austin, JH Meek, JS Wyatt. Three-dimensional optical tomography of the premature infant brain. *Physics in Medicine and Biology* 47:4155–66, 2002.
71. SR Arridge. Optical tomography in medical imaging: Topical review. *Inverse Problems* 15:R41–R93, 1999.
72. J Nygren, J Hult, M Richter, M Alden, M Christensen, A Hultqvist, B Johansson. Three-dimensional laser induced fluorescence of fuel distributions in an HCCI engine. *Proc. Combust. Inst.* 29:679–85, 2002.
73. F Großmann, PB Monkhouse, M Ridder, V Sick, J Wolfrum. Temperature and pressure dependences of the laser-induced fluorescence of gas-phase acetone and 3-pentanone. *Appl. Phys. B* 62:249–53, 1996.
74. D Frieden, V Sick. A two-tracer LIF strategy for quantitative oxygen imaging in engines applied to study the influence of skip-firing on in-cylinder oxygen contents of an SIDI engine. SAE Technical Paper 2003-01-1114, 2003.
75. JD Koch, RK Hanson. Temperature and excitation wavelength dependencies of 3-pentanone absorption and fluorescence for PLIF applications. *Appl. Phys. B* 76:319–24, 2003.
76. J Reboux, D Puechberty, F Dionnet. Study of mixture inhomogeneities and combustion development in a S.I. engine using a new approach of laser induced fluorescence (FARLIF). SAE Technical Paper 961205, 1996.
77. D Frieden, V Sick. A two-tracer LIF strategy for quantitative oxygen imaging in engines applied to study the influence of skip-firing on in-cylinder oxygen contents of an SIDI engine. SAE Technical Paper 2003-01-1114, 2003.
78. D Frieden, V Sick, J Gronki, C Schulz. Quantitative oxygen imaging in an engine. *Applied Physics B: Lasers and Optics* 74:137–41, 2002.
79. C Weaver, S Wooldridge, S Johnson, V Sick, G Lavoie. PLIF measurements of fuel distribution in a PFI engine under cold start conditions. SAE Technical Paper 2003-01-3236, 2003.
80. CA Garcia-Stewart, N Polydorides, KB Ozanyan, H McCann. Image reconstruction algorithms for high-speed chemical species tomography. *3rd World Congress on Industrial Process Tomography,* Banff, Canada, 2003, pp. 80–85.
81. ÖL Gülder, GJ Smallwood. Do turbulent flame fronts in spark-ignition engines behave like passive surfaces? SAE Technical Paper 2000-01-1942, 2000.
82. DA Rothamer, JB Ghandhi. Determination of flame-front equivalence ratio during stratified combustion. SAE Technical Paper 2003-01-0069, 2003.
83. DL Reuss, V Sick. Investigation of HCCI combustion with combined PLIF imaging and combustion analysis. *3rd Joint Meeting of the U.S. Sections of the Combustion Institute,* Chicago, IL, 2003.
84. Y Ikeda, J Kojima, H Hashimoto. Local chemiluminescence spectra measurements in a high-pressure laminar methane/air premixed flame. *Proc. Combust. Inst.* 29:1495–501, 2002.
85. B Kim, M Kaenko, Y Ikeda, T Nakajima. Detailed spectral analysis of the process of HCCI combustion. *Proc. Combust. Inst.* 29:671–77, 2002.
86. B Stojkovic, TD Fansler, MC Drake. Quantitative high-speed imaging of soot temperature and relative soot concentration in a spark-ignited direct injection engine. *3rd Joint Meeting of the U.S. Sections of the Combustion Institute.* The Combustion Institute, Chicago, IL, 2003.
87. H Zhao, N Ladommatos. Optical diagnostics for soot and temperature measurements in Diesel engines. *Prog. Energy Combust. Sci.* 24:221–55, 1998.

88. RW Pitz, TS Cheng, JA Wehrmeyer, CF Hess. Two-photon predissociative fluorescence of H2O by a KrF excimer laser for concentration and temperature measurement. *Applied Physics B* 56:94–100, 1993.
89. BJ Kirby, RK Hanson. Infrared PLIF imaging of CO and CO_2. *Proc. Combust. Inst.* 28:253–59, 2000.
90. WG Bessler, C Schulz, T Lee, JB Jeffries, RK Hanson. Carbon dioxide UV laser-induced fluorescence in high-pressure flames. *Chem. Phys. Lett.* 375:344–49, 2003.
91. V Sick. Exhaust-gas imaging via planar laser-induced fluorescence of sulfur dioxide. *Applied Physics B* 74:461–63, 2002.
92. GT Kalghatgi, P Snowdon, CR McDonald. Studies of knock in a spark ignition engine with CARS temperature measurements and using different fuels. SAE Technical Paper 950690, 1995.
93. RP Lucht, RE Teets, RM Green. Unburned gas temperatures in an internal combustion engine. 1. CARS temperature measurements. *Combust. Sci. Technol.* 55:41–61, 1987.
94. J Jonuscheit, A Thumann, M Schenk, T Seeger, A Leipertz. One-dimensional vibrational coherent anti-Stokes Raman-scattering thermometry. *Optics Letters* 21:1532–34, 1996.
95. CF Kaminski, J Engström, M Aldén. Quasi-instantaneous two-dimensional temperature measurements in a spark ignition engine using 2-line atomic fluorescence. *Proc. Combust. Inst.* 27:85–93, 1998.
96. S Einecke, C Schulz, V Sick, R Schießl, A Dreizler, U Maas. Two-dimensional temperature measurements in the compression stroke of an SI engine using two-line tracer LIF. SAE Technical Paper 982468, 1998.
97. NP Tait, DA Greenhalgh. PLIF imaging of fuel fraction in practical devices and LII imaging of soot. *Ber. Bunsenges. Phys. Chem.* 97:1619–25, 1993.
98. UE Meier, D Wolff-Gaßmann, W Stricker. LIF Imaging and 2D temperature mapping in a model combustor at elevated pressure. *Aerosp. Sci. Technol.* 4:403–14, 2000.
99. M Thiruchengode, S Nair, S Prakash, D Scarborough, Y Neumeier, T Liewen, J Jagoda, JM Seitzman, B Zinn. An active control system for LBO margin reduction in turbine engines. *41st Aerospace Sciences Meeting & Exhibit.* AIAA, Reno, NV, 2003.
100. MR Morell, JM Seitzman, M Wilensky, E Lubarsky, J Lee, B Zinn. Interpretation of optical emissions for sensors in liquid fueled combustors. *39th Aerospace Sciences Meeting & Exhibit.* AIAA, Reno, NV, 2001.
101. JM Seitzman, R Tamma, BT Scully. Broadband infrared sensor for active control of high pressure combustors. *36th Aerospace Sciences Meeting & Exhibit.* American Institute of Aeronautics and Astronautics, Reno, NV, 1998.
102. F Schuler, F Rampp, J Martin, J Wolfrum. TACCOS: A thermography-assisted combustion control system for waste incinerators. *Combustion and Flame* 99:431–39, 1994.
103. W Meienburg, J Wolfrum. *In situ* measurement of ammonia concentration in industrial combustion systems. *Proc. Comb. Inst.* 23:231–36, 1990.
104. W Meienburg, H Neckel, J Wolfrum. *In situ* measurement of ammonia with a $^{13}CO_2$-waveguide laser system. *Appl. Phys. B* 51:94–98, 1990.
105. V Ebert, R Hemberger, W Meienburg, J Wolfrum. *In situ* gas analysis with infrared lasers. *Ber. Bunsenges. Phys. Chem.* 97:1527–34, 1993.

9 Multiphase Flow Measurements

Tomasz Dyakowski and Artur J. Jaworski

CONTENTS

9.1 INTRODUCTION

A description of multiphase flow starts with the concept of flow regime. For a given flow system, defined by the particular pipeline geometry and the material flowing through it, various spatial and temporal patterns (flow regimes) arise through self-organization of the multiphase systems. These patterns are the result of forces acting on the flowing phases as well as on the interfaces between them. For example, in gas–liquid systems, the interfaces are free to deform, break up, or coalesce. In horizontal pipes, the resulting flow regimes have been described as bubbly, intermittent, stratified, or annular flow [1]. If the interfaces coalesce, then the resulting phase separation may lead to a stratified flow, where there is minimal contact between the liquid phases. In a chemical reactor, this situation would have a detrimental effect on the reaction and production rates. If, on the other hand, the interfaces between the phases break up, then a bubbling flow might result wherein the enhanced mixing would increase the reaction rate. It can be seen that the type of flow regime encountered in multiphase processes has a significant economic impact.

Identification of the flow regime is therefore important in assessing the effects of multicomponent flow on process system performance. Examples include a large variation in oil–gas separator performance caused by slugging in the feed pipes and the superior performance of heat exchanger pipes when exposed to an annular liquid flow rather than partially exposed to the gaseous component in a two-phase flow. Since much of the behavior depends on these flow patterns, their prediction is a critical issue in process design.

Spatial and temporal flow patterns can be characterized from the macroscopic point of view. This description may specify the position or diameter of a flow channel, the type of flowing phases, and their superficial velocities. On the other hand, to measure the volumetric flow rates, much more detailed information is needed, such as the size and trajectories of individual bubbles and particles, or the wavelength and amplitude of an interface's oscillatory motion.

The proper instrumentation is necessary to reveal flow patterns, and numerical simulations are essential to interpret what the experiments reveal. Ideally, what is learned from one approach should be available for use in the other. A number of experimental techniques are known under "flow visualization," which is a well-established field [2]. Likewise, countless journals and books have been devoted to discussions of direct numerical simulations and other theoretical models of multiphase behavior. However, a complete empirical description of multiphase flow requires three-dimensional (3D) or even four-dimensional (4D) imaging.

Cross-sectional imaging of multiphase flow can be accomplished through electrical tomography (ECT; see Chapter 4), where a set of electrical sensors is installed around the pipe circumference [3]. Electrical tomography is beginning to make promising contributions to control systems and is well suited for flow pattern identification in opaque pipes or conduits. This technology can generate a cross-sectional image at a particular time or reveal the flow pattern from the temporal sequence of cross-sectional images at some defined location in a pipe or process vessel. An electrical property represented by an image, such as dielectric permittivity or resistivity, can also represent a property such as density or chemical composition. It is possible to use the image to visualize the changing properties of the flow streams in the process. Imaging the flow process at two cross-sections and cross-correlating the results obtained opens a new way for on-line measurement of the volumetric flow rate. This chapter discusses applications of process imaging (with an emphasis on electrical tomography) for studying the dynamics of two-phase flows in freight pipelines (i.e., pneumatic and hydraulic conveyors) and various types of heat and mass exchangers.

9.2 FLOW PATTERN RECOGNITION

Flow patterns have been identified for steady, fully developed, adiabatic flows in pipes [2, 3]. Such observations are critical for modelling the dynamics of transient and 3D flows. The flow patterns were obtained by using both the direct and the indirect methods. The first method identifies the flow regimes by direct visual inspection, high-speed photography, or the radial attenuation method based on nucleonic radiation. The disadvantage of the direct methods is, very often, a subjective judgment of the researcher, who sets arbitrary criteria in order to establish the boundaries between various flow patterns. On the other hand, the indirect methods provide a unique identification based on a more objective inspection of the flow behavior, typically based on mathematical approaches. These approaches are frequently based on Fourier data analysis [4] or more advanced concepts such as deterministic chaos theory [5].

In terms of the interaction of the sensors with the flow, the techniques of gathering data can be either invasive or noninvasive. The first kind of measurement is usually realized by probes introduced into the flow. The measurement techniques applied may vary, and typical examples include electrical probes for conductivity or dielectric permittivity measurements [6], fiber-optic probes [7], Pitot tubes [8], hot wire anemometers [9], and others. Of course, the main disadvantage of these techniques is that they only provide a local measurement; additionally they introduce some disturbances in the flow field, which invariably will affect the flow behavior.

Flow identification through the use of distributed noninvasive electrical sensors (such as those used in electrical tomography) has already been demonstrated [10, 11]. It should be emphasized that although process tomography is associated with imaging, the image itself can very seldom be used directly for control purposes. Omitting an explicit image reconstruction phase, which is time consuming, is advantageous where on-line control is concerned. Therefore, it is useful to consider means of recognizing the flow patterns from the raw projection data.

Extracting concise and useful flow feature data from a huge raw data set (on the order of a few gigabytes) is a challenging task. Here the fundamentals of two methods of flow recognition based on processing a raw data set are discussed. The first method applies *a priori* knowledge of the flow pattern [11], whereas the second uses statistical analysis [12]. The fundamentals of these methods and the results obtained are discussed.

9.2.1 *A Priori* Knowledge Approach

The method described in this section was developed for rapid flow pattern identification of a transient gas–liquid (air–kerosene) flow in a horizontal pipe. The type of likely flow pattern was known beforehand, and the identification was to be taken at a precisely defined location, in order to study the separation ability of a T-junction used as a preseparator module. The local nature of the requirement, linked to flow pattern identification, ruled out the use of pressure probes and flow maps to determine the flow pattern.

The general idea for flow pattern recognition without image reconstruction relies on finding the geometrical properties hidden in a frame corresponding to a set of capacitance measurements. It is important to emphasize that the decision is made on a "one frame only" basis to ensure the process is as quick as possible. This point is discussed later with the description of the measurement apparatus.

The two principal identifiers are proposed as follows:

1. If the flow is annular, all adjacent electrode pairs are likely to give similar capacitance measurements. This similarity can be related to the variance of these measurements.

2. The balance between electrode pair measurements at the upper and lower section of the pipe provides information related to the classification of a stratified flow.

Before considering the means of identifying the two types of flow pattern introduced above, it is necessary to consider the limitations of the sensing equipment used to collect the necessary data. There are limitations related to speed and spatial resolution, even though image reconstruction is not the primary aim. Another limitation to be considered is the fact that even if the current frame is given as a 2D representation (as an image or as an identified flow pattern), it is related to a physical sensor that is 3D in nature. The capacitance measurements are actually an average taken over the electrode length. The electrode length introduces a paradox: on the one hand it must be short to allow for approximating a small measurement volume as a 2D slice; on the other hand, the electrode length L, shown in Figure 9.1, has to be long enough to convey the energy required for sensing purposes and allow for a uniform field distribution to take place. The data acquisition system computes the capacitance from the time needed to discharge the "capacitor" defined by a given electrode pair. The electronics impose a maximum charge per surface unit and hence a minimum electrode surface area. The field uniformity is also helped by the use of guard electrodes, shown in Figure 9.1.

Because the data collected from the sensor is related to the average state of the flow passing through the pipe taken along the electrode length, very rapid changes in the flow pattern cannot be detected. The ECT system can achieve at least 60 frames per second (fps). Taking 50 fps as conservative estimate, it takes 20 ms to record a sequence of measurements. Since the length of a typical electrode set is 35 mm, a flow pattern at least 35 mm long will be recorded reliably up to a velocity of 1.75 m/s. If an event (a slug for instance) has a characteristic length shorter than the electrodes or moves faster than the data acquisition time, only an average value is recorded.

Taking these limiting factors into account, two options are open for study:

1. Flow pattern identification on one frame with a number of frames kept in a buffer to confirm the decision.
2. Time series approach, where several frames are stored and the data are processed in batches for the flow pattern recognition by averaging over a period $n\Delta t$ with n being the number of buffered frames and Δt the time needed to capture one frame.

Assuming the phenomenon being observed by the sensor is detectable (according to the acquisition time and electrode length restriction), both options have their advantages and disadvantages. The first option has problems if the flow oscillates between annular and stratified every other frame. This situation would correspond to very thin slices, or plugs, perturbing an annular flow pattern or even an erratic flow pattern that cannot be matched by the decision-making speed. The second option would give a smoother answer, over a period of time,

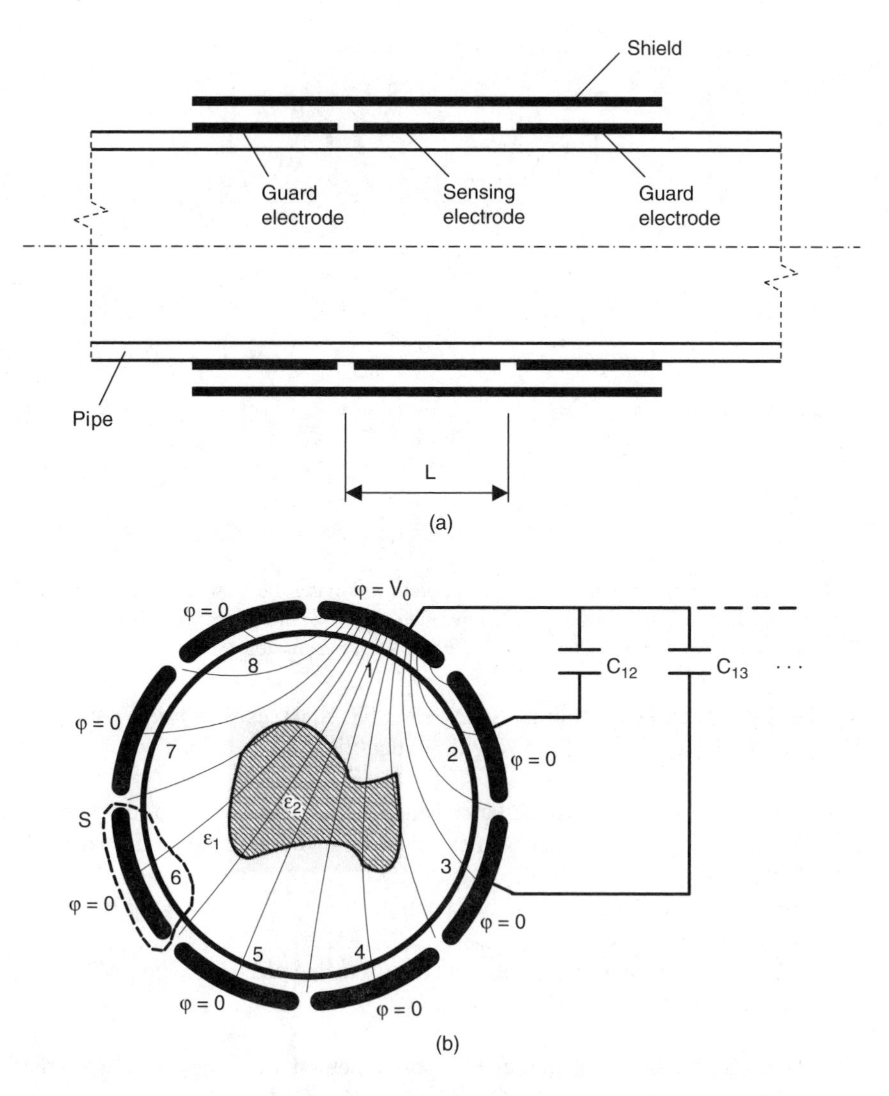

FIGURE 9.1 Longitudinal (a) and transverse (b) cross-sections through a single-plane sensor for electrical capacitance tomography (ECT) measurements. (b) also shows the measurement principle; here, for clarity, the external shield has not been shown.

thus giving the average flow pattern corresponding to the sample taken. The drawback is that any change in flow pattern during the sampled time would be ignored.

The first option (identifying a flow pattern from one frame) allows for a better slug detection system since the time needed for flow recognition is far closer to the data acquisition time than in the second option. A classification algorithm has been designed using three simulated annular flow patterns (AF1, AF2, and AF3)

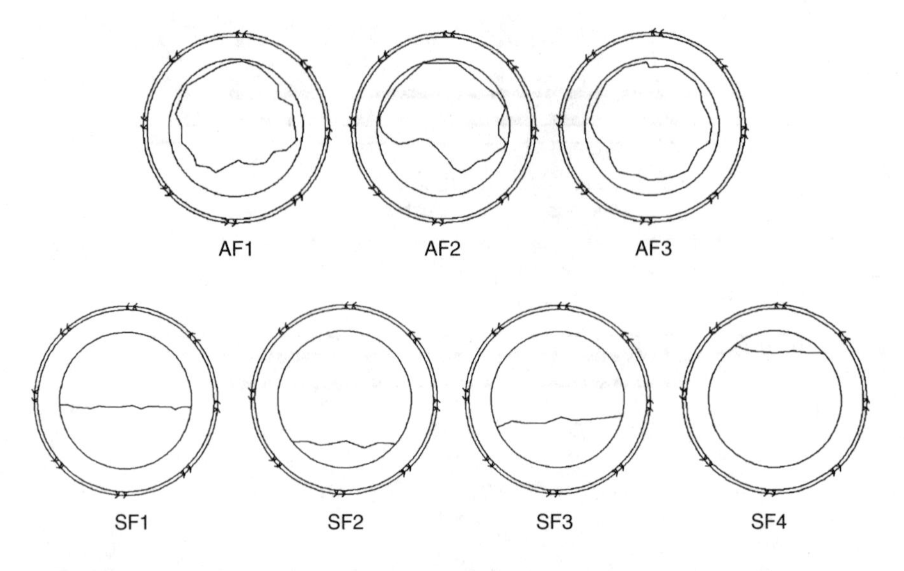

FIGURE 9.2 Examples of flow patterns applied for solving the forward problem. (From Jeanmeure, L.F.C., Dyakowski, T., Zimmerman, W.B.J., and Clark, W., *Experimental Thermal and Fluid Science*, 26, 763–773, 2002. With permission.)

and four separated flow patterns (SF1, SF2, SF3, SF4), as shown in Figure 9.2. For these flow patterns the following flow identifiers were defined:

RTB—ratio of average capacitance values taken from top electrode-pairs over average capacitance values taken from bottom electrode pairs; $\mathrm{RTB} = (C_{35} + C_{46} + C_{36})/(C_{17} + C_{28} + C_{27})$

VSA—variance on the second adjacent electrode pairs; $\mathrm{VSA} = \mathrm{var}\ (C_{13} + C_{24} + C_{35} + C_{46} + C_{57} + C_{68} + C_{71} + C_{82})$

AFE—average capacitance values taken on facing electrode pairs; $\mathrm{AFE} = (C_{15} + C_{26} + C_{37} + C_{48})/4$

The reader might note that, in the proposed method, capacitances between the adjacent pairs of electrodes (C_{12}, C_{23}, C_{34}, C_{45}, C_{56}, C_{67}, C_{78}, and C_{81}) are disregarded. These electrode pairs have a very high gradient of electrical field between them and therefore require an extremely fine mesh for numerical simulation. To simplify this approach, these pairs are not considered.

Capacitance measurements were simulated by solving the forward problem under the MATLAB environment and using the associated PDE Toolbox. The proposed classification parameters were evaluated for the test flow patterns indicated above. Categorizing annular flow patterns works best when using the second adjacent electrodes, whereas stratified flow correlates with the grouping of top and bottom electrode-pairs. Adjacent electrodes are more sensitive to what happens at the periphery of the tube and far less to the permittivity distribution in the center area.

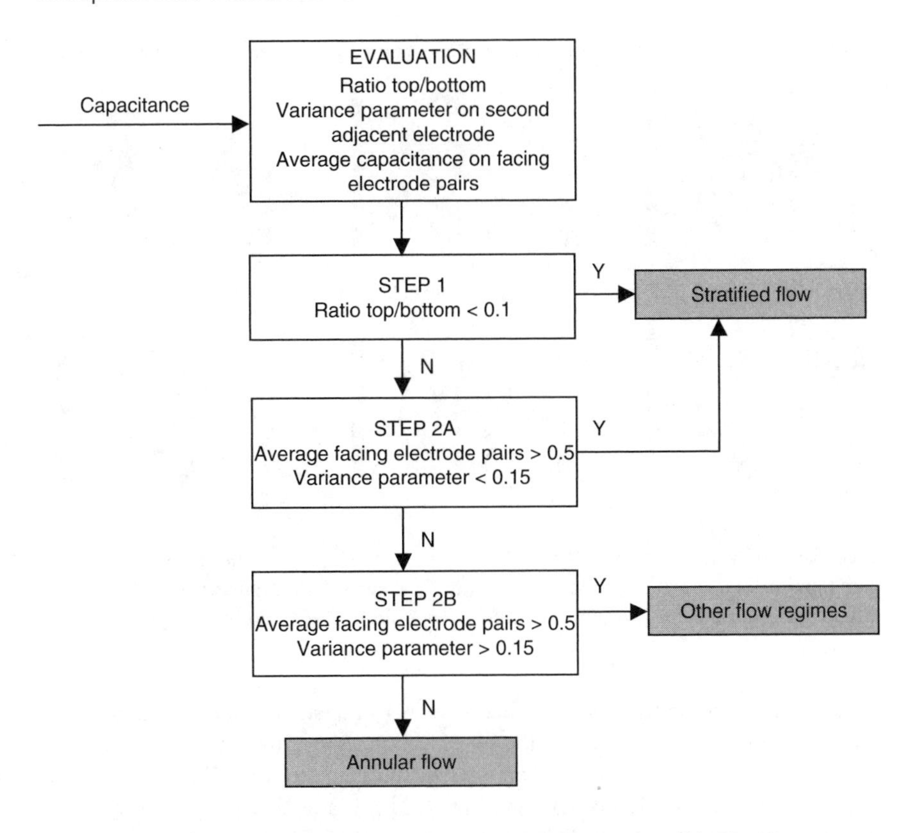

FIGURE 9.3 Decision tree. (From Jeanmeure, L.F.C., Dyakowski, T., Zimmerman, W.B.J., and Clark, W., *Experimental Thermal and Fluid Science*, 26, 763–773, 2002. With permission.)

A decision tree for identifying the flow regime (see Figure 9.3) has been constructed from such observations. The various thresholds shown in Figure 9.3 are given as conservative estimates. Because the identification process is not based on mutual exclusion (i.e., "nonannular flow" does not induce "stratified flow"), it is always possible to end up with a flow pattern belonging to other types of flow regimes. The reader might note that, in the proposed method, a slug flow would be detected when the average capacitance value taken on facing electrodes is greater than 0.95, which would correspond to a pipe 95% full, taking into account measurement errors.

The image sequence in Figure 9.4 corresponds to the formation and collapse of a slug flow with superficial gas velocity 5 m/s and superficial liquid velocity 0.2 m/s. These images were obtained at 25 fps (0.04-s interval). The associated classification results are given in Table 9.1. The black and white images are interpreted as follows: the air phase is shown in black or dark grey, and the kerosene phase appears slightly lighter. The interface, or pixels that are undetermined due to limitations of the image reconstruction, is shown in white. Frame 216 to Frame 218 are evaluated as slugs

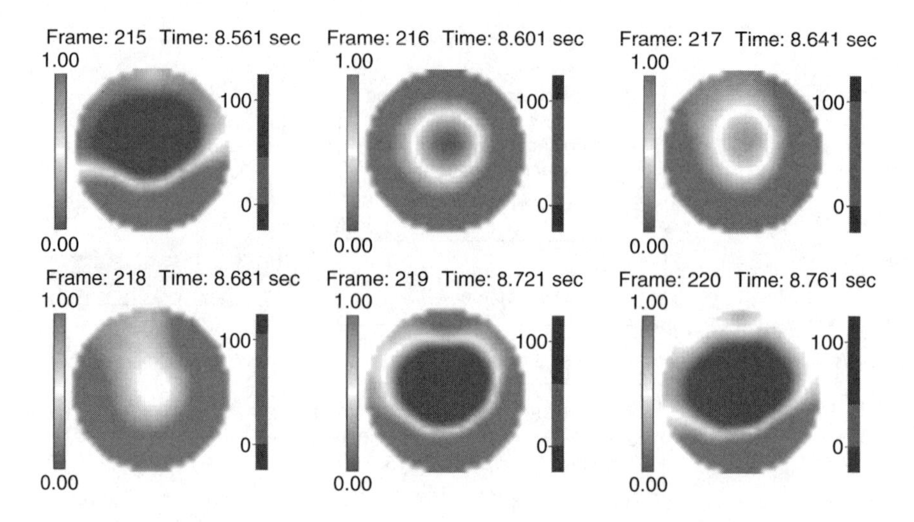

FIGURE 9.4 A sequence of images illustrating the passage of a slug through the measurement volume. (From Jeanmeure, L.F.C., Dyakowski, T., Zimmerman, W.B.J., and Clark, W., *Experimental Thermal and Fluid Science*, 26, 763–773, 2002. With permission.)

because of the phase fraction. Frame 216 shows an unstable pattern, collapsing in the following frames and filling the pipe section. Slug flow appears as follows: Frame 215 and Frame 216 show the liquid creeping up the wall. Then the kerosene phase fills the pipe on Frame 217 and Frame 218. Frame 219 shows a very short annular flow pattern. Frame 220 returns to a stratified flow.

It should be noted that the slug formation process presents a short-lived annular flow pattern in front of and behind the passing slug. In fact, it appears as if a bubble of air is trapped ahead and behind the slug under consideration.

TABLE 9.1
Decision Parameter Evaluation by Jeanmeure et al. [11]

Image	Ratio Top/Bottom	Variance (2nd adj. elec.)	Average Facing Electrodes	Classification
Frame 215	0	0.164	0.203	Stratified
Frame 216	0.790	0.006	0.872	Slug
Frame 217	0.734	0.015	0.907	Slug
Frame 218	0.904	0.004	0.967	Slug
Frame 219	0.172	0.103	0.219	Annular
Frame 220	0.046	0.115	0.116	Stratified

Source: From Jeanmeure, L.F.C., Dyakowski, T., Zimmerman, W.B.J., and Clark, W., *Experimental Thermal and Fluid Science*, 26, 763–773, 2002. With permission.

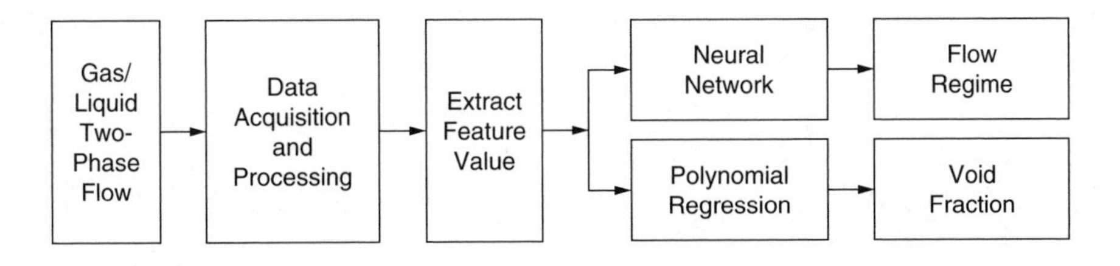

FIGURE 9.5 Schematic of measurement data analysis method. (From Dong, F., Jiang, Z.X., Qiao, X.T., and Xu, L.A. *Flow Measurement and Instrumentation*, 14, 183–192, 2003. With permission.)

This observation concurs with the notion of a 3D effect on the flow inside a pipe, as expressed by Benjamin [13]. Imaging at a higher frame rate of 100 fps (with off-line image reconstruction) gives more evidence of these phenomena. The results obtained by ECT imaging are also in accordance with the work of Fukano and Inatomi [14] on the formation of annular flow. Their analytical simulations describe the formation of annular flow with the liquid creeping up symmetrically along the side of the pipe, as observed from the tomographic data.

9.2.2 Statistical Approach

A concept for flow pattern recognition based on the statistical analysis of raw data from electrical resistivity tomography (ERT) is reported by Dong et al. [12]. They used air and water as working fluids and collected 208 measurements from a 16-electrode system by modifying the commonly applied measurement strategy. Next the data set was transformed into the feature set, which holds the most essential information of the original data content. The results from this set (see Figure 9.5) were applied as the input to a three-layer radial basis function neural network for flow pattern identification. Table 9.2 shows the criteria used to distinguish among five basic flow patterns (flow of water without air, bubble flow,

TABLE 9.2
Identification of the Flow Regimes According to the Output from the Neural Network Applied in Study by Dong et al. [12]

Flow Pattern	Output from Neural Network
Single phase, pipe fully filled by water	0 ± 0.3
Bubble flow	1 ± 0.3
Slug flow	2 ± 0.3
Multibubble flow	3 ± 0.3
Annular flow	4 ± 0.5

Source: From Dong, F., Jiang, Z.X., Qiao, X.T., and Xu, L.A. *Flow Measurement and Instrumentation*, 14, 183–192, 2003. With permission.

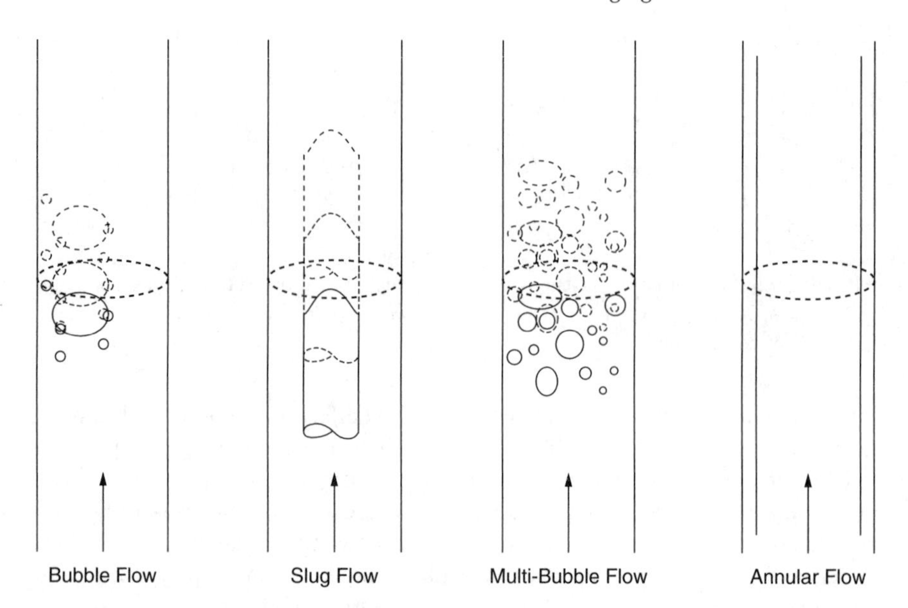

FIGURE 9.6 Schematic of data acquisition for various flow regimes. (From Dong, F., Jiang, Z.X., Qiao, X.T., and Xu, L.A. *Flow Measurement and Instrumentation*, 14, 183–192, 2003. With permission.)

slug flow, multibubble flow, and annular flow). The flow patterns investigated are also schematically shown in Figure 9.6.

The authors presented some preliminary results of the flow regime identification. Table 9.3 shows results of five experiments and the results of decisions taken by the neural network. It can be seen that except for Experiment 2 for annular flow and Experiment 3 for slug flow, a satisfactory agreement was obtained between the actual flow regime and the neural network predictions.

TABLE 9.3
Output from Neural Network for the Flow Regimes Studied by Dong et al. [12] (Each output is given for five different experiments.)

Flow Pattern	Exp.1	Exp. 2	Exp. 3	Exp. 4	Exp. 5
Single phase, pipe fully filled by water	0.0004	–0.0750	0.0023	0.0922	–0.1002
Bubble flow	0.9993	0.9991	0.9989	0.9980	1.0198
Slug flow	2.1360	1.7452	2.6983	2.1023	1.8932
Multibubble flow	2.8563	3.0760	3.1094	3.0291	3.2490
Annular flow	4.0983	3.2613	3.9272	3.7983	4.1230

Source: From Dong, F., Jiang, Z.X., Qiao, X.T., and Xu, L.A. *Flow Measurement and Instrumentation*, 14, 183–192, 2003. With permission.

9.3 FLOW PATTERN IMAGING

The preceding section discussed two approaches for recognizing the flow pattern based on raw projection data. The considerable difference between the relative dielectric permittivity (or resistivity) of flowing fluids enabled the introduction of some qualitative criteria to distinguish between various flow patterns. However, application of the concept of tomographic measurement, which involves a rotating electrical field (as shown in Figure 9.1b), offers an opportunity for producing a cross-sectional distribution of bulk electrical properties within a pipe or channel cross-section, as explained in Chapter 4. The images produced (tomograms) show the relative changes of electrical properties as a result of changes in the flow pattern.

A calibration procedure is needed to relate the measured electrical parameters (dielectric permittivity or resistivity) to the bulk mechanical properties (density or concentration). It should be emphasized that the measured bulk electrical properties not only are a function of mechanical properties (density and concentration of flowing phases) but they also depend on the temperature, air humidity, water salinity, and many other factors. To avoid errors caused by the electrode geometry, the calibration procedure has to use the same sensors as the tomographic measurement. Another method of relating the electrical values to mechanical ones is based on an analytical approach that involves solving the forward problem combined with an appropriate set of constitutive equations; this is fully discussed by Lucas et al. [15] and Norman et al. [16].

9.3.1 2D INSTANTANEOUS AND AVERAGE PHASE DISTRIBUTION

The measurement of phase distribution in a pipe or apparatus is vital for understanding the complex interaction between flowing phases. Images obtained from electrical tomography show the cross-sectional spatial variation of dielectric constant or resistivity. With the proper calibration, these images depict the distribution of phase density or phase concentration in the process. Tomographic measurements can be used to validate various flow models applied in computational fluid dynamics or to construct the constitutive equations describing interactions between flowing phases [17], as discussed below.

Pugsley et al. [7] studied the dynamics of gas–solids flow within a circulating fluidized riser with an internal diameter of 14 cm. The experimental results were obtained for a mixture of placebo pharmaceutical granules with mass density of 1100 kg/m^3 (1.1 g/cm^3) and particle size ranging from 40 μm to greater than 3 mm in diameter. They validated the results from the ECT system by comparing the reconstructed images to the images obtained from a fiber optic probe. This study showed that the time-averaged results from both imaging techniques (as shown in Figure 9.7) are in agreement for the bubbling mode for gas superficial velocities above 0.25 m/s. The authors state that their findings are consistent with earlier results [18], which show that the ECT system works better in dense beds. On the other hand, the instantaneous measurements obtained by both techniques show poor

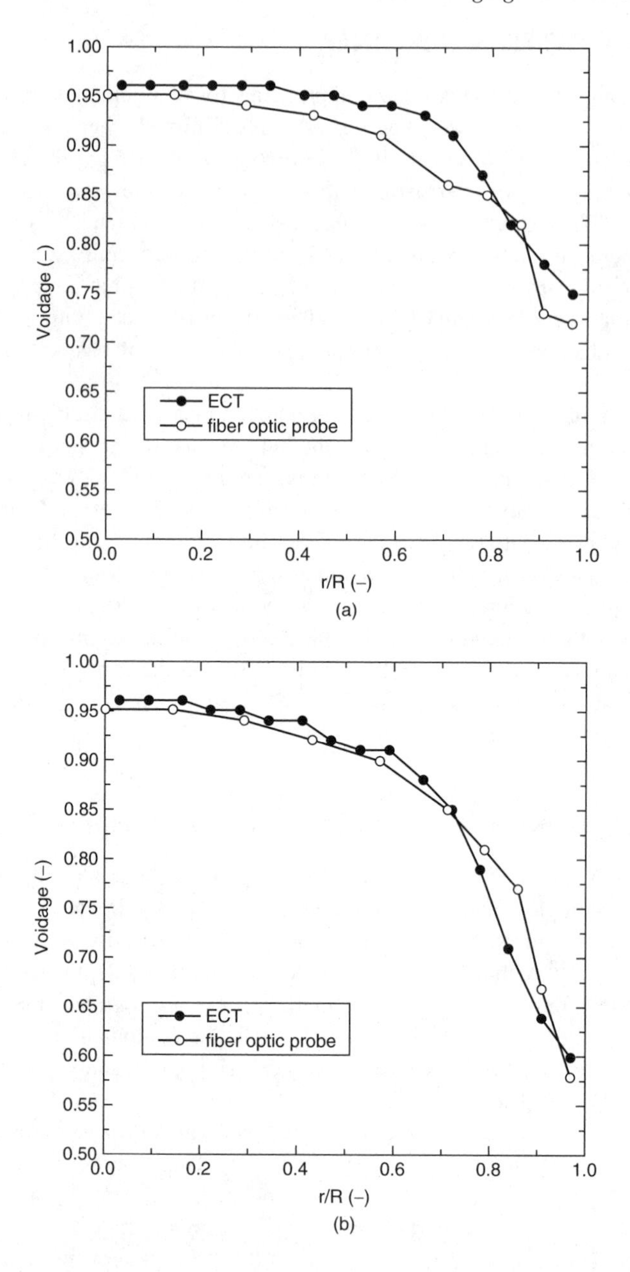

FIGURE 9.7 Comparison of radial voidage profiles measured with electrical capacitance tomography (ECT) and fiber optic probe in the circulating fluidized riser containing FCC catalyst (diameter 80 μm). (a) Riser solids mass flux 148 kg/m^{2}s and superficial gas velocity 4.7 m/s; (b) Riser solids mass flux 264 kg/m^{2}s and superficial gas velocity 4.7 m/s. (From Pugsley, T., Tanfara, H., Malcus, S., Cui, H., Chaouki, J., and Winters, C., *Chemical Engineering Science*, 58, 3923–3934, 2003. With permission.)

agreement, which the authors suggest might be caused by the difference in measured volumes: 0.002 cm^3 for the optical fiber and 0.77 cm^3 for the ECT. The volumes through which capacitance measurements are taken are much larger than the voxel volume, but it should be emphasized that the voxel size does not affect the accuracy of the reconstructed image. For a given electrode arrangement, the number and size of voxels can be varied somewhat independently of the number and size of the electrodes. Therefore, the spatial resolution is mainly a function of the geometry of the electrodes, as discussed by Wang [19].

Also, Lucas et al. [15] have applied a local intrusive probe to validate the results from electrical resistivity tomography. They used a dual plane electrical tomography system to collect data for upward solids–liquid flows in vertical and inclined channels with an internal diameter of 80 mm. The results were obtained for 4-mm diameter plastic beads flowing in water. The authors reported good qualitative agreement between the profiles obtained using these two techniques, as shown in Figure 9.8 and Figure 9.9.

George et al. [20] studied the dynamics of gas–liquid churn-turbulent flows in a vertical column with an inner diameter of 19 cm. The measurements were taken at distance of 97 cm from the gas inlet. Variations in conductivity were minimized by keeping the liquid at nearly constant temperature (±0.2 °C). Churn-turbulent measurements were taken for five gas flow rates (420 cm^3/s, 830 cm^3/s, 1250 cm^3/s, 1670 cm^3/s, and 2500 cm^3/s) through a vertical column filled with water; the corresponding superficial gas velocities were within the range of 1.5 to 8.8 cm/s. The results from electrical impedance tomography (EIT) were compared with gamma-densitometry tomography (GDT) measurements. An example of the comparison of time-averaged gas volume fractions between these two techniques is shown in Figure 9.10. The authors claim that the average cross-sectional values and radial profiles from both methods agreed to within 1% of gas void fraction. Such a good agreement was achieved despite the large difference in collection times for both methods (about 23 min for GDT but less than 20 s for EIT).

Bubble diameters and rise velocities within a bubbling fluidized bed have been measured by Halow et al. [21, 22], Wang et al. [23], Makkawi and Wright [24], White [25], and McKeen and Pugsley [26]. Knowledge of these two parameters is critical for a better modelling of gas–solids interactions, as discussed by McKeen and Pugsley [26]. They applied their results, shown in Figure 9.11, to modify the gas–solids drag term used in a numerical two-fluid model. The authors claim that predictions from their modified two-fluid model are in good agreement with experimental results, as shown in Figure 9.12 and Figure 9.13.

The application of an ECT system for imaging gas–solids flow patterns within a dipleg is discussed by Wang [19]. The results showed that solids falling down from a cyclone distribute themselves into an annular shape around the wall of the dipleg and also into the center of the dipleg. Except for the region near the wall, the time and cross-sectional average solids volume fraction distributions compared well with those obtained from the literature correlations, as shown in Figure 9.14.

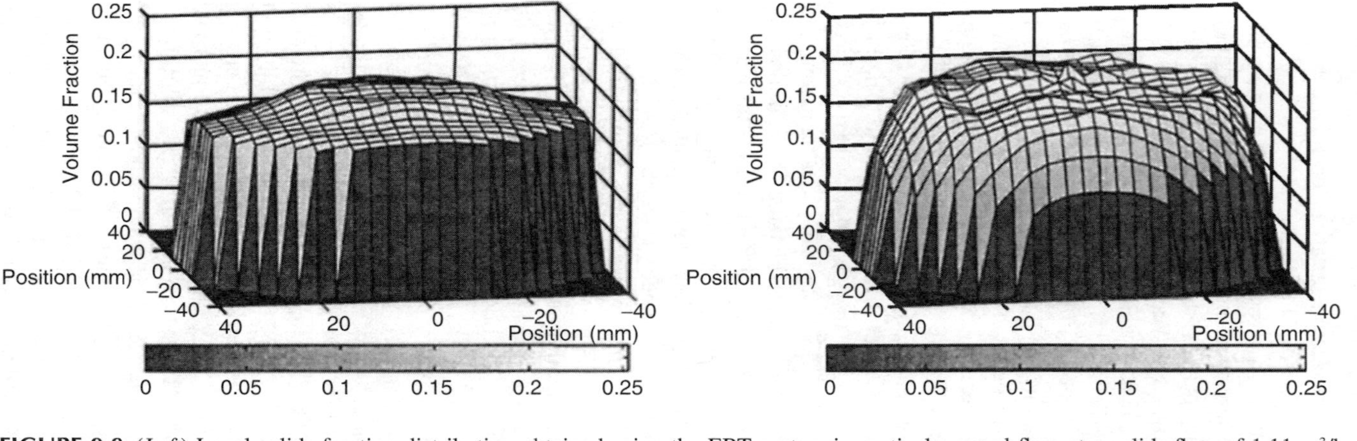

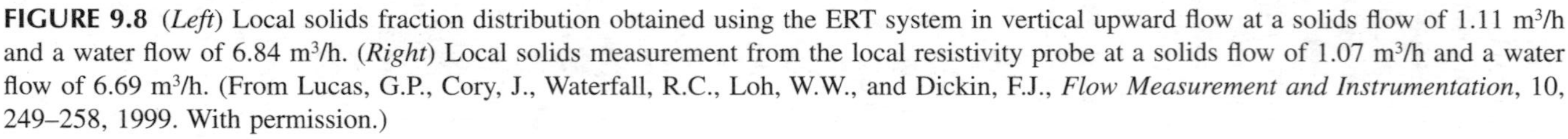

FIGURE 9.8 (*Left*) Local solids fraction distribution obtained using the ERT system in vertical upward flow at a solids flow of 1.11 m^3/h and a water flow of 6.84 m^3/h. (*Right*) Local solids measurement from the local resistivity probe at a solids flow of 1.07 m^3/h and a water flow of 6.69 m^3/h. (From Lucas, G.P., Cory, J., Waterfall, R.C., Loh, W.W., and Dickin, F.J., *Flow Measurement and Instrumentation*, 10, 249–258, 1999. With permission.)

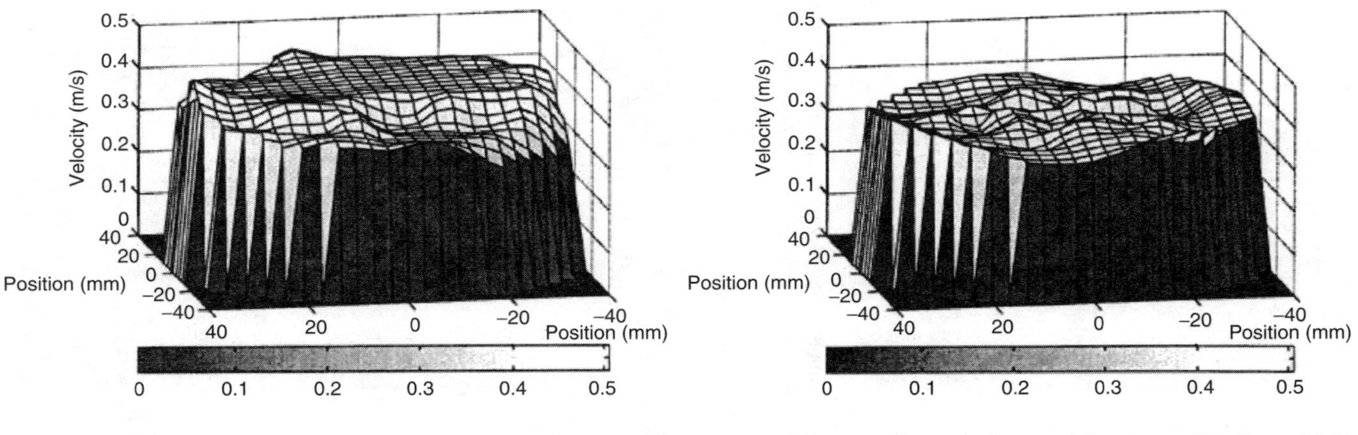

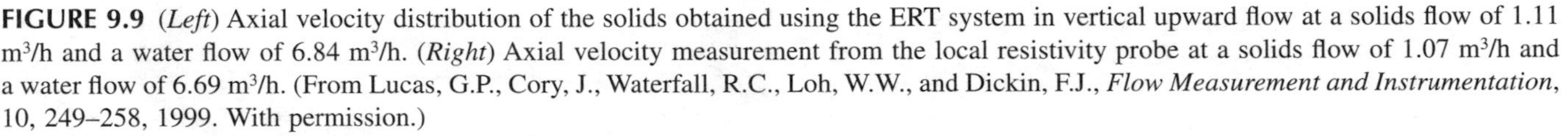

FIGURE 9.9 (*Left*) Axial velocity distribution of the solids obtained using the ERT system in vertical upward flow at a solids flow of 1.11 m^3/h and a water flow of 6.84 m^3/h. (*Right*) Axial velocity measurement from the local resistivity probe at a solids flow of 1.07 m^3/h and a water flow of 6.69 m^3/h. (From Lucas, G.P., Cory, J., Waterfall, R.C., Loh, W.W., and Dickin, F.J., *Flow Measurement and Instrumentation*, 10, 249–258, 1999. With permission.)

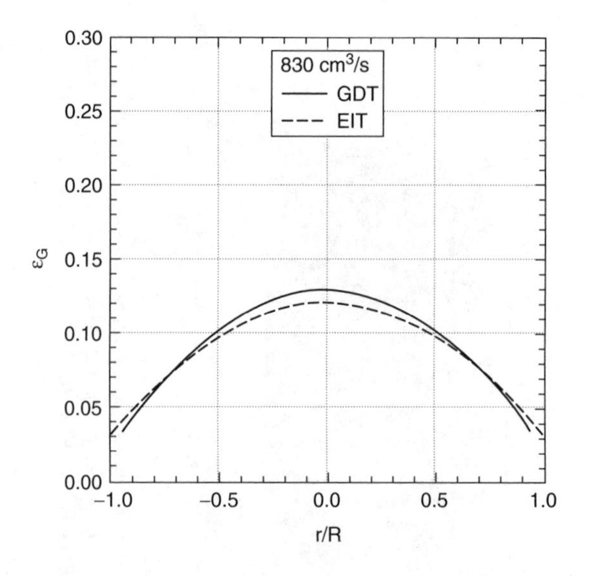

FIGURE 9.10 Comparison of symmetric radial gas volume fraction profiles from GDT and EIT; gas volumetric flow rate 830 cm^3/s. (From George, D.L., Torczynski, J.R., Shollenberger, K.A., O'Hern, T.J., and Ceccio, S.L., *International Journal of Multiphase Flow*, 26, 549–581, 2000. With permission.)

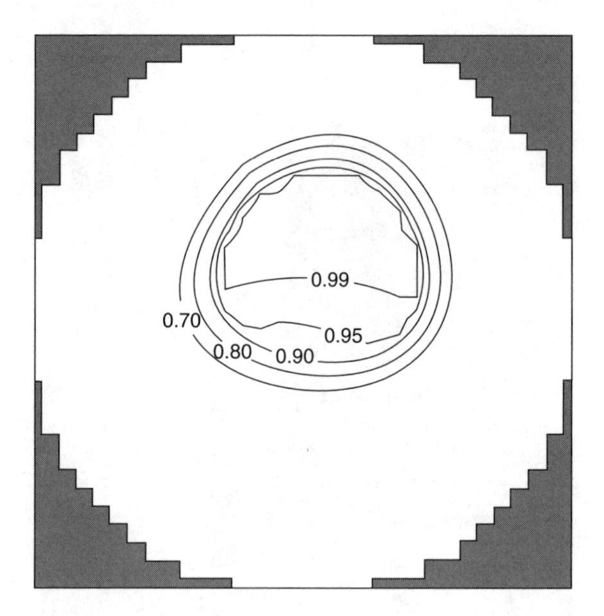

FIGURE 9.11 Voidage contours showing a typical bubble within the cross-section of the ECT sensor. (From McKeen, T., and Pugsley, T., *Powder Technology*, 129, 139–152, 2003. With permission.)

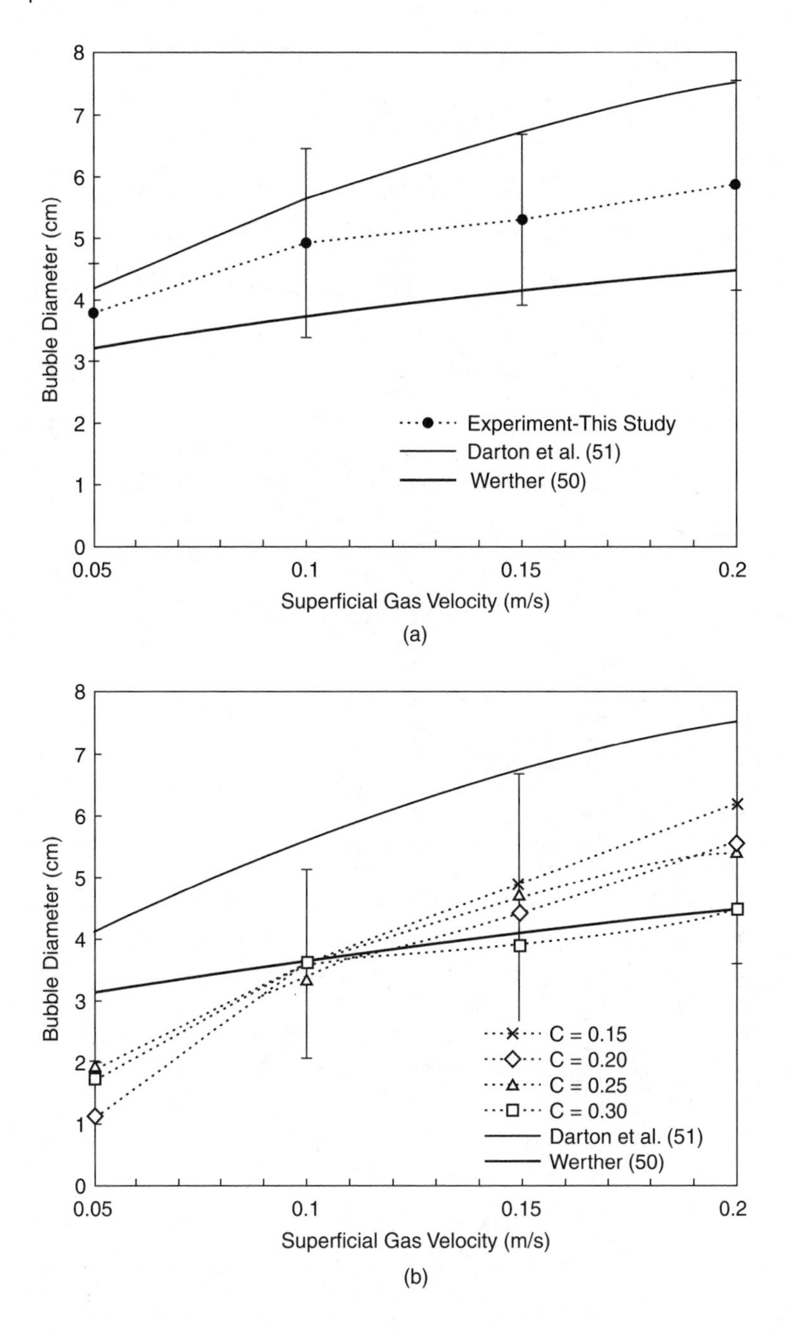

FIGURE 9.12 Comparison of experimental (a) and simulated (b) bubble diameters. (From McKeen, T., and Pugsley, T., *Powder Technology*, 129, 139–152, 2003. With permission.)

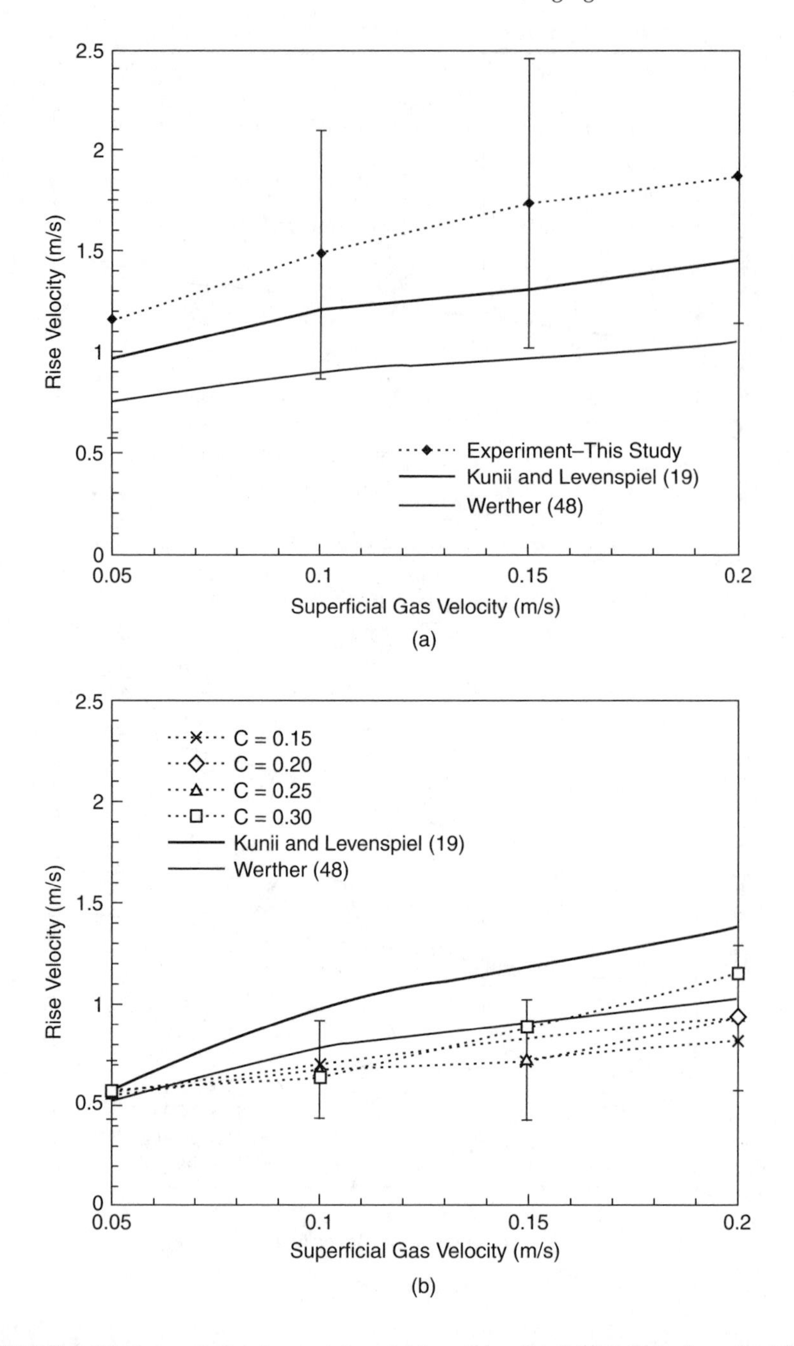

FIGURE 9.13 Comparison of experimental (a) and simulated (b) bubble rise velocities. (From McKeen, T., and Pugsley, T., *Powder Technology*, 129, 139–152, 2003. With permission.)

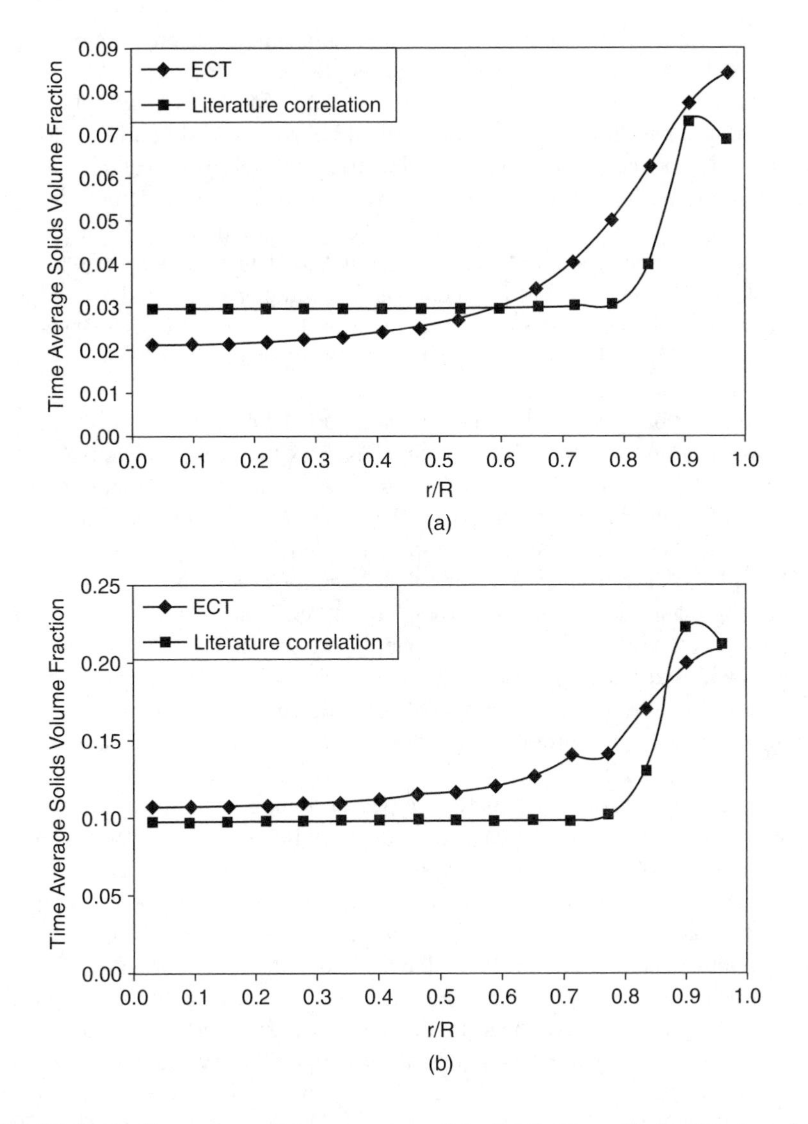

FIGURE 9.14 Comparison of time average solids distribution with the literature data (ECT, electrical capacitance tomography). (From Wang, S.J., Ph.D. dissertation, UMIST, Manchester, U.K., 1998. With permission.)

9.3.2 3D Macro-Flow Structures

Successive images, taken at a single location, represent how the phase distribution varies with time. Assuming that the fluid flows with constant velocity along the channel axis, a 3D representation of the phase distribution along the channel can be constructed. This representation can be constructed either on an arbitrary length scale or by fixing the scale to the product of fluid velocity and elapsed time.

The latter approach needs a method of measuring fluid velocity, which can be provided (for example) by a twin-plane tomographic system.

The effect of air velocity and the presence of solids on a 3D air bubble flow structure in a gas–liquid column was studied by Warsito and Fan [27, 28]. The results were obtained for two types of dielectric fluids (Norpar 15 and Paratherm) and polystyrene beads, whose permittivity is close to that of Paratherm. The measurements were taken from a vertical column of 0.1 m internal diameter and 1 m height. The gas distributor was a single nozzle with a diameter of 0.5 cm. A twin-plane ECT sensor using 12 electrodes for each plane was used; the distance of Plane 1 and Plane 2 from the distributor was 10 cm and 15 cm, respectively. A modified Hopfield dynamic neural network algorithm (see Chapter 4) was applied for data reconstruction.

The effect of air velocity on bubble break-up and coalescence for Norpar 15 is shown in Figure 9.15a. Figure 9.15b illustrates the effects of liquid density and viscosity on bubble size and trajectory. The bubbles in Paratherm liquid (density 870 kg/m^3 and viscosity 0.317 $Pa \cdot s$) are larger and more concentrated close to the column axis than for Norpar 15 (density 770 kg/m^3 and viscosity 0.253 $mPa \cdot s$). The tomograms show there was no significant difference in the gas bubble shape and trajectory when polystyrene particles were added. The authors conclude that the imaging capability of the ECT system is limited to bubbles with sizes larger than the spatial resolution within a vertical column. For smaller bubbles, the tomograms provide information on the amount of the dispersed phase (e.g., solids concentration or gas voidage, or gas hold-up within a gas–liquid system).

Three-dimensional flow structures based on tomographic measurements can be validated by using direct imaging, as discussed elsewhere [2, 29]. Of course, the information obtained from flow visualization is fundamentally different from that obtained by combining tomographic images. Whereas the photographs show the spatial information at a given instant, the tomographic results represent the temporal changes at a given spatial location (that of the sensor). The two approaches would be equivalent only if the flow structures were "frozen" while moving along the pipe. Although this is not necessarily true, the apparent similarities between the photographs and the tomographic data are worth noting. Using a twin-plane tomographic system allows one to determine the time delay between the appearance of slugs in respective planes and, therefore, the propagation velocity of slugs. The length of the slug can then be calculated as the time that the slug was present on one of the planes multiplied by the propagation velocity.

Figure 9.16a gives a series of photographs illustrating the motion of a slug of solids in a pneumatic conveying system. Figure 9.16b shows a time series of cross-sectional ECT images of the same system. The first seven images in Figure 9.16b show the transition between a half-filled pipe and a full pipe that corresponds to the passage of the slug front. Similarly, the last four images show the passage of the slug's tail through the measurement plane. All images in between correspond to the slug passing through the sensing plane (images between $t = 0.12$

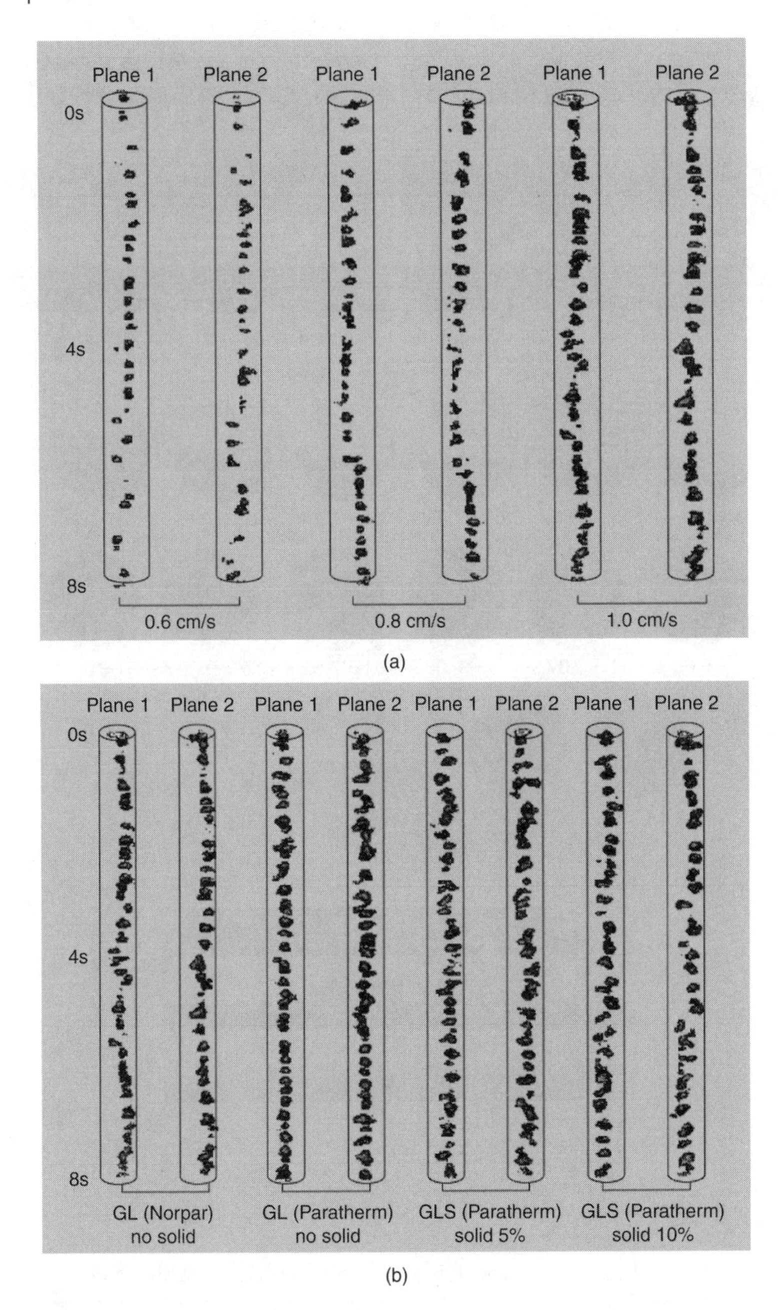

FIGURE 9.15 (a) The effect of air velocity on the bubble size and trajectory for bubbles ascending through Norpar 15; (b) the effect of adding polystyrene beads on the bubble size and trajectory for Norpar 15 and Paratherm. (From Warsito, W., and Fan, L.-S., *Chemical Engineering Science*, 56, 6455–6462, 2001. With permission.)

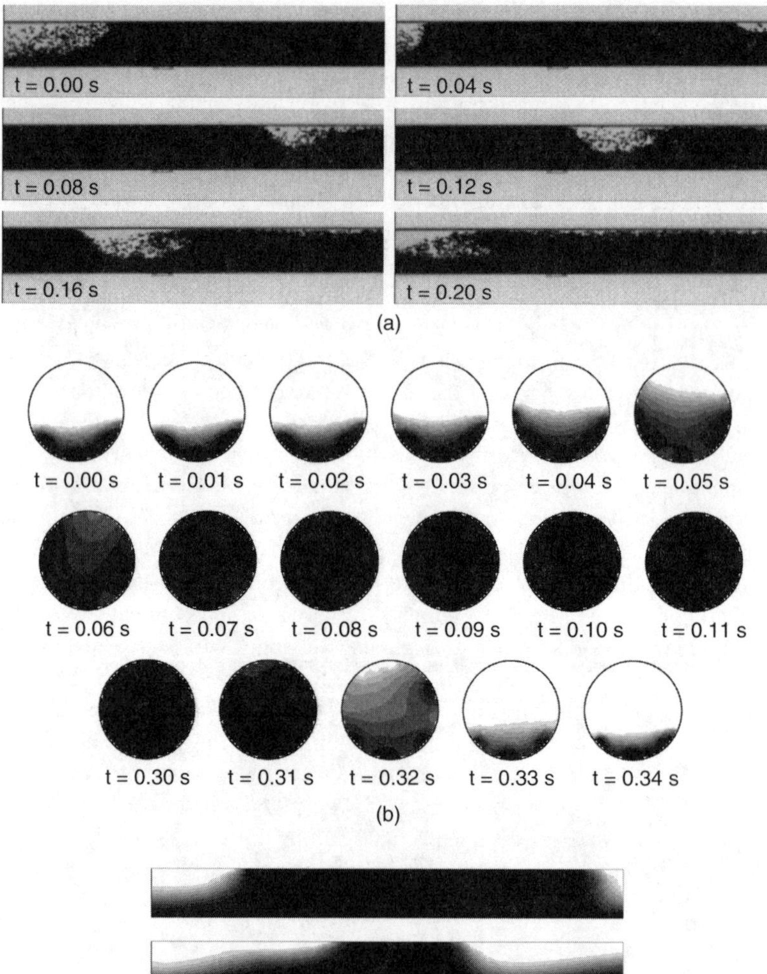

FIGURE 9.16 (a) High-speed camera visualization of the slug flow in a horizontal pipe. (b) A series of ECT images corresponding to the passage of a slug through the sensor. Images between t = 0.12 s and t = 0.29 s are omitted since they show fully filled pipe. (c) Longitudinal cross-section of the slug flow estimated from transaxial tomograms; the image was obtained by extracting pixels from the vertical axis of each tomogram and combining them as a time series. (From Jaworski, A.J., and Dyakowski, T., *Measurement Science and Technology*, 12, 1–11, 2001. With permission.)

s and $t = 0.29$ s are omitted to save space). Use of a twin-plane ECT system enables an approximate reconstruction of the shape of the slugs, as presented in Figure 9.16c. Here, the pixels along the vertical line passing through the center of the pipe are selected from each frame. These pixels are combined to give a longitudinal cross-section of the slug. This figure illustrates the problems encountered by the ECT measurement: limited spatial resolution, averaging of information along the electrode length, and blurring of the sharp boundaries between the phases.

Figure 9.17a illustrates the flow pattern directly observed in a vertical pipe. The conveying of solids typically consisted of two distinctive phases. A "train" of a few slugs, each approximately 10 to 20 cm long, appeared on average every three or four seconds. It could be seen that some particles "rain down" from the preceding slug onto the following slug. At the end of each passage, some granular material (most probably from the tail of the train) dropped downward under the gravity. ECT images, shown in Figure 9.17b, depict the internal structure of the plugs, which cannot be seen with photographic techniques. It is worth noting that the ECT reconstruction reflects the changes in porosity of the material within the slugs. It can be inferred that the density increases toward the pipe wall, which most probably corresponds to an increase in interparticle stresses. The center of the slug, on the other hand, seems more porous, probably due to the passage of the gas through the slug. Finally, Figure 9.17c shows the longitudinal cross-section of the flow structures obtained in a manner analogous to that explained with reference to Figure 9.16c.

The results presented in Figure 9.16 and Figure 9.17 show that there are similarities between the images produced by direct and indirect techniques. This is not necessarily true for other flow patterns, as fully discussed by Jaworski and Dyakowski [29]. It is apparent, however, that despite the limitations of ECT (i.e., reconstruction algorithm errors and averaging along the electrodes), the technique can provide unique information about the structure and evolution of 3D and unsteady gas–solids flows.

9.4 SOLIDS MASS FLOW MEASUREMENTS

An attractive idea for solids mass flow measurement is based on the use of twin-plane tomography systems and cross-correlation techniques. Figure 9.18 shows a schematic of the idea behind this method. The instantaneous solids mass flow rate at time t through the cross-section A of a pipe can be written as

$$\dot{m}(t) = \iint_A \rho(x,y,t)v(x,y,t)dx\,dy \tag{9.1}$$

where $\rho(x, y, t)$ stands for the instantaneous density at point (x, y) of cross-section A and $v(x, y, t)$ denotes an instantaneous velocity at point (x, y) in the direction

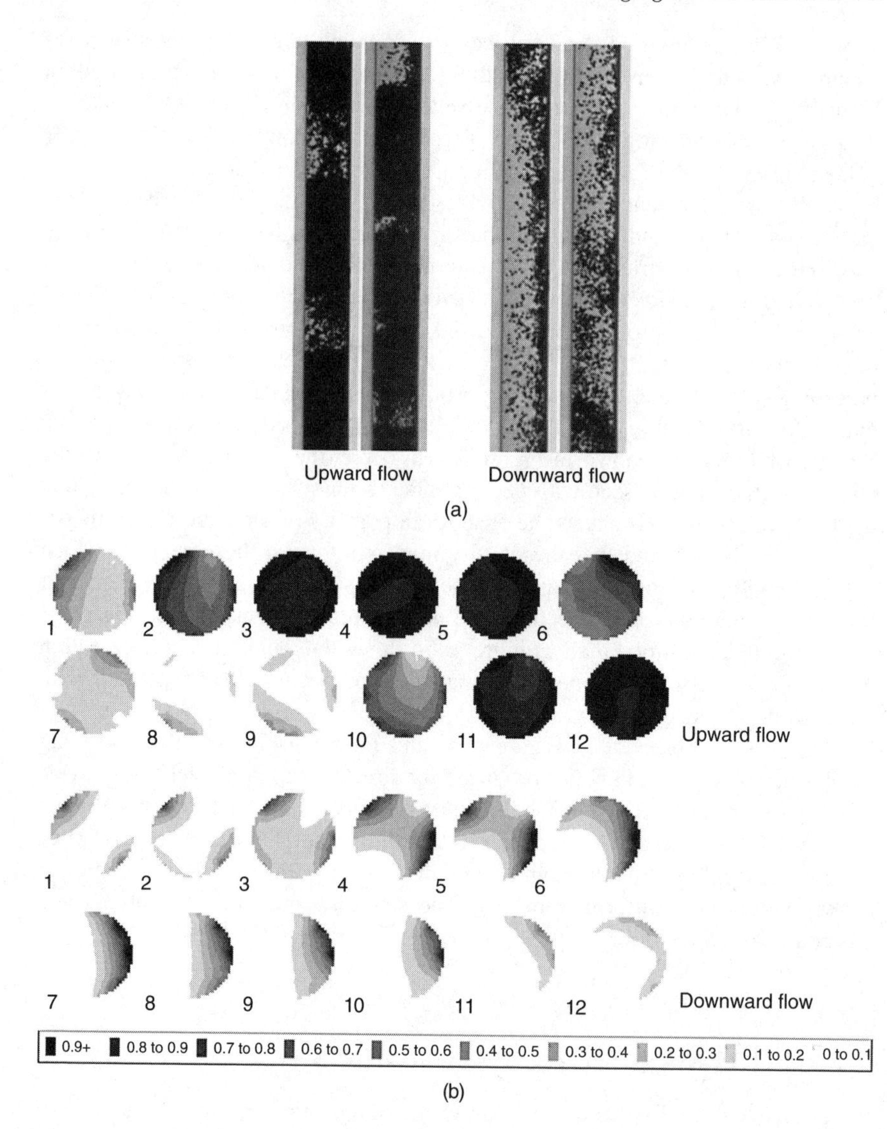

FIGURE 9.17 (a) Photographs showing slugs travelling upward and material dropping downward in between the trains of slugs. (b) The cross-sectional distribution of dielectric permittivity for upward slug flow (the first and second rows) and downward return of material (third and fourth rows). (c) Temporal changes in the permittivity distribution across the diameter of the pipe: left column for upward travelling slugs and right column for material falling downward. (From Jaworski, A.J., and Dyakowski, T., *Measurement Science and Technology*, 12, 1–11, 2001. With permission.)

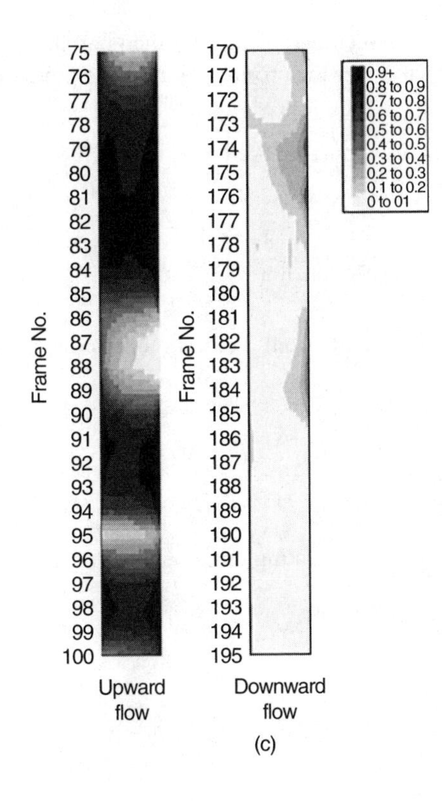

FIGURE 9.17 (Continued).

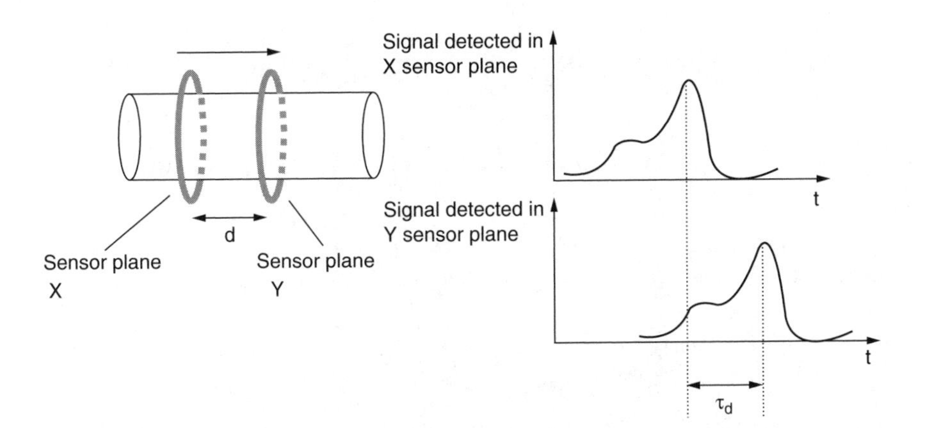

FIGURE 9.18 A typical tomography system for velocity measurements based on the correlation function. (From Mosorov, V., Sankowski, D., Mazurkiewicz, L., and Dyakowski, T., *Measurement Science and Technology*, 13, 1810–1814, 2002. With permission.)

perpendicular to the cross-sectional plane. It is then possible to define averages of both density and velocity over the cross-section of the pipe A as follows:

$$\bar{\rho}^A(t) = \frac{1}{A}\iint_A \rho(x, y, t)dxdy \tag{9.2}$$

$$\bar{v}^A(t) = \frac{1}{A}\iint_A v(x, y, t)dxdy \tag{9.3}$$

The instantaneous density and velocity can now be expressed in the following way:

$$\rho(x, y, t) = \bar{\rho}^A(t) + \rho'(x, y, t) \tag{9.4}$$

$$v(x, y, t) = \bar{v}^A(t) + v'(x, y, t) \tag{9.5}$$

where, by definition of the fluctuating components:

$$\iint_A \rho'(x, y, t)dxdy = 0 \tag{9.6}$$

$$\iint_A v'(x, y, t)dxdy = 0 \tag{9.7}$$

Using the above equations, it is easy to show that:

$$\begin{aligned}\dot{m}(t) &= \iint_A [\bar{\rho}^A(t) + \rho'(x, y, t)]\cdot[\bar{v}^A(t) + v'(x, y, t)]dxdy \\ &= \iint_A \bar{\rho}^A(t)\bar{v}^A(t)dxdy + \iint_A \rho'(x, y, t)\bar{v}^A(t)dxdy \\ &+ \iint_A \bar{\rho}^A(t)v'(x, y, t)dxdy + \iint_A \rho'(x, y, t)v'(x, y, t)dxdy\end{aligned} \tag{9.8}$$

Of course, the second and third terms on the right-hand side are zero by definition. The first term can be easily integrated, and, therefore,

$$\dot{m}(t) = A\bar{\rho}^A(t)\bar{v}^A(t) + \iint_A \rho'(x, y, t)v'(x, y, t)dxdy \tag{9.9}$$

Equation 9.9 highlights the reason why tomography is required to calculate the correct mass flow rate of solids in highly nonuniform flows. It clearly shows that the more spatially non-uniform the flow (either in velocity or in density), the larger the contribution to mass flow by the second term on the right-hand side. Therefore, considering the flow on a pixel-by-pixel basis becomes essential.

A useful approach for analyzing flow between two cross-sectional planes is to cross-correlate the images. The concept of using a cross-correlation technique is not new. In fluid mechanics, it has been for many decades a standard technique for investigating the flows within a boundary layer [30] and for tracking the movement of coherent structures shed by aerodynamic bodies [31]. In the area of multiphase flows, the velocity field can be measured by cross-correlating the time-varying signal arising from one phase being dispersed in another (e.g., solids in gas or liquid) [32]. Some obvious (but often tacit) assumptions made while measuring the velocity field by cross-correlation techniques are that

- The sensors' size is small relative to their separation.
- There are measurable disturbances in the flow field being investigated.
- The velocity field propagation (convection velocity) can be associated with the propagation velocity of these disturbances.

The cross-correlation function is defined by the following equation:

$$R_{12}(\tau) = \lim_{T\to\infty} \frac{1}{T} \int_0^T S_1(t) S_2(t+\tau)\,dt \tag{9.10}$$

where S_1 and S_2 are the detected signals from sensor Planes 1 and 2, τ is a time delay, and T is a time period for which the cross-correlation is calculated. The correlation function reaches the maximum for time delay $\tau_{\max}$. This time delay is an estimation of the transit time of solids between the two planes.

For the reconstructed images, the cross-correlation at each pixel can be calculated as described by Etuke and Bonnecaze [33]:

$$R_{12}[p] = \sum_{k=0}^{T-1} V_{1,n}[k] V_{2,n}[k+p] \tag{9.11}$$

In the above equation, p stands for the shift number in time sequence of frames, n is the pixel index (i.e., position), and T is the number of images for which the cross-correlation is calculated. $V_{1,n}[k]$ and $V_{2,n}[k+p]$ are numerical values associated with pixel n from the kth image obtained from Sensor 1 and from the $(k + p)$th image from Sensor 2, respectively.

The knowledge of transit time and distance between the two sensors enables the calculation of the solids velocity from the following equation:

$$V = \frac{d}{p_\tau \cdot T_F} \tag{9.12}$$

where T_F is the time period of a single frame, p_τ is the shift (in number of frames) for which the correlation in Equation 9.11 reaches the maximum, and d is the distance between sensor planes. From the reconstructed images we can calculate the axial velocity at each pixel in the pipe cross-section, as shown in Figure 9.19.

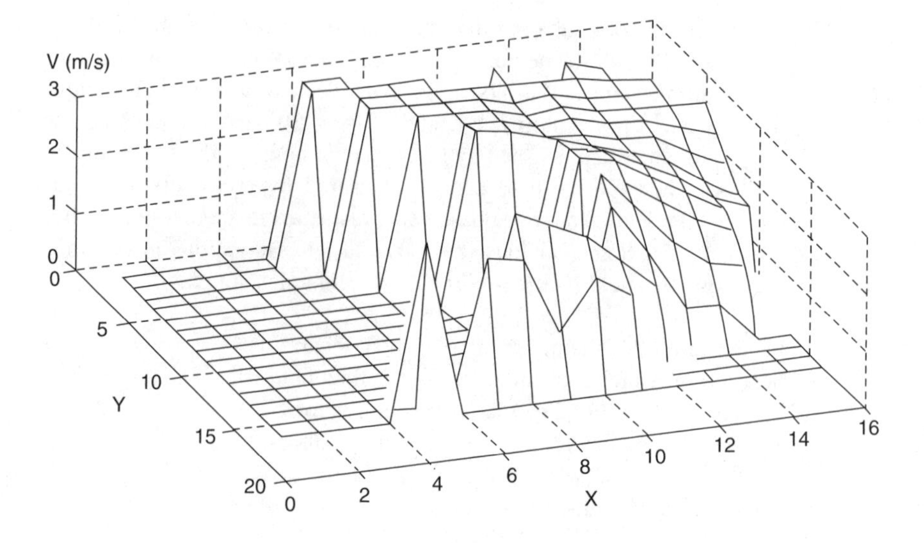

FIGURE 9.19 An example of the velocity profile (vertical axis is velocity, horizontal axes are pixels coordinates in 2D space). (From Mosorov, V., Sankowski, D., Mazurkiewicz, L., and Dyakowski, T., *Measurement Science and Technology*, 13, 1810–1814, 2002. With permission.)

The solids mass flow rate is calculated as an integral of solids concentration and velocity profiles along the cross-sectional area of the pipe.

Figure 9.19 and Figure 9.20 show representative distributions of the axial component of velocity obtained for pneumatic and hydraulic conveying systems. The results were obtained for upward flows in vertical channels. The results for the pneumatic conveyor (see Figure 9.19) were obtained just after a bend connecting horizontal and vertical sections of the pipe; therefore, the solids velocity distribution is not uniform. The effect of pipe inclination on the solids velocity profile in a hydraulic conveyor is illustrated in Figure 9.20. The results were obtained for spherical plastic beads with a mean diameter of 4 mm and a density of 1340 kg/m^3 (1.34 g/cm^3) hydraulically conveyed in a pipe with an inner diameter of 80 mm [15].

The classic cross-correlation technique for solids velocity measurement is based on the assumption that particle trajectories are parallel to each other and perpendicular to the sensor plane. Such a method is therefore valid only for certain flow patterns. A more sophisticated tomographic technique, discussed below, offers a new method to calculate the solids velocity without requiring assumptions about the solids trajectory. Relaxing this assumption is important since in many cases the solids trajectories are very complex.

In the classical method [17], corresponding pixels in sensing Plane 1 and Plane 2 are correlated, but the method can be extended by cross-correlating each pixel from Plane 1 with many pixels from Plane 2, as illustrated in Figure 9.21.

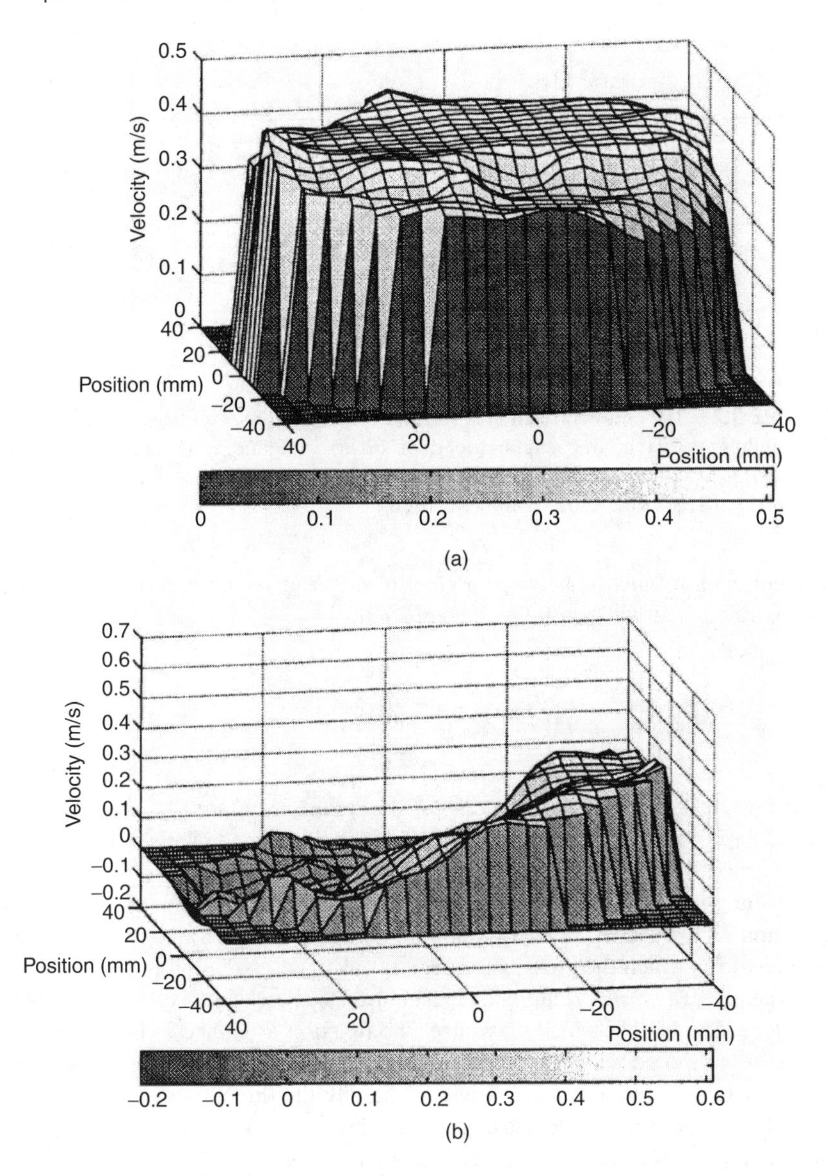

FIGURE 9.20 Axial velocity distribution of solids measured with ERT: (a) Vertical upward flow at a solids flow rate of 1.11 m^3/h and a water flow rate of 6.84 m^3/h. (b) Upward flow inclined at 5° to the vertical, at a solids flow of 0.41 m^3/h and a water flow of 4.04 m^3/h. (From Lucas, G.P., Cory, J., Waterfall, R.C., Loh, W.W., and Dickin, F.J., *Flow Measurement and Instrumentation*, 10, 249–258, 1999. With permission.)

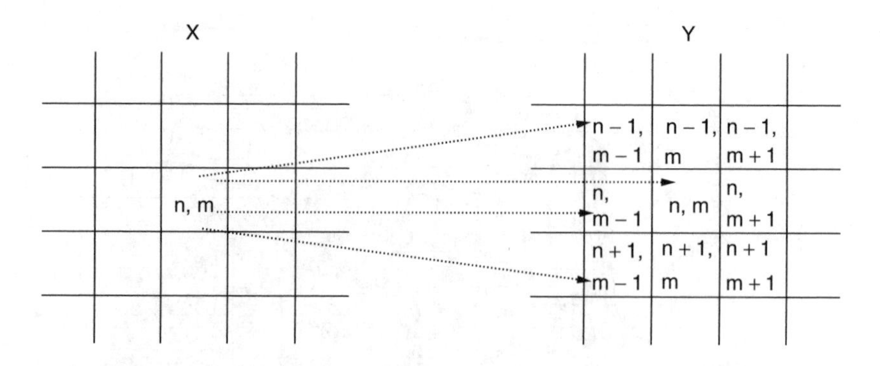

FIGURE 9.21 The cross-correlation is calculated between the pixel (n,m) in Plane 1 and surrounding pixels in the neighborhood of (n, m) in Plane 2. (From Mosorov, V., Sankowski, D., Mazurkiewicz, L., and Dyakowski, T., *Measurement Science and Technology*, 13, 1810–1814, 2002. With permission.)

The correlation function between a pixel from Plane 1 with pixels from Plane 2 can be calculated with the following equation:

$$R_{1,[n,m]2[n-i,m-j]}[p] = \sum_{k=0}^{T-1} V_{1,[n,m]}[k]V_{2[n-i,m-j]}[k+p], \quad (i, j) \in B \qquad (9.13)$$

where (n, m) are the coordinates of the pixel in Plane 1, $V_{1,[n,m]}[k]$ and $V_{2,[n-1,m-j]}[k + p]$ are the numerical values of the designated pixels in Plane 1 and Plane 2, and **B** is the neighborhood of pixel (n, m) on Plane 2. Offset vectors (i, j), which lie within **B**, are used to select pixels in the neighborhood of (n, m). In Equation 9.13, p stands for the time shift in number of frames, and T is the number of images for which the cross-correlation is calculated.

The pixel $(n - i,m - j)$ in Plane 2 that correlates best with pixel (n, m) in Plane 1 is defined to be that which maximizes the function R defined in Equation 9.13. Thus, the proposed technique does not require assumptions concerned with the solids motion within a sensor volume. Generally the direction of solids velocity in a 3D volume can be calculated from the pixels offsets i and j. The knowledge of transit time for which the cross-correlation function reaches its maximum value and distance between the two planes enables the calculation of the solids velocity from Equation 9.12.

An application of this concept is illustrated by the measurement of tangential velocity in a swirl flow. Both the radial profile of the solid concentration distribution and the tangential component of the total velocity vector may be obtained from the best-correlated pair of pixels method. The flow rig used in these experiments consisted of a small hopper connected to a vertical pipe (I.D. 30 mm) around which an ECT sensor was installed. A specially designed device induced swirl in the solids flow, with the amount of swirl determined by an adjustment on

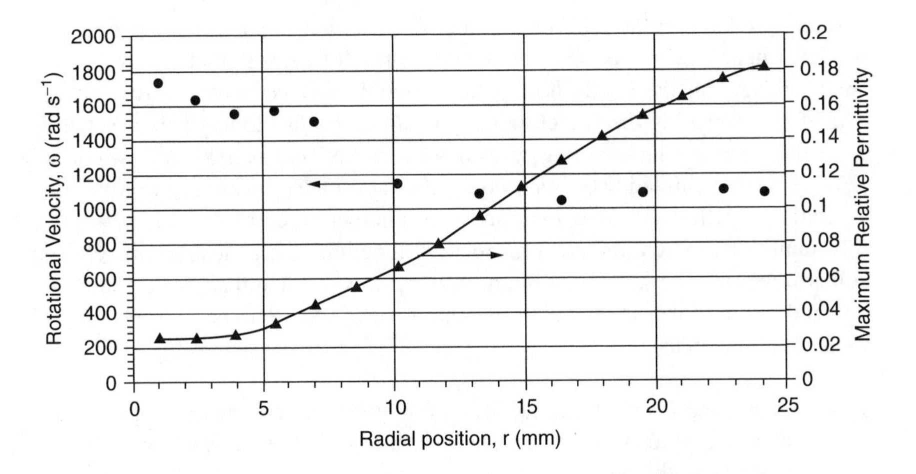

FIGURE 9.22 Results of the measurements using best correlated pair of pixels principle: solids concentration on the left vertical axis, swirl velocity on the right axis; both are shown as functions of radial position.

the blower (M. Byars, personal communication, 2003). Figure 9.22 shows the results of such measurements on 4 mm polypropylene beads.

As a summary of this section, the results obtained both for the pneumatic and the hydraulic conveying systems are presented in Table 9.4. The solids mass flow rates were estimated using the "classical" approach of Equation 9.1 for the hydraulic conveyor and the "best correlated pair of pixels" method for the pneumatic conveyor. These estimates are compared to direct measurements of the solids flow rate. The results presented here indicate that tomographic sensors can provide solids flow rate data with an accuracy of 20%.

The results are encouraging, because they prove that a formalized approach to calculating the mass flow rate from process imaging can give correct answers to within an order of magnitude. It should be emphasized that there are some inherent limitations in applying the cross-correlation technique, as discussed by Mosorov et al. [34]. The first is concerned with the velocity discrimination, which

TABLE 9.4
Comparison of Solids Flow Rates Measured by Electrical Tomography with Those Obtained by the Weighing Method

Flow Type	Estimated Flow Rate from Process Tomography	Value Obtained by Direct Weighing Method
Hydraulic conveying	1.31 m^3/h (volumetric flow)	1.11 m^3/h
Pneumatic conveying	736 kg/h	789 kg/h

is directly related to the ratio of the sampling interval to the sensor spacing. On the other hand, the sensor spacing should be small to ensure that only relatively small changes in the solids flow pattern would exist between the two planes. Therefore, increasing the velocity discrimination of a fast tomographic system is needed to decrease the sampling interval, as discussed by Jaworski and Dyakowski [29]. The other limitation is imposed by the ratio of the maximum to minimum velocities, which determines the maximum number of correlation delays. Since this number is limited, the cross-correlation function cannot distinguish between solids velocities below the minimum measurable value and their zero velocity.

In the area of mass flow measurement, it is expected that the future work will focus on three areas:

1. Increasing the capture rate of the tomographic equipment to approximately 500 to 1000 fps to provide a more accurate estimate of solids velocity in the flow.
2. Improving the models that relate measured electrical parameters (dielectric permittivity or conductivity) to solids concentration.
3. Investigating the correlation techniques in more detail.

REFERENCES

1. M Annunziato, G Girrardi. Horizontal two phase flow: a statistical method for flow pattern recognition. *Proc. Third International Conference on Multi-Phase Flow,* The Hague, 1987.
2. AJ Smits, TT Lim, eds. *Flow Visualization: Techniques and Examples.* Singapore: World Scientific Publishing, 2000.
3. A Plaskowski, MS Beck, R Thorn, T Dyakowski. *Imaging Industrial Flows.* London: Institute of Physics Publishing, 1995.
4. P Seleghim Jr., FE Milioli. Improving the determination of bubble size histograms by wavelet de-noising techniques. *Powder Technol.* 115:114–123, 2001.
5. A Tsutsumi, R Kikuchi. Design and scale-up methodology for multi-phase reactors based on non-linear dynamics. *Appl. Energy* 67:195–219, 2000.
6. J Werther, O Molerus. The local structure of gas fluidized beds: 1. A statistically based measuring system. *Int. J. Multiphase Flow* 1:123–138, 1973.
7. T Pugsley, H Tanfara, S Malcus, H Cui, J Chaouki, C Winters. Verification of fluidized bed electrical capacitance tomography measurements with a fiber optic probe. *Chemical Eng. Science* 58:3923–3934, 2003.
8. Guo Q, Yue G, J Zhang, Z Liu. Hydrodynamic characteristics of a two-dimensional jetting fluidized bed with binary mixtures, *Chemical Eng. Science* 56:4685–4694, 2001.
9. C Lorencez, M Nasr-Esfahany, M Kawaji, M Ojha. Liquid turbulence structure at a sheared and wavy gas-liquid interface. *Int. J. Multiphase Flow* 23:205–226, 1997.
10. J-E Cha, Y-C Ahn, M-H Kim. Flow measurement with an electromagnetic flowmeter in two-phase bubbly and slug flow regimes. *Flow Meas. Instrum.* 12:329–339, 2002.

11. LFC Jeanmeure, T Dyakowski, WBJ Zimmerman, W Clark. Direct flow pattern identification using electrical capacitance tomography. *Experimental Thermal and Fluid Science* 26:763–773, 2002.
12. F Dong, ZX Jiang, XT Qiao, LA Xu. Application of electrical resistance tomography to two-phase pipe flow parameters measurement. *Flow Meas. Instrum.* 14:183–192, 2003.
13. T Brooke, T Benjamin. Gravity currents and related phenomena. *J. Fluid Mech.* 31:209–248, 1968.
14. T Fukano, T Inatomi. Numerical analysis of liquid film formation in an annular flow in a horizontal tube. *Proceedings of the 4th International Conference on Multiphase Flow,* ICMF 2001, New Orleans, LA, 2001.
15. GP Lucas, J Cory, RC Waterfall, WW Loh, FJ Dickin. Measurement of the solids volume fraction and velocity distributions in solids-liquid flows using dual-plane electrical resistance tomography. *Flow Meas. Instrum.* 10:249–258, 1999.
16. JT Norman, HV Nayak, RT Bonnecaze. Non-invasive imaging of pressure-driven suspension flows with electrical resistance tomography. *Proceedings of 3rd World Congress on Industrial Process Tomography,* Banff, Canada, 2003, pp. 312–317.
17. RA Williams, MS Beck. *Process Tomography: Principles, Techniques and Applications,* Oxford, U.K.: Butterworth-Heinemann, 1995.
18. T Dyakowski, RB Edwards, CG Xie, RA Williams. Application of capacitance tomography to gas-solids flows. *Chemical Eng. Science* 52:2099–2110, 1997.
19. SJ Wang. Measurement of fluidization dynamics in fluidized beds using capacitance tomography. Ph.D. dissertation, UMIST, Manchester, U.K., 1998.
20. DL George, JR Torczynski, KA Shollenberger, TJ O'Hern, SL Ceccio. Validation of electrical-impedance tomography for measurements of material distribution in two-phase flows. *Int. J. Multiphase Flow* 26:549–581, 2000.
21. JS Halow, P Nicoletti. Observation of fluidized bed coalescence using capacitance imaging. *Powder Technol.* 69:255–277, 1992.
22. JS Halow, GE Fasching, P Nicoletti. Observation of a fluidized bed using capacitance imaging. *Chemical Eng. Science* 48:643–659, 1993.
23. SJ Wang, T Dyakowski, CG Xie, RA Williams, MS Beck. Real time capacitance imaging of bubble formation at the distributor of a fluidized bed. *Chemical Eng. Science* 56:95–100, 1995.
24. YT Makkawi, PC Wright. Tomographic analysis of dry and semi-wet bed fluidization: the effect of small liquid loading and particle size on the bubbling behavior. *Chemical Eng. Science* 59:201–213, 2004.
25. RB White. Using electrical capacitance tomography to investigate gas solid contacting. *Proceedings of 3rd World Congress on Industrial Process Tomography,* Banff, Canada, 2003, pp. 840–845.
26. T McKeen, T Pugsley. Simulation and experimental validation of a freely bubbling bed of FCC catalyst. *Powder Technol.* 129:139–152, 2003.
27. W Warsito, L-S Fan. Measurement of real-time flow structures in gas–liquid and gas–liquid–solid flow systems using electrical capacitance tomography (ECT). *Chemical Eng. Science* 56:6455–6462, 2001.
28. W Warsito, L-S Fan. ECT imaging of three-phase fluidized bed based on three-phase capacitance model. *Chemical Eng. Science* 58:823–832, 2003.
29. AJ Jaworski, T Dyakowski. Application of electrical capacitance tomography for measurement of gas-solids flow characteristics in a pneumatic conveying system. *Meas. Science Technol.* 12:1109–1120, 2001.

30. BA Kader. The second moments, spectra and correlation functions of velocity and temperature fluctuations in the gradient sublayer of a retarded boundary layer. *Int. J. Heat Mass Transfer* 39:331–346, 1996.
31. BHK Lee, S Marineau-Mes. Investigation of the unsteady pressure fluctuations on an F/A-18 wing at high incidence. *J. Aircraft* 33:888–894, 1996.
32. EA Hammer. Three component flow measurement in oil/gas/water mixtures using capacitance transducers. Ph.D. dissertation, University of Manchester, U.K., 1983.
33. EO Etuke, RT Bonnecaze. Measurement of angular velocity using electrical impedance tomography. *Flow Meas. Instrum.* 9:159–169, 1998.
34. V Mosorov, D Sankowski, L Mazurkiewicz, T Dyakowski. The 'best-correlated pixels' method for solid mass flow measurements using electrical capacitance tomography. *Meas. Science Technol.* 13:1810–1814, 2002.

10 Applications in the Chemical Process Industry

David M. Scott

CONTENTS

10.1 INTRODUCTION

The chemical process industry has traditionally relied on the measurement of simple quantities (such as temperature and pressure) to control its operations, but as manufacturing processes become increasingly complex, additional types of information are required. Common process measurement needs now include contamination detection, particulate size and shape, mixture uniformity, concentration profile, thermal profile, amount of fluidization, and various factors related to product quality. Process imaging can provide such data in a wide array of applications; indeed, in many cases these techniques offer the only viable measurement option. This chapter provides a variety of specific examples of process imaging applications in the chemical process industry.

Process instruments are often categorized according to their placement. "In-line" sensors are installed directly into a vessel or pipe and provide a continuous readout. When it is not technically feasible or cost-effective to use an in-process sensor, a side-stream or other sampling technique can be used to present samples to a nearby ("at-line") instrument. In some cases, the instrumentation is too big, too complicated, or too delicate to place near the manufacturing process, so samples are carried to another location (usually a quality control lab) for analysis with an "off-line" device. The term "on-line" is widely used to differentiate sensors that function automatically from those that require human assistance, but it blurs the distinction between "in-line" measurements made directly in the process (and therefore under process conditions) and "at-line" measurements made on sampled material. Since the physical environment of the process material often affects the very quantities to be measured, this distinction can be an important one. For this reason, the term "on-line" is avoided in the present discussion.

The data generated by these process sensors can be used for closed-loop process control, process monitoring (used for "open-loop control"), and product or process research and development. Each one of these end uses carries a specific economic advantage. Process control maximizes product quality and minimizes waste. The ability to control complex chemical process plants directly can also be used to implement new production operations that would not have been possible in the past. Process monitoring capability provides chemical plant operators with direct information about the current status of the constituent unit operations. These data allow operators to maximize asset productivity by identifying and mitigating potential problems before they impact production. Finally, information obtained during the initial experimentation and pilot-plant scale-up stages can greatly enhance the research and development effort aimed at creating new products or new manufacturing processes.

The process imaging examples discussed in this chapter are listed in Figure 10.1, where they are arranged according to their end use and placement in the process. This arrangement is a bit arbitrary, since many of these applications entail both monitoring and control. These examples were chosen to illustrate the breadth of applications in the chemical process industry. Since the primary theme of this book is process control, most of the discussion will focus on the applications in the upper-left region of the figure. Two additional applications are

	Process Control	Process Monitoring	Process R&D
In-Line	Polymer Extrusion (D) Pneumatic Conveying (T)	Crystallization (D)	Polymerization (T)
At-Line	Granulation (D)	Particle Morphology (D)	
Off-Line		Compounding (D)	Media Milling (T)

FIGURE 10.1 Examples of process imaging in the chemical industry. Applications of direct imaging are denoted with (D); tomographic applications are denoted with (T).

included to demonstrate the use of process imaging in the development of new processes and products. It should be noted that research applications tend to be either in-line on real processes (to take advantage of real-world process conditions not found in the lab) or off-line. Examples of both direct imaging (see Chapter 3) and tomographic imaging (see Chapter 4) are included in Figure 10.1.

10.2 APPLICATIONS RELATED TO PROCESS CONTROL

10.2.1 Polymer Melt Extrusion

The manufacture of nylon and other thermoplastic polymers generally involves casting or spinning the molten material following the polymerization cycle. Occasional manufacturing problems arise due to particulate contamination in the polymer, which causes filament breakage during spinning, dyeability variations, and other performance problems. The most common contaminates are gels (lumps of cross-linked polymer), bits of thermally degraded polymer, and extraneous particles entrained in the process during charging of the autoclave (where polymerization takes place).

By monitoring the contamination level in a polymer production line, it is possible to identify the onset of process upsets that lead to production of substandard material. The root cause of the problem can often be diagnosed by observing the timing of the onset and duration of the upset; this timing information is also vital for proper disposition of the product.

Since the contaminants typically found in molten polymer streams are larger than 10 μm in diameter yet occur at very low (ppm) levels, process imaging can be used to measure their quantity and size. A variety of optical probes (both commercial and customized) have been developed to detect the presence of gel particles and other contaminants in polymer processes [1]. Some of these probes have been used to study mixing and residence time distributions in extruders. These camera systems must withstand the harsh environment of polymer extruders that operate around 320 °C and pressures of about 55 MPa (>540 atmospheres).

An early example of a commercial instrument used for polymer applications was the Kayeness FlowVision. It detects and measures the size of contaminants by analyzing the images obtained by transmitting visible light (via fiber optic bundles) through the polymer flow. Two opposing "optical bolts" (hollow bolts containing sapphire windows) are installed in the process to provide optical access. Due to significant light scattering within the polymer, the maximum optical bolt separation is limited to approximately 1 cm for unpigmented polymer. Contaminants in the flow block a portion of the light, casting a shadow on the opposite sapphire window. A fiber optic imaging bundle, which preserves the spatial relationship among fibers in the bundle, carries the image to a CCD camera. The video signal is connected by a fiber optic video link to a video digitizer and image processor. A region of interest (ROI) is defined within the camera's field of view, and contaminants appearing in that ROI are analyzed using

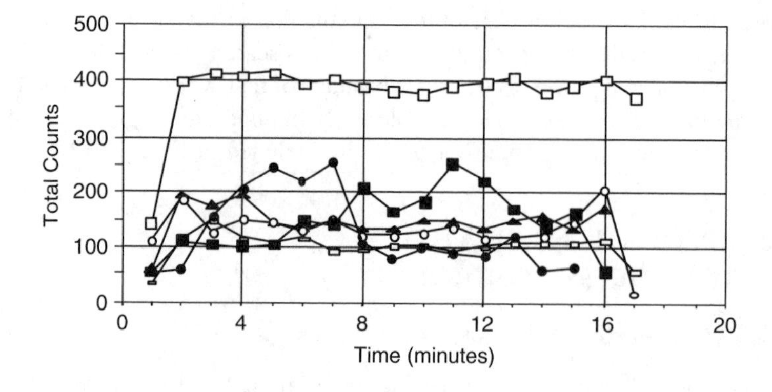

FIGURE 10.2 Contamination levels (total counts) vs. time for six autoclave batches of polymer.

edge detection and other feature extraction techniques discussed in Chapter 5. The system generates particle size distributions and total count rates for the observed particles.

One application of this technology has been to detect contamination at a production autoclave flake-casting nozzle [2]. The autoclave produces nearly 1.4 tonnes of polymer during a 2-hour cycle, but casting takes only 18 minutes; the polymer contamination level is monitored during this final casting stage. Typical data are shown in Figure 10.2. The level of contamination is observed to vary from batch to batch. Data from some batches also show increased contamination at the beginning and end of the casting step. This effect is probably due to pressure fluctuations that jolt degraded polymer residue loose from the autoclave walls and internal piping. By monitoring the level of contamination, manufacturing personnel can identify which production lines need to be cleaned before major problems arise.

Another application of in-line imaging is to detect particulate releases in continuous polymerizer (CP) operations [3]. CP lines feed spinning cells where the molten polymer is spun into fibers. In such applications, particulate impurities in the polymer lower the luster of the fiber surface and increase the possibility of thread breakage, which disrupts the manufacturing operation. Contamination increases have been tied to specific events; for example, a sheared pin in a transfer line metering pump caused a major particulate release, as shown in Figure 10.3. The increase in total particulate counts coincided with a 30% loss in fiber luster (measured with a separate in-line sensor), and the upset condition lasted nearly 48 h. When the particulate contamination decreased to an acceptable level, normal production was resumed. This example demonstrates how imaging technology enables automatic separation of off-quality material from normal product.

Figure 10.4 shows one concept for a complete thermoplastic extrusion and pelletization operation controlled with feedback from process imaging [4]. There, the polymer melt flowing through the extruder is checked for contaminants and gels with an in-line microscope. After extrusion, the polymer strands pass through

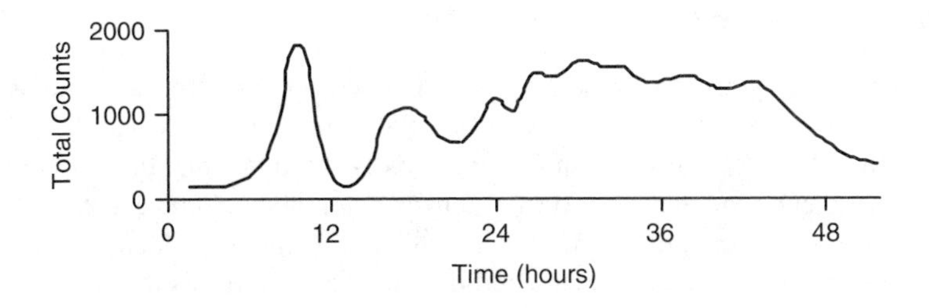

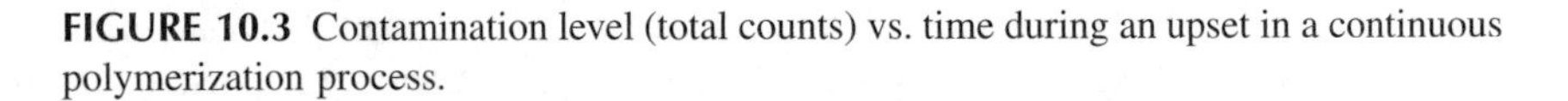

FIGURE 10.3 Contamination level (total counts) vs. time during an upset in a continuous polymerization process.

a water-filled quench tank (which quickly cools the polymer below its glass transition temperature) and into a pelletizer, where the material is cut into pellets by a rotating blade. A camera-based sensor monitors the polymer strands and sounds an alarm if one of them breaks. The size and shape of the product pellets are measured by a third imaging system, which provides continuous feedback to the process controller. These devices therefore provide the information necessary to maintain the quality and uniformity of the product.

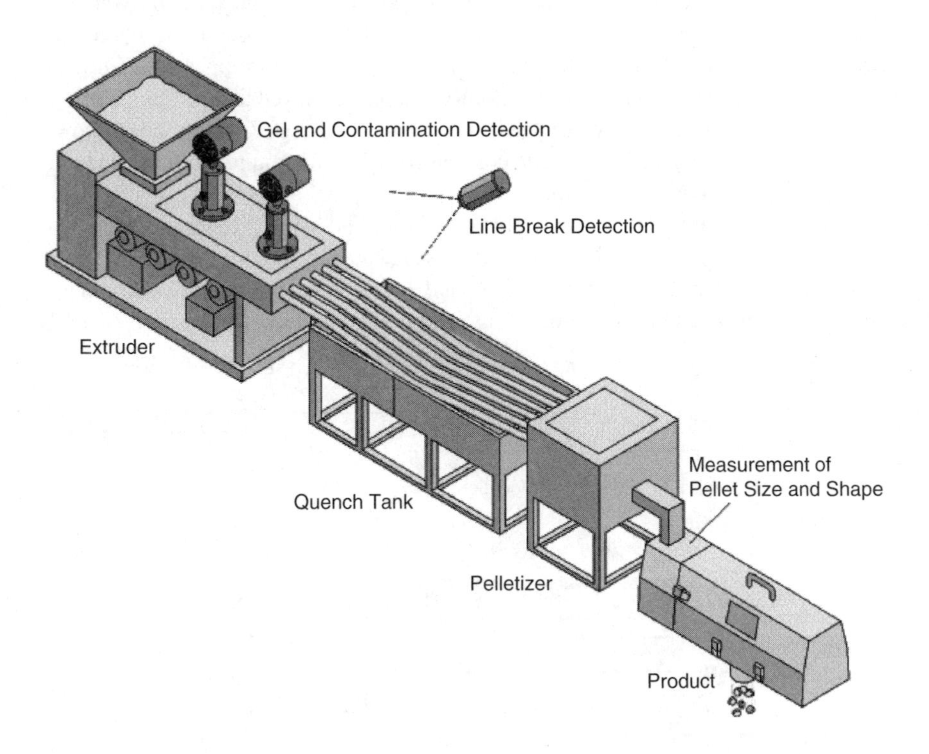

FIGURE 10.4 Concept for automatic control of a polymer extrusion process. (Courtesy of J.M. Canty, Inc.)

10.2.2 Pneumatic Conveying

Pneumatic conveying is a deceptively simple operation used in the manufacture and transportation of bulk solids, e.g., particulate solids, powders, and flakes. The basic concept is that a gas stream moving rapidly through a pipeline will entrain particulate matter and deposit it further downstream. It should be evident that the control of such an operation presents a challenge: if the air velocity is too low, the solids drop out of the air stream and plug the line, but if it is too high, energy is wasted (and the product may also suffer attrition or other damage). In many cases, pneumatic conveyors are used to transport material between unit operations within a manufacturing plant, so if a problem develops it generally impacts other areas as well.

The behavior of a pneumatic conveyor depends on the density of the particles and the air flow conditions. At sufficiently high air velocities, nearly all of the particles will be entrained in the gas flow (assuming little or no cohesive force between the particles). This regime is called dilute phase flow [5]. In horizontal pipes, if the solids mass flow rate is constant and the air velocity is reduced to a critical point (the saltation velocity), particles will start to drop out of the air stream and form a settled layer on the bottom of the pipe. In vertical pipes, a similar effect called "choking" occurs at low air velocity. The saltation velocity marks an abrupt transition from dilute phase flow to dense phase flow, where the settled material travels along the bottom of the pipe in the form of dunes. Further reduction in the air velocity may eventually result in a packed bed. If this situation occurs, increasing the air velocity will not necessarily restore dilute phase flow. The dynamic behavior of pneumatic conveyors is therefore quite nonlinear, which compounds the control problem.

Pneumatic conveying processes have been studied using the electrical tomography techniques described in Chapter 4. Capacitance tomography (ECT), for example, can identify the flow regime and the degree of particle entrainment [6, 7]. One study used a pneumatic rig as shown in Figure 10.5, where several

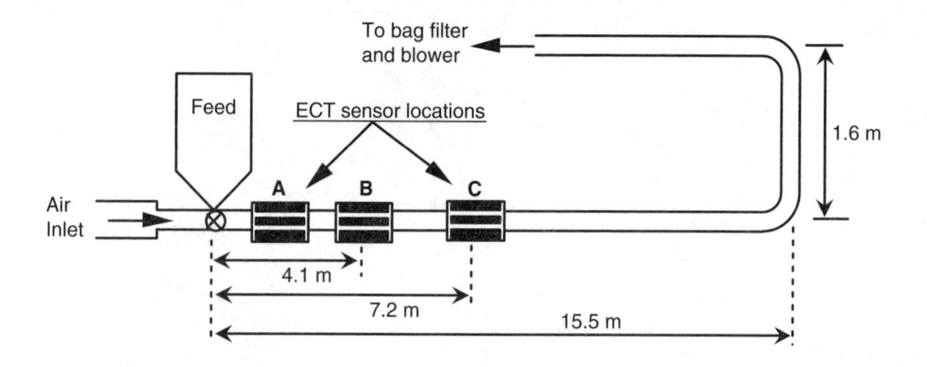

FIGURE 10.5 Layout of pneumatic conveying line used in tests (ECT, electrical capacitance tomography).

observation points are indicated [8]. The ECT sensor was moved from one location to another as required, and the tests were repeated under the same nominal operating conditions. Two materials were conveyed: acetal resin pellets (with a mean particle diameter of 2.85 mm and density of 1350 kg/m^3) and sea salt (with an initial mean diameter of 6.1 mm and density of 2200 kg/m^3). Air inlet velocities of 15 to 40 m/s were used with mass loading factors ranging from 1:1 to 15:1 to obtain dilute-phase conveying conditions.

The acceleration length of a horizontal pipe is the minimum length required for the particles to reach a steady flow regime. Within this length, unstable flow patterns can be encountered, with suspended particles redepositing along the bottom of the pipeline. Such behavior was observed with the ECT sensors located at position B (see Figure 10.5) during the conveying of sea salt. The deposition was seen in the tomographic images, which showed a layer of settled particles at location B; this settled layer was not as prevalent during conveyance of acetal resin pellets under the same flow operating conditions. No image of a settled layer was observed at location C, so the particles were fully entrained in the air flow at that point.

The effect of loading factor on the dispersion of solids (in this case, sea salt) can be seen in Figure 10.6, which shows ECT images at locations A and B under two operating conditions. Figure 10.6a and Figure 10.6b were taken with a loading factor of 13.6, whereas Figure 10.6c and Figure 10.6d were taken with a loading factor of only 4.8. Figure 10.6a and Figure 10.6c were taken at location A, which is close to the injection point; the settled layer is evident in both images.

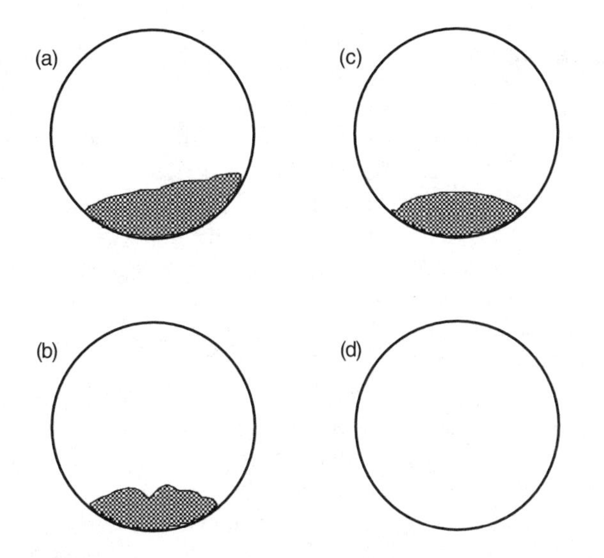

FIGURE 10.6 Electrical capacitance tomography (ECT) images of dilute-phase pneumatic conveying show the effect of loading factor on solids dispersion. (a) is at location A with a loading factor of 13.6; (b) is at location B with a loading factor of 13.6; (c) is at location A with a loading factor of 4.8; and (d) is at location B with a loading factor of 4.8.

Figure 10.6b and Figure 10.6d are the corresponding images at location B. The settled layer was only observed in the case of the larger loading factor; the absence of such a layer in Figure 10.6d indicates a fully entrained flow at that point. Additional work on this application has been described elsewhere [9]. These results demonstrate that dune formation in a pneumatic conveying process can be monitored by a tomographic imaging system.

Deloughry and coworkers have taken this application a step further: they have used a similar ECT sensor operating at 10 frames per second (fps) to implement a closed-loop control scheme for a pneumatic conveyor running under dilute phase flow [10]. The control variable is the number of pixels in the image with gray levels that exceed a certain threshold value; this count corresponds to the amount of saltation present at the sensor's location. A block diagram of their automatic feedback control system is shown in Figure 10.7.

In this application, a minimum air velocity has to be maintained to maintain mass flow. Therefore, a hard limiter was included on the control input to prevent it from going below the minimum value required for operation of the conveyor. This limiter operates as part of a computer-based control algorithm. Several trials were required to determine the optimum integration time (<5 s) and gain setting (<0.3) for the controller; different settings would be needed on other conveyors. It is recognized that the slight time delay introduced by the imaging system would require a predictive controller for the best performance [10]. However, the results of the study clearly demonstrate that the feedback provided by an imaging system can be used for automatic control of a pneumatic conveyor.

10.2.3 Granulation

The purpose of a granulation process is to transform powder (i.e., fine particles) or powder mixtures into larger particles with a desired size distribution and bulk density. Since granules are convenient to transport, store, measure, and apply, many intermediate chemicals and consumer products are granulated. Typical formation processes are pan or fluidized bed granulation (where a binder solution is sprayed onto the powder) and paste extrusion. For many applications, a uniform shape and size distribution are required. Therefore, a high-throughput measurement system is needed to control the granulation process.

It is difficult to measure the size of extruded granules with conventional instruments (such as laser diffraction) due to the cylindrical shape of the particles. To control the size distribution of such granules, it is necessary to use an approach based on imaging techniques. DuPont developed such an instrument in the mid-1990s to control an extrusion granulation process [2]. This instrument acquires images of the granules as they slide down an inclined plane and determines the length and width of each one. Granules are measured at a rate of over 1000 per minute, so a complete size distribution with good counting statistics can be determined in a few minutes. Of course, the actual production rate is much higher than the measurement rate, so this instrument is used as an at-line device

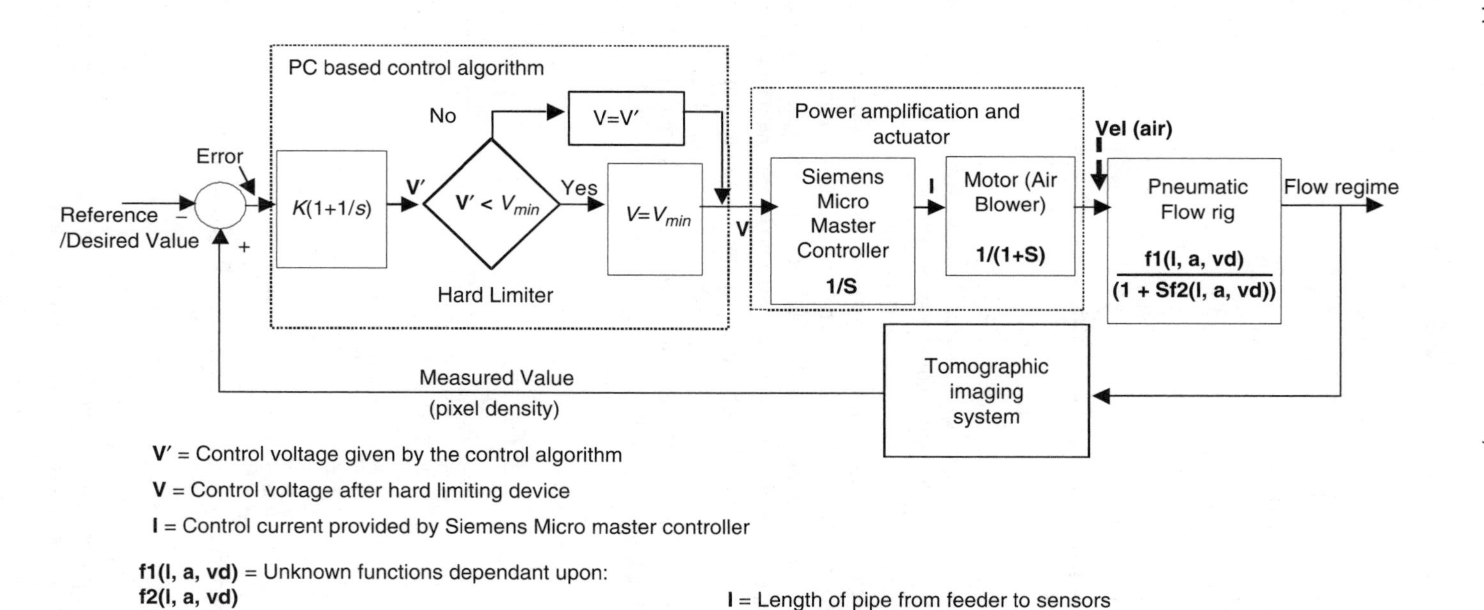

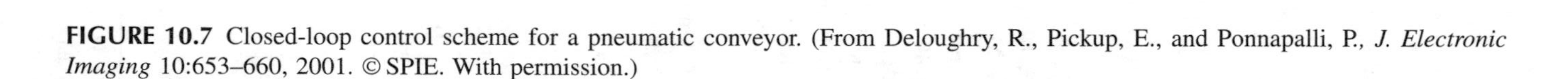
FIGURE 10.7 Closed-loop control scheme for a pneumatic conveyor. (From Deloughry, R., Pickup, E., and Ponnapalli, P., *J. Electronic Imaging* 10:653–660, 2001. © SPIE. With permission.)

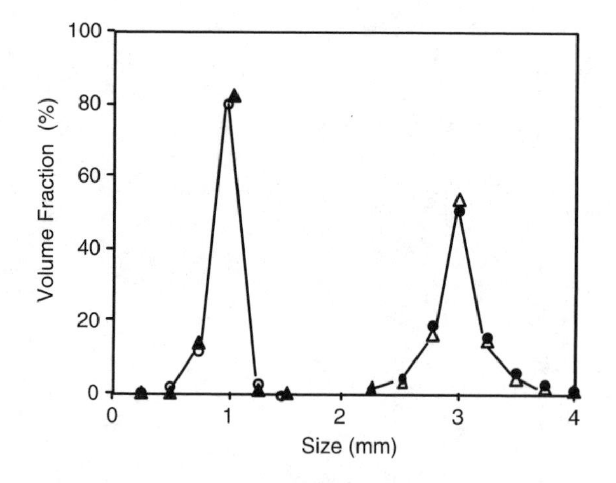

FIGURE 10.8 Length and width distributions of a Teflon® granule reference sample. (From Scott, D. M., Process imaging for automatic control, *Proc. of SPIE,* 4188:1–9, 2001.)

that looks at a side-stream of product. In recent years, Retsch, Microtrac, Canty, and others have independently introduced similar commercial instruments.

Typical granule size data obtained by the DuPont instrument are shown in Figure 10.8 and Figure 10.9. Figure 10.8 shows the granule length and width distributions for a reference sample consisting of Teflon® strands 1.0 mm in diameter that have been chopped to a length of 3.0 ± 0.2 mm. More than 1500 individual "granules" are in the sample, and all of them are measured by the imaging system. In Figure 10.8, the volume-weighted granule width distribution is depicted by filled triangles

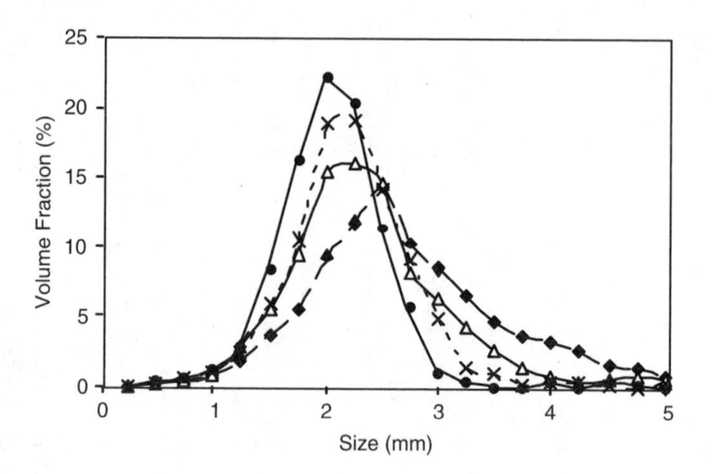

FIGURE 10.9 Length distribution of product granules produced under various operating conditions. (From Scott, D. M., Process imaging for automatic control, *Proc. of SPIE,* 4188:1–9, 2001.)

on the left, and the granule length distribution is depicted by the open triangles on the right. Lines have been added to guide the eye. The observed width (1.0 mm) and length (3.0 mm) have the expected values. A repeat measurement (depicted by open circles and filled circles in Figure 10.8) shows that the reproducibility is excellent.

This instrument is used in a granulation process, the data in Figure 10.9 show the volume-weighted length distributions for product granules that were generated by changing a single process operating parameter; lines have been added for clarity. This parameter has a profound effect on granule length, as shown in the figure. Sample uniformity is indicated by narrow length distributions; here, the most uniform sample is the one designated by circles. A much broader distribution (indicated by diamonds) was obtained with the worst process condition. Thus the image-based technique provides a quick and reliable measurement of granule uniformity. This granulation process is now routinely controlled and optimized via the granule size measurements provided by this instrument.

10.2.4 Crystallization

The morphology of product crystals is an important characteristic that can seriously affect the operation of centrifuges and dryers, not to mention handling characteristics such as flowability. Crystal morphology is influenced by a number of factors, including the vessel design, the solids concentration, and temperature gradients. Conventional (i.e., off-line) techniques for characterizing morphology are not well suited for monitoring or controlling crystallization processes. Removing crystals from the mother liquor causes additional growth, and fine particles tend to stick to larger particles. These effects change the particle surface and shift the apparent median size, making a representative sample difficult to obtain.

To solve the problem of sampling, an in-line camera probe can image and characterize the product crystals in the slurry exiting a full-scale industrial crystallizer [11]. One example of a simple camera built for an industrial crystallizer application is shown in Figure 10.10. This camera is based on an industrial borescope (Schott model 10RS455D) used for inspecting the interiors of closed vessels. The camera probe is a 2.5-cm diameter stainless steel tube that protrudes into the

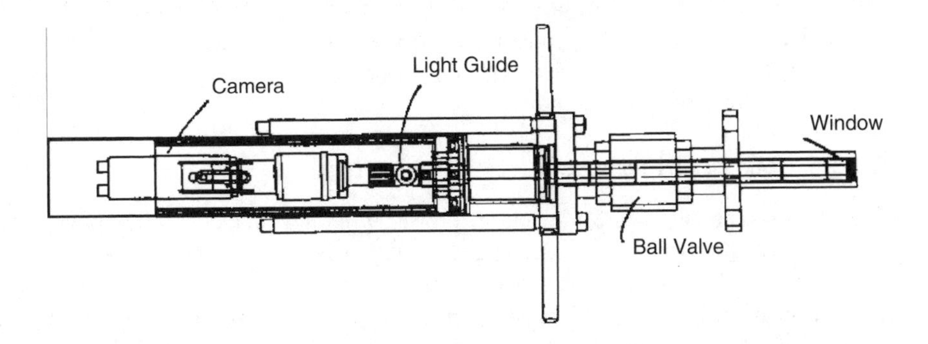

FIGURE 10.10 Outline of a camera probe for monitoring industrial crystallizers.

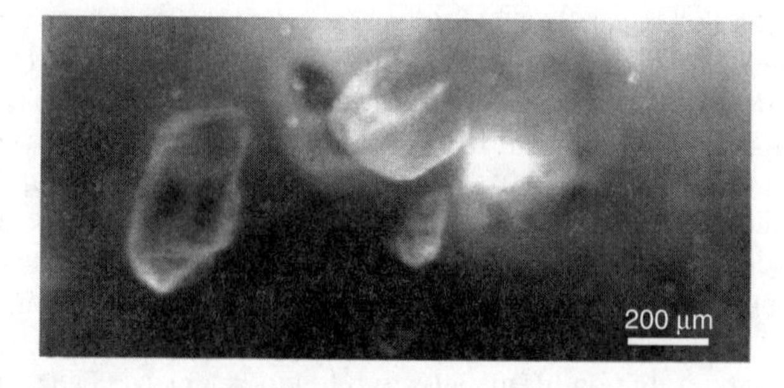

FIGURE 10.11 In-line image of crystals produced in a plant during start-up. (From Scott, D. M., Boxman, A., and Jochen, C. E., *Part. Syst. Charact.* 15:47–50, 1998. © Wiley-VCH. With permission.)

process stream through a ball valve. A sapphire window at the end of the probe provides optical access to particulate material inside the process. A strobe light (mounted in the control box) is used to "freeze" the motion of the moving particles. Light is carried from the strobe to the crystals via a fiber optics bundle, and the image is relayed to a CCD camera by the borescope optics. The camera is optically coupled to the eyepiece of the borescope via a lens. The video signal is transmitted to a frame digitizer via a fiber optic link, so the computer and operator interface can be mounted in any convenient location (such as the control room).

This camera was designed to be compatible with standard ball valves that were already installed at various locations in the crystallization plant. Thus the camera can easily be moved from one location to another. This instrument has been used at production sites to monitor the crystal morphology as part of a process optimization program. In recent years, several vendors (e.g., Lasentec, Canty, and Sympatec) have introduced similar types of in-process imaging probes.

Figure 10.11 shows an in-line image taken during start-up of the crystallizer. The crystal edges have been rounded by attrition occurring within the crystallizer. Other crystal features that have been observed with this camera probe include crystal twinning, inclusions, and relative transparency (an indication of crystal purity). These characteristics are used to diagnose and correct problems in the crystallizer operation. In-line imaging provides a direct means of monitoring crystallization processes; coupled with image processing, it could be used for automatic control of size and morphology.

10.2.5 Particle Morphology

Particle-based processes and products are prevalent throughout industry. The physical characteristics (including size, shape, and other factors) of the particles can have a profound impact on the process, so particle characterization is an essential capability. Image analysis is often used in conjunction with microscopy to measure the size and shape of individual particles [12]. In fact, a number of morphological parameters can be measured, including projected area and projected diameter, length, width, and various "shape factors." One disadvantage of traditional microscopy is that the number of particles that can be feasibly measured is rather small; with small sample sets, the poor counting statistics lead to a large variance in the collected data.

An automated indexing microscope system can measure the individual size and shape of thousands of particles without human supervision. By creating such a large data set for each sample, one can determine the full distribution of morphological parameters instead of estimating their mean values from a few observations.

A suitable system consists of a computer-controlled X-Y positioning sample stage, a light source, a CCD camera (with a macro lens), and a computer with an image digitizer. A good macro lens is adequate to enable measurement of particles in the 100 to 800 μm range, but one can also use microscope objectives to inspect smaller particles (down to 10 μm). The top of the stage should be transparent so the particles can be illuminated from either the back (by placing the light source beneath the stage) or the front. In operation, the X-Y stage is moved stepwise in a raster pattern that allows the entire surface to be imaged. Many thousands of particles can be distributed across the stage and automatically imaged in this manner.

A useful technique that simplifies the image processing is to affix the particles in a regular array so that they do not touch each other. An array of sticky spots can be created on a glass or polymer substrate [13]. These spots are designed to be large enough to hold the particles, but small enough to prevent more than one particle from sticking at a given location. Small arrays of these dots are available commercially [14]. One populates the array by repeatedly sprinkling the particles on it and gently removing the excess. This approach certainly has the potential for introducing sampling error [12]. However, for the purpose of the present discussion, it should suffice to state that with the proper technique nearly all of the particles in a small sample are captured on the array.

This type of instrument has been used to optimize a droplet polymerization reaction used to produce polymer spheres about 0.4 mm in diameter. Figure 10.12

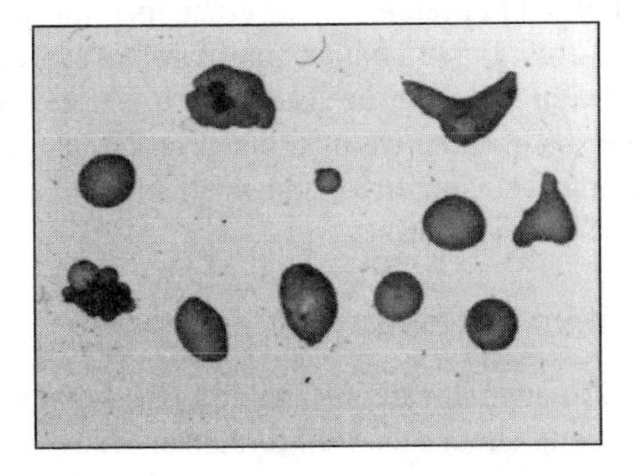

FIGURE 10.12 An example of the variety of particle shapes produced by a droplet polymerization process. (From Scott, D. M., Process imaging or automatic control, *Proc. of SPIE*, 4188:1–9, 2001.)

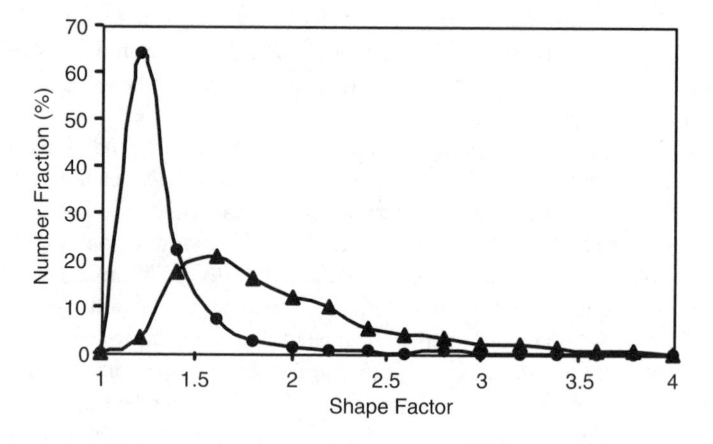

FIGURE 10.13 Shape factor distributions of particles produced by two different process conditions. (From Scott, D. M., Process imaging for automatic control, *Proc. of SPIE*, 4188:1–9, 2001.)

illustrates the variety of shapes and particle sizes that have been produced under various process conditions. The final application for these beads required reasonably spherical particles, and morphology measurements provided the quantitative feedback necessary for process optimization.

One simple measure of particle sphericity is the shape factor S, defined as

$$S = P^2/4\pi A \tag{10.1}$$

where P is the perimeter of the particle's image and A is the projected area. For a circle, $S = 1$, whereas S is larger for other shapes. For large aspect ratios (length/width), S is proportional to the aspect ratio. When a collection of particles is analyzed, the shape factor S generally has a distribution of values.

Figure 10.13 shows a number-based distribution of shape factors observed for two polymer samples produced under different operating conditions. It is clear from these data that one sample contains a substantially greater population of near-spherical beads than the other. In distinction to visual inspection, this automated system provides a quantitative measurement that can easily be tailored to the specific application. The simplicity and relatively low cost of this approach make it a useful at-line instrument.

10.2.6 Compounding of Reinforced Polymers

Reinforced plastic is used for injection-molded parts that require dimensional stability and resistance to chemicals. Reinforcement materials (such as chopped glass or carbon fibers) are compounded with a polymer matrix to improve the tensile strength and strength-to-weight ratio. One example is glass-reinforced nylon, which features high tensile strength, stiffness, and impact resistance and is widely used in automotive applications.

The ultimate mechanical properties of an injection-molded part depend not only on material properties but also on how well the reinforcing material is dispersed throughout the part, how much total reinforcing material is present ("loading"), and the length of the reinforcing fibers [15]. A part with poor dispersion (or inadequate loading) will have abnormally low mechanical strength. During product development, the dispersion and loading of an injection-molded test sample must therefore be quantified to assess the success or failure of a new compound.

Destructive techniques have generally been used to determine loading. The most common method is to weigh and ash the samples. The ash is washed so that the residue is essentially glass. The glass is then weighed, and the ratio of the glass weight to the weight of the original sample is the approximate glass loading. This process takes as long as 40 min and is not suitable for fluoropolymers because those matrices react with the glass at elevated temperatures. In the past, there was no satisfactory method for quantifying dispersion.

A fast and quantitative method of analyzing these reinforced polymers has been developed through the combination of real-time x-ray imaging technology with rudimentary image processing techniques [16]. This system measures the concentration and dispersion uniformity of glass in the polymer and provides a quantitative means of monitoring the compounding process.

When x-rays pass through a material, the transmitted intensity I is reduced from the incident intensity I_0 according to Beer's Law:

$$I = I_0 \exp(-\mu t) \tag{10.2}$$

Here t is the thickness of the sample (in cm), and μ is the linear attenuation coefficient of the sample (in 1/cm). Measurements of the absorbance (I/I_0) can be used to estimate the linear attenuation coefficient. For a two-phase system such as glass-reinforced nylon, the loading (volume fraction) α of the glass phase can be determined from measurements of the linear attenuation coefficients of the reinforced polymer sample (μ), the polymer matrix (μ_n), and the glass (μ_g):

$$\alpha = (\mu - \mu_n)/(\mu_g - \mu_n) \tag{10.3}$$

Such measurements can be made with radioscopic (i.e., real-time radiographic) imaging systems, which essentially record the transmitted intensity $I(x, y)$ at each point (x, y) in the projected image. Any internal structure in the object (such as clumps of poorly dispersed glass) will introduce variability into the transmitted radiation pattern. By examining the statistics of the distribution of gray levels within the image, it is possible to measure the local and average glass content (loading) as well as the fiber dispersion (how well the fibers are distributed).

Figure 10.14 shows a diagram of the instrument [16]. A microfocus x-ray source is mounted above a computer-controlled X-Y table, which holds an array of samples. An x-ray image intensifier mounted underneath the X-Y table converts

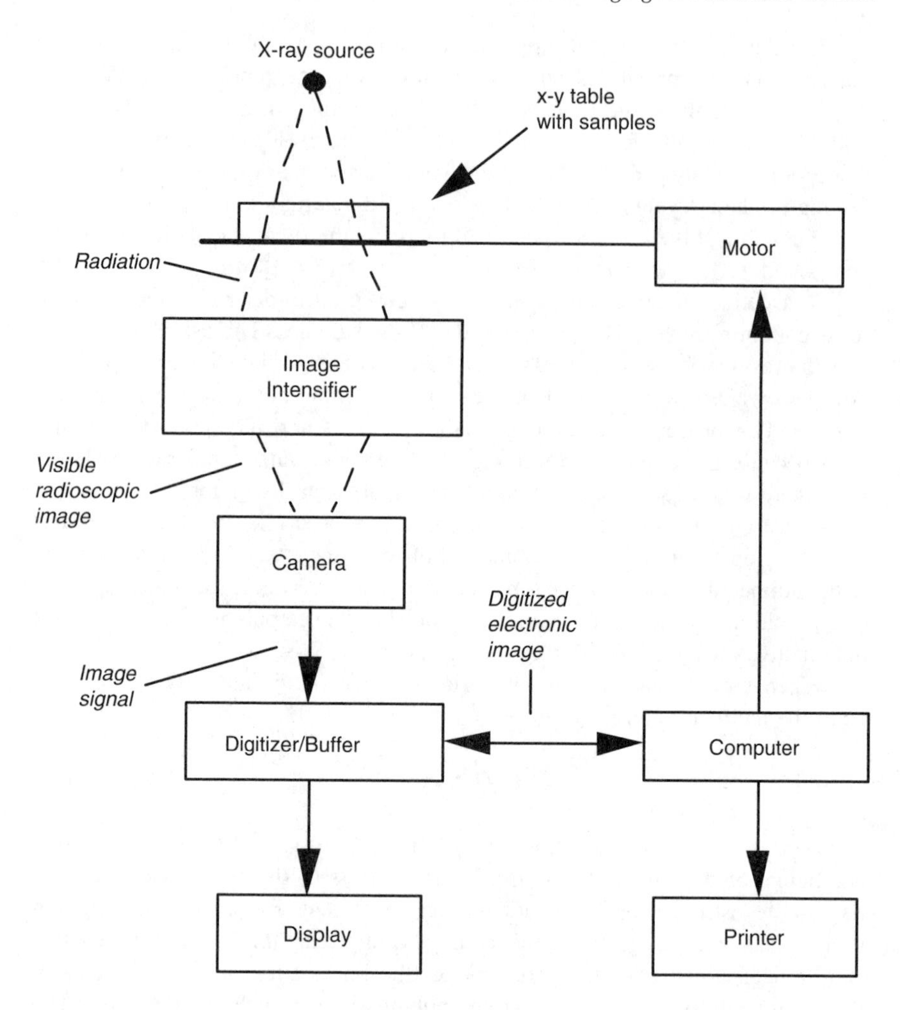

FIGURE 10.14 Block diagram of instrument used to monitor a polymer compounding process.

the latent image of the radiation passing through the sample into a visible image. This image is recorded by a CCD video camera and digitized by a video frame buffer residing in the computer. Due to noise in the intensified images, it is necessary to average 100 frames in order to achieve acceptable signal-to-noise ratios, and each pixel in the digitized image must be independently corrected (normalized) for variations in baseline and gain [17]. The average loading of each sample is calculated from Equation 10.3, and the quality of fiber dispersion is indicated by the standard deviation of the gray levels in the normalized radiographic image.

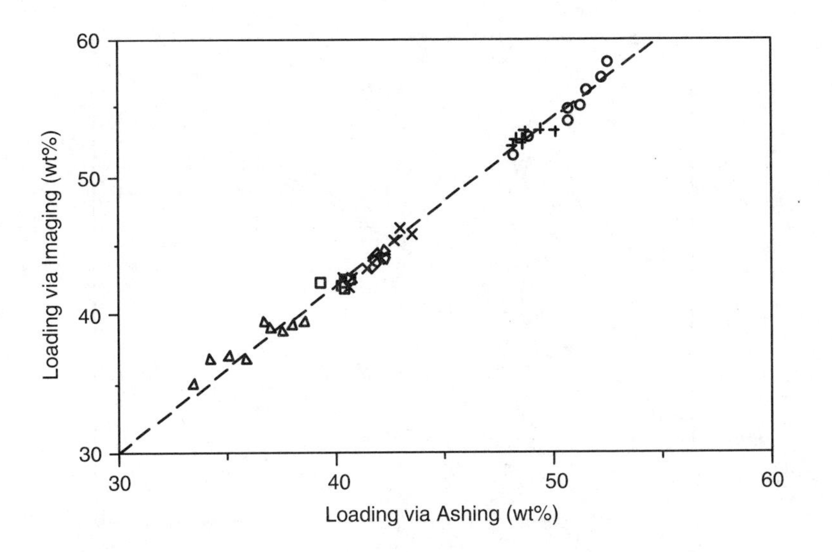

FIGURE 10.15 A comparison of glass loading measurements determined by x-ray imaging vs. data on the same samples subsequently obtained via the conventional ashing test. (From Scott, D. M., U.S. Patent 5,341,436, 1994.)

Figure 10.15 compares loading measurements made with this instrument to the results obtained by the conventional ashing method. Samples of both commercial and experimental grades of glass-reinforced polymer were studied, and the agreement between predicted and observed values is remarkable. Although there is a small systematic error of a few percent (probably due to errors in the linear attenuation values for glass and polymer), the relative error is only 0.5%. Since the ashing method can easily underestimate the loading if any glass is washed away during the rinsing step, these data may be closer to the actual values than is indicated in the graph.

Since inhomogeneity in the composite material generates large gray-level variations in the radiograph, this instrument has been used to improve compounded material by providing a quantitative measure of fiber dispersion. Melt viscosity, a factor long suspected to influence dispersion, has in fact been shown to correlate quite strongly with the dispersion as measured by the standard deviation of gray-level values in the normalized radiographic image. Measurements of the standard deviation vs. the melt viscosity for three different resins are shown in Figure 10.16. The improved dispersion at lower melt viscosity is thought to be the result of better wetting of the glass fibers by the resin. Also, resins that contained plasticizer generally exhibited better glass dispersion due to the lowered viscosity. Processing details such as glass fiber length, screw design, and compounding conditions were also observed to affect the dispersion. This off-line instrument is therefore quite useful for monitoring the inspection-molding process.

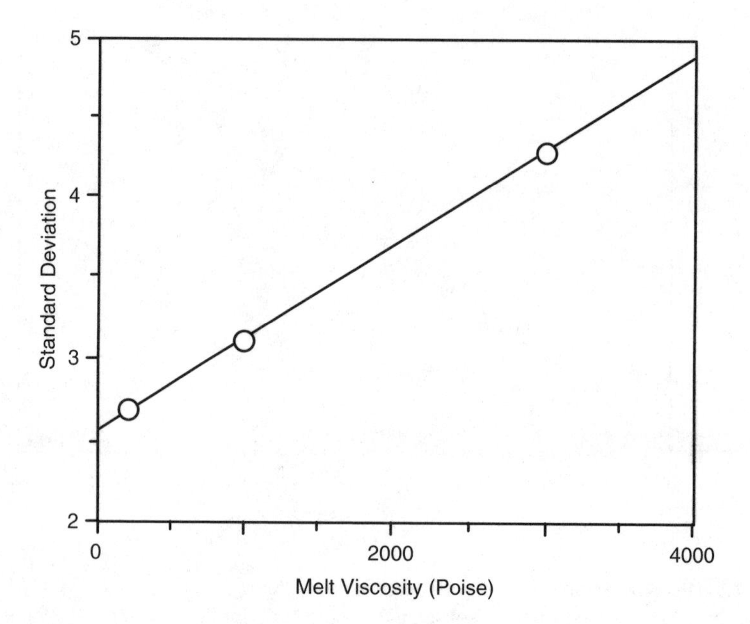

FIGURE 10.16 Variability in gray-level vs. melt viscosity. (From Scott, D. M., *Composite Part A,* 28A:703–707, 1997 © Elsevier. With permission.)

10.3 APPLICATIONS RELATED TO PROCESS AND PRODUCT R&D

10.3.1 POLYMERIZATION MONITORING

Many common polymers are produced in autoclaves, which are closed reaction vessels operating at high temperature and pressure. Nylon, for instance, is formed by heating nylon salt (precipitated from a mixture of hexamethylene diamine and adipic acid) with some water to a temperature of 300 °C and a pressure of 1.72 MPa (about 17 atmospheres). Most process sensors are not compatible with this environment, and consequently very little information is available about the properties of the polymer while it is still in the process. As in the case of most batch processes, these polymers are made according to a recipe with no closed-loop control to correct for variations in reactor loading, moisture content, or other process variables.

A three-year research collaboration between DuPont and the University of Manchester Institute of Science and Technology (UMIST) in the United Kingdom was able to address the problem of making in-line measurements during polymerization. Electrical resistance tomography (ERT) was deemed the only feasible sensing method for this application, so the main program objective was to visualize changes and spatial variations in material properties occurring during various stages of nylon polymerization. The measurements were made on a small-scale autoclave at the DuPont Nylon production plant in Wilton, U.K. Although making

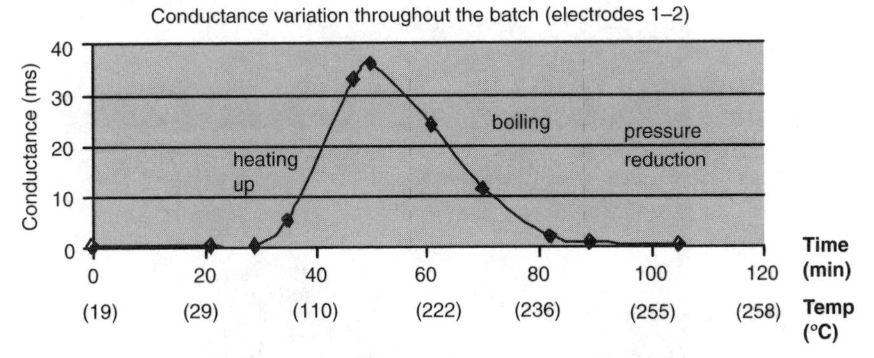

FIGURE 10.17 Conductance measured between two adjacent electrodes in the ERT sensor vs. time.

sensitive electrical measurements in an autoclave posed significant technical challenges, this project provided the world's first tomographic images of the nylon polymerization process [18].

The ERT technique requires that the materials to be imaged have finite conductance. Figure 10.17 shows the measured conductance between a pair of adjacent ERT electrodes as a function of time and temperature. Initially, nylon salt is fed into the autoclave with a small amount of water on top. At this initial stage, the electrical conductivity of the nylon salt and water mixture is too low to generate images. It takes time for the water to penetrate the salt, but the rising temperature in the autoclave accelerates the process. As the salt dissolves, that region becomes conductive. For the observed electrode pair, measurable conduction occurs after about 30 min. Once the entire mixture becomes conductive, tomographic images can be obtained.

The conductance continues to rise until the temperature reaches about 205 °C, at which point the nylon polymerization starts. Figure 10.18 shows a sequence of representative tomographic images; these were obtained at a temperature of 210 °C and pressure of 1.1 MPa. At a temperature of about 220 °C, the pressure is released, causing the volatile components in the mixture to expand rapidly.

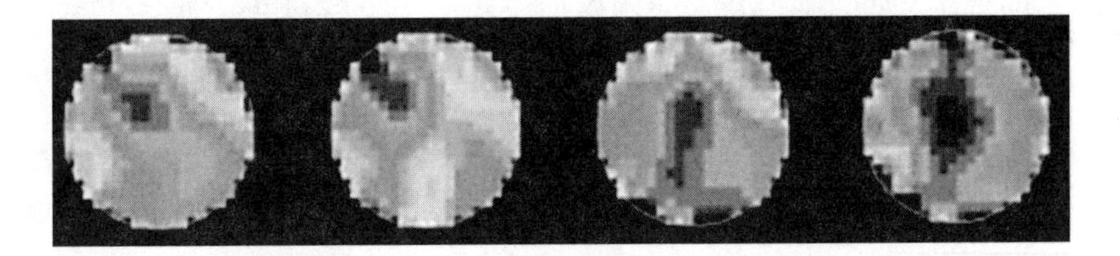

FIGURE 10.18 Sequential ERT cross-sectional images (at 1s intervals) of nylon undergoing polymerization in an autoclave. (From Dyakowski, T. et al., *Chem. Eng. J.* 77:105–109, 2000.)

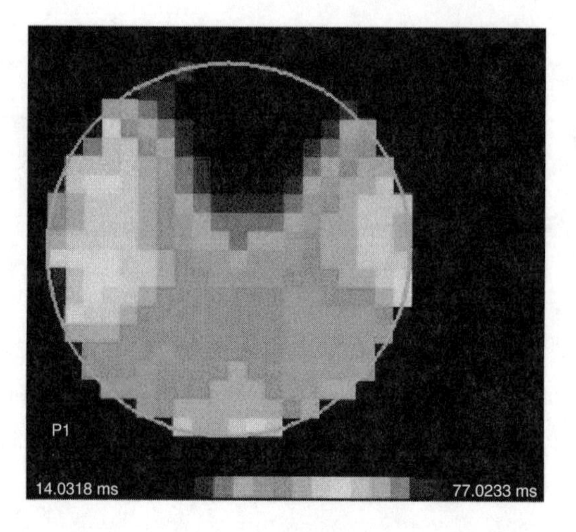

FIGURE 10.19 Zone of low conductivity (due to incomplete melting of the nylon salt) observed during the early stages of nylon production. (From Dyakowski, T. et al., *Chem. Eng. J.* 77:105–109, 2000.)

In the final stages of the process, the resulting foam reduces the conductance to the extent that ERT images cannot be captured.

Figure 10.19 shows a tomographic image of the reaction vessel where some of the nylon salt has not completely melted. The gray-level of such images can be interpreted to imply the amount and spatial distribution of moisture in the autoclave, and the growth of inhomogeneous regions can be detected and quantified. This information can be used to improve autoclave designs and operating procedures. These results demonstrate the feasibility of monitoring polymerization reactions under typical process conditions with ERT.

10.3.2 Media Milling

Agitated media mills are used throughout industry for size reduction and dispersion of a variety of particles, such as pigments, polymers, pharmaceuticals, and agricultural chemicals. Such mills use rotating agitators to stir and thereby fluidize a bed of grinding beads, which typically fill 80% of the volume of the grinding chamber. The nominal size of these beads (called grinding media) ranges from 0.2 to 3 mm, and the beads may be glass, ceramic, or metallic. The particles to be milled are made into a slurry to fill the remaining volume in the chamber. Energy is transmitted from the agitator to the liquid phase, which accelerates the beads; particles in the slurry break when they are nipped between colliding beads [19]. Some mills pump slurry through the chamber continuously; others operate in batch mode, with no net flow of slurry through the mill.

Optimum grinding and energy utilization occur when the grinding media are uniformly distributed throughout the mill [20]. Experience shows that when the flow rate of the slurry through a continuous mill exceeds a critical value, grinding beads begin to pack at the retainer screen, causing screen wear, media wear, an increase in power consumption, and an overall decrease in grinding efficiency. Likewise, if the beads are not fully fluidized in a vertical (batch) mill, they remain on the bottom of the mill and wear grooves into the chamber walls. Grinding under such conditions is ineffective. Successful grinding requires both effective particle capture (between grinding beads) and sufficient impact intensity. Bead fluidization influences both capture statistics and collision intensity; therefore, it is a prerequisite for optimum grinding.

Process parameters such as bead size, bead density, bead filling, fluid viscosity, and rotational speed affect the state of fluidization, but traditionally there has been no direct way to measure it. An interesting off-line application of the electrical capacitance tomography (ECT) technique discussed in Chapter 4 is to measure axial and radial bead distributions as well as overall bead fluidization in vertical media mills [21]. As demonstrated below, this technique provides a versatile tool to determine the optimum operating conditions for agitated mills.

Figure 10.20 depicts the vertical mill used in this study. Due to the type of imaging system used, the mill is constructed entirely of nonconducting materials. The mill consists of a Plexiglas® tube with outer diameter (OD) 14.6 cm and inner diameter (ID) 13.3 cm. The total length of the tube is about 30 cm. A Plexiglas plate is cemented inside the tube at a distance of 11.4 cm from the bottom. This plate defines the bottom of the milling chamber and provides physical access for the sensing plane of the ECT sensor, as shown in the figure. The ECT system [22] has a sensor with an ID of 15.2 cm, allowing it to fit around the vertical mill housing and slide along the axis of the mill. This mill is small by industrial standards, but it exhibits the same operational behavior seen in larger mills. As a research tool it provides a platform for quickly measuring the amount of fluidization produced by a variety of operating conditions and agitator designs.

The mill is closed at the top, and the agitator shaft protrudes from a small opening in the lid. The purpose of the agitator is to fluidize the grinding beads and to supply them with the kinetic energy needed to break the particles in the slurry. The shaft of the agitator is connected to a variable speed motor via a torque transducer. This transducer measures torque and rotational speed so that the power input to the mill can be determined. The mill is completely filled to avoid the formation of a vortex. Corn oil is used as a surrogate for the particle slurries, and the grinding media are 1 mm ceramic beads.

Several different agitators have been studied. They are similar to typical designs actually used in industrial mills, but to be compatible with the ECT sensor, they were constructed of Delrin® engineering polymer. Two common designs are the "disk" and "pin" agitators. The "disk" agitator has five removable solid disks, each 12.8 cm in diameter and 0.6 cm thick. The disks are centered on a 3.2-cm diameter shaft, with a gap of 2.5 cm between the disks. The "pin" agitator has a shaft 2.5 cm in diameter, with six 1.3-cm diameter pins (i.e., rods) installed in mounting holes drilled through the shaft at right angles. The pins are 10 cm long, and they are mounted in an alternating pattern (E-W, N-S, E-W, etc.) along the shaft at a spacing of 2.5 cm.

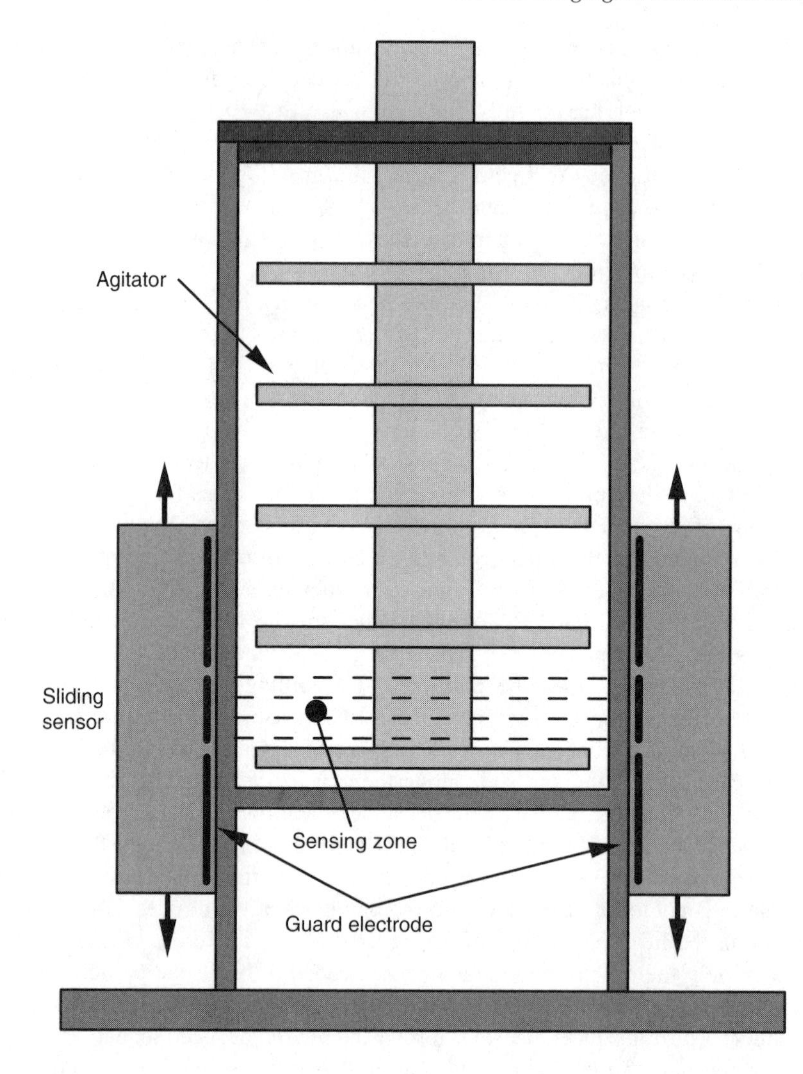

FIGURE 10.20 Cross-section of the vertical mill.

As described in Chapter 4, the tomography system measures the electrical capacitance between every possible pair of electrodes in the array and uses these data to determine the local dielectric constant (as a function of position in the cross-sectional plane) via a tomographic reconstruction. Since the dielectric properties of the fluid and media are fixed, the local bead fraction can be calculated from Equation 10.3 if the linear attenuation coefficients there are replaced by the respective dielectric constants. After the system is calibrated against a fully packed media bed (which by definition has a bead fraction of 1), the cross-sectional image in effect shows the bead fraction at every point in that plane.

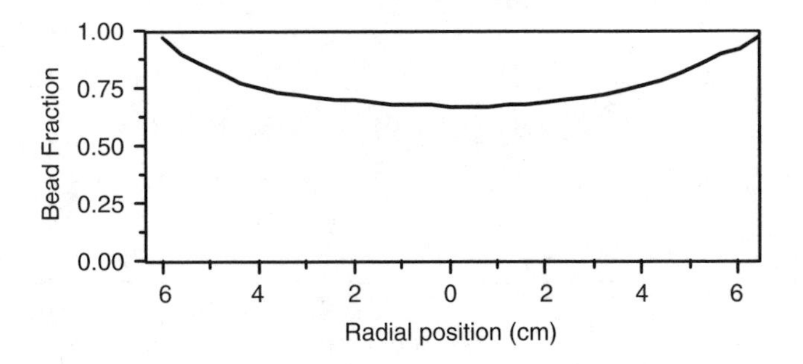

FIGURE 10.21 Radial distribution of beads in the mill. (From Scott, D. M. and Gutsche, O. W. *Proc World Cong. Industrial Process Tomography,* Buxton, U.K., pp. 90–95, 1999.)

Cross-sectional ECT images of the mill provided the information to determine the radial distribution of the grinding beads. Figure 10.21 shows a typical result for the disk agitator rotating at 600 rpm. The centripetal force imparted by the agitator tends to push the beads outward, thus increasing the bead fraction near the wall of the mill. This effect is clearly seen in the figure.

The axial distribution is determined by integrating the bead fraction across the sensing plane at a given axial position. Figure 10.22 shows the axial distribution of beads for the pin agitator at a speed of 586 rpm. The bead fraction is approximately constant at 85% over the lower 8 cm of the milling chamber and then decreases to about 50% near the top. This result is due both to poor mixing at the top of the chamber and to gravity, whose downward pull must be overcome by momentum transferred in collisions between the beads. Higher agitation speeds increase the collision rate and send material higher in the vertical mill. Likewise, the "weight" of material already fluidized tends to compress the beads at the bottom of the mill; therefore the fluidization is lowest there. Averaging the bead fraction shown in Figure 10.22 along the axis of the mill gives a value of about 80%, the value expected from conservation of bead volume.

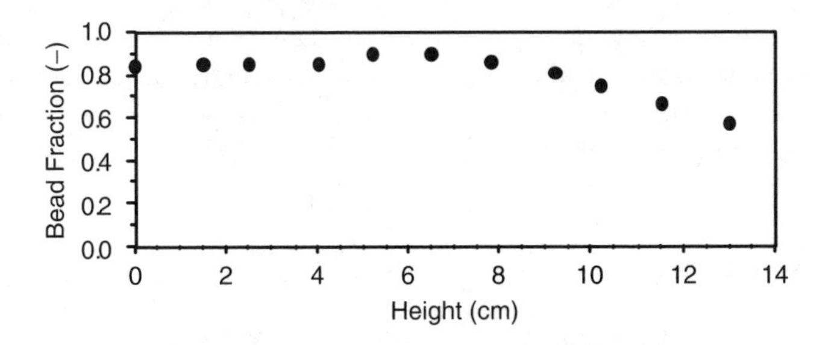

FIGURE 10.22 Axial distribution of beads in the mill. The bead volume fraction is a function of axial position (height). (From Scott, D. M. and Gutsche, O. W. *Proc World Cong. Industrial Process Tomography,* Buxton, U.K., pp. 90–95, 1999.)

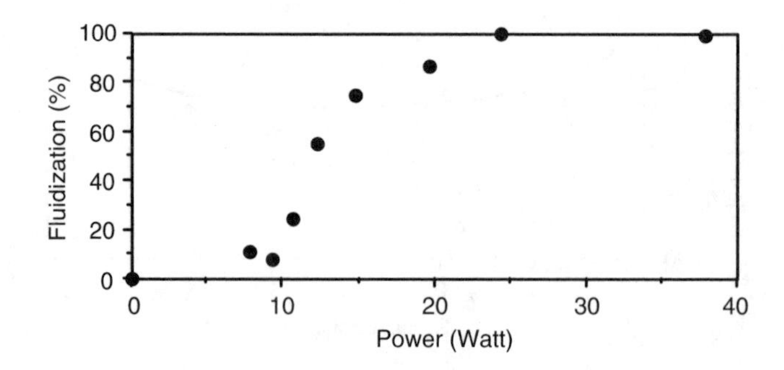

FIGURE 10.23 Measured fluidization of the grinding beads as a function of input power for the pin agitator. (From Scott, D. M. and Gutsche, O. W. *Proc World Cong. Industrial Process Tomography,* Buxton, U.K., pp. 90–95, 1999.)

A simple connection between fluidization (f) and bead fraction (p) can be expressed as

$$f = 5(1 - p) \quad (10.4)$$

Note that this equation assumes the overall bead volume is 80%. Using this definition, fluidization is a number ranging from 0 to 100%, where 0% corresponds to the packed state (bead fraction is 1) and 100% corresponds to the fully fluidized state. Figure 10.23 shows the fluidization produced by the pin agitator as a function of input power, which is calculated from the torque and speed of the agitator shaft. This approach provides the only direct means of quantifying fluidization of media in mills, and it has clear implications for process improvements.

10.4 CONCLUSION

A wide variety of imaging techniques have been applied to real processes in the chemical industry. The applications listed here demonstrate the versatility of process imaging, whether it is direct or reconstructed (tomographic). In many cases, such as media mills or crystallizers, these techniques represent the only viable option for obtaining the required data. The virtually instantaneous feedback provided by these instruments has already been used to implement control schemes in extruders, conveyors, crystallizers, and other unit operations. It is clear that this broad technology field will find additional control applications in other processes.

REFERENCES

1. WE Wolf. Optical probes for polymer process stream measurements. *DuPont Polymer Eng. J.* 1:25–35, 1994.
2. DM Scott, G Sunshine, L Rosen, E Jochen. Industrial applications of process imaging and image processing. In: H McCann, DM Scott, eds. *Process Imaging for Automatic Control, Proceedings of SPIE,* Vol. 488, pp. 1–9, 2001.

3. CE Jochen, DM Scott. On-line contamination detection in polymer processes. *Proceedings of Control of Particulate Processes IV,* Delft, 1997, pp. 177–182.
4. JM Canty, Inc., (Buffalo, New York), 2003 catalog.
5. D Mills. *Pneumatic Conveying Design Guide.* London: Butterworths, 1990.
6. A Plaskowski, MS Beck, R Thorn, T Dyakowski. *Imaging Industrial Flows.* Bristol: Institute of Physics, 1995.
7. SL McKee. On-line measurements of particle-fluid transport processes using tomographic techniques. M.Sc. thesis, University of Manchester Institute of Science and Technology, Manchester, U.K., 1992.
8. SL McKee, T Dyakowski, RA Williams, TA Bell. Tomographic imaging of pneumatic conveying processes. *Proceedings of the 1993 Workshop of the European Concerted Action on Process Tomography (ECAPT),* Karlsruhe, Germany, 1993, pp. 89–92.
9. T Dyakowski, RA Williams. Pneumatic conveying and control. In: RA Williams, MS Beck, eds. *Process Tomography.* Oxford, U.K.: Butterworth-Heinemann, 1995, pp. 433–445.
10. R Deloughry, E Pickup, P Ponnapalli. Closed loop control of a pneumatic conveying system using tomographic imaging. *J. Electronic Imaging* 10:653–660, 2001.
11. DM Scott, A Boxman, CE Jochen. In-line particle characterization. *Part. Part. Syst. Charact.* 15:47–50, 1998.
12. T Allen. *Particle Size Measurement,* 5th ed. London: Chapman Hall, 1997, pp. 112–148.
13. A Cairncross, U Klabunde. Method and Product for Particle Mounting. U.S. Patent No. 5,356,751 (1994).
14. Tacky Dot™ microscope slides are sold by Structure Probe, Inc., West Chester, Pennsylvania.
15. MJ Carling, JG Williams. Fiber length distribution effects on the fracture of short-fiber composites. *Polymer Composites* 11:307–313, 1990.
16. DM Scott. Nondestructive analysis of dispersion and loading of reinforcing material in a composite material. U.S. Patent No. 5,341,436 (1994).
17. DM Scott. Density measurements from radioscopic images. *Mater. Eval.* 47:1113–1119, 1989.
18. T Dyakowski, T York, M Mikos, D Vlaev, R Mann, G Follows, A Boxman, M Wilson. Imaging nylon polymerization processes by applying electrical tomography. *Chem. Eng. J.* 77:105–109, 2000.
19. A Kwade. Wet comminution in stirred media mills: research and its practical application. *Powder Technol.* 105:14–20, 1999.
20. H Weit. Betriebsverhalten und Maßstabsvergrößerung von Rührwerkskugelmühlen. Dissertation, Technical University of Braunschweig, 1987.
21. DM Scott, OW Gutsche. ECT studies of bead fluidization in vertical mills. *Proc. World Congress on Industrial Process Tomography,* Buxton, U.K., 1999, pp. 90–95.
22. Process Tomography Ltd. (Wilmslow, Cheshire, U.K.).

11 Mineral and Material Processing

Richard A. Williams

CONTENTS

11.1 MOTIVATION FOR DEVELOPMENT OF IMAGE-BASED TECHNIQUES

Primary and secondary processing of minerals and metals involves handling materials in particulate form of variable size, shape, chemical composition, and mineralogical composition. (A particle is simply defined as something that is small in relation to its surroundings.) Such complex process systems demand careful assessment of the nature of the feedstocks and quality of the products. In *De Re Metallica* of 1556 [1], Agricola describes the first recorded image-based control system based on hand sorting of ores to pick out the high grade pieces of ore from lower grade material. The use of the discerning human eye to assess color, hue, and luster coupled with size and shape was effective as an ore-sorting and quality control mechanism—indeed, the same protocol and variations on it have been in continuous use since Agricola's time. With optical sensors and discriminating intelligence beginning to emulate human capabilities, at least for the simplest of tasks, it is not surprising, therefore, that the minerals industry has been an early adopter of process imaging, as this chapter will seek to describe.

The need to process large quantities of materials (thousands of tonnes per day) while measuring properties of particulates that may be on the 10-μm length scale means that up to many billions of individual measurements may have to be taken. This need demands a methodology that is rapid, easily deployed, and amenable to automation. An example is the development of optical sorters for industrial minerals (by color detection) and diamond sorting (by fluorescence). These devices have also found applications in the food and other industries, in which each particle has to be inspected. Of course, even if the whole process stream is not to be analyzed, it is evident that the issue of sampling from any high mass flow rate stream is of critical importance. The application of appropriate statistical methods of analysis forms an important part of the measurement and assessment procedure. If one broadens the sensor domain beyond the visible optical field to include any property that can be mapped into space to form an image, then a myriad of techniques need to be scrutinized. Justifications for the selection of the techniques discussed in this chapter will become evident, based on the industrial and commercial drivers relating to the sector.

This chapter considers the role of on-line imaging methods that have been developed for industrial applications in systems that deal with handling particulate suspensions of solids, gases, and liquids in liquids or gases. Often these systems are complex and the suspensions are present as multiphase mixtures. The mineral and material processing communities have been among the pioneer developers and first adopters of key platform technologies in analytical methods, chemistry, physical separation, and advanced modeling. These capabilities are based on the application of fundamental research and continue to have a significant impact, driven by the ongoing need for development of sustainable practices. These practices include mitigating or avoiding long-term consequences for the future human population, including topics such as energy, air quality, water use and discharge, stewardship of land and biodiversity, product stewardship (i.e., through the product's entire life cycle), and waste management. The precise targets for these areas depend on the specific metal, mineral, quarrying, or material recycling operation. Process imaging methodologies have an important role in delivering solutions to the emerging challenges for the mineral industry. Examples of these challenges include:

- Reduction in energy uses in processing
- Discovery of new energy materials
- Adoption of green chemistry or alternative processing routes
- Safeguarding of water resources
- Demonstration of adherence to best practice at all times

With increasingly global mining enterprises, the provision of internationally accepted benchmarks for performance will continue to stimulate the invention and adoption of new sensing technology. The innovation routes through which novel sensors and sensor systems are invented and subsequently implemented are complex, driven by serendipity, legislative push, and sound commercial payback, as discussed elsewhere [2–4]. This chapter illustrates the emergence of a variety of measurement methods on particulate-based systems that can be deployed to deliver practical benefits in the following areas:

- Validation and development of process models (theoretical, phenomenological, empirical)
- Quantitative flow measurement of solids and liquids
- Measurements for fault diagnosis
- Process optimization
- Quality control of products and by-products

These topics will be described in the order that follows the mineral processing route, namely: ore breakage (comminution) to achieve liberation; concentration through means of particle classification and separation; water recovery and wastewater treatment; and finally the detailed analysis of products. The focus is on the application of the method rather than detailed consideration of the sensor physics or measurement principles, which can be found elsewhere in this volume and in the reference citations. The chapter also highlights major opportunities around a new paradigm of *multiscale measurement coupled with multiscale modeling*, based on advanced sensors, communications, and computing. Connecting real-time measurement sensors with dynamic simulation models opens up new horizons for visualizing the process behavior from an experimental and theoretical basis.

11.2 DESIGN OF COMMINUTION EQUIPMENT

The reduction of particle size is ubiquitous in primary processing of ores in order to liberate mineral components from associated unwanted (gangue) components. The matrix of mineral phases is unlocked by physical disintegration through crushing (at coarse sizes down to particle diameters on the order of centimeters) followed by grinding (down to fine sizes of a few microns). The process of comminution is the most energy demanding and expensive operation in any industry and in fact accounts for a significant portion of global energy consumption. Hence small improvements in comminution efficiency can yield major cost savings and bring remission of the associated environmental carbon dioxide burdens. The equipment used ranges from very large tumbling mills, of order 7 m in diameter, to small batch stirred bead mills, of order 0.5 m or less. These are often operated in a wet state, although this is not universal. In the latter case, the mill is charged with media (e.g., ceramic beads) that nip together to compress and abrade the mineral particles, causing breakage and fragmentation. Information is sought on the motion of the grinding media, the particulate solids being milled, and the fluid carrying the solids through the mill. Design variables can include the form of the internals of the mills (liner design, media size distribution) and operation (rotation speed, stirring, etc.).

A range of imaging methods has been used to interrogate such systems. The measurement environment is aggressive but not necessarily in the chemical sense. The intensity of mechanical energy expended inside the devices makes it extremely difficult to derive information from sensors placed inside them. In some exceptions, sacrificial cameras have been positioned to look into mills or, more commonly, high-speed video photography has been used to monitor media motion through transparent walled test equipment. Two types of imaging inspection are considered here, based on measurements of bead milling behavior deduced from (a) external tracking at an interface (the upper surface in a bead mill) using high-speed digital photography and (b) internal tracking of tracer beads using positron emission particle tracking tomography. These methods are complementary to a third method, namely the use of dielectric (capacitance) tomography to visualize the distribution of beads in the mill [5], as described in Chapter 10.

11.2.1 External High-Speed Video Imaging of Bead Movement

Direct video imaging is perhaps the most commonly used method of quantitative analysis when optical access is available. Even when it is not possible to optically match the grinding beads with the background fluids (in order to see through the mill), examination of bead motion at a wall or interface is often feasible. For example, analysis of bead motion in the attritor (vertical mill) shown in Figure 11.1 using a high-speed digital camera has been reported [6]. Here a 2.25 dm^3 mill was used; the impeller had four arms, with two arms parallel to one another and the other two set perpendicular (see Figure 11.1a). The mill contained 2475 stainless steel balls of 9 mm diameter. Information was obtained for three rotation speeds (60 rpm, 120 rpm, and 180 rpm). Through the high-speed video (Kodak HS 4530), it was possible to record images at 500 images per second from which the surface velocity of balls could be measured (using Optimas 6.5 software). Figure 11.1b and Figure 11.1c show typical images for rotation at 60 rpm (at which speed the impellers cannot be seen) and at the maximum speed of 180 rpm (in which the first and second impellers can be seen). The results of computer simulations shown in Figure 11.1 (d, e, f, and g) indicate how the balls are lifted around the circumference of the attrition mill with increasing rotational speeds due to the increasing centrifugal force.

In Figure 11.2, (a) and (b) show the measured velocity field of the balls at the surface, noting the difficulty in recording the velocity for balls close to the side of the wall closest to the camera viewpoint (hence data are shown here for the far wall). The ball velocity is low near the impeller shaft and increases with radial distance, with a maximum value near the tip of the impeller arm. These data enable quantitative comparison [see (c) and (d) in Figure 11.2], with the bead motion predicted from independent simulations based on the application of discrete element models (DEM). Figure 11.2e shows the relation between the balls' mean velocity and the radial location from the impeller shaft. There is close agreement between the experiment and simulation results of the balls' velocities. The objective of this study was to

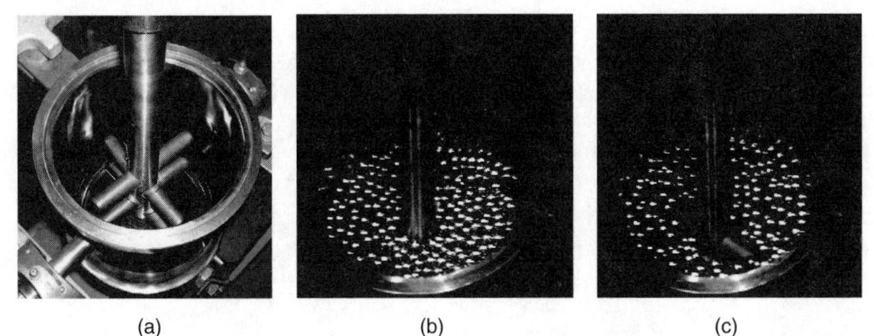

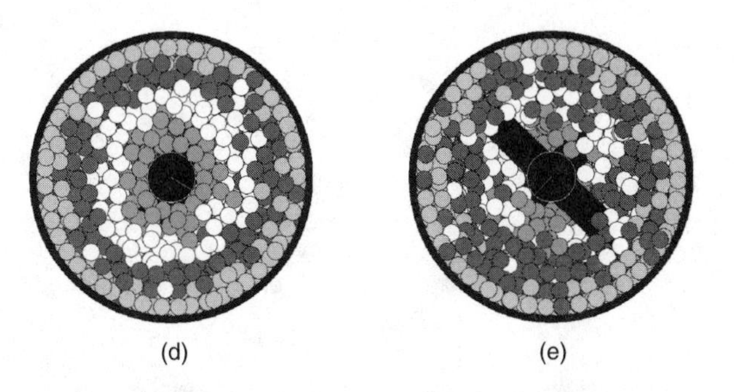

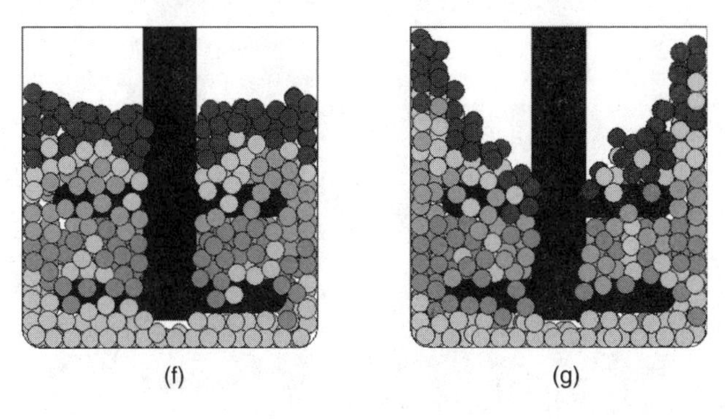

FIGURE 11.1 Tracking individual stainless steel grinding beads in a vertical mill (attritor) at rotor speeds of (a) 60 rpm and (b) 180 rpm. The rotor geometry can be seen in an empty attritor (c). High speed digital photography of the attritor provides data to validate simulations viewed from the top (d, e) and side (f). (From Mio, H., Saito, F., Ghadiri, M., Eastman, R., and Stephen, R., *2nd Asian Particle Technology Symposium,* Penang, 2003. With permission.)

validate such a model, in which constants relating to the mechanical properties of the balls are needed to generate dynamic simulations of bead mixing effects. In this case, image analysis was performed on a limited region of interest in the process (the upper surface of the attritor) to verify the model. The results gave confidence in other outputs of the simulation to predict optimal conditions for good powder mixing and best practice in attritor design.

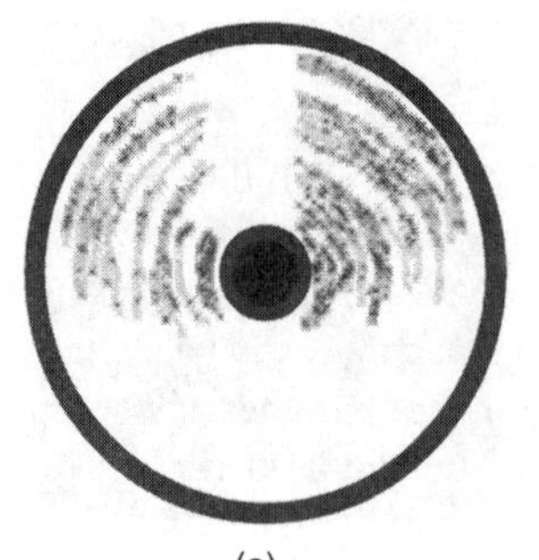

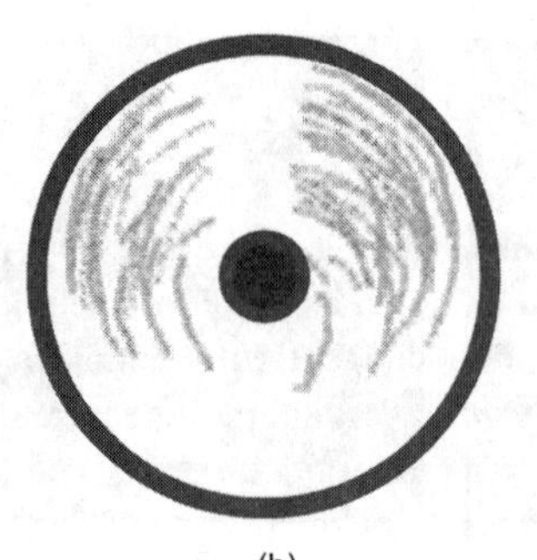

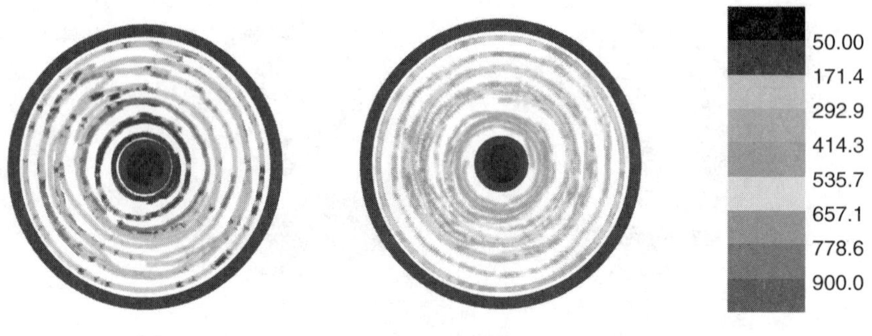

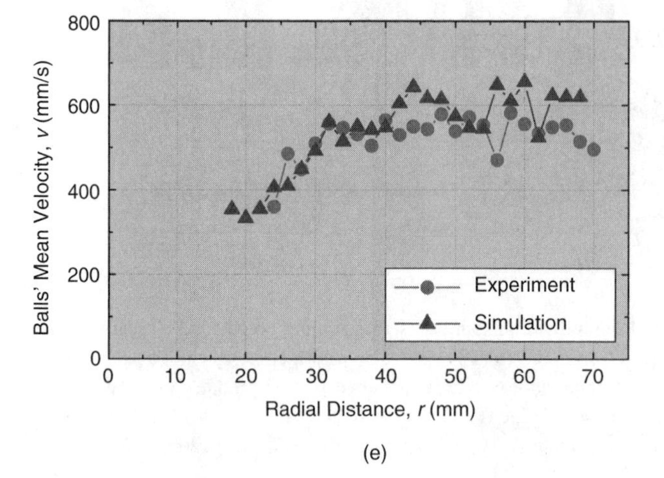

FIGURE 11.2 Comparison of velocity fields from high speed video imaging of balls at (a) 60 rpm and (b) 180 rpm with the corresponding flow modeling prediction (c, d). The velocity scale is shown in mm/sec (e). Measurements of quantitative mean velocity as a function of radius at 180 rpm are compared with results of DEM simulation in (e). (From Mio, H., Saito, F., Ghadiri, M., Eastman, R., and Stephen, R., *2nd Asian Particle Technology Symposium,* Penang, 2003. With permission.)

11.2.2 Internal Bead Tracking Using Positron Emission Particle Tracking

Industrial applications of positron emission particle tracking (PEPT, described in Chapter 4) have been developed at the University of Birmingham and reviewed in detail by Parker et al. [7]. The dielectric tomographic method cited above [5] is well suited for mapping concentration variations, but it cannot easily be applied to track the motion of an individual particle or an assembly of particles of a given type. The ability to map the trajectory and velocity of individual particles is a requirement of verification for many particle flow codes such as the DEM. When specific particles are labeled with a positron emitting species, the β^-decay yields back-to-back gamma-rays that can be used to pinpoint the location of the source (i.e., particle) along each line of sight within the field of two planar detectors.

Figure 11.3 shows a photograph of a small attritor used in bead milling, positioned between two planar detectors in a positron emission camera. By using

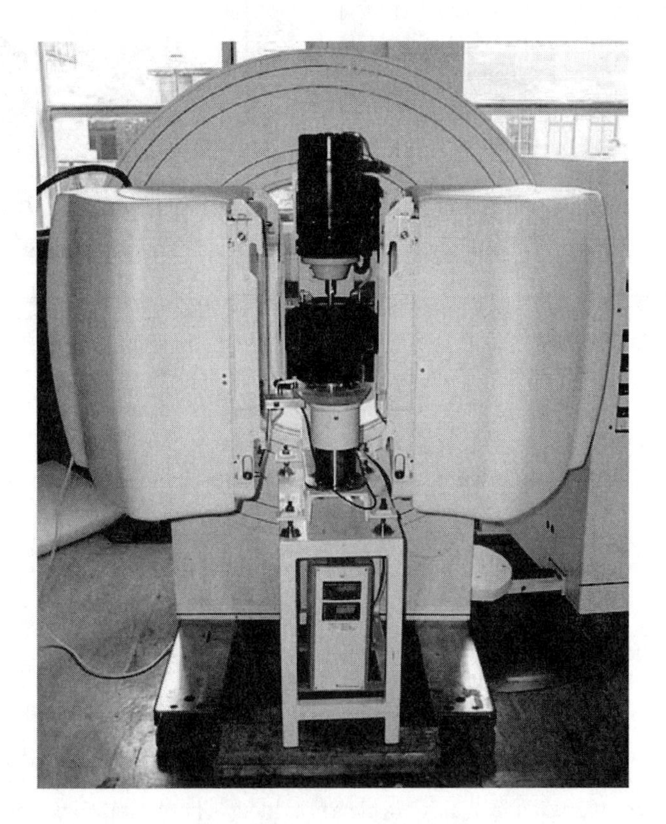

FIGURE 11.3 Positron emission particle tracking (PEPT) camera at University of Birmingham applied to the measurement of bead motion in a 2 dm^3 laboratory scale mill. (From Conway Baker, J., Barley, R.W., Williams, R.A., Jia, X., Kostuch, J., McLoughlin, B., and Parker, D.J., *Mineral Engineering,* 15, 53–59, 2002. With permission.)

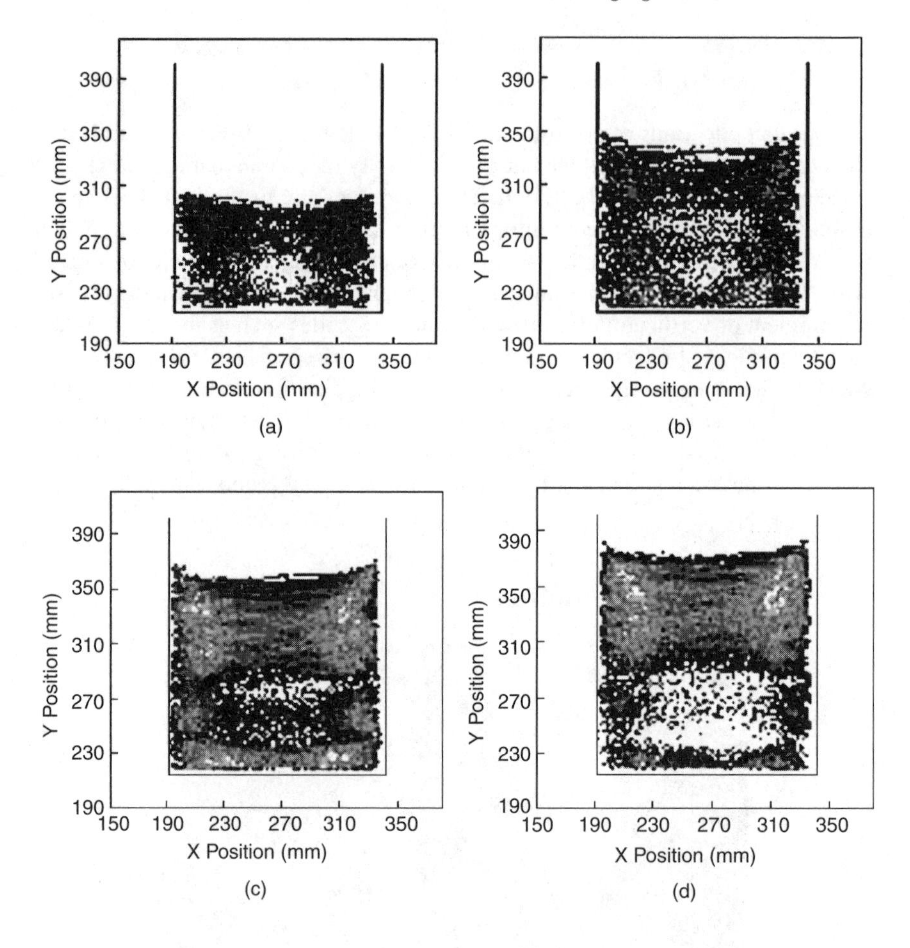

FIGURE 11.4 Example of occupancy and fill information as a function of mill stirring rate (a) 200 rpm, (b) 400 rpm, (c) 600 rpm, and (d) 800 rpm. Such data together with detailed velocity maps can be obtained to compare with distinct element simulation. (From Conway Baker, J., Barley, R.W., Williams, R.A., Jia, X., Kostuch, J., McLoughlin, B., and Parker, D.J., *Mineral Engineering,* 15, 53–59, 2002. With permission.)

a labeled bead, it is possible to follow the trajectory of the media, thus producing detailed velocity maps, and to compute the occupancy of the bead at different locations in the mill. As shown in Figure 11.4, a three-dimensional (3D) DEM of the attritor provides the occupancy information (i.e., the bead fraction discussed in Chapter 10) as a function of stirring rate [8]. The effect of fill on stirring rate and the regions of highest occupancy are clearly evident. These data were used to assess the extent of media fluidization, confirming previous observations reported on dielectric imaging studies of similar mills (see Section 10.3.2). Velocity maps for each condition can also be obtained, again providing quantitative validation of the simulations from DEM models. These data provide invaluable insights for process optimization and control strategies. For example, ongoing work seeks to identify the best strategy for controlling the bead size distribution in order to maximize the milling efficiency. Additional interesting examples of PEPT methods used to assess and validate models for segregation in rotating devices have also been reported [9].

Such information is also useful for estimating the energy dissipation occurring in the mill, to be used in conjunction with estimates for particle–particle collisions from DEM. In practice, it still remains challenging to distinguish energy contributions to interactions ranging from direct contact to abrasion, especially in wet systems where the quantification of lubrication effects remains to be fully solved. However, such coupled measurement and modeling approaches provide the best available means for assessment of mill performance. Other complementary methods, such as thermal imaging and mapping (e.g., using the thermochromic liquid crystals described in Chapter 3), can also be used, but they are experimentally more demanding to implement.

11.3 GRANULAR FLOW AND BULK TRANSPORTATION

The transport of dry and damp minerals via hoppers, belts, and conveying lines is important, as plant operability is often poor due to pluggage of conveying lines resulting from subtle changes in the feedstock (e.g., moisture content, size distribution, particle shape). The difference in dielectric constant between most powders and the dispersed fluid can be sensed through a capacitance measurement, provided that the fluid is electrically nonconducting. If a number of sensing electrodes are placed around the process pipelines in which the granular mixture is flowing, sufficient information can be obtained by rapid sequential determination of the capacitance between pairs of electrode plates to enable an image of the dielectric distribution to be captured. This technique, known as electrical capacitance tomography (ECT), is described in detail in Chapter 4. ECT can be used to image the dielectric constant inside conveying lines for dilute- and dense-phase processes.

11.3.1 PIPELINE CONVEYING

Pneumatic conveying offers many advantages over other methods of granular solids transport, such as low routine maintenance and manpower costs, dust-free transportation, and flexible routing. The main disadvantage is the reliance upon empirical procedures for conveyer design, which often result in an unnecessarily high or variable wear rate and power consumption. In addition, product degradation and particle segregation can be major problems. Slug flow (discussed in Chapter 9), when it occurs in a conveyor, has the advantage of low air flow and hence energy demand, low pipeline erosion, and low product degradation. However, the control requirements of such a transportation system are clearly far more

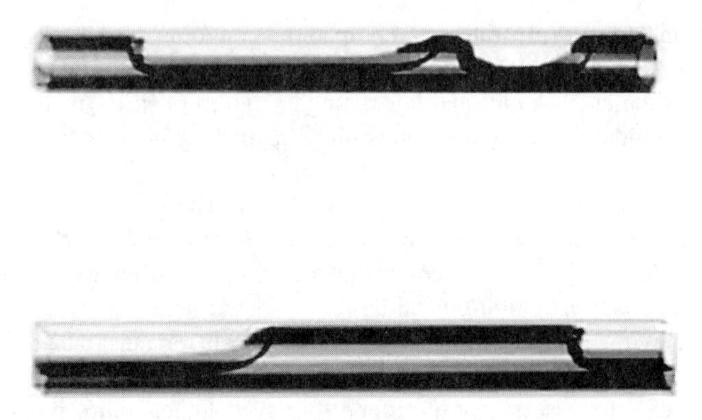

FIGURE 11.5 Real-time visualization of dense granular flows in pneumatic conveying using electrical capacitance tomography, showing here two different flow types exhibiting different slug lengths and frequencies. (From Ostrowski, K.L., Luke, S.P., Bennett, M.A., Williams, R.A., *Journal of Powder Technology,* 102, 1–13, 1999. With permission.)

acute in respect to the maintenance of flow regime and the prevention of blockage. Some practical issues involved in control of such systems based on tomographic image data are considered here in some detail. These issues are relevant to the conveying of silicas, coal, and other nonmineral materials.

Figure 11.5 shows the first real-time 3D visualization of powder slugs in a pneumatic conveyor, described elsewhere [10], from which details on the geometric form and velocity of the slugs and moving beds can be obtained. More detailed studies have sought to elucidate the formation and transport properties of different slugging characteristics in horizontal and vertical pipes, as well as to provide comprehensive information on the shapes of the slugs' nose and tail to interpret granular flow models [11].

In this analysis, consideration is given to statistical parameters such as perturbations of the average value of the normalized dielectric constant $\langle e \rangle$ in the pipe cross-section. Figure 11.6 shows examples of three different slug types and the corresponding normalized dielectric constant-time plots; the corresponding visualizations have been included for clarity. Note the increasing distance separating slugs from (a) to (c). The horizontal arrows indicate regions from which data were extracted for further analysis. The typical information that can be obtained includes the average volume occupied by solids and the height of the solid–gas interface (if present). The sequences of images may be used to extract the following data:

- Slug length and distribution of slug lengths
- Slug velocity and velocity distribution
- Slug frequency and distribution
- Correlation analysis of the above parameters

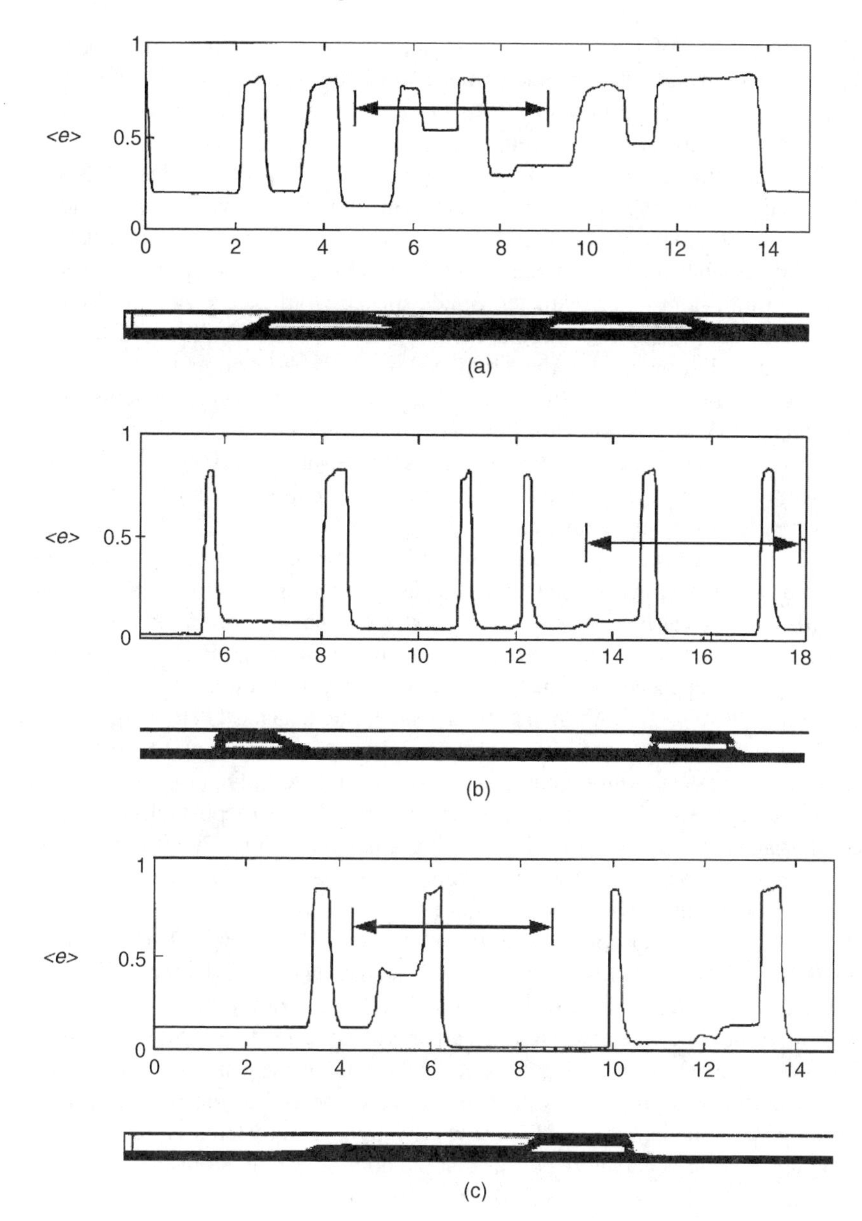

FIGURE 11.6 Measurements of fluctuations in dielectric constant due to changes in the relative volume fraction of solids crossing a single sensor plane during dense phase conveying for three different slugging flow structures. (From Ostrowski, K.L., Luke, S.P., Bennett, M.A., Williams, R.A., *Chemical Engineering Journal,* 3524, 1–8, 1999. With permission.)

Perturbations of the averaged cross-section normalized dielectric constant <*e*> were obtained as the arithmetic mean value for all pixels (814 in this case). Such analysis is sufficient to estimate the flow pattern and so provide the necessary information for the control procedure.

The experimental vacuum conveying system used [10] was designed to enable different types of powder flows to be generated in a controlled manner. It was a closed loop for solids and an open system for air. In order to facilitate visual observations of the flow regime, the horizontal test section, 3 m long, was made from transparent pipe with a 52-mm inner diameter. Solids sucked from a tank were transferred through a standpipe 11 m in length and stored in the vacuum conveyor. Induced air was pushed through a filter and removed by the vacuum pump to atmosphere. The vacuum conveying system can generate a wide range of flow patterns, from dense slug flow up to dilute, fast flow. The materials conveyed were nylon plastic pellets having a bulk density of 750 kg m^{-3}, a solid density of 1120 kg m^{-3}, and a size of 2 to 3 mm with an aspect ratio between 1 and 2. For this medium, the maximum mass transferred during a single run was about 25 kg.

The ECT sensor was based on 12 sensing electrodes, 100 mm long, with guard electrode assemblies at each end. Since the distance required to develop a particular perturbation typical for the tested flow regime was unknown *a priori*, the sensor was designed to move along the test section. Images from the ECT systems were reconstructed using a conventional linear back-projection, as explained in Chapter 4. These data were used in the statistical and stochastic analysis [12].

Within the range of dense flow, some subregimes may be recognized and defined. Even under constant air inlet conditions, it was possible to distinguish between slow and fast (or dense and less dense) slug flows. Additionally, by keeping inlet conditions constant and periodically closing and opening another injection valve, it was possible to obtain a very regular slug structure (called here *injected* flow). Examples of <*e*> plotted vs. time (in seconds) for these three slug structures are shown in (a), (b), and (c) of Figure 11.7. For comparison, Figure 11.7(d) shows a typical signal for the dilute flow. Table 11.1 presents a set of basic statistical parameters for each signal. Table 11.2 gives a summary of flow conditions for the fast and slow slug flows.

Differences between particular slug patterns are clearly visible in these data. The slow slug flow is more regular than the fast slug flow; generally, the fast slug flow represents the more disordered structure compared to the slow slug flow. The top diagram in Figure 11.7 shows the trend to blockage typical for this structure. Another characteristic was that the bottom level of solids for all signals was not constant but rather a random variable. The flow parameters (i.e., air and solid flow rates and their superficial velocities) shown in Table 11.2 do not provide clear information for distinguishing a particular slug pattern and are insufficient for control purposes. Such information, however, is provided by the statistical parameters: the normalized mean dielectric constant and its standard deviation and mean. The standard deviation of the signal provided a clear differentiation between slug flow and dilute flow (or blockage). The median calculation was a convenient tool to differentiate between slow and fast slug flow patterns. Thus, an on-line control procedure could be based on these two estimators alone.

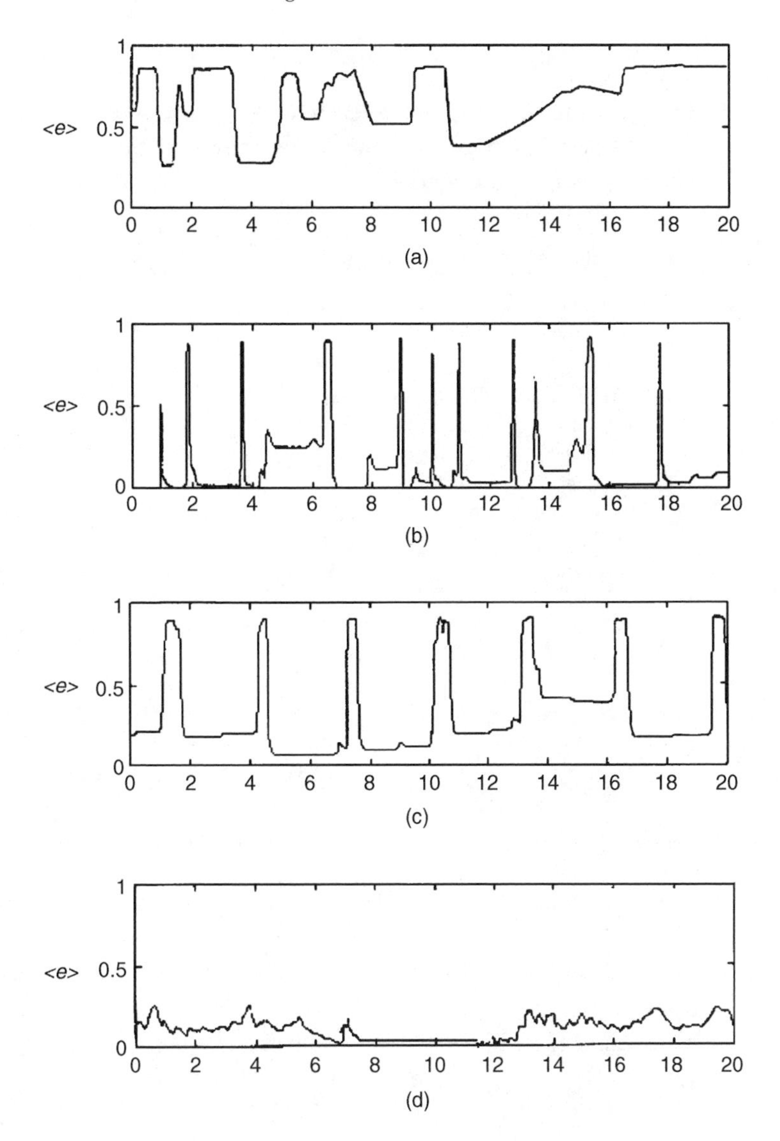

FIGURE 11.7 Typical plots of <*e*> vs. time (in sec) for three slugging flow regimes corresponding to (a) slow slug flows, (b) fast slug flows, and (c) slug flows where extra gas is injected. For comparison, (d) shows the signature for a dilute flow. (From Ostrowski, K.L., Luke, S.P., Bennett, M.A., Williams, R.A., *Chemical Engineering Journal,* 3524, 1–8, 1999. With permission.)

TABLE 11.1
Flow Regimes Corresponding with the Signals <e> vs. Time Shown in Figure 11.7 and Associated Statistical Parameters of These Signals

No	Flow Regime	Mean Value of <e>	Median of <e>	Standard Deviation of <e>
1	slow slug	0.70	0.77	0.16
2	slow slug	0.66	0.72	0.20
3	slow slug	0.53	0.59	0.25
4	slow slug	0.51	0.38	0.22
5	slow slug	0.51	0.44	0.24
6	fast slug	0.17	0.08	0.24
7	fast slug	0.16	0.08	0.23
8	fast slug	0.13	0.04	0.21
9	fast slug	0.12	0.04	0.21
10	fast slug	0.10	0.04	0.19
11	injected	0.30	0.20	0.26
12	injected	0.11	0.04	0.21
13	injected	0.10	0.04	0.19
14	injected	0.08	0.04	0.16
15	dilute	0.11	0.11	0.09

By using off-line analysis, it is possible to complete a more comprehensive estimation including auto-correlation, power spectra, and histograms of the signals, as shown in Figure 11.8 for the slow and fast slugging responses. From an examination of the entire data sets for all flows, it is clear that the macroscales

TABLE 11.2
Air and Solid Flow Rates and the Superficial Velocities for Selected Number of Fast and Slow Slug Flows Listed in Table 11.1

No.	Air Flow Rate* $[m^3\ s^{-1}] \cdot 10^3$	Air Superficial Velocity $[m\ s^{-1}]$	Solid Flow Rate $[kg\ h^{-1}]$	Solid Superficial Velocity $[m\ s^{-1}]$
1	0.89	0.42	1370	0.16
2	0.48	0.23	1210	0.14
3	0.70	0.33	1320	0.15
4	0.79	0.37	1370	0.16
6	0.87	0.41	1410	0.16
7	0.89	0.42	1440	0.17
8	0.93	0.44	1480	0.17

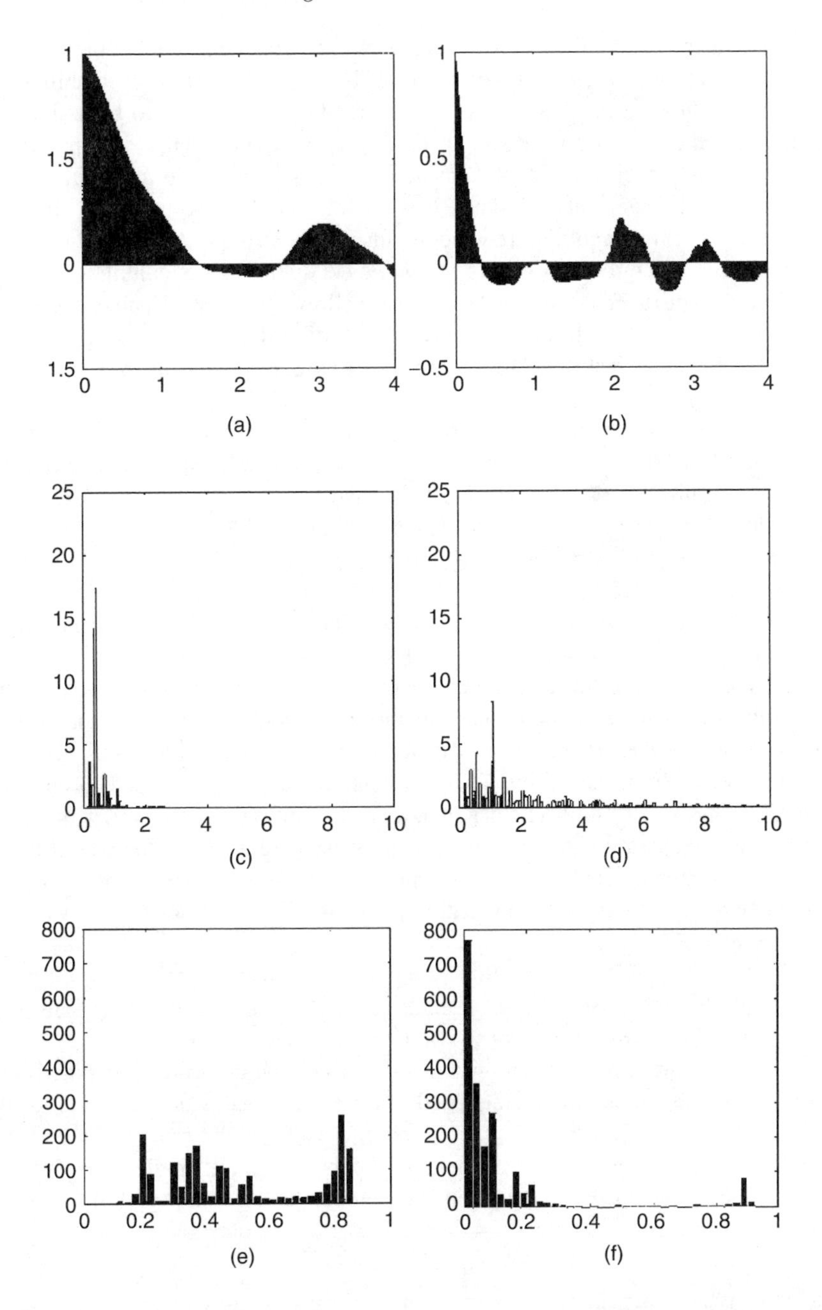

FIGURE 11.8 Statistical analysis showing the autocorrelation of the <*e*>(*t*) signal for (a) slow and (b) fast slugs; the power spectra for (c) slow and (d) fast slugs; and histograms for (e) slow and (f) fast slugs. (From Ostrowski, K.L., Luke, S.P., Bennett, M.A., Williams, R.A., *Chemical Engineering Journal,* 3524, 1–8, 1999. With permission.)

as well as microscales are larger for slow slug flow by about one order of magnitude. The injected flow could also be characterized by relatively high repeatability of the signal. Thus, calculation of the auto-correlation was shown to be a simple, effective, and robust tool to distinguish between particular dense flow patterns. If the slug frequencies are to be considered, it is more suitable to calculate the power spectral density of the signal. It is common practice to taper the original signal (<*e*>) with an appropriate window function before transformation, thereby reducing any discontinuities at its edges. Power spectra were calculated using windowing and a Fast Fourier Transform. Slow slug flow is characterized by a narrow spread in the power spectrum, with a distinctive strong peak at about 0.3 Hz. The fast slug flow represents a wider spread, up to 3 Hz, with few maxima.

Examples of the histograms for the flow regimes being tested are also shown in Figure 11.8. The range of <*e*> equal to <0, 1> was subdivided into 100 quantization levels. For the slow slug flow, distribution of probability was more symmetrical; however, the random character of the solid levels is clearly visible. The <*e*> distributions were close to bimodal (i.e., representing the minimum and maximum solid levels) as plotted in histogram form in Figure 11.8e. It is seen that the slow flow represents a more symmetrical distribution than the fast flow.

In conclusion, these simple estimators were found to be effective tools for on-line recognition of flow; additional techniques are introduced in Chapter 9. It is possible to extend the number of estimators and include, for example, skewness or kurtosis; however, there is no evidence that they would significantly clarify the flow recognition capability. These principles are being extended to on-line control of conveyors with special regard to the prompt identification of flow regime and the quantitative estimation of statistical parameters for particular flow patterns. These data can also be used for optimization and control based on imaging, for example, by using genetic algorithms [13, 14]. This technique is relevant to coal conveying.

At the microscale of inspection, the passage of discrete, micron-sized particles can also be sensed using miniaturized wall-mounted dielectric sensors. The application of these sensors to determine particle velocities and the shape of individual particles has been reported [15]. Such methods are well suited for design of sensors that can be embedded in pipe or surface wall to provide local microscale information. Use of macro- and microscale measurements in tandem provides a means to probe phenomena that occur at different length scales simultaneously. Such practices are not yet in widespread use, but their potential has been identified [16].

11.3.2 Belt Conveying

Of particular interest is the use of ECT to assess *mass flow rate* of material being conveyed on belt conveyors [17]. Figure 11.9 shows the concept and a measurement example, where the contour profile presented by particles on the belt might be used to infer information about the particle properties themselves (such as

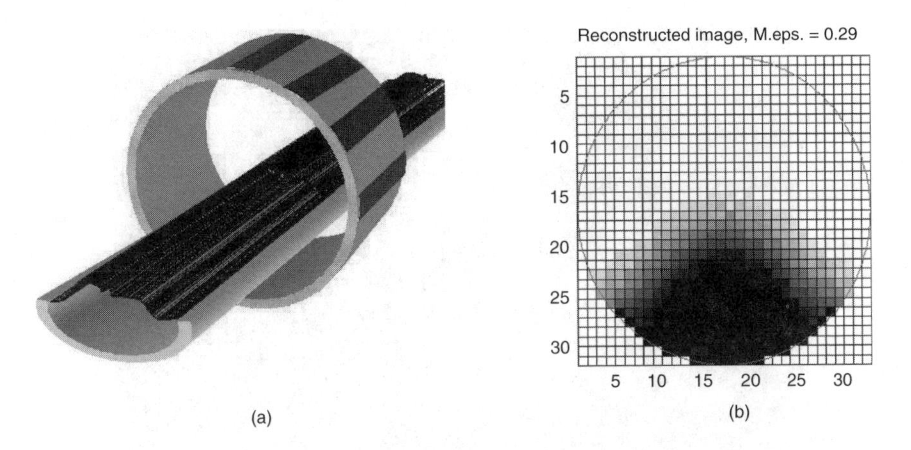

FIGURE 11.9 (a) Concept of on-belt measurement using electrical capacitance tomography to estimate solids flow and (b) details of profile measurement from which particle properties could be inferred. (From Williams, R.A., Luke, S.P., Ostrowski, K.L., Bennett, M.A., *Chemical Engineering Journal,* 77, 57–63, 2000. With permission.)

shape or size distribution). The proposition is that the fractal dimension of this profile is related to the size and shape distributions of the particulates. For example, in the production of high quality road stone, measurement of crushed product on a belt is distinguishable since well-shaped particulates, obtained by running the crusher in an overfed mode, pack differently than products in which the shape is not controlled. Such approaches are under investigation with a view to industrial implementation.

In addition, optical techniques are used extensively to observe and quantify particle size and in some cases particle reflectivity or luster. Measurements can range from simple on-line image analysis of particle size or shape to determination of specific quality parameters (e.g., impurity content in processed silica sand streams). Figure 11.10 shows an infrared imaging device used to estimate moisture content in a mineral powder.

Such measurements are made either directly from above the belt or on a free-falling sheet of materials, the latter being commonplace. These devices are in routine use in sorting of industrial minerals (such as marble, calcite, and diamonds) and recycled mineral, aggregate, and scrap metals products. An example of an optical sorting machine that operates in conjunction with conveyer belt feed is shown in Figure 11.11a. The figure shows a view of the inside free-fall section of the machine (b) and examples of the feed materials (e) separated by it into two streams of product (c) and reject (d). Sorting of product streams by size or shape using process imaging is less common due to the difficulty of designing reliable systems for removal of out-of-specification material. However, very high speed systems do exist for real-time analysis in a laboratory environment. Secondary recycling of metals provides a major impetus to the development of these systems for municipal recycling. A variety of "property imaging" tools have been developed

FIGURE 11.10 Example of on-belt analysis of mineral product using infrared sensors for moisture detection and other properties. (NDC InfraRed Engineering Ltd.)

for these applications, as reviewed elsewhere [18]. For sorting different metals for recycling, these tools have included combinations of dual energy x-ray mapping with video imaging and simultaneous use of laser inductor breakdown spectral analysis (LIBS).

11.4 PARTICLE CLASSIFICATION IN CYCLONES

The topic of separation in cyclone fields raises the issue of the link between simulated visualization of fluid properties and actual measurements from tomographic and imaging methods. There are two primary reasons for the relatively slow development of computational fluid dynamics and its application to real multiphase flow problems in industrial mineral systems. The first is the inherent complexity of multibody interactions and the effort to describe turbulence. The second is the paucity of real-time measurements that can work *in situ* on real processes, thus providing experimental validation of theoretical results.

It is a fact that relatively little research has been reported where equal vigor has been expended on both measurement and modeling. The problem is complex since a robust model of a polydisperse mineral system must describe particles that are affected by a variety of forces acting over a range of length scales. The net result is that the simulation must embrace information that is both species specific and position specific (in the process). Furthermore, the phenomena may

(a) (b) (c) (d) (e)

FIGURE 11.11 (a) Industrial optical on-belt sorter. (b) Illumination and particle ejection systems and examples of (c) product and (d) reject material separated from the (e) feed stream in a mineral application. (RHEWUM Datasort GmbH)

occur over very different time scales (on the order of 10^{-9} to 10^{3} sec). This is exceedingly demanding computationally, but examples of such multiscale modeling approaches do exist [4]. Future developments will give rise to methodologies where the models resident in process simulators have submodels for processes that could be empirical, phenomenological, or statistical in nature. These models could then be used alone or in combination, according to the data sets available. Turning to measurement, the deployment of different types of sensors simultaneously to measure phenomena that are occurring at different length scales and at different time scales is extremely rare. The use of soft sensors (whose outputs are interpreted empirically via multivariate statistics or related methods) is more common, but these do not generally provide a happy companion for multiscale simulations. Multiscale measurement sensors that measure at the molecular, micrometer, millimeter, and macro scales are needed but are still largely unknown. The advent of tomographic sensors [19] that can be used on industrial pipes and processes has provided a further stimulus to this view, since in some instances 3D and real-time measurement capabilities can be utilized.

11.4.1 Hydrocyclones

Hydrocyclones are used for particle separation, separating species on the basis of their size or density relative to the suspending fluid. They produce either a specific sized product (classification) or a product with maximum solids content (thickening/dewatering), and they can be used for washing products [20]. Here an example based on hydrocyclones is used to demonstrate the knowledge that can be gained by combining the measurement and model in a fully integrated system. By integrating the model used to interpret the measurements with the simulation, it is possible to increase confidence levels in both. For example, the combination of tomographic measurements with computational fluid dynamics fused around a common set of model-based assumptions provides a sound methodology [21].

Figure 11.12 shows sets of tomographic measurements made using electrical impedance tomography (EIT), ultrasound, and x-rays to image solids, air core, and process dynamics in an industrial hydrocyclone used to separate clays. These images have been used with a physical model of the hydrocyclone to deduce

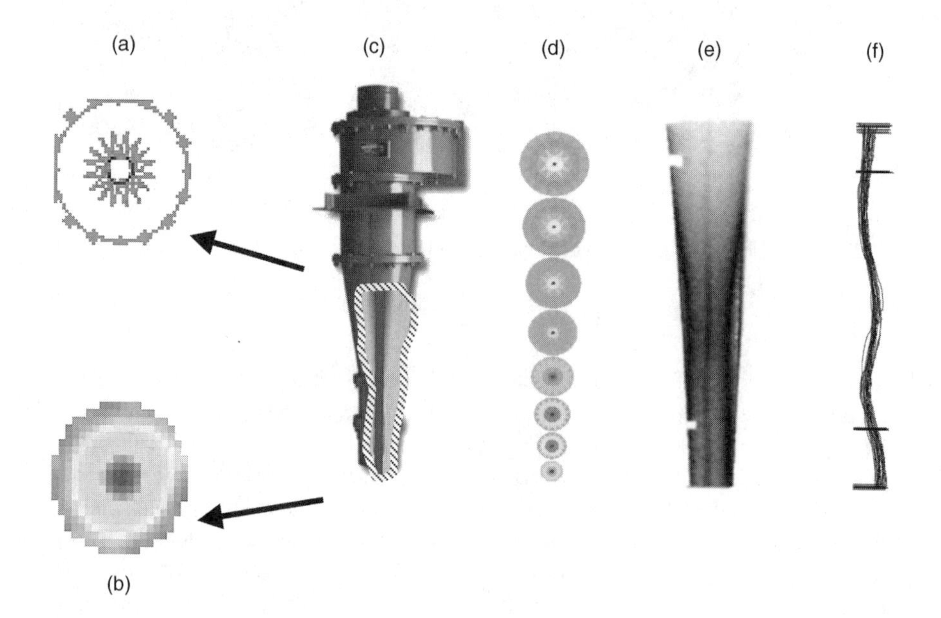

FIGURE 11.12 Combining different tomographic sensors based around a process flow simulation model to extract the air core diameter in an operating industrial hydrocyclone. Data here show (a) ultrasound tomography and (b) impedance tomography cross-sections showing the air core; (c) a cut-away view showing conductivity profiles inside the hydrocyclone; (d) cross-sectional images of conductivity along the length of the cyclone; (e) an x-ray photograph; and (f) a 3D image of the air core tortuosity derived from x-ray data. (From Cullivan, J.C, Williams, R.A., and Cross, C.R., *J. Trans. I. Chem Eng. Part A: Chemical Engineering Research and Design,* 81, 455–466, 2003. With permission.)

(quantitatively) the air core size and motion for comparison with a rigorous and fully 3D flow simulation [22]. The existence of a persistent nonsymmetric air core observed inside operating industrial separators has implications for the flow, since hitherto symmetry had been assumed. Incorporation of this new information into flow simulations revealed new asymmetric structures and phenomena within the hydrocyclone that have consequences for the mechanism of separation through the occurrence of local zones of enhanced radial transport (see Figure 11.13). Other imaging methods that have also been used with hydrocyclones include acoustic mapping, digital image analysis of the morphology of the external discharge stream (underflow), and video imaging of the internal air core oscillations by a camera mounted on a probe arm.

Two further examples are considered here that have been deployed in industrial environments for control based on image analysis. These involve video analysis of hydrocyclone underflow and EIT analysis of the hydrocyclone air core.

Optical inspection of the underflow discharge from a single cyclone or a bank of cyclones (see Figure 11.14a) provides great insight into operation. In the case of a hydrocyclone performing a size classification function (closing the circuit on a mill discharge, for example), it is desirable to operate the hydrocyclone so that the underflow is just spraying, as in Figure 11.14(b) and not quite going into a roping state. If the underflow exhibits roping, the cut size is dramatically increased and the proportion of fines leaving the overflow is changed. Hence, close control of such cyclones is desirable. Optical image analysis has been used for this purpose, along with alternative measurements of passive acoustic sound and vibration meters. Optical methods have some attractions due to their simplicity; consequently, several independent groups have proposed control methods based on video analysis of the exterior of the cyclone underflow [23]. The methods range from examination of the discharge angle through to detailed analysis of the oscillations and use of swirl velocity estimates obtained by tracking fluid motion on the periphery of the cone.

From Figure 11.12 it is clear that detailed mapping of the air core can be undertaken. This mapping has been used as a means of identifying fault conditions in an operating cyclone [20, 21]. For example, Figure 11.15 shows the electrical resistivity image taken close to the hydrocyclone underflow for three operating states: blockage (no discharge flow is visible), rope discharge, and spray discharge. The colors and contours of the reconstructed resistivity images can be used to infer the size of the air core when it exists.

In addition to this practical tool for identifying states of operation, a more careful analysis and reconstruction procedure can be used to measure the size of the air core as a function of operating conditions at any chosen axial distance along the cyclone. Figure 11.16a shows a sequence of simple resistivity images obtained by simple back-projection reconstruction. It is seen that as the flow rate is increased (raising the inlet pressure from 1 bar to 3.2 bar), the size of the central region also increases, indicating a larger air core; this finding is consistent with independent measurements. Quantitative measurement from such blurred images is not satisfactory; hence, an alternative form of data analysis is needed.

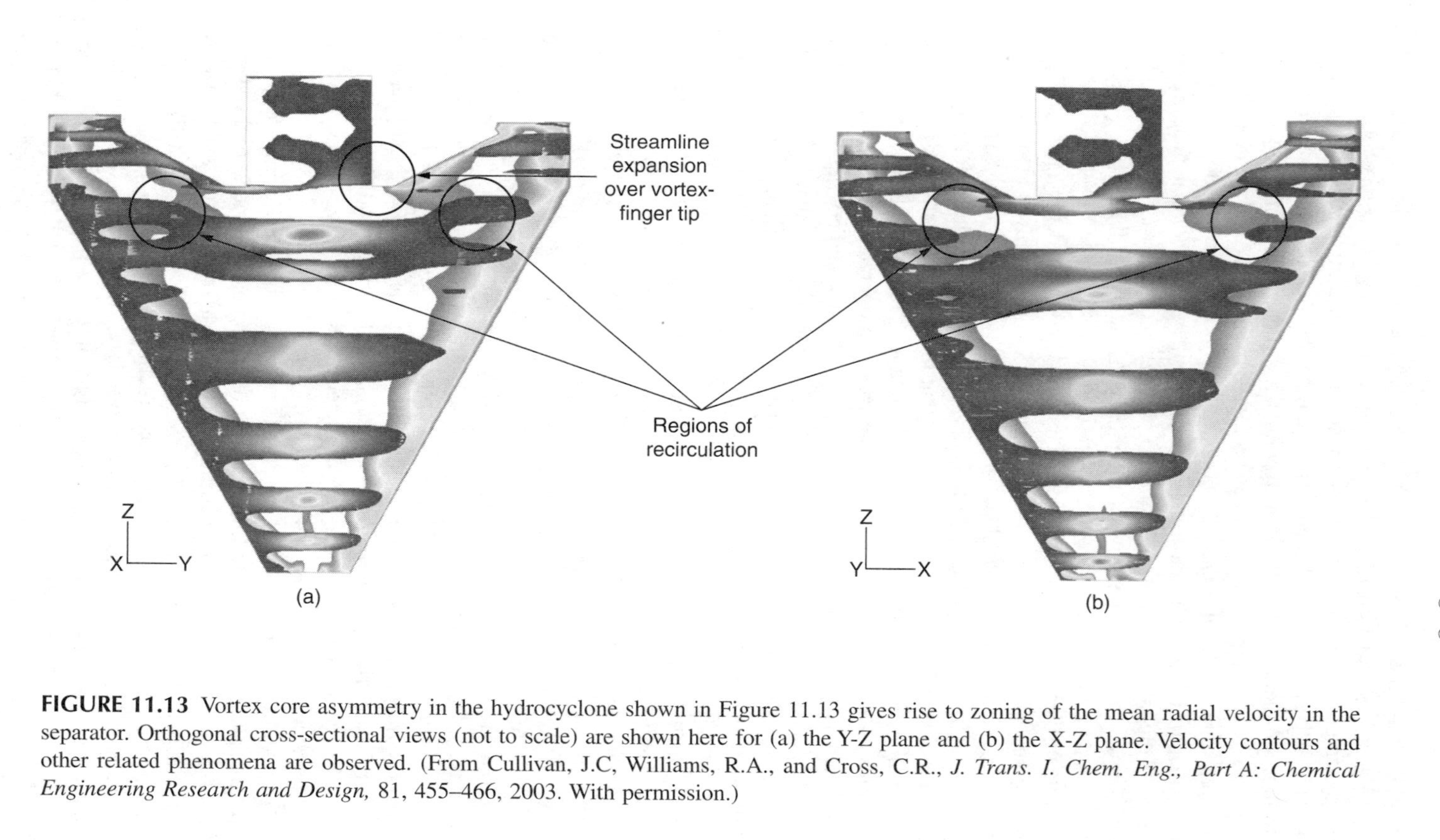

FIGURE 11.13 Vortex core asymmetry in the hydrocyclone shown in Figure 11.13 gives rise to zoning of the mean radial velocity in the separator. Orthogonal cross-sectional views (not to scale) are shown here for (a) the Y-Z plane and (b) the X-Z plane. Velocity contours and other related phenomena are observed. (From Cullivan, J.C, Williams, R.A., and Cross, C.R., *J. Trans. I. Chem. Eng., Part A: Chemical Engineering Research and Design,* 81, 455–466, 2003. With permission.)

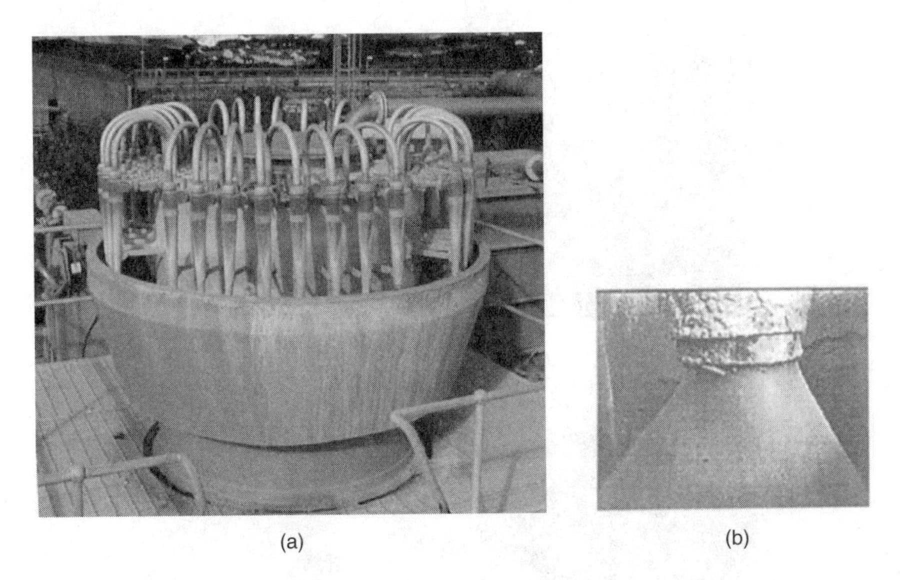

(a) (b)

FIGURE 11.14 (a) Industrial carousel/bank of 50 mm diameter hydrocyclones used in clay refining. (From Williams, R.A., Jia, X., West, R.M., Wang, M., Cullivan, J.C., Bond, J., Faulks, I., Dyakowski, T., Wang, S.J., Climpton, N., Kostuch, J.A., and Payton, D., *Minerals Engineering,* 12, 10, 1245–1252, 1999. With permission.) (b) Spray discharge from a hydrocyclone.

A model-based reconstruction [21] relies on the presence of a nonconducting and centrally located air core and low conductivity at the cyclone walls (due to higher solids concentration there). The problem is reduced to fitting values for three parameters (a, b, c) in a quadratic equation for the conductivity, which is a function of radial distance (r) defined as

$$(r) = a + b(3r - 2) + c(10r^2 - 12r + 3) \quad (11.1)$$

Here the value of r ranges from 0 (the air core) to 1 (the wall of the hydrocyclone). Figure 11.16b shows the image obtained from the parametric fit to the projection data (resistivity).

Use of this method greatly improves the information content. The radius can be resolved more precisely and with a known degree of confidence. If necessary, an image can be provided, this being now a parametric image rather than one obtained through back-projection inversion. The resolution of air core measurement is excellent, enabling quantitative measurements to be made to ± 0.05 mm, as seen in Figure 11.17. This method has been used for on-line control and to identify the onset of defect conditions such as spigot wear (which invariably results in a larger air core diameter) [23, 24].

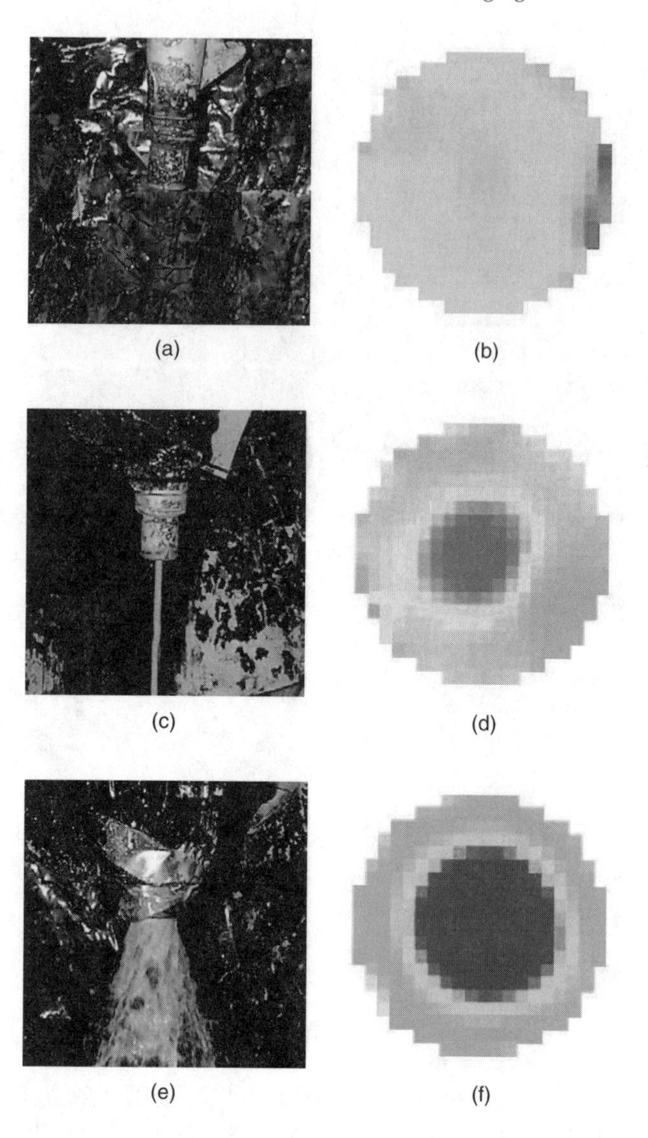

FIGURE 11.15 Photographs of hydrocyclone underflow discharge and corresponding EIT (resistivity) images close to the underflow port used to identify three states of operation: (a, b) blockage, no flow; (c, d) thickened rope discharge; and (e, f) dilute spray discharge. (From Williams, R.A., Jia, X., West, R.M., Wang, M., Cullivan, J.C., Bond, J., Faulks, I., Dyakowski, T., Wang, S.J., Climpton, N., Kostuch, J.A., and Payton, D., *Minerals Engineering,* 12, 10, 1245–1252, 1999. With permission.)

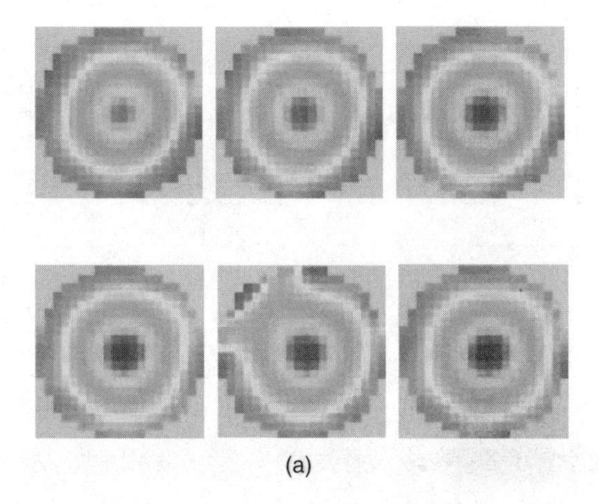

(a)

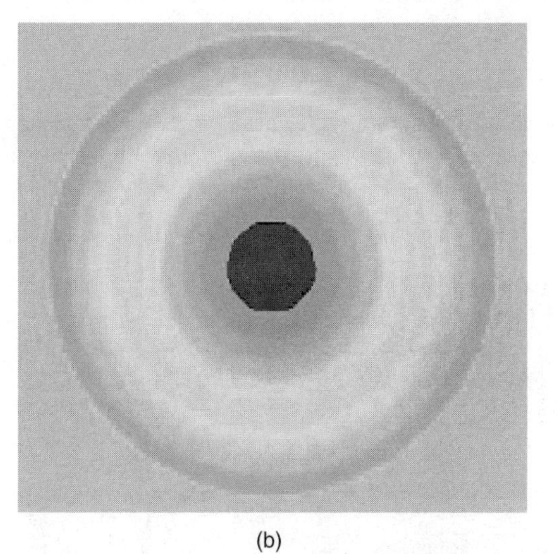

(b)

FIGURE 11.16 (a) Sequence of restitivity images reconstructed using a back-projection algorithm for an air core midway along a 50 mm diameter hydrocyclone as the inlet pressure is increased. (b) A model-based parametric interpretation for the radial conductivity profile, based on Equation 11.1. (From West, R.M., Jia, X., and Williams, R.A., *Chemical Engineering Journal,* 77, 1–2:31–26, 2000. With permission.)

11.4.2 De-Oiling Cyclones

Electrodes installed within the industrial separators used in de-oiling industrial wastes and offshore oil platforms have been used to obtain electrical images for equipment diagnosis and optimization [25]. In such processes, fluid droplets are separated rather than solid particulates. Figure 11.18 shows impedance images across the head of a 2 m long by 40 mm diameter de-oiling cyclone revealing

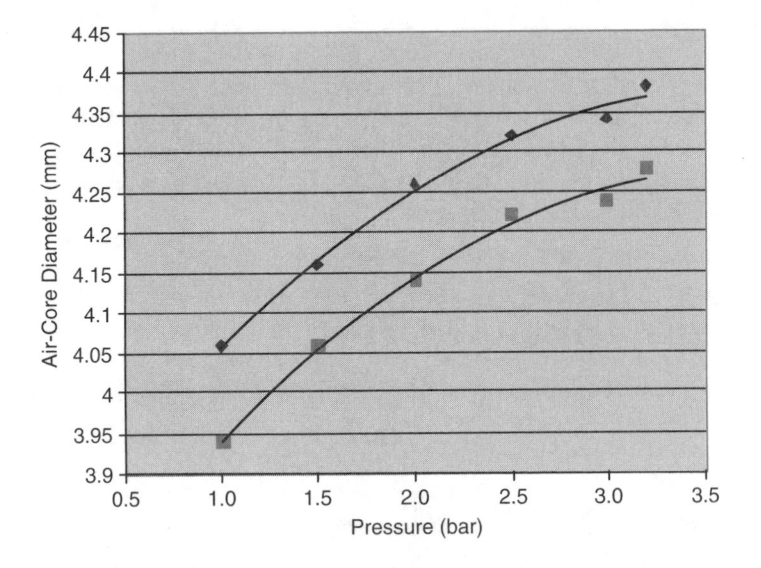

FIGURE 11.17 Example of data reconstruction using the model based approach on data from Figure 11.18 (*lower line*) showing quantitative measurements for the air core as function of applied pressure drop and a second trial where the underflow spigot was worn resulting in a larger air core size (*upper line*).

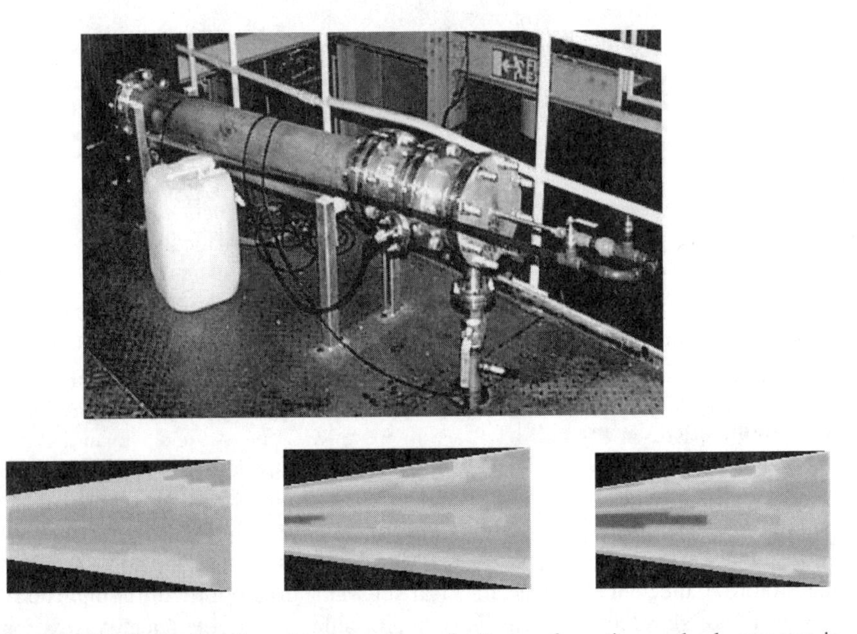

FIGURE 11.18 On-line impedance imaging of oil core formation as back pressure is increased in the head section of an industrial water producing oil–water separator (*photograph*). Evolution of an oil core can been seen in sequence of tomograms. (From Bennett, M.A., and Williams, R.A., *Minerals Engineering,* 17, 605-614, 2004. With permission.)

the evolution of an oil core. Such information can be used to identify appropriate operating conditions for different feed emulsions. The detailed transients in oil–water profiles within the separator can be observed and compared with physical simulations. This capability has implications for control of the separator and start-up procedures. Direct observation of the image can help the operators to tune the process. Like the application to hydrocyclones, this provides a practical tool for industrial and pilot plant deployment.

11.5 PERFORMANCE OF FLOTATION CELLS AND COLUMNS

Flotation separation is ubiquitous in mineral, water, and food processing for selective removal of hydrophobic solids (via attachment to gas bubbles that rise up to form foam) from nonhydrophobic solids that remain dispersed in the process fluid. Here two important image-based approaches for on-plant analysis of performance are considered.

11.5.1 EIT Analysis of Bubble Columns and Foams

Tomographic imaging has been used to visualize the dispersion of air in the slurry zone of flotation equipment and to map gas concentration in the foam zone. For example, EIT has been used to map out gas profiles within mixer tanks and (coupled with positron emission tomography, PET) to extract solids concentration data [26, 27]. PET enables direct observation of size segregation effects in suspension mixing when small populations of sieved radioactively labeled particles are used to visualize the homogeneity of the mixing process as a function of particle size.

Figure 11.19 shows the effect of increasing the gas sparging rate for constant agitation at 200 rpm provided by a centrally mounted impeller (not shown in the image) and in the absence of solids. The extent of gas distribution can be deduced from such images by averaging over time many of these instantaneous snapshots. The addition of solids and surfactant to initiate the flotation separation gives rise to an overlying foam structure. If this process is performed in an industrial flotation column (up to 30 m height and 2 m diameter) rather than a tank, a very deep froth layer can form. The behavior of froths is of significant interest and has been examined through various imaging methods based on inserted probes and top-mounted cameras [28].

Electrical imaging of the slurry bubble structure has been undertaken in laboratory studies; this technique makes it possible to visualize directly the nature of the bubbles [29, 30]. Figure 11.20 shows two images for different gas rates and the corresponding photomicrographs of bubble structure observed through the transparent wall of the column. Image data provides a means to assess the homogeneity and details of the structure of the gas–liquid·mixture in a unique manner. Hence, it is possible to conceive of using EIT to enable control of a process based on measurement of microstructure. For example, all the spatial tomographic data can be reduced to a single statistical homogeneity index; plots of this value vs. gas rate and

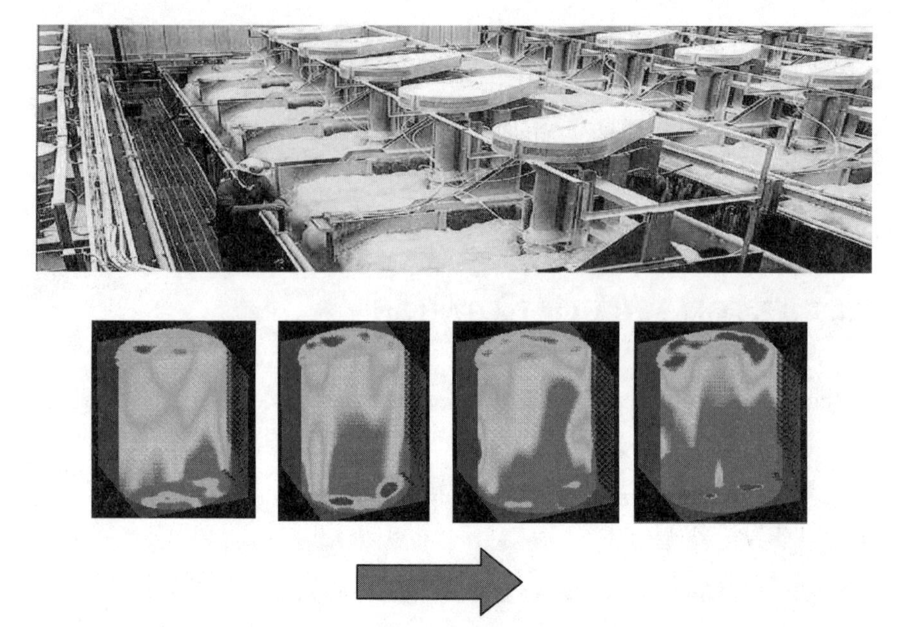

FIGURE 11.19 EIT imaging of gas in flotation equipment (shown) enables visualization of the instantaneous gas hold-up, shown here as a function of increasing sparging rate for a constant impeller speed (200 rpm). (From Williams, R.A., and Beck, M., *Process Tomography: Principles, Techniques, and Applications.* Oxford: Butterworth-Heinemann, 1995. With permission.)

surfactant dosage show how the onset of transition from a uniform to an unstable churning flow can be monitored. Alternatively, tomograms can be used to control the addition of surfactant to ensure adjustment of surface tension to produce the desired bubble size distribution. Such data are not accessible from external observation of the bubble column, and this application demonstrates a powerful property of tomographic visualization based on microstructure and texture.

Cross-correlation of pixelated data between sensor planes and within a plane can yield estimates of axial, angular, and tangential velocity, as discussed in Chapter 9. This technique has been used extensively in the auditing of bubble and flotation columns [29, 30]. Detailed analysis of the upper foam structure as it drains is also possible with EIT methods, based on analyzing the hydraulic pathways that give rise to electrical conduction paths.

11.5.2 Direct Imaging of Flotation Foams

The use of high-speed image cameras has been developed to a sophisticated level in the analysis of the state of mineral-laden foams as they overflow flotation tanks. The goal has been to use machine vision to translate operator experience gained from visual inspection of bubble size, distortion, color, and luster into robust

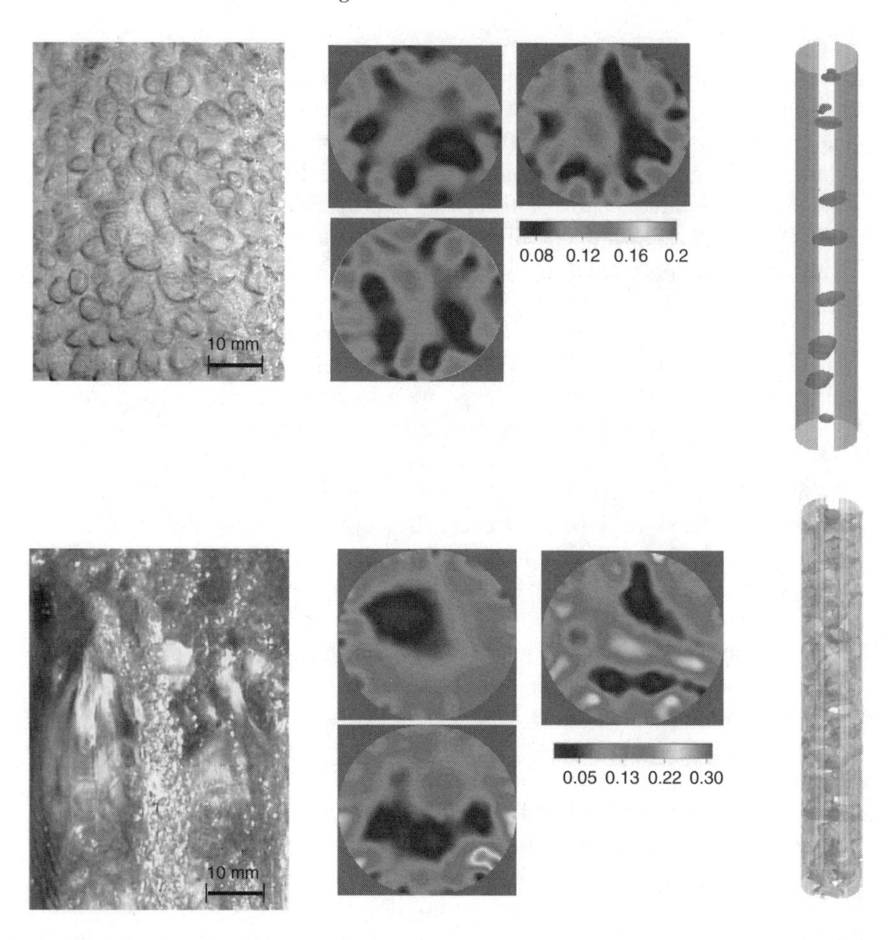

FIGURE 11.20 Identification of bubbly (*top*) and churning flow (*bottom*) from EIT images. Stacked images can be used to produce a pseudo-3D view (*right*). Video imaging (*left*) can be used to confirm structure for laboratory columns with transparent walls. (From Bennett, M.A., West, R.M., Luke, S.P., Jia, X., and Williams, R.A., *Journal of Chemical Engineering Science,* 54, 21, 5003–5012, 1999, and Bennett, M.A., West, R.M., Luke, S.P., and Williams, R.A., *Minerals Engineering,* 15, 225–234, 2002. With permission.)

control systems. One of the most common sources of disturbances in flotation is the changing ore quality, since the ore is fed to the concentrator plant as a mixture of many types of ore (coming from different parts of the mine). This variety gives rise to variable performance, with some flotation tanks yielding good recovery with a simultaneously high concentrate grade, whereas others produce a lower recovery and yield. Since ore quality is difficult to measure on-line, continuous feedback control of flotation conditions through the adjustment of reagents is needed. Also, under- or overdose of reagent can produce a secondary source of disturbances, and due to long process delays the effect of corrective control actions cannot always be seen immediately. This area has received much attention since the early pioneering work [28, 31] that led to the development of new kinetic models for flotation based on detailed bubble properties [32–34]. The technical challenge is to obtain images and then use different image-based analysis methods to extract bubble statistics. Various instrumental vision-based methods have been developed to classify froth. The extraction of texture features (see Chapter 5) is accomplished through statistical methods, Fourier transforms, power spectrums, or more advanced methods coupled with fluid motion models.

Figure 11.21 shows some photographs of a flotation cell in which the 3D nature of the froth can be seen (a) from the side and (b) from the top. Distinct variations in lighting present challenges to standard image analysis protocols. Various software approaches have been used to segment the images (Figure 11.21c shows one example) and then analyze changes to the images with time [35–39]. Different operational conditions (gas rates, surfactant loading, mineral compositions and sizes) give rise to very different shallow and deep froth structures. Figure 11.22 shows a froth in which windows develop, devoid of mineral matter, whereas other regions may have variable luster. All these characteristics provide important information and have been used to develop commercial software tools and devices (e.g., FrothMaster, Smartfloat, JKFloat) that have been used in industrial flotation plants, notably relating to copper, zinc, and platinum recovery in South Africa, Chile, and Australia. The translation from laboratory testing to pilot testing and now full industrial implementation has occurred over a period of 10 years in tandem with the availability of advanced low-cost optical sensors and smart statistical and artificial intelligence algorithms. Several case studies are available in academic and commercial literature describing benefits to enhanced recovery and improved reagent and water usage through adoption of these methods. In practice, they are not always easy to operate, since many flotation cells are located outdoors and changes in lighting conditions must be taken into account.

More sophisticated methods have been developed and deployed with success using optical spectroscopy and image analysis from which detailed mineralogical data can be estimated [40]. These methods normally require well-controlled lighting conditions in a covered flotation tank. An example of such an arrangement is given in Figure 11.23 based on work on the Outokumpu's Pyhäsalmi concentrator [39]. Here, various statistical froth classification methodologies were tested using data from a digital camera mounted above the tank. Color analysis was used for composition analysis to obtained grade information with side-mounted light sources and detectors. An industrial research program run for a prolonged period has demonstrated the viability of this method and quantified cost savings through overall improved grade, principally achieved through judicious control of copper sulphate additions. Closed loop control was based on four image-derived measurements: froth speed, bubble collapse rate, bubble transparency, and mean bubble size. In some variations of the controller, the mean value of the red channel in the color image was also used.

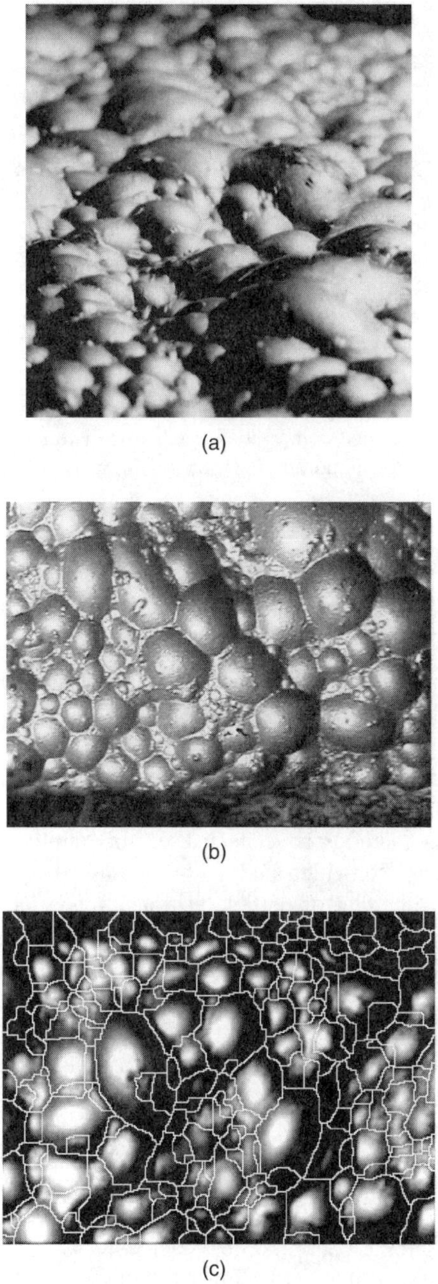

(a)

(b)

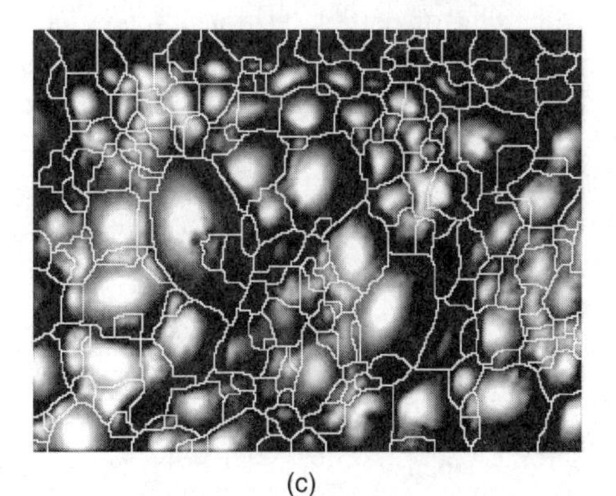

(c)

FIGURE 11.21 Photographs from industrial flotation cells showing the 3D nature of froths viewed from the side (a) and the top (b). Plan views are used to extract bubble statistics and velocities through various segmentation and analysis methods (c). (Images from chalcopyrite flotation at Pyhasalmi Mine, Finland, courtesy of J. J. Cilliers.)

FIGURE 11.22 Froth photograph showing bubbles exhibiting mineral-free caps. Information on mineral loading and composition needs to be extracted from image analysis. (Images from chalcopyrite and bornite flotation at Northparkes Mine, Australia, courtesy of J. J. Cilliers.)

11.6 SOLIDS SETTLING AND WATER RECOVERY

The settling of solids in mineral slurries is a major design issue in mineral processing. For transportation it is desirable to build a stable (nonsettling) mixture through control of the particle size distribution and concentration and occasionally through use of viscosity modifying agents. Routinely, information is sought on the particle size, shape, and concentration of such mixtures. For water recovery, a nonstable fast settling mixture is required.

Optical imaging methods have been used to examine such systems using in-process sensors based on video analysis [41] or laser-based inspection methods [42]. Direct optical imaging of particle size and shape has developed rapidly; several vendors offer in-process microscope probes, such as the one shown in Figure 11.24 (see also the example in Chapter 10). Multiple deployments of these devices in mineral process systems are of interest to enable mapping of flocculation and other allied processes within a process separator.

EIT has been used to visualize the extent of solids segregation in hydraulic conveying. From early studies [43], a more detailed analysis has been performed to optimize plant design using pipe-mounted systems with multiple planes of electrode sensors. For example, Figure 11.25 shows data in which EIT methods were used to map out the solid concentration profiles at different distances downstream of swirl inducing elements and for different flow rates. Hence, the proficiency of the solids suspensions induced by swirl could be assessed [44]. This study demonstrates how such methods can be used routinely to abstract quantitative measurements of solids disposition in moving flows. Such data are invaluable to verify fluid dynamics simulations and have resulted in new practical correlations for solids suspension. Recent advances in the ability to undertake fast EIT imaging up to 1000 frames per second offer new opportunities in using

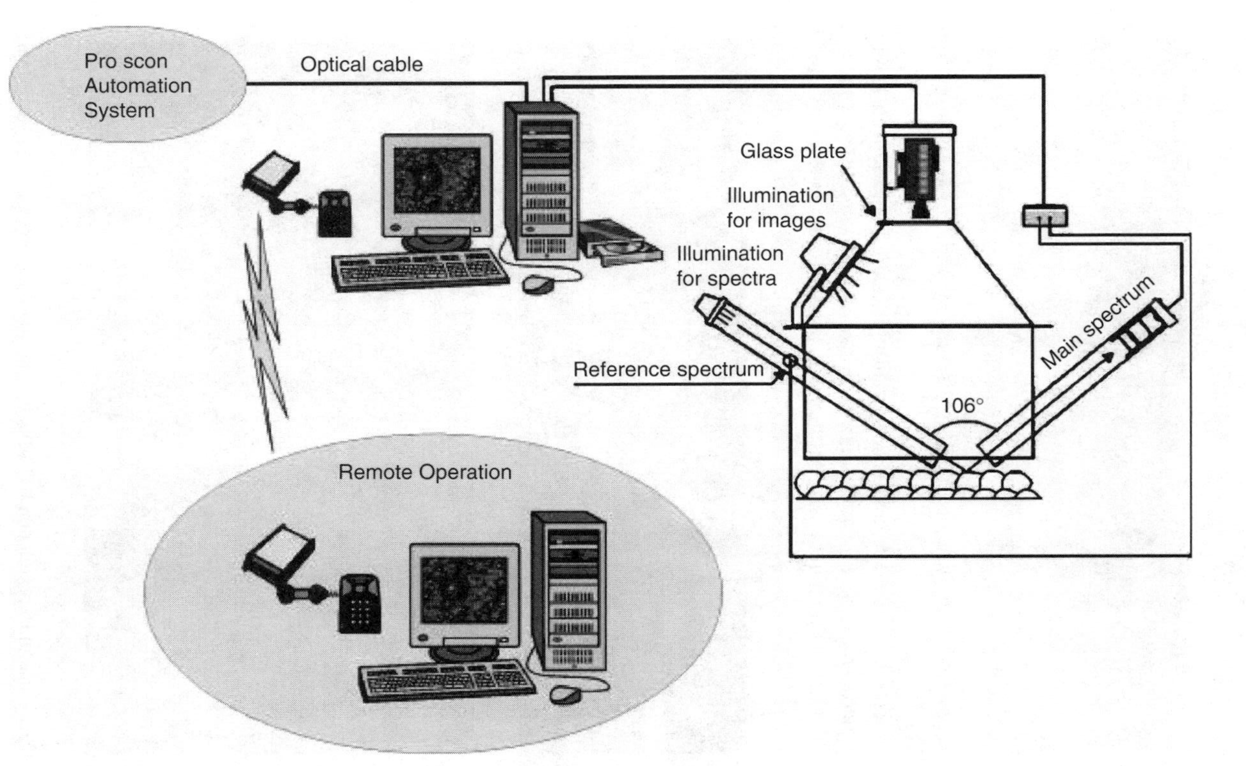

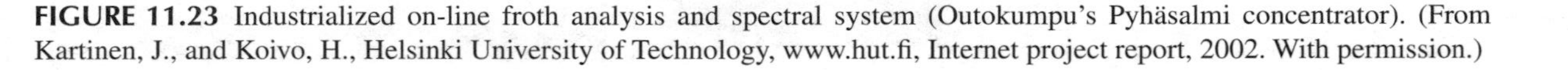

FIGURE 11.23 Industrialized on-line froth analysis and spectral system (Outokumpu's Pyhäsalmi concentrator). (From Kartinen, J., and Koivo, H., Helsinki University of Technology, www.hut.fi, Internet project report, 2002. With permission.)

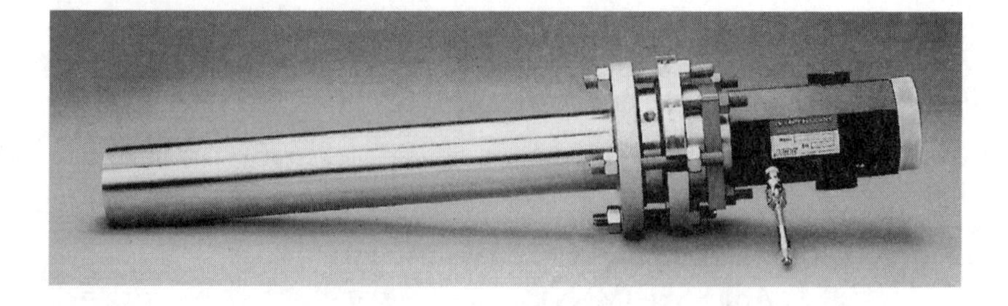

FIGURE 11.24 An in-process camera. (Courtesy of J.M. Canty Ltd., Dublin.)

impedance cameras to analyze fast flowing streams (Wang, personal communication, 2004).

In water removal processes, solids segregation is desired, and settling equipment is designed to maximize the rate of gravitational settling. In industrial practice, very large diameter thickeners and settlers are used. Commercially available

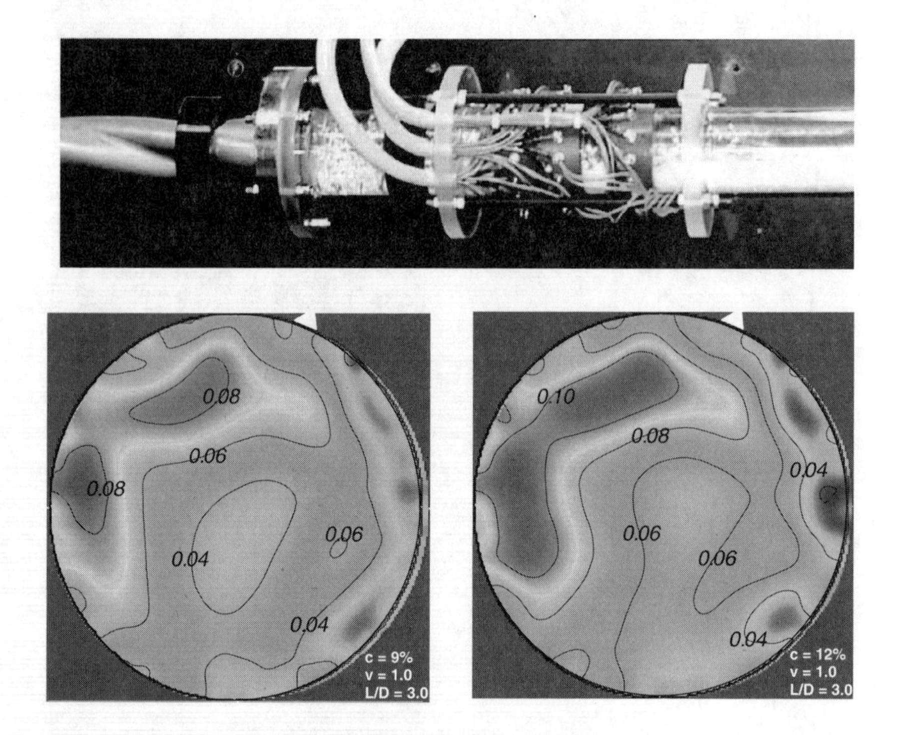

FIGURE 11.25 (*above*) Experimental EIT sensors auditing the behavior of a highly swirling slurry flow in a pipe. (*below*) Examples of the instantaneous solid concentration image at a given plane for solid loadings of 6.4% v/v and 8.5% v/v solids. Contours show the volume fraction of solids. (From Wang, M., Jones, T.F., and Williams, R.A., *Trans. I. Chem. Eng.*, 81(A), 854–861, 2003. With permission.)

electrical visualization methods have been deployed in such systems based on tomographic probes and ring arrays [45]. For example, there has been much interest in the design of the feed well in industrial thickeners using EIT, since the addition of slurry and flocculants in this zone affects overall efficiency [46].

11.7 MICROSCALE ANALYSIS OF GRANULES, FLOCS, AND SEDIMENTS

Use of tomographic techniques to measure at smaller length scales within a complex structure is obviously desirable. For some time synchrotron work has enabled fully 3D compositional mapping of, for instance, minerals in a rock specimen to reveal the chemical and mineralogical composition. This information has been obtained with considerable effort and at significant cost. Considered below are some alternative approaches to the challenge of microstructure determination. The key development has been the linking of new measurement capability with suitable modeling methods to enable property prediction.

High-resolution CT, or x-ray microtomography (XMT), has become practical with the advent of suitable commercial systems, such as the Skyscan system with resolution down to micrometers. Miller et al. have reported significant work on macro and micro x-ray methods [47,48]. XMT methods can be used to assess filtration media (see Figure 11.26), comminuted ores, pastes, and other structured entities such as flocculated minerals [49]. When applied to liberation analysis, it overcomes severe limitations associated with stereological inversion of data gathered using two-dimensional (2D) methods (polishing, sectioning, and image analysis).

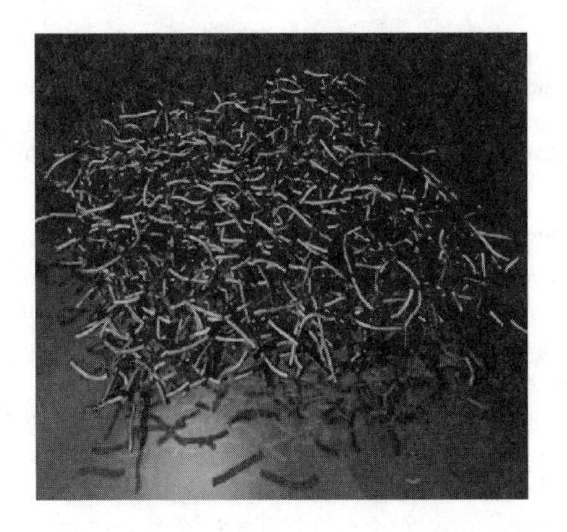

FIGURE 11.26 X-ray microtomography data reconstructed to visualize filtration media in three dimensions. The fiber diameter is on the order of 500 μm. (From Williams, R.A., and Jia, X., *Advanced Powder Technology,* 14, 1, 1–16, 2003. With permission.)

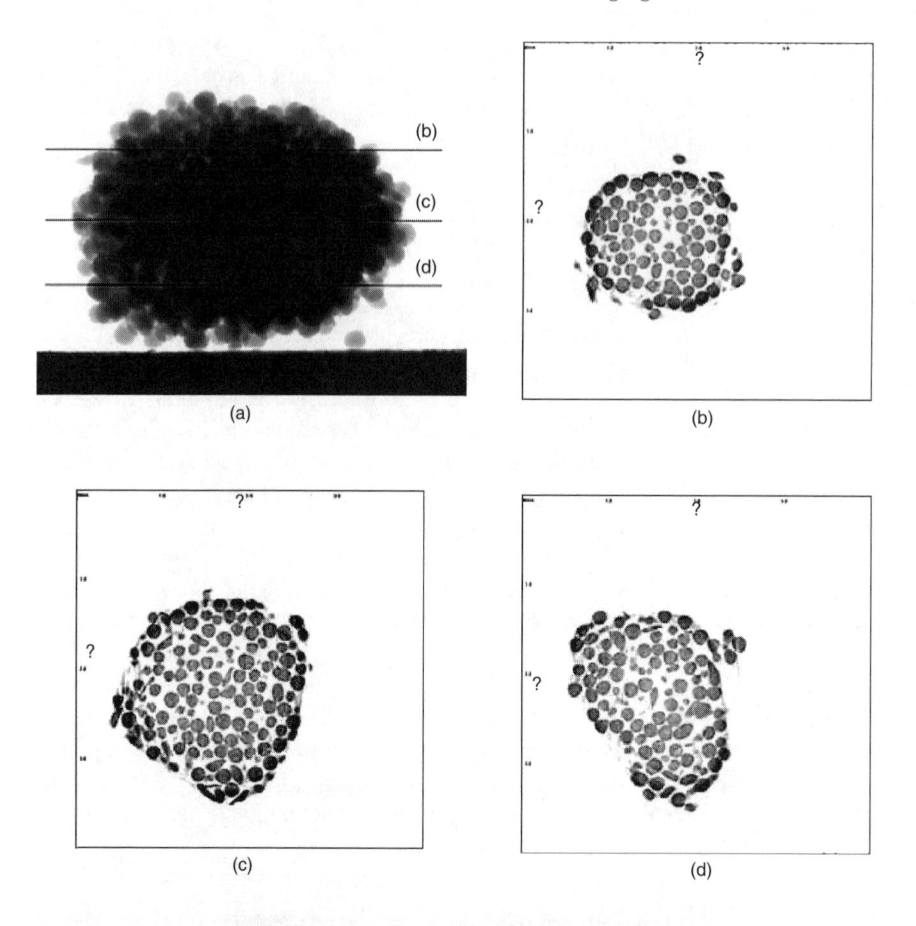

FIGURE 11.27 X-ray microtomographs of a silica granule made with PVA binder, showing (a) side view and views of cross-sections at (b) Z = 2.342 mm, (c) Z = 1.874 mm, and (d) Z = 1.445 mm. (From Golchert, D.J., Moreno, R., Ghadiri, M., Litster, J., and Williams, R.A., *Advanced Powder Technology,* 15, 447–458, 2004. With permission.)

Figure 11.27 shows analysis of friable granules of products based on x-ray photography and selected cross-sectional views derived from the tomogram. Detailed information on the structure and coordination number of the granule can be obtained and compared with theoretical models to predict the formation and breakage properties of these structures [50].

By combining digital structure analysis with property simulation, it is possible to gain detailed information on complex properties using Lattice Boltzman methods for detailed flow and pressure drop simulations [49, 51]. For example, consider the comparison of images obtained from (optical) confocal scanning microscopy (of individual flocs) with x-ray microtomographic analysis of sedimented flocs [52]. In Figure 11.28, silica flocs were made by flocculation at different levels of solids concentration. These were visualized using confocal microscopy

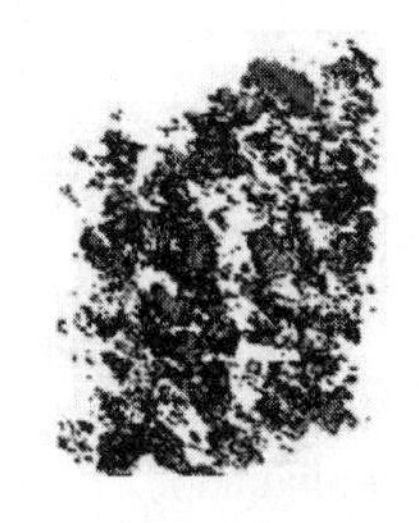

(a)

(b)

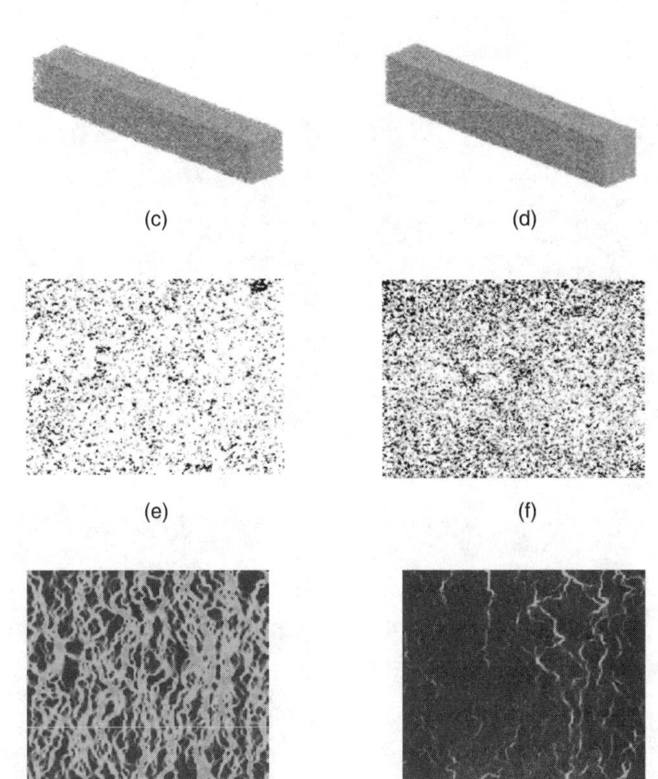

(c) (d)

(e) (f)

(g) (h)

FIGURE 11.28 Analysis of individual polymer flocculated silica aggregates using confocal scanning for two levels of silica loading, (a) 5 g/l and (b) 10 g/l. Sedimented flocs of this material (c, d) are analyzed with x-ray microtomography; examples are shown (e, f) of tomographic images through the sediment. (g, h) Predicted liquid flux within the settled sediment based on the tomographic data. (From Selomulya, C., Jia, X., and Williams, R.A., *3rd World Congress on Process Tomography,* Banff. Manchester: VCIPT, 2003. With permission.)

(*top row*) and then subjected to sedimentation. Portions of the sediment were then examined using XMT *in situ* from which the detailed fluid network could be estimated. The digitized 3D structure was used in a Lattice Boltzman simulation to estimate liquid flux in the sediment, shown in Figure 11.28 in a 2D slice through the sediment (*bottom row*). The effect of solids loading is shown. Clearly, such methods are set to provide a powerful means of detailed analysis relevant to materials and waste handling in the minerals industry. Prediction of filtration behavior in 3D can be obtained with additional computation (see Figure 11.29) and application of a 3D flow model [52]. This example demonstrates the value that can be gained by applying microscale 3D imaging coupled with a microscale flow model.

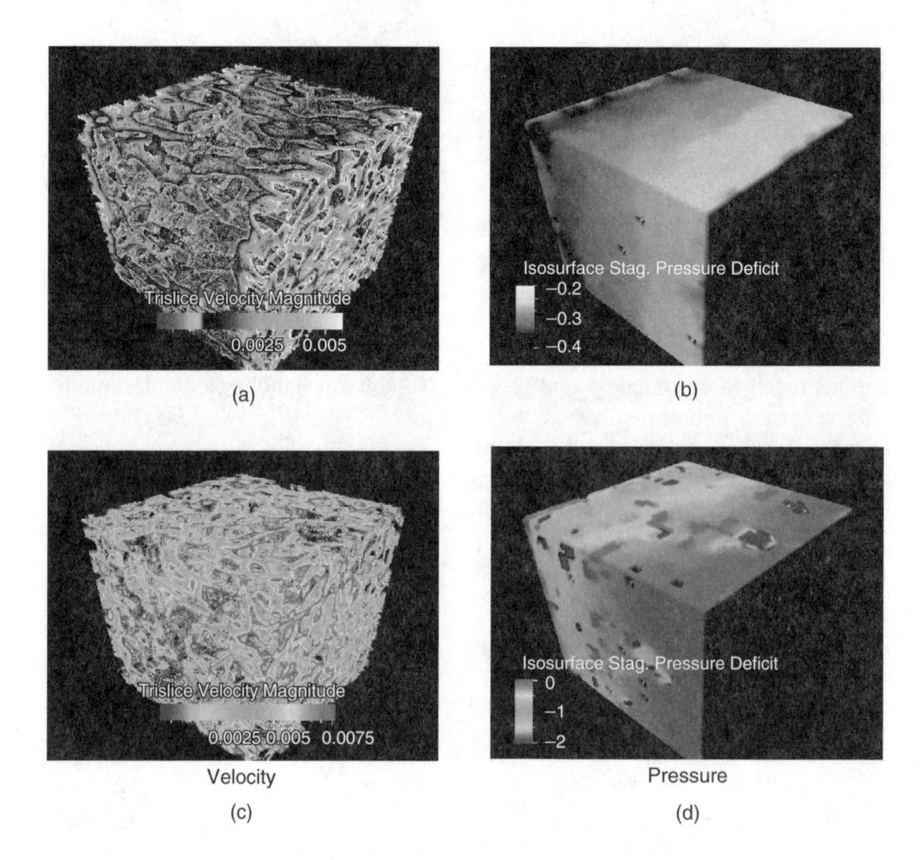

FIGURE 11.29 Direct prediction of filtration behaviour from x-ray microtomography using 3D microscale measurement and microscale Lattice Boltzman modeling showing (a) velocity profiles and (b) pressure drop for sediment derived from a 5 g/l flocculated silica slurry and (c, d) for higher solids slurry (10 g/l). (From Selomulya, C., Jia, X., and Williams, R.A., *3rd World Congress on Process Tomography,* Banff. Manchester: VCIPT, 2003. With permission.)

11.8 CONCLUDING REMARKS

Imaging methods have important applications in mineral and material processes for the advancement of fundamental knowledge and for industrial measurement and control. For fundamental application, the fusion of multiscale process knowledge with measurements offers a way forward to devise more robust phenomenological models. Ultimately the proper integration of (a) sensors that simultaneously measure over different length and time scales with (b) models that predict the different phenomena over those time and length scales will provide a powerful diagnostic tool.

For industrial measurement, imaging can be used to provide advisory information, or the data can be integrated with operator-based decision-making algorithms. Simple vision-based systems (such as a camera or line scanner viewing the upper surface of mineral products, suspensions, or froths) can be challenging to use as part of a routine control strategy, since spurious interpretations can occur. Rapid advances continue to be made in the area of on-line optical analysis by probes and cameras, with increasingly sophisticated analysis and lower cost hardware. These advances will further stimulate use of these methods in mineral processing operations.

Tomographic tools provide a complementary route to direct process imaging by permitting the detailed internal behavior of the process to be visualized. To date, their use has largely been for process optimization rather than as part of an operating control system. However, this position is likely to change as coupling between various types of heuristic or statistical dynamic process models renders the approach more practical and reliable. Notably, in handling complex multimineral and multiphase mixtures, they will increasingly offer practical routes for auditing stream composition and mass flow rates. For some areas of industrial needs, such as measurement and control of sedimentation, applications are emerging as a result of adoption of simpler sensing systems.

REFERENCES

1. G Agricola. *De Re Metallica* (1556). Translated by HC Hoover and LH Hoover. New York: Dover, 1950, p. 638.
2. RJ Batterham, SH Algie. The role of technology in the minerals industry. *Proc. XIX International Mineral Processing Congress.* San Francisco: Society of Mining Engineers, 1996, pp. 19–15.
3. TJ Napier-Munn. Invention and innovation in mineral processing. In: *Innovation in Physical Separation Technologies* (Richard Mozley Memorial Volume). London: The Institution of Mining & Metallurgy, 1998, pp. 11–31.
4. RA Williams. The impact of fundamental research on minerals processing operations of the future. *XXII International Mineral Processing Congress.* Marshalltown: SAIMM, 2003, vol. 1, pp. 2–13.
5. DM Scott, OW Gutsche. ECT studies of bead fluidization in vertical mills. *Proc. 1st World Congress on Industrial Process Tomography.* Buxton, U.K., 1999, pp. 90–95.

6. H Mio, F Saito, M Ghadiri, R Eastman, R Stephen. Analysis of particle mixing behaviour in attrition ball milling by discrete element method. *2nd Asian Particle Technology Symposium,* Penang, 2003.
7. DJ Parker, RN Forster, P Fowlers, PS Taylor. Positron emission tracking using the new Birmingham positron camera. *Nucl. Inst. & Meth.* A477:540–545, 2002.
8. J Conway-Baker, RW Barley, RA Williams, X Jia, J Kostuch, B McLoughlin, DJ Parker. Measurement and motion of grinding media in a vertically stirred mill using positron emission particle tracking (PEPT). *Miner. Eng.* 15:53–59, 2002.
9. JPK Seville. Tracking single particles in process equipment. *Proc. Part. Syst. Anal. 2003,* Harrogate, U.K., 2003.
10. KL Ostrowski, SP Luke, MA Bennett, RA Williams. Real time visualisation and analysis of dense phase powder conveying. *J. Powder Technol.* 102:1–13, 1999.
11. A Jaworski, T Dyakowski. Application of electrical capacitance tomography for gas-solid flow characterisation in a pneumatic conveying system. *Meas. Sci. Technol.* 12:1109–1119, 2001.
12. KL Ostrowski, SP Luke, MA Bennet, RA Williams. Application of capacitance tomography for on-line and off-line analysis of flow patterns in horizontal pipelines of pneumatic conveyors. *Chem. Eng. J.* 3524:1–8,1999.
13. D Neuffer, A Alvarez, DH Owens, KL Ostrowski, SP Luke, RA Williams. *Proceedings of the 1st World Congress on Industrial Process Tomography,* Buxton, U.K. 1999, 71–77.
14. RA Williams. Process tomography. In: Masuda, Higashatani, eds. *Handbook of Powder Technology.* 1st ed., Utrecht: VSP, 1997, pp. 869–877.
15. TA York, I Evans, Z Pokusevski, T Dyakowski. Particle detection using an integrated capacitance sensor. *Sensors and Actuators A:Physical,* 2001, 92: 103, 74.
16. A Malmgren, G Oluwande, G Riley, Multiphase low in coal fired power plant, DTI Project Report COALK R252, DTI Publication URN 03/1635, p. 84, 2003.
17. RA Williams, SP Luke, KL Ostrowski, MA Bennett. Measurement of bulk particulates on a belt conveyor using dielectric spectroscopy. *Chem. Eng. J.* 77:57–63, 2000.
18. WL Dalmijn, MA Reuter, TPR de Jong, UMJ Boin, The optimization of the resource cycle impact of the combination of technology, legislation and economy. *Proc. XXII Inernational Mineral Processing Congress,* SAIMM, 2003, Volume 1: pp. 80–105.
19. RA Williams, M Beck. *Process Tomography: Principles, Techniques, and Applications.* Oxford: Butterworth-Heinemann, 1995, p. 550.
20. RA Williams, X Jia, RM West, M Wang, JC Cullivan, J Bond, I Faulks, T Dyakowski, SJ Wang, N Climpon, JA Kostuch, D Payton. Industrial monitoring of hydrocyclone operation using electrical resistance tomography. *Miner. Eng.* 12, 10:1245–1252, 1999.
21. RM West, X Jia, RA Williams. Parametric modelling in industrial process tomography. *Chem. Eng. J.* 77, 1–2:31–26, 2000.
22. JC Cullivan, RA Williams, CR Cross. Understanding the hydrocyclone separator through computational fluid dynamics. *J. Trans. I. Chem. Eng., Part A: Chem. Eng. Res. Design* 81:455–466, 2003.
23. RA Williams, OM Ilyas, T Dyakowski. Air core imaging in cyclonic separators using electrical resistance tomography. *Coal Preparation* 15:149–163, 1995.
24. RA Williams, FJ Dickin, JA Gutierrez, T Dyakowski, MS Beck. Using electrical impedance tomography for controlling hydrocyclone underflow discharge. *Control Eng. Practice* 5:253–256, 1997.

25. MA Bennett, RA Williams. Monitoring the operation of an oil-water separator using impedance tomography. *Miner. Eng.* 17:605–614, 2004.
26. SL McKee, DJ Parker, RA Williams. In: DM Scott, RA Williams, eds. *Frontiers in Industrial Process Tomography.* New York: AIChE and Engineering Foundation, 1995, p. 249.
27. RM West, X Jia, RA Williams. Quantification of solid-liquid mixing using electrical resistance and positron emission tomography. *Chem. Eng. Comms.* 175:71–97, 1999.
28. JJ Cilliers. Flotation froths: visualisation, characterisation and interpretation. *Chemical Technol.* 13–15, 1997.
29. MA Bennett, RM West, SP Luke, X Jia, RA Williams. Measurement and analysis of flows in a gas-liquid column reactor. *J. Chem. Eng. Sci.* 54, 21:5003–5012, 1999.
30. MA Bennett, RM West, SP Luke, RA Williams. The investigation of bubble column and foam processing using electrical capacitance tomography. *Miner. Eng.* 15:225–234, 2002.
31. DW Moolman, C Aldrich, JSJ Van Deventer, WW Stange. Digital image processing as a tool for on-line monitoring of froth in flotation plants. *Miner. Eng.* 7:1149–1164, 1994.
32. N Sadr-kazemi, JJ Cilliers. An image processing algorithm for measurement of flotation froth bubble size and shape distributions. *Miner. Eng.* 10:1373–1383, 1997.
33. SJ Neethling, JJ Cilliers, ET Woodburn. Prediction of water distribution in flowing foams. *Chem. Eng. Sci.* 55:4021–4028, 2000.
34. W Xie, SJ Neethling, JJ Cilliers. A novel approach for estimating average bubble size for foams flowing in vertical columns. *Chem. Eng. Sci.* 59:81–86, 2004.
35. H Hyötyniemi, R Ylinen. Modeling of visual flotation froth data. *Control Eng. Practice* 8:313–318, 2000.
36. DW Moolman, C Aldrich, JSJ Van Deventer, DB Bradshaw. The characterisation of froth surfaces and relation to process performance by using connectionist image processing techniques. *Miner. Eng.* 8:23–30, 1995.
37. DW Moolman, C Aldrich, JSJ Van Deventer, DB Bradshaw. The interpretation of flotation froth surfaces by using digital image analysis and neural networks. *Chem. Eng. Sci.* 50:3501–3513, 1995.
38. JM Oestreich, WK Tolley, DA Rice. The development of a colour sensor system to measure mineral composition. *Miner. Eng.* 8:31–9, 1995.
39. J Kartinen, H Koivo, Machine vision based measurement and control of zinc flotation circuit, Helsinki University of Technology, www.hut.fi, Internet project report, 2002.
40. G Bonifazi, P Massacci, A Meloni. Prediction of complex sulfide flotation performances by a combined 3d fractal and colour analysis of the froths. *Miner. Eng.* 13:737–746, 2000.
41. J Jennings, MJ Wilkinson, K Wood-Kaczmar, J Hayler, M Forth, G Bret, H Singh. Imaging drug particles and crystallisation processes with an in-line video technique. *Particulate Systems Analysis 2004,* Royal Society of Chemistry, 5 pp., 2004.
42. C Selomulya, RA Williams. In-process measurement of particulate systems. Chapter 7 in *Advances in Granular Materials: Fundamentals and Applications.* SJ Antony (ed.). London: Royal Society of Chemistry, 2004.

43. YS Fangary, RA Williams, WA Neil, SP Luke, J Bond, I Faulks. Application of electrical resistance tomography to detect deposition in hydraulic conveying systems. *Powder Technol.* 95: 61–66, 1998.
44. M Wang, TF Jones, RA Williams. Visualisation of asymmetric solids distribution in horizontal swirling flows using electrical resistance tomography. *J. Trans. I. Chem. Eng.* 81(A): 854–861, 2003.
45. http://www.itoms.com.
46. JB Farrow, PD Fawell, L Mittoni, TV Nguyen, M Rudman, K Simic, JD Swift. Techniques and methodologies for improving thickener performance. *Proceedings of XXII International Mineral Processing Congress,* SAIMM, 2:827–837. 2003.
47. CL Lin, JD Miller, A Cortes. X-ray computed tomography applied to mineral processes. *KONA.* 10: 88 pp., 1992.
48. JD Miller, CL Lin, A Garcia, H Arias. Ultimate recovery in heap leaching operations as established from mineral exposure analysis by x-ray microtomography. Littleton: Society of Mining Engineers. Preprint No 02-170 from SME Annual Meeting, Phoenix, AZ, 2002.
49. RA Williams, X Jia. Tomographic analysis of particulate systems. *Advanced Powder Technol.* 14, 1:1–16, 2003.
50. DJ Golchert, R Moreno, M Ghadiri, J Litster, RA Williams. Application of X-ray microtomography to numerical simulations of agglomerate breakage by distinct element method. *Advanced Powder Technol.* 15:447–458, 2004.
51. X Jia, N Gopinathan, RA Williams. Modelling complex packing structures and their thermal properties. *Advanced Powder Technol.* 13:55–71, 2002.
52. C Selomulya, X Jia, RA Williams. Tomographic analysis of flocs and flocculating suspensions. *3rd World Congress on Process Tomography,* Banff, Canada. Manchester: VCIPT, 2003, pp. 145–151.

12 Applications in the Metals Production Industry

James A. Coveney, Neil B. Gray, and Andrew K. Kyllo

CONTENTS

12.1 INTRODUCTION

The metals production industry is concerned with the extraction of metals from mineral ores. This is achieved through a variety of different process paths that depend on the metal that is being produced. In general, ores are processed to remove impurities and increase the concentration of the valuable mineral. This concentrate then undergoes smelting to produce a crude metal which is then refined further to achieve the required purity for the final product. This chapter does not address the processing of ores prior to their smelting, but a widely used overview of metallurgical processing is given by Rosenqvist [1].

The high temperatures inherent in extractive metallurgy, and the aggressive materials involved, present difficulties in directly observing many of the relevant processes by conventional sensor technologies. The industry has turned to process imaging and other sensing technologies that are able to overcome the hostile processing environment. Now these sensors are finding application in various steps of metal production: metallurgical furnaces, ladles, continuous casters, and inspection of the final product. These technologies provide a better understanding of process conditions and enable advanced process control schemes. The demands for product quality, improved process safety, and reduced emissions are driving the development of new sensors in the metallurgical industry.

The application of process imaging and other sensing technology can be considered in two parts: first, the sensing system employed to take the measurements and, second, the analysis of those measurements to provide the required information regarding the process. The sensing technique must be sufficiently sensitive to detect the relevant change in process conditions in the often "noisy" environments. Positioning of sensors and the number of sensors that can be installed are often compromised due to physical restrictions and the need to protect the sensors from the surrounding environment. These limitations result in the collected data set being less than ideal, presenting a challenge in the processing of the data.

Several considerations determine how the data are processed. These include the quality of data, the goal of the measurement, and the amount of time available to carry out calculations. Often a simple data processing option is chosen to minimize on-line computation time to satisfy these requirements. A single number output rather than a full image may be sufficient in many cases to provide the required information regarding the process, making the output easier to feed into a control loop and reducing ambiguity for operator interpretation. In other cases, a detailed image of the system will be required, involving more complex and time-consuming computation.

The harsh environments encountered in the metals production industry have caused the development of imaging techniques to lag behind the more sophisticated applications in other industries. This situation presents opportunities for development of new systems, as the information available regarding a process may be very limited. In many cases, information that would be seen as relatively simple in other industries would be enthusiastically welcomed in the metals processing industry.

Systems that have been developed are frequently based on operator knowledge of the process, such as infrared detection of entrainment. Experienced operators are able to detect subtle changes in the color of a stream when slag had become entrained in the metal. Infrared systems have subsequently been developed to detect the change in appearance of the metal stream. Systems developed from operator knowledge aim to automate the detection process, which in turn improves reliability compared to operator judgment. Furthermore, the sensitivity of the system is superior to that of the operators. Automation of the process makes it possible to produce a log of the measured data, which can be analyzed to provide improved understanding of the process.

This chapter reviews applications within the metals processing industry of the imaging techniques discussed in the previous chapters. Applications that are covered include slag entrainment detection, monitoring of furnace refractory wear, flow monitoring, and imaging solid state phase change. Several competing techniques are detailed for each application, and a comparison between the techniques is made. Developing technologies are discussed, highlighting the potential for future development of process imaging in the metals production industry.

12.2 DETECTION OF SLAG ENTRAINMENT

In many stages of metal production, vessels will contain metal (or matte) and slag as separate phases. The subsequent stages of production require the metal phase to be removed from the vessel separately from the slag phase. Tapping metals, mattes, and slags from furnaces; pouring from ladles; and casting of metals are all common processes in the metals production industry. Entrainment occurs when one phase is mixed with the other. Slag entrainment in a metal complicates the downstream operations, and the purity of the final product may be compromised. The presence of slag increases the wear rate of the refractories of downstream operations that have to deal with the contaminated metal. Entrainment of metal in the slag phase either leads to a loss of product or necessitates reprocessing of the slag to recover lost product.

Figure 12.1 shows a vessel containing separate metal and slag phases, where the metal is being removed from the vessel via a nozzle or taphole. The most apparent cause of entrainment occurs when the level of the slag–metal interface approaches

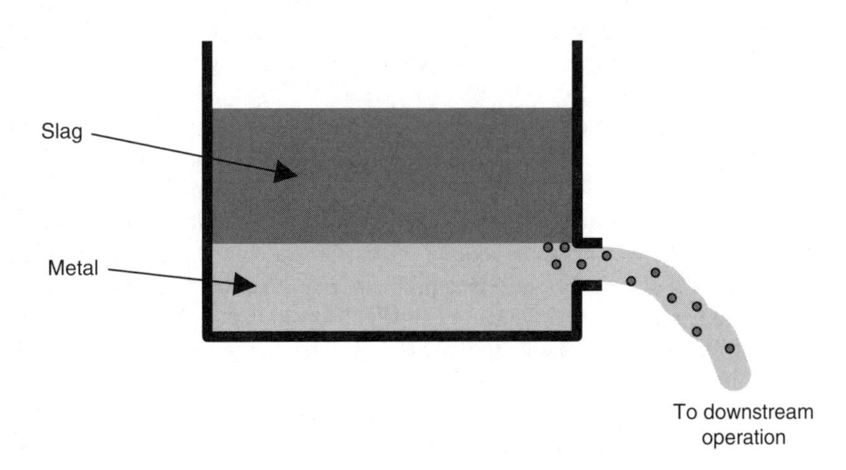

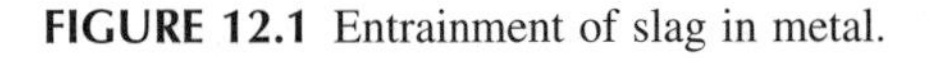
FIGURE 12.1 Entrainment of slag in metal.

the level of the taphole and the slag is drawn into the stream of metal exiting the vessel [2]. Entrainment can also be caused by fluid dynamical effects such as vortexing due to the outflow of the metal, causing mixing of the two phases. Such mixing may also result from injection of gases or solids into the vessel [3]. The process of opening the taphole has also been known to result in entrainment in the early stages of the operation.

The most rudimentary form of entrainment detection (still used in many situations) is visual detection by the process operators, who observe a subtle color change in the molten stream. However, this change in color is difficult to detect as a result of the intense radiation from the hot, molten stream. Overall, observation of slag entrainment by operators is subjective and inconsistent, and it does not allow low levels of slag entrainment to be detected.

Accurate detection of the level of entrainment enables the use of a feedback loop to control the pour/cast/tap operation and determine when the flow should be stopped. Monitoring the level of entrainment in conjunction with the vessel's operation will provide understanding of the causes of entrainment for the particular process. This understanding may allow these causes to be addressed by modifications to the operation, and thus reduce the level of entrainment.

12.2.1 Infrared Detection

Infrared detection is based on similar principles to the operator observing a color change that indicates entrainment, but with an IR camera (see Chapter 3) sensing the radiation emitted from the surface of the molten stream. As the slag has a significantly higher emissivity than the metal (ε_{slag} = 0.15 to 0.22 whereas ε_{steel} = 0.077 at 10 μm [4]), slag of the same temperature as the metal appears "hotter" to the IR camera. The camera is positioned to focus on the molten stream. The air between the stream and the camera will often contain smoke, dust, and gases that absorb some of the IR radiation before it reaches the camera. This attenuation, along with the ability of the camera's optics to focus on the stream, limit the maximum distance at which the camera can be separated from the steam.

On the other hand, a difficulty associated with positioning the IR camera too close to the stream is the effect of the stream in heating the camera itself. The detector is kept at least 5 m from the stream to protect it from the intense heat of the molten stream, as uncooled cameras cannot be operated in ambient temperature above 50 °C. Air- or water-cooling of the camera may extend this range to 150 °C. The field of view can also be a problem. For the case of monitoring steel being poured from a ladle, Goldstein et al. [5] observed that when the IR camera is aimed at the mouth of a ladle, the intensely bright slag in the ladle affects the clarity of the image.

The choice of the camera's wavelength sensitivity range is important. For the detection of slag in steel, operating the camera over the range 8 to 12 μm has been shown to minimize the effects of attenuation caused by suspended particles and gases in the air, thus providing a clear contrast between slag and steel [6].

Rau and von Röpenack [7] applied an IR detection system produced by Applied Measuring and Process Control Systems (AMEPA GmbH) to slag detection in steel being tapped from a converter. Application of this system resulted in lower impurities, with a 33% reduction in phosphorus and a 19% reduction in the silicon content of the steel.

The required image analysis, as described in the U.S. patent by Stofanak et al. [8], can be performed by digitizing the acquired image and converting it to gray scale. The gray scale of each pixel (assuming 8-bit digitization) can vary from 0 (black) to 255 (white). A range on this scale can be taken to represent steel and another range to represent slag; Stofanak et al. suggested ranges of 60 to 160 for steel and 230 to 255 for slag. Each pixel in the image can thus be classified as representing steel or slag, or it is left unclassified. The number of pixels in each class can be counted in order to calculate the slag-to-steel ratio. This ratio is used to determine when the flow should be stopped, either by passing the information to a control system, or by sounding an alarm for human operator action. The system also records the cumulative level of slag in the stream.

An alternative image analysis technique was proposed by Goldstein et al. [5], where the average emissivity for the image is measured and compared with values obtained from calibration runs under two known conditions: first, when the stream is known to be all steel and, second, when the stream is contaminated with a known level of slag.

12.2.2 Electromagnetic Detection

This technique involves a pair of coils (excitation and detection) positioned near the nozzle or taphole, as shown in Figure 12.2. An alternating current is passed

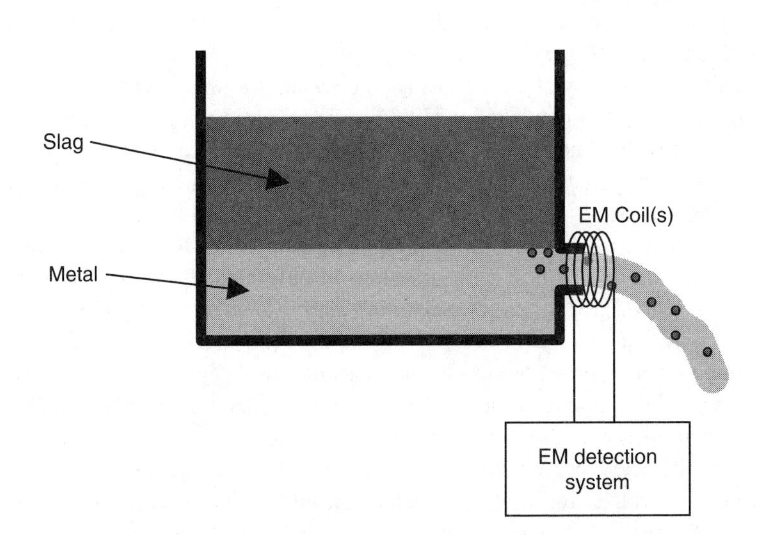

FIGURE 12.2 Electromagnetic detection of entrained slag.

through the excitation coil, inducing eddy currents in the molten material within the nozzle. These eddy currents depend on the composition of the molten material. Thus, the level of the signal induced in the detection coil is sensitive to the stream composition. Specifically, the detected signal is dependent on the electrical conductivity of the material within the stream. Steelmaking slags have conductivities in the range 500 to 1500 S/m [9], to be compared with those of molten metals, which are about 10^6 S/m [10]. Therefore, the presence of slag in the stream can have a large effect on the level of the signal induced in the detection coil.

The coils can be arranged in a number of different configurations. The systems produced by AMEPA [11] have both coils surrounding the nozzle so that the coils are coaxial to the stream, while the Molten Process Control AB (MPC) system [12] has the excitation and detection coils on opposite sides of the nozzle. Coils positioned coaxially to the stream provide better sensitivity to changes of conductivity within the stream, but installation or replacement of the coils is more difficult compared to coils adjacent to the nozzle. Coil systems must also be designed to withstand the high temperatures inherent in this environment. This is achieved by choosing appropriate wire to construct the coils, careful positioning of the coils, and possible inclusion of some form of cooling (e.g., compressed air).

Choosing an appropriate frequency (f) at which to excite the coils is important: As f is decreased, the magnitude of the response detected due to the presence of slag also decreases. The penetration of the magnetic field from the coil through the nozzle wall into the stream is governed by the electromagnetic skin depth, which is inversely proportional to $\sqrt{f}$. If the frequency is too high, the skin depth will not be sufficient for the magnetic field to penetrate the nozzle wall and into the stream. The choice of excitation frequency is thus a compromise between achieving adequate skin depth and sufficient sensitivity, and it is typically in the range 100 Hz to 5 kHz.

Analysis of the signal is similar in some respects to that of the IR technique, although no image is displayed. Calibration is required with a slag-free stream and a stream with a known level of slag, to provide reference measurements. The system can then present the level of slag as a percentage, and the information may be passed on to a control loop or an alarm system. In an application of electromagnetic slag detection in steel being tapped from a basic oxygen furnace (BOF) used in combination with a system to automatically close the taphole, Fritz and Grabner [13] report a reduction in slag content from 10 to 15 kg/t steel to 1 to 3.5 kg/t steel.

If the nozzle is prone to significant wear due to the aggressive nature of the molten materials passing through it, the profile of the stream will change over time. This in turn will affect the eddy currents induced in the stream and, hence, the signal detected by the coils, resulting in an inaccurate prediction of the entrainment level. Using this dependence, an eddy current detection system is being developed [14] to monitor the condition of tapholes as they wear.

12.2.3 Vibrational Detection

Vibrations are usually a result of the flowing stream, and their characteristics change as slag becomes entrained in the stream or when vortexing commences within the pouring vessel. Some experienced operators have been able manually to detect these subtle changes in vibration. However, this approach is inconsistent and operator dependent [15]. Automated detection relies on an accelerometer detecting the change in vibration from the stream, and such systems are now emerging onto the market. These systems have been applied to pouring steel from a ladle through a nozzle into a tundish, where the nozzle is held in place by a holding arm, as shown in Figure 12.3 [16, 17]. The accelerometer is mounted on the holding arm so that it is adequately separated from the stream in order to protect it from the high temperatures. The vibration caused by the flow of the stream is transmitted to the accelerometer via the holding arm. As with the IR and electromagnetic detection systems, the vibration detection system must be calibrated under desirable and undesirable flow conditions. The system can also be connected to a control loop or alarm system.

12.2.4 Comparison of Techniques

Each of the systems described above uses a simple analysis of the sensor readings. This approach has the advantage of being computationally fast, providing real-time feedback to control the process. However, each system also has limitations. Most of these systems require calibration to be performed for each specific application. Vibrational detection requires that the accelerometer be mounted in a position where the vibrations from the stream will be transmitted to the accelerometer while being sufficiently separated from the intense heat of the stream. Access to the nozzle or taphole region of the vessel is required for implementation of the electromagnetic detection system. For applications where the time between opportunities to access

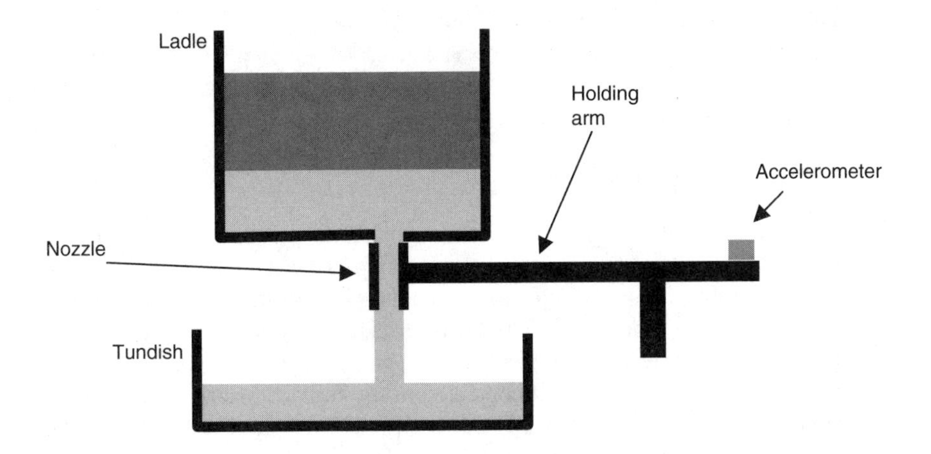

FIGURE 12.3 Vibrational detection of entrained slag.

this area is unacceptably long, electromagnetic detection may not be an appropriate technique.

Slag entrainment can be caused by vortexing, and the slag can be drawn into the center of the stream [18]. IR detection will be unable to detect such slag, as it is only sensitive to the surface of the stream. In cases where entrainment due to vortexing is a problem, such as bottom tapping [19], electromagnetic detection or vibrational detection should be considered.

12.3 MEASUREMENT OF FURNACE REFRACTORY WEAR

Furnaces contain large amounts of molten materials and hot gases. To ensure these are effectively contained, furnaces are lined with refractory material that is able to withstand the high temperatures and aggressive nature of these materials. Over time, exposure to these harsh conditions causes the refractory to erode away. The possibility for a catastrophic accident exists if excessive erosion of the refractory material occurs. This may result in a loss of containment of the molten materials or may allow the molten materials to contact water channels in furnace cooling and cause an explosion. Such accidents will put the operators in serious danger, cause significant damage to the furnace and nearby equipment, and require a long shutdown to repair the damage.

To avoid these consequences, furnaces are periodically relined with new refractories. This depends critically on the type of furnace: relines may occur several times a year for converters, or every 10 to 20 years in the case of iron blast furnaces. Relining a furnace is a major undertaking, requiring significant capital expenditure and a lengthy shutdown of the furnace.

Monitoring the wear profile of the furnace refractory is necessary to determine when the furnace is approaching an unsafe condition. This will allow the maximum safe campaign life to be gained from the refractory lining while ensuring furnace integrity is maintained. Monitoring will also show the wear rate associated with various operating practices. The wear rate may then be controlled by modifying operating practices to prolong refractory life.

12.3.1 THERMAL IMAGING

Thermal imaging is based on the measurement of temperature at many points within the refractory, with an array of thermocouples embedded in it at various depths. The set of measured temperatures can then be used to model the thermal profile of the furnace, which can be used to determine the extent of refractory wear.

The thermal profile is related to the wear profile using the isotherm of the temperature at which the material contained in the furnace begins to melt. This technique has been developed mainly in the steel industry for the monitoring of the refractory profile of blast furnace hearths. The isotherm used in the case of the steel industry is 1423 K, the iron–carbon eutectic [20]. The 1423 K isotherm is used to define the boundary of the melt. Allowance is made for the formation

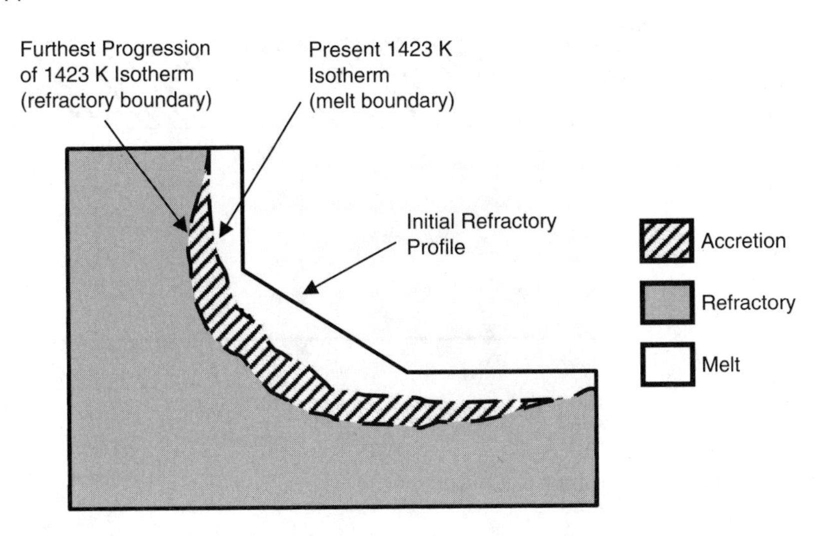

FIGURE 12.4 Section of furnace showing progression of 1423 K isotherm indicating refractory wear and accretion buildup within the furnace.

of a layer of accretion on the refractory material, which can have a significant influence on the life of the refractory. The position of the 1423 K isotherm is monitored over the campaign life of the furnace. The furthest progression of the 1423 K isotherm is defined as the wear line, or the refractory boundary. The region between the refractory boundary and the melt boundary is the accretion layer, as shown in Figure 12.4.

The heat conduction calculation methods employed to determine the position of the 1423 K isotherm range from simple one-dimensional (1D) approaches up to the complex three-dimensional (3D) case, using finite element methods [21]. Conduction of heat is governed by Fourier's Law of heat conduction [22]:

$$\mathbf{q} = -k\nabla T \tag{12.1}$$

where $\mathbf{q}$ is heat flux (W/m^2), k is thermal conductivity (W/mK), and T is temperature (K).

Simple 1D heat conduction may be assumed for the central region of the furnace base, where the heat flux will be in the axial direction, and for the region of the furnace walls sufficiently remote from the base such that the heat flux will be purely in the radial direction. In one dimension, Fourier's Law becomes

$$q_z = -k\frac{dT}{dz} \tag{12.2}$$

Thermocouples (e.g., A and B in Figure 12.5) are positioned at different depths in the refractory. Knowing the thermal conductivity allows the line of constant heat flux to be calculated, which can then give the position of the 1423 K

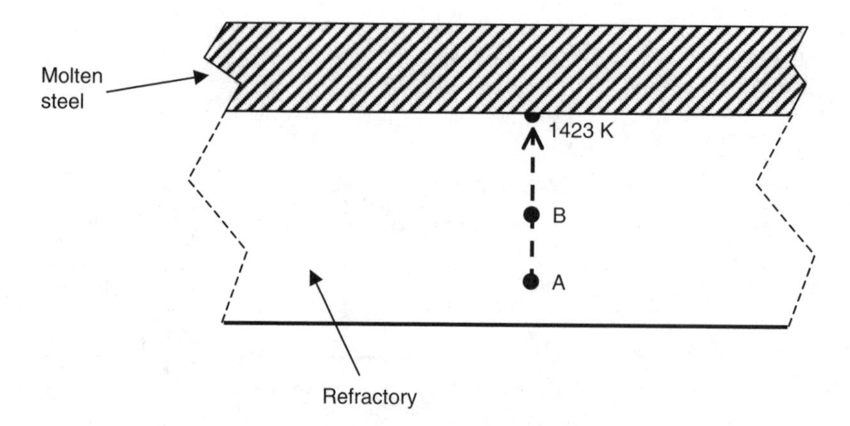

FIGURE 12.5 One-dimensional calculation of 1423 K isotherm location.

isotherm. These calculations are done by numerical means, as there may be layers of different refractory material as well as the accretion layer, each with its own thermal conductivity which may be a function of temperature. Three-dimensional FEM modeling of the blast furnace hearth by Shibata et al. [21] found that very little of the furnace could be modeled with 1D heat conduction. Rex and Zulli [20] employed a method using sets of three thermocouples to calculate heat conduction in two dimensions. Thermocouples are positioned at points A, B, and C, in Figure 12.6. The temperatures at A and B allow the heat flux in the x direction

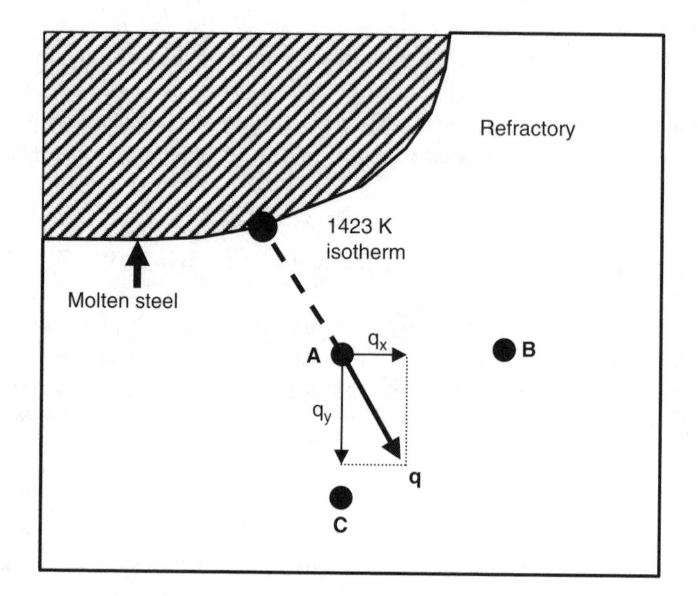

FIGURE 12.6 Two-dimensional calculation of 1423 K isotherm location.

(q_x) to be calculated, and the temperatures at A and C allow the heat flux in the y direction (q_y) to be calculated. These give the vector heat flux $\mathbf{q}$. It is then possible to calculate back along $\mathbf{q}$ to find the position of the 1423 K isotherm.

The set of thermocouples should be arranged to ensure the walls and the base of the furnace are covered sufficiently so that local changes in the refractory profile can be detected. In the instrumentation of a blast furnace hearth, Rex and Zulli used more than 200 thermocouples [23]. The requirements of the intended calculation method should also be considered in the layout of the thermocouples. Accurate knowledge of the position of each thermocouple is necessary for the calculated profile to be realistic. When thermocouples are installed at multiple depths, care must be taken to avoid heat conduction along the multiple thermocouples, which will give a misrepresentation of the thermal gradient. For this reason, insulating materials should be used for the thermocouple protective sheaths rather than metallic sheaths.

These calculations assume the heat transfer in the refractory to be at steady state. Under- or overestimation of the wear profile will occur with temperatures measured when the hearth is not at steady state. Rex et al. [23] observe that because of the huge thermal inertia of a furnace and its contents, it may take up to a week for steady state conditions to be reached.

Thermal imaging of the furnace profile provides continuous monitoring of the wear profile of the refractory and the profile of accretion deposited on the refractory. This gives an indication of when a reline is required and allows the effect of various operating practices on the refractory profile to be monitored. The practices can then be controlled or modified to protect the furnace refractory to maximize the time between furnace relines.

12.3.2 Ultrasonic Inspection

Ultrasonic techniques have been employed to monitor refractory thickness of metallurgical furnaces [24–28]. The measurements are made using pulse–echo mode, where the ultrasonic wave is transmitted into the refractory wall and reflected waves are detected by a receiver on the same side of the wall as the transmitter.

A typical setup for the pulse–echo mode is shown in Figure 12.7 [29]. The transmitter and the receiver are positioned on the surface of the refractory wall with a separation (w) between the transmitter and the receiver. The transmitter is driven to produce a short tone burst, causing a bulk wave and a surface wave to be produced. The receiver detects the bulk wave as it is partially reflected at the interface. Knowing the velocity of the wave through the bulk (v_b) and the time between transmission and detection of the bulk wave (t_b) allows the thickness (h) to be calculated with Equation 12.3 [24]. An image of the furnace refractory profile can then be constructed by taking measurements of the refractory thickness over an array of points.

$$h = \sqrt{\left(\frac{v_b t_b}{2}\right)^2 - \left(\frac{w}{2}\right)^2} \tag{12.3}$$

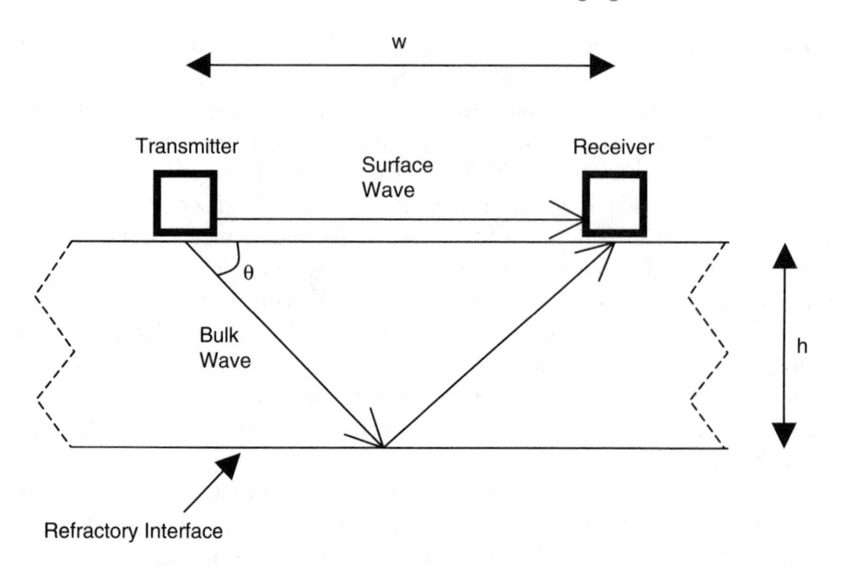

FIGURE 12.7 Pulse–echo arrangement for ultrasonic thickness measurement.

The separation between the transmitter and the receiver should be sufficiently large such that the bulk wave reaches the receiver before the surface wave. Fiorillo et al. [24] have shown that the best results were obtained when the separation was chosen such that the transmission angle (θ) was between 40° and 50°. If w is not sufficiently large, the higher order derivatives of the measured signal must be analyzed to identify the arrival of the bulk wave, which will be masked by the surface wave [29].

Given access to the refractory material, the ultrasonic technique may be used to take periodic measurements, where the transmitter and receiver are moved over various positions on the vessel wall to take an array of point measurements. Alternatively, a fixed array of transmitters and receivers may be permanently installed on the vessel's walls.

An alternative to using an ultrasonic transmitter is the rebound hammer method, shown in Figure 12.8. A hammer is struck against the surface of the refractory to produce an impact elastic wave. This wave is reflected off the interfaces in the material and detected using a receiver, just as in the pulse–echo technique. A second receiver is positioned to detect the time the hammer strikes.

Fujiyoshi et al. [28] found that an impact elastic wave was able to penetrate 1.2 m into a carbon block with little damping, whereas ultrasonic waves of 1 MHz and 200 kHz were unable to penetrate into the carbon block further than 0.3 m and 0.7 m respectively. Bell et al. [25] report that with the rebound hammer technique, refractory thickness can be measured to an accuracy of ±15%. Fujiyoshi et al. [28] performed refractory thickness measurements on an operating blast furnace (BF) using the impact elastic wave method and compared the results with thickness calculations from thermocouple measurements in the BF wall. The impact elastic wave method underestimated the thickness of the BF wall by 100 to 200 mm

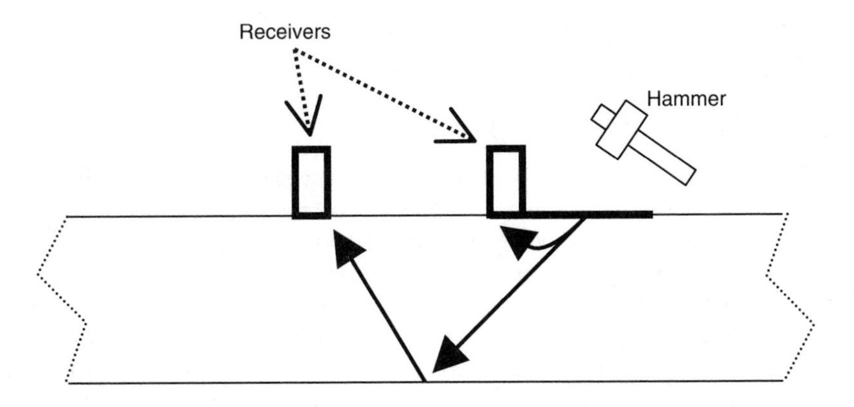

FIGURE 12.8 Rebound hammer arrangement for ultrasonic thickness measurement.

compared with the thickness calculated by the thermocouple measurements. A bore of the BF was taken when it was blown out (at the end of its campaign life), revealing that it comprised a 60 mm thick mantle outer layer, followed by a 110 mm thick layer of ramming material, a 730 mm thick carbon refractory brick, and finally a 200 mm thick layer of cracked carbon brick intermixed with accretions. The presence of this layer of cracked carbon brick and accretion accounts for the difference between the thicknesses calculated by the ultrasonic and thermocouple techniques. The ultrasonic wave detected would have been a reflection from the interface between the cracked and uncracked carbon refractory, causing the thickness of the cracked carbon and accretion layers to be ignored. The thickness of the cracked carbon and accretion layers would have been taken into account using the thermocouple measurements.

Since there is a significant temperature gradient across the refractory, dependence of the bulk wave velocity on temperature must be known for accurate thickness calculations to be made. The ultrasonic velocities through zircon brick and soda–lime glass have been shown to have a temperature dependence [26], whereas the velocity through carbon brick was found to have a negligible dependence on temperature over the range 20 to 1200 °C [28].

Refractories constructed from multiple layers increase the complexity of these measurements, as there is reflection of ultrasound from each interface. This also serves to attenuate the signal. The geometry of the layered media and the velocity of the bulk wave through each of the materials must be known in order to identify the signals reflected from each of the interfaces and calculate the overall thickness of the refractory. Obtaining sensible measurements for multiple layer systems is notoriously difficult. Fiorillo et al. [30] were able to apply this technique to measure the thickness of refractory lining of torpedo cars. The cars consisted of a steel shell lined with two different layers of refractory material. The thickness of the refractory could be measured using this technique with an associated error of less than ±10%. Bell et al. [25] were able to measure the thickness of a range of refractory slabs which were 15 to 25 mm thick and made from high-alumina brick, kiln bat, or fired magnesite. With the ultrasonic method, the measured thicknesses were all within 4 mm of the true values.

12.3.3 Laser Furnace Inspection

Laser monitoring of refractory lining thickness has been used since the early 1980s [31] and is a widely used technique for batch type furnaces [32]. Laser monitoring can be divided into two types: point measurement systems and scanning type systems. The point measurement system, as illustrated in Figure 12.9, involves a laser directed toward an individual point on the refractory surface of an empty vessel. A short laser pulse is transmitted, and the time for the pulse to reach the refractory and then reflect back to the detector is measured. From this propagation time, the distance between the sensor and the refractory can be calculated, and, hence, the thickness of refractory at that point. An image of the furnace refractory profile can then be established by taking thickness measurements over an array of points. Industrial applications of such systems have reported accuracies of ±5 mm for these point measurements on BOF steel converters [33, 34].

To measure the refractory profile over the entire vessel, many point measurements must be made. A scan of a converter at Rautaruukki Steel using two measurement heads takes 5 minutes, producing <2000 measurements [33]. The scanning type systems employ a line of multiple points, allowing measurements to be acquired more quickly, with hundreds of thousands of point measurements being taken in a 2-minute scan [33]. More recent systems with a fixed sensor head and rotating mirror systems to scan an array of points over the refractory surface are capable of taking 200,000 measurements in 20 seconds [33, 35]. The ability to acquire such a large number of data points quickly allows detailed images of the refractory profile to be reconstructed.

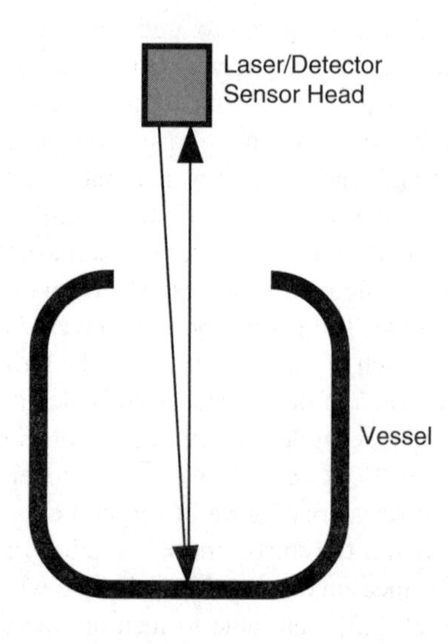

FIGURE 12.9 Laser point measurement system.

Laser systems require a direct line of sight between the sensor head and the refractory surface. For a scan of the surface of the refractory, the vessel must be empty. Such opportunities are common for vessels such as BOF converters, ladles, and torpedoes where this technique has been widely applied [32, 33, 35]. Laser systems have also been used to measure the refractory or gunning material thickness of blast furnace shafts during shutdowns (P. Zulli, personal communication, 2003).

Vessels with small openings may not afford the sensor head a direct line of sight to the whole surface of the refractory, requiring the sensor head to be moved to several positions in order to complete a scan. Smoky and dusty environments also present difficulties to the transmission of the laser light. The measurement system can identify when smoke or dust causes unacceptable interference to individual measurements. In this case, the system interpolates between usable data points [33].

12.3.4 Comparison of Techniques

In this application, ultrasonic, laser, and thermal imaging techniques have similar accuracy, of the order of 10 to 15% of the refractory thickness. The laser technique allows a great number of point measurements to be taken in a short space of time, allowing very detailed wear profiles to be reconstructed. This ensures that localized regions of wear will be detected. The major drawback for the laser method is that it requires the vessel to be empty. The opportunity to scan the empty vessel does not occur for a significant number of important metallurgical vessels.

Neither thermocouples nor ultrasonic monitoring require an empty vessel and so are more practical techniques for measuring the wear profile of vessels. Both techniques, however, acquire relatively few point measurements compared to the laser technique. Ultrasonic inspection does not necessarily rely on a steady state heat transfer through the vessel, but it may require knowledge of the thermal profile. Monitoring the refractory temperature with thermocouples provides not only the wear profile of the vessel but also valuable information about the processes occurring within it. Thermocouple data can also be used to validate heat transfer models of the process [36].

12.4 FLOW MEASUREMENT

Many examples of flow can be observed throughout various stages of the production of metals. Flow regimes influence mixing, dissolution, chemical reaction, and heat transfer, as seen in processes such as casting, pouring, injection, and operation of burners. The flow regime may involve only a single phase, as in pouring of steel from a ladle to a tundish, or it may involve multiple phases, as in the case of injection of solid feed into a melt.

The ability to monitor such flow regimes provides understanding of the mixing, dissolution, chemical reaction, and heat transfer. Understanding of the process may be used to make modifications to the process equipment design, feed rate, feed specification, and operating procedures that can result in improved productivity, increased life of equipment, improved product quality, and reduced energy consumption.

Many of the important flow regimes encountered are very difficult to monitor directly with existing sensor technologies due to the fluids being opaque and the high temperatures present. To gain an understanding of these flows, lab-scale water models are often constructed. These may be used as a stand-alone investigation or used to validate computational models of the process. Particle image velocimetry (PIV), described in Chapter 8, has been extensively used to measure 2D flow fields of such models [37, 38]. Experimental modeling employing PIV has been used to investigate continuous casting of steel and aluminium [37, 39–41], hot dip galvanizing [42], flash smelting burners [43], and oxygen lancing.

12.4.1 Monitoring of Flow Regimes in the Continuous Caster

Continuous casting of steel is a key process in the steel industry; the majority of the world's annual production of 635 million tonnes of steel is produced via this process [44]. As shown in Figure 12.10, molten steel in the tundish flows though a submerged entry nozzle (SEN) into the water-cooled copper mold. The steel exits from the bottom of the mold with the edges of the slab having solidified to form a solid strand. The slab then undergoes further cooling and rolling to solidify the remainder of the steel and form the slab into the required shape.

Substantial work has been undertaken in the study of the continuous casting process with the aim to optimize the process and to improve the quality of the

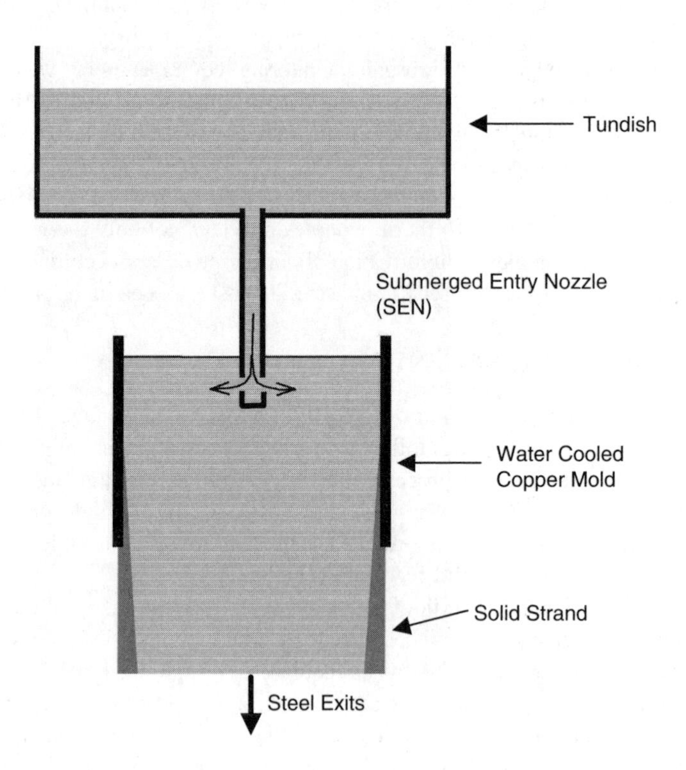

FIGURE 12.10 Schematic of a continuous steel caster.

steel cast. Unfavorable flow regimes in the continuous caster can cause entrainment of the mold flux or of argon bubbles. These inclusions cause defects in the steel that is cast. Computer models of the flow patterns in the mold have been validated with PIV measurements (see Chapter 3) performed on water models of the system. A comprehensive review of continuous casting, and modeling that has been performed to improve understanding of its operation, is provided by Thomas [45].

A PIV water modeling study of the continuous casting mold was undertaken at LTV Steel [46]. In the course of the study, measurements were also taken on an industrial continuous casting mold using a mold flow control sensor (MFC), which is being developed by AMEPA. A magnetic field is set up across the mold, using permanent magnets, as shown in Figure 12.11. Flow of the electrically conductive steel through the magnetic field causes currents to be induced in the steel [47]. The effects of these induced currents are detected by pairs of magnetic field sensors, which are separated by a known distance. Inhomogeneities in the flow cause peaks in the sensor measurements. The velocity of the steel flow is determined by the time taken for the inhomogeneities to move from one sensor to the next, using software to identify matching peaks from the sensors' measurements [48]. The MFC sensors are capable of measuring velocities of 10 to 100 cm/s, though no claims as to their accuracy have been reported in the literature.

Results from PIV measurements performed on the water model compared well with MFC measurements on the continuous caster's mold when there was no gas injection with the liquid flowing through the SEN into the mold [46]. In practice, argon is injected into the SEN to reduce nozzle clogging [49]. At high argon flow rates, the steel flow patterns measured using the MFC sensor were shown to be unstable, while the water model, which used either air or argon injection, yielded reproducible results showing no such instability in the flow regime [46]. In this case, the water modeling was not adequate to capture the complexity of this process. Other PIV water modeling and computational simulations have been able to capture such instabilities in the flow regimes in the continuous caster mold [50]. Plant measurements using sensors, such as the MFC sensor, measure what is actually happening in the process.

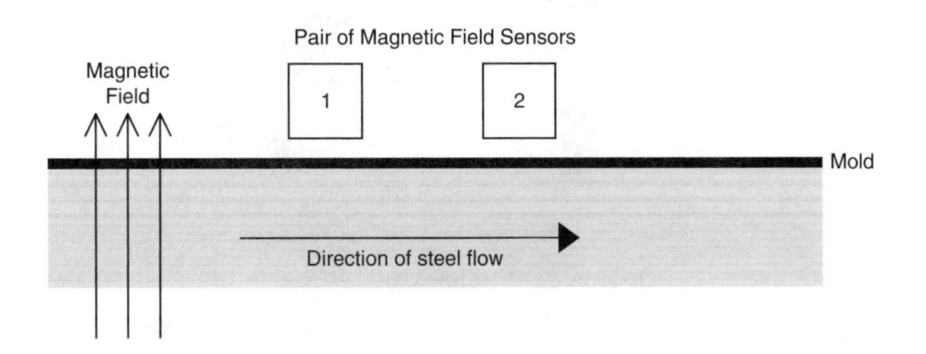

FIGURE 12.11 Arrangement for the mold flow control (MFC) sensor.

MFC measurements have allowed the identification of different flow regimes within the mold. This has enabled research to correlate the different flow regimes to the quality of steel cast to identify favorable flow regimes in the mold. The relation between the different flow regimes to the control variables (cast speed, slide gate position, argon injection rate, SEN depth, etc.) has also been examined. Thus the MFC can be used to provide feedback to control the continuous caster, where control variables are adjusted to promote a favorable flow regime in the mold, resulting in better quality of the cast steel.

The operation of the continuous caster is also influenced by the flow pattern within the SEN itself. The flow of the steel from the tundish through the SEN into the mold is controlled by a stopper or a slide gate at the top of the SEN that restricts the flow. Other factors that influence the flow pattern within the SEN are the level of steel in the tundish, the amount of wear the SEN has experienced, the argon injection rate, and any clogging the SEN may be experiencing [49]. Potential flow patterns within the SEN include annular flow, a central stream, bubbly flow, or flow down one side of the nozzle. The bottom section of the SEN may be completely filled with molten steel while the top section is only partially filled, or the SEN may be completely filled with steel.

There has been substantial physical (water) and computational modeling of the flow regimes within the SEN [50–52], but information about the flow of molten steel through a SEN is still lacking. In the past, it was not possible to measure flow directly in the process. A collaboration between Lancaster University, Metal Process Control, and Corus Ltd. is developing an electromagnetic tomographic (EMT) system to image the distribution of steel within the SEN [53]. Ultimately, this system could be used to control the slide gate or stopper to achieve the desired steel distribution.

The EMT system uses an array of coils surrounding the SEN, as shown in Figure 12.12. Cold modeling has been undertaken to show that such an array of coils is able to detect different distributions of steel within the SEN. In these experiments, Woods metal (an alloy with a melting point of 70°C) was used to

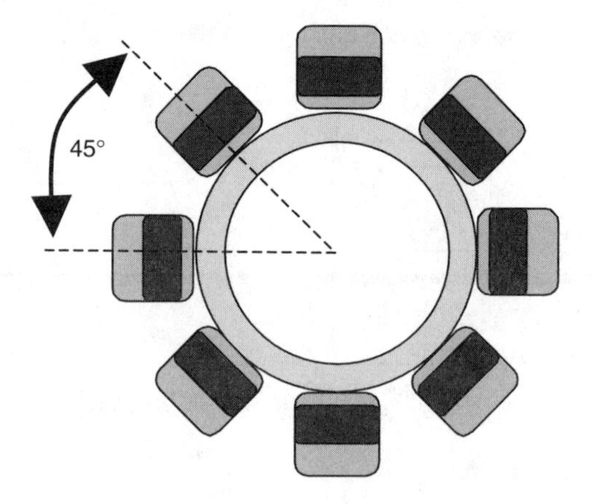

FIGURE 12.12 Arrangement of the eight coils around the SEN.

represent the molten steel. By measuring the coil responses to excitation frequencies over a range of 100 to 1000 Hz, it was possible to distinguish between annular flow, bubbly flow, a central stream, and an empty SEN [53].

Further cold work was performed to produce sensitivity maps for each pair of coils in the array. The area within the SEN was mapped with a grid of 10 mm × 10 mm pixels. The impedance change for each coil pair was measured when a copper rod was positioned over each pixel. The image reconstruction problem is detailed by Ma et al. [54]. This work falls into the category of iterative techniques based on linear back-projection, as discussed in Chapter 4. Images produced from measurements taken with metal bars (19 mm, 25 mm, and 38 mm in diameter) showed it is possible to distinguish between bars of different sizes.

Hot trials were conducted with a quartz tube placed off-center within an SEN. The coil array positioned around the SEN was contained within a nonmetallic air-cooled box to protect the coils from the intense heat. Four tonnes of molten steel was poured through the quartz tube, with the flow halted twice during the pour. The measured signals clearly showed these two halts to the flow (as shown by the two black images in Figure 12.13) but also showed significant drift, which

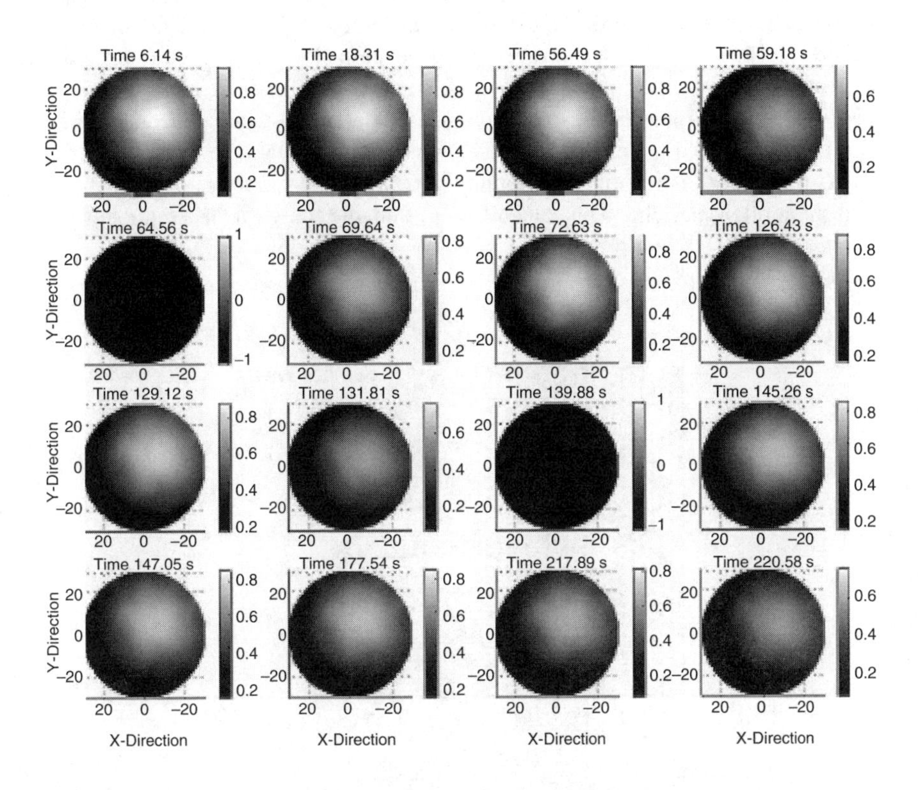

FIGURE 12.13 Images of molten steel flow within SEN. (From Ma X., Peyton, A.J., Binns, R. and Higson, S.R. *3rd World Cong. Industrial Process Tomography,* 736–742, 2003. With permission.)

was attributed to a combination of sensor electronic drift and the changing temperature of the support materials. The thermal drift effect was approximated by an exponential curve that was subtracted from the measured signal. Reconstructed images of the steel distribution, using this compensated signal, are shown in Figure 12.13; the steel stream is clearly off-center.

The EMT system can monitor the flow regimes that actually occur in the SEN of continuous casters. This approach will help to establish the impact on flow regimes of changes in process variables such as stopper or slide gate position, argon injection rate, and the metal height in the tundish. These data will provide a valuable comparison with results of computational and water models. Ultimately the system could be used to control the stopper or slide gate of the continuous caster.

12.4.2 Electromagnetic Monitoring of Flow through Tapholes

In another application of EMT, a tomographic system is being developed at the University of Melbourne to monitor the flow of molten materials though tapholes [14, 55–57]. Tapholes are regions of metallurgical furnaces that are periodically opened and closed to facilitate the removal of molten products, as depicted in Figure 12.14. This technique was originally developed to monitor the freeze layer of a molten metal solidifying within a cooled taphole. The shape of the boundary between the solid and liquid metal phase is determined from impedance measurements using coils positioned in an array around the taphole.

Current work is locating the boundary between the refractory of the taphole and the molten material. This capability will enable the wear of a taphole to be

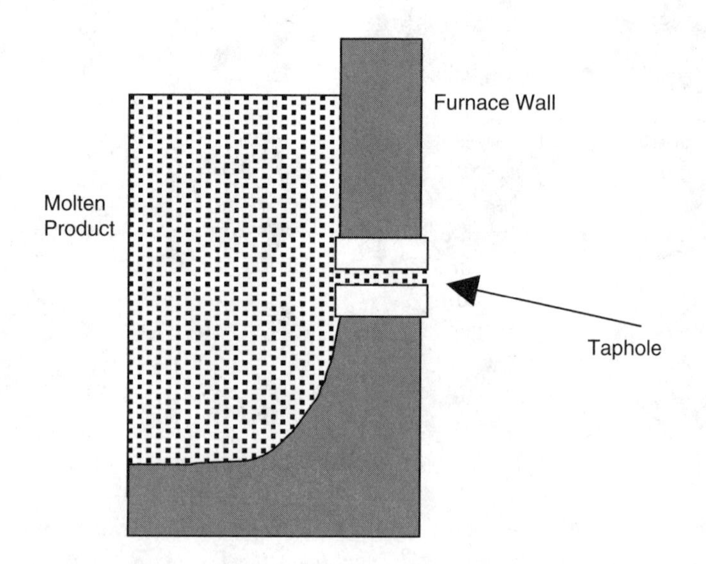

FIGURE 12.14 Taphole region of a metallurgical furnace.

monitored and will give a clear indication of when it requires rebuilding. The process of rebuilding is labor-intensive and costly, since it requires the furnace to be idle. Taphole rebuilds must be performed on a regular basis to prevent failure, which presents a significant hazard to operator safety and furnace integrity [58]. At present, the decision to rebuild a taphole is based on a "rule of thumb" such as a set throughput or a certain number of openings of the taphole. However, there is great variation in the wear rate. The ability to monitor the condition of the refractory itself improves safety and reduces furnace downtime. Quantitative wear rate information could be utilized to modify operating practices to extend taphole life.

Pham et al. [57] have developed a 3D mathematical model to describe the relationship between the geometry within the taphole and the response induced in electromagnetic sensing coils. The model uses integral equations that are solved numerically with the moments method. An iterative algorithm is used to solve the inverse problem of predicting the boundary geometry from the measured responses.

A measurement system for this application has been developed, based around a data acquisition board controlled by a National Instruments LabVIEW™ program. Cold measurements taken using this system have indicated that it would be sufficiently sensitive to detect a 2.5-mm change in diameter of a taphole [14]. The cold measurements are used in conjunction with a commercially available finite element method package (Maxwell 3D™ from Ansoft Corp.) to validate Pham's mathematical model. This system was installed in an industrial taphole to demonstrate its operation. This wear monitoring system may also offer the potential for entrainment detection.

12.4.3 Imaging of Flow in Ladle Stirring

Ladles serve not only as transfer vessels for molten products but also as reaction vessels for further metallurgical processing, especially in the steel industry. Fluxes and alloying materials may be added to achieve the required chemistry and homogeneity of the metal and slag phases. To allow the necessary chemical reaction to take place, ladles are stirred with argon (or sometimes nitrogen). Stirring is achieved by argon either being injected though a porous plug at the bottom of the ladle (see Figure 12.15) or injected through a lance from the top of the vessel [59]. The raised region where the gas discharges is known as the spout and the area within the spout that is not covered by slag is known as the spout eye. Stirring must be sufficient to enable the required chemical reactions and mass transfer processes to occur. The stirring of ladles with argon requires a considerable input of energy. Overstirring can result in increased oxide inclusions (slag in steel) due to surface turbulence, reoxidation of steel due to increased air–steel contact, loss of product due to splashing, excessive argon consumption, and difficulties in heating with electrodes. Furthermore, ladle life is decreased due to arc flaring and decreased reheating ability resulting from arc instability [60].

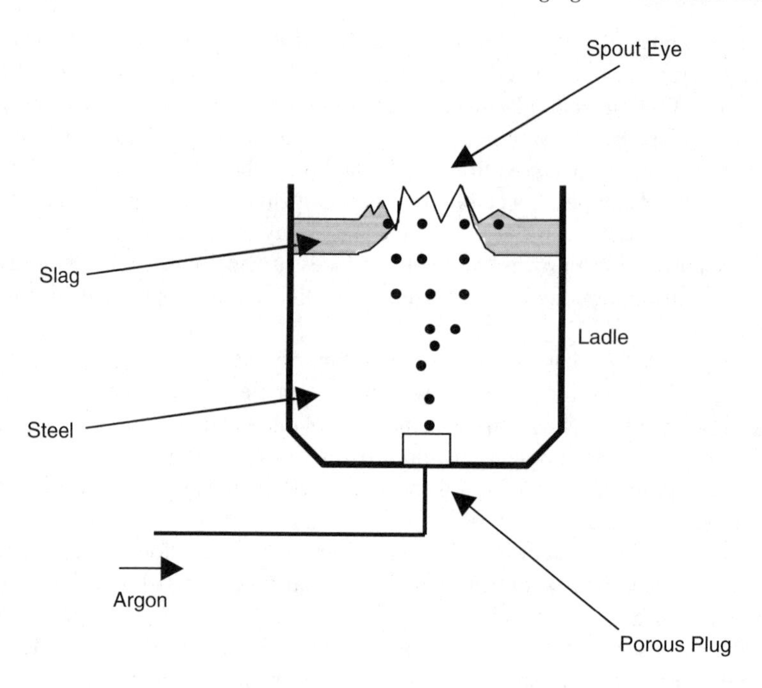

FIGURE 12.15 Bottom stirring of a ladle.

The argon flow rate is used to control the rate of stirring. Control of flow rate has previously been achieved by either monitoring the back-pressure in the argon line or by visual observation of the spout eye. The first of these measures is unreliable since accretions that may build up on the porous plug will partially block the plug, increasing the resistance to flow of gas and possibly redirecting the flow [60]. Additionally, the pores of the plug may wear over time, reducing the resistance of the plug to the flow of gas [61]. Control of flow rate by operators observing the spout eye is subjective, and their view of the spout eye is often hampered by smoke generated during reheating [60]. Operators tend to be more concerned about the steel leaving the ladle meeting the required specifications than minimization of stirring, which leads to a culture of over-stirring the vessel.

Yonezawa and Schwerdtfeger [62, 63] investigated the relationship between the volumetric flow rate of gas (Q_b) into the ladle and the average area of the spout eye (A_{es}) and height of the spout (h). Cold laboratory scale experiments were carried out using a mercury bath with an oil layer to represent the slag, along with industrial measurement on 120-tonne and 350-tonne steel ladles. In the cold experiments, the area of the spout eye was measured with a CCD camera [62] and the height of the spout was measured with a resistivity probe [63]. The geometry of the eye was found to be highly dynamic, and so an average size of the eye was used. Nondimensional analysis of the measurements produced correlations relating a dimensionless number (A_{es}/hH), involving the spout eye area, to a function of the Froude number (Q_b^2/gH^5) or (Q_b^2/gH^4h), where H is the height of the slag layer and g is the gravitational acceleration.

These correlations gave a reasonable fit to the measured data, although there was some discrepancy between the hot and cold data. Subagyo et al. [64] analyzed the same data to give the relationship

$$\left(\frac{A_{es}}{(h+H)^2}\right)=(0.02\pm0.002)\left(\frac{Q_b}{gH^5}\right)^{0.375\pm0.0136} \tag{12.4}$$

This empirical result gives a better fit to the measured data (with a standard deviation of 0.169 compared with 0.311) than Yonezawa's correlation involving the term (Q_b^2/gH^5). These correlations are limited to ladles of similar geometry to those used in the experimental work. Subagyo and Brooks [65] are currently using this relationship to develop an image analysis system that will measure the spout eye area to control the argon flow rate into the ladle.

The area of the spout eye can be determined by multivariate image analysis, as detailed by Geladi [66], Bharati and MacGregor [67], and Subagyo and Brooks [65]. The image of the spout eye is captured with a CCD camera. The image of $n \times m$ pixels is represented in the RGB (red/green/blue) color format, wherein each pixel is specified by three values. The image is unfolded to a 2D ($m.n \times 3$), and a singular value decomposition is performed [68]. This operation produces vectors that classify each pixel in the image as bare metal, thin slag, fluid slag, or crusty slag, where the total number of bare metal pixels is proportional to the spout eye area.

Equation 12.4 is highly sensitive to the thickness of the slag layer (H), which is poorly known in practice. H varies with time due to addition of fluxes and alloying materials and due to chemical reaction occurring within the ladle. Overestimating the slag thickness by 32% leads to the volumetric flow rate being overestimated by 100% in Equation 12.4. In the experimental work on which this correlation is based, Yonezawa and Schwerdtfeger [62] state that "the slag layer thickness was about 50 mm" and explain that the slag thickness may have been 40 mm thick, which would make the correlation they put forward match the hot data more closely. Meszaros et al. [9] state that the method used for slag height measurement has an associated error of ±30 mm. None of the authors investigating the spout eye area as a means of controlling the gas flow rate has suggested a means of measuring slag thickness, so extreme caution must be adopted when using these correlations.

Yonezawa and Schwerdtfeger [63], following the relationship between the spout height and the stirring intensity, developed a simpler correlation:

$$\Delta h_{\max}=1.94Q_b^{0.4} \tag{12.5}$$

where $\Delta h_{\max}$ is the maximum spout height (in cm), and Q_b is in l/min. The great advantage of this relationship is that the flow rate is dependent only on the spout height and not on poorly known variables such as the slag height. There is good agreement between this relationship and cold measurements taken using a mercury bath, as well as measurements taken on 120-tonne and 350-tonne industrial ladles.

However, at low argon flow rates, the relationship overestimates the spout height. In this case, the Froude number is defined as (Q_b^2/gd^5), where d is the diameter of the porous plug; the Froude number is <0.1 for low argon flow rates.

Alternatively, the long-term average spout height (Δh_{LT}) could be used in place of Δh_{max}. In cold modeling by Yonezawa and Schewerdfeger [63] using the mercury bath, it was found that $h_{max}/h_{LT} \approx 3.16$. Substitution into Equation 12.5 yields the following correlation:

$$\Delta h_{LT} = 0.61 Q_b^{0.4} \tag{12.6}$$

This result compares well with the correlation of Guo and Irons [69] based on their water modeling:

$$\Delta h_{LT} = (0.63 \pm 0.022) Q_b^{0.491 \pm 0.012} \tag{12.7}$$

The spout height was determined by suspending a steel tube above the ladle prior to the commencement of argon stirring, the end of the steel bar being level with the surface of the fluid. The maximum value of spout height was determined by the length of pipe dissolved by the spout. Though useful in these experiments, this technique clearly could not be applied to measuring spout heights for control of stirring during normal operation of a ladle. This measurement is a potential application for the laser systems discussed earlier.

Vibrational monitoring offers an alternative means of monitoring ladle stirring. Measurements using accelerometers have shown that vibrations in the frequency range 50 to 90 Hz are proportional to the stirring intensity in the ladle [60, 61], with the accelerometer being mounted either directly on the ladle (such as the ladle lid hinge) or on a structure pressed against the ladle (such as a side wall sensor arm). The frequency range suggests that the vibrations are not due to bubbles bursting at the surface, which would produce signals in the 10-Hz range [60], but rather they are the characteristic harmonic vibrations of the individual ladle [60, 61]. The control system must be made insensitive to events such as the bursting of bubbles on the surface, to avoid erroneous changes in stirring intensity. Systems using the vibration technique of monitoring ladle stirring have been developed separately by Nupro Corp. and Stelco Inc. Both systems are used commercially [60, 61].

12.5 IMAGING SOLID STATE PHASE CHANGE

Rolled steel sheet is a significant commodity; the U.S. market consumed more than 51 million tonnes in 2001 [70]. Rolled steel sheet is formed in a hot strip steel mill, where steel slab is heated to approximately 1200 °C before it is passed through a series of rolling mills to produce steel sheet of the required thickness (see Figure 12.16). The steel sheet, now at around 900 °C, passes out onto a runout table equipped with an array of top and bottom cooling sprays, which reduce the steel's temperature to 500 °C before the sheet is fed onto the coiler. The cooling of the steel sheet along the runout table controls the microstructure and the resulting physical properties of the final product.

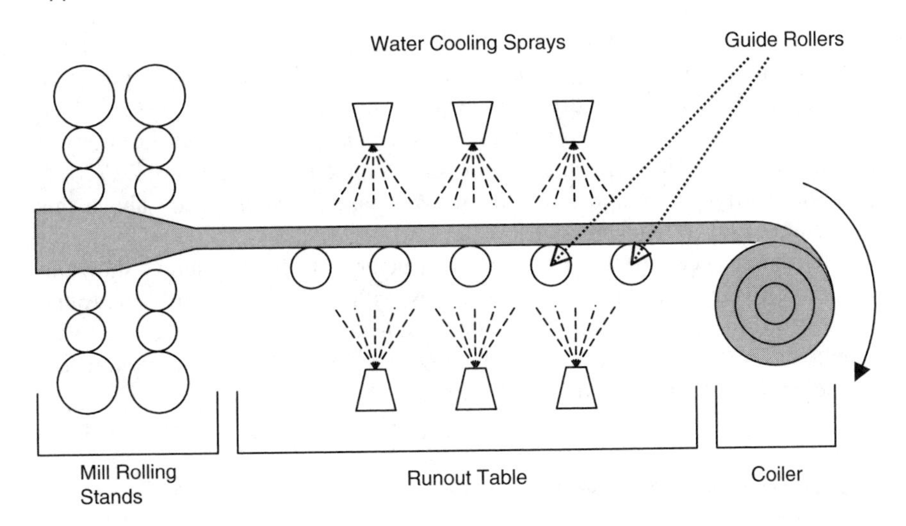

FIGURE 12.16 Hot strip mill.

Steel type is determined by the carbon content in the iron as well as by the level of alloying components (i.e., Mn, Ni, Mo, Cr, Ti, V, etc.). Steel sheet passing out onto the runout table at around 900 °C is in the form of austenite (see Figure 12.17), a high-temperature paramagnetic face-centered cubic phase. As the steel strip is cooled, the steel undergoes a solid-state phase transformation to form

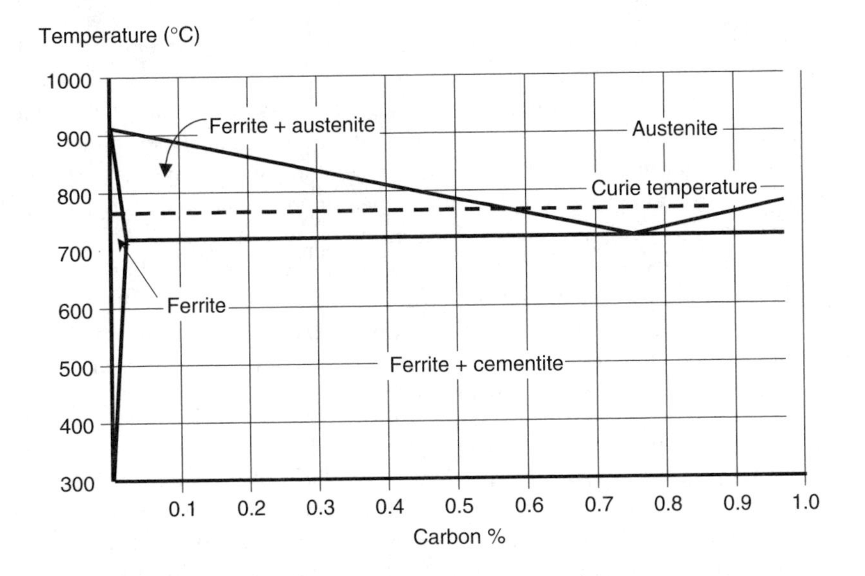

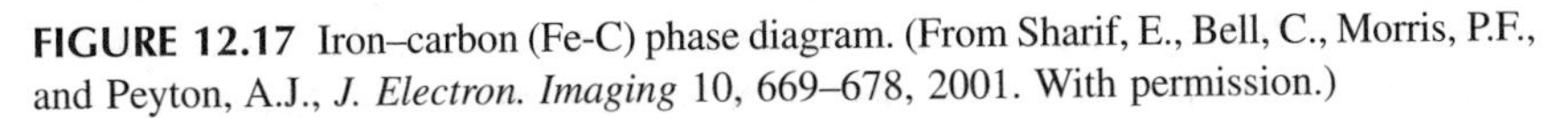
FIGURE 12.17 Iron–carbon (Fe-C) phase diagram. (From Sharif, E., Bell, C., Morris, P.F., and Peyton, A.J., *J. Electron. Imaging* 10, 669–678, 2001. With permission.)

a mixture of a body-centered cubic ferrite phase and a cementite (Fe_3C) phase, with the ferrite phase accounting for the majority. The cooling rate and composition control the dispersion of the cementite phase through the ferrite phase, which influences the physical properties of the steel such as yield stress, tensile strength, and ductility [71]. The Curie temperature for this material is approximately 770 °C [72], depending on composition; the ferrite phase is paramagnetic above this temperature and ferromagnetic below it.

To obtain steel with the desired microstructure, the cooling sprays shown in Figure 12.16 are divided into numerous zones, with each zone carefully controlled to achieve the necessary cooling rate and cooling uniformity across the steel strip. The control of these zones of cooling sprays relies on the ability to monitor the temperature distribution of the steel strip as it cools.

12.5.1 Infrared Pyrometry

Infrared (IR) pyrometers have been used to monitor the cooling of steel strip since the late 1970s [73], and their use is now commonplace. Pyrometers measure the steel temperature at a single point. Initially they were used only to monitor the temperature of the steel strip passing into and out of the cooling section of the runout table, but over time it has become common to include additional pyrometers along the cooling section.

Pyrometers operate by collecting the IR radiation, using either an optical lens system focused on the object of interest or an optical fiber system (see Chapter 3). The IR radiation is transmitted to the detector, which operates in either single-color or dual-color mode. A single-color detector measures the intensity of a narrow band of the IR spectrum. This intensity is related to the temperature of the object of interest through a calibration function, which requires the emissivity of the object to be known. The dual-color detector measures the intensity of the IR radiation at two distinct wavelengths. The ratio of these intensities, rather than the absolute intensities, is related to the temperature of the object of interest. Changes in emissivity significantly affect measurements taken in single-color mode but have less effect on dual-color measurements [74].

External factors such as airborne dust, water droplets, or steam in the line of sight between the pyrometer and the object of interest absorb the IR radiation and adversely affect measurements. This limits the effectiveness of pyrometers that are installed along the cooling section of the runout table, due to the large quantities of water droplets and steam obscuring the surface of the steel.

12.5.2 Electromagnetic Imaging

Various techniques to monitor the solid state transformation of steel have been used on a laboratory scale, including laser ultrasonics, conventional ultrasonics [75], x-ray diffraction, and electromagnetic techniques [76]. Of these methods, only the electromagnetic techniques have been taken beyond the laboratory scale;

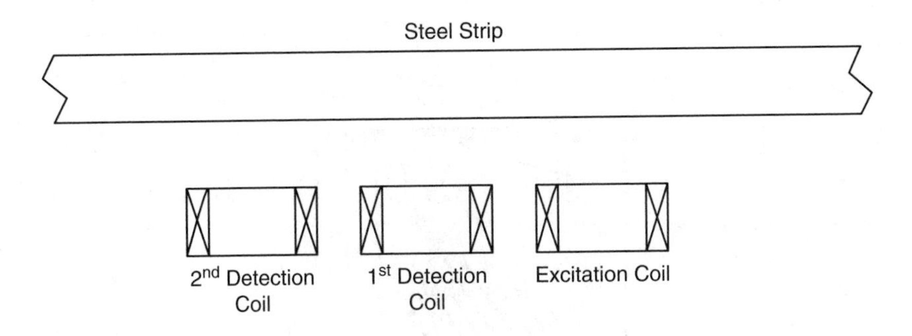

FIGURE 12.18 Sensor element arrangement used by Yahiro et al.

they are sensitive to the changing magnetic condition of the material as it cools. The electromagnetic techniques have several potential advantages: the sensors are noncontacting, and the presence of water, steam, or dust has no effect on the measurements. To achieve sufficient penetration of the magnetic field into the steel, electromagnetic sensors are operated at relatively low frequencies, of the order of 2 Hz [76].

At a production scale, electromagnetic sensors were installed at the Mizushima Works Hot Strip Mill [77]. A linear array of eight electromagnetic sensors was positioned between the guide rollers of the runout table and used in conjunction with the existing pyrometer system to monitor the cooling of the steel strip and to control the water sprays of the 24 cooling zones along the runout table. The electromagnetic sensors consisted of one excitation coil and two detection coils (see Figure 12.18), which allowed compensation to be made for the variation in distance between the hot strip and the sensor (lift-off). The use of the electromagnetic sensors in conjunction with pyrometry allowed the temperature of the strip to be controlled to ±20 °C for 97.8% of the time, compared with 90% when pyrometry alone was used. This improved control resulted in a 15% reduction in the deviation in the physical properties of the final product.

The use of arrays of electromagnetic sensors has been extended to two dimensions by Sharif et al. [71]. This allows variations in strip temperature to be detected not only along the length of the strip but also between the edge of the strip and the center. Each sensor consists of a 1500-turn excitation coil wound around a silicon steel core with Hall effect field detectors at each end of the core (see Figure 12.19). The Hall sensors have a sensitivity of 2 V/mT to changes in the magnetic field, with a background signal of approximately 2 V [78]. The design of the sensor did not allow for lift-off compensation to be made. The steel sample used was sufficiently flat to keep variations in the lift-off to less than 1 mm, which equated to less than 5% of the total change in signal between the untransformed and completely transformed case. The thermocouple included in the sensor allowed the temperature of the coil to be monitored, which made it possible to compensate for the varying coil temperature. This reduced the drift

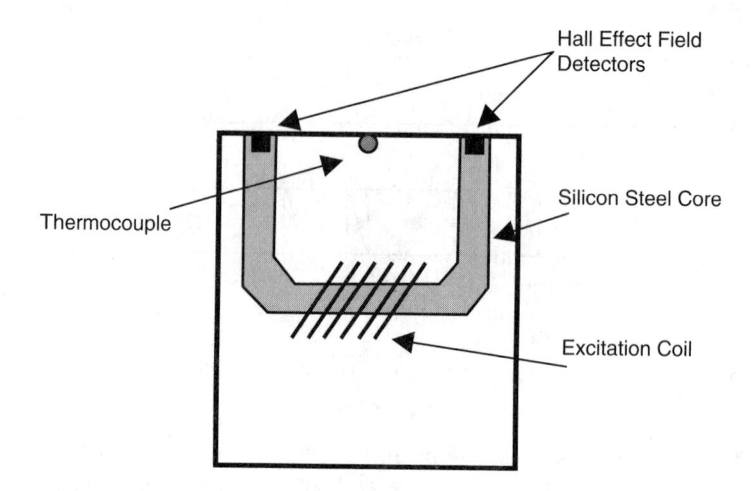

FIGURE 12.19 Sensor element used by Sharif et al.

in the signal due to the changing coil temperature to less than 5% of the total change in signal between the untransformed and completely transformed case.

An array of seven sensors was positioned between the guide rollers of the runout table. A sample of low carbon steel sheet (C 0.077%, Si 0.27%, Mn 1.64%, Ti 0.111%) heated to >1100 °C was passed over the sensor array. The sample was then quenched with water from one end and repeatedly passed over the array until the sample fully transformed. The image of the steel sheet was produced using five of the seven sensors, arranged with one row of three sensors and one row of two sensors. The sensors could sample at a rate of 6 samples per second [71], allowing 11 samples to be taken for each pass of the steel sheet and giving a 5×11 matrix of measurements for each pass. Each of these measurements was then scaled relative to the measurements taken for the untransformed steel sample and the fully transformed steel sample to provide an estimation of the transformed fraction. Each of the 5×11 transformed fractions could then be mapped to its position on the steel sheet. A cubic spline interpolation was used to produce an image of the steel sample. The images produced showed profiles of the steel sheet transformation consistent with the cooling applied to the sheet. It should be noted that the above method of scaling may not be practical in a production situation, unless it is possible to perform calibration runs beforehand.

The use of electromagnetic transformation detectors in combination with pyrometry has provided a more accurate measurement of the solid state phase change of the steel strip as it passes through the cooling section. This improvement in monitoring the transformation of hot rolled steel strip has made it possible to achieve higher quality. Although the use of IR pyrometers alone to monitor the cooling of steel strip is well established and widespread, the improved control afforded by the electromagnetic methods discussed here makes it attractive to use both techniques.

12.6 CONCLUSION

This chapter has outlined a number of process imaging systems and associated instrumentation systems that have been successfully applied to monitor processes in the metals production industry. In some cases, these systems have been incorporated into process control systems; others are used for intermittent inspection or monitoring. Systems that are under development for new applications in the field were discussed. The industry's need for imaging technologies that are able to withstand harsh environments was highlighted, along with some promising new methods. As better sensors and faster computers become available, progress will be made toward meeting additional applications. It is clear that the metals production industry presents considerable engineering challenges, many of which can be met by image-based control systems.

REFERENCES

1. T Rosenqvist. *Principles of Extractive Metallurgy.* 2nd ed. Singapore: McGraw Hill, 1983.
2. G Assaad. Entrainment in two-layer fluid flowing over a weir. Ph.D. thesis, University of Melbourne, Australia, 2001.
3. SW Ip, JM Toguri. Entrainment of matte in smelting and converting operations. *39th Annual Conference of Metallurgists of CIM,* Ottawa, 2000, pp. 291–302.
4. D Goldstein, A Sharan, J Stofanak. Infrared imaging for BOF slag detection. *Iron Steelmaker* 27:331–38, 2000.
5. D Goldstein, A Sharan, E Fuchs. System and method for minimizing slag carryover during the tapping of a BOF converter in the production of steel. U.S. Patent No. 6,129,888. October 10, 2000.
6. R Selin. *Vanadium in Ore-based Iron and Steelmaking.* LKAB monograph, November 2000.
7. H Rau, I von Röpenack. Applications of thermographic slag detection. *Metall. Plant Technol. Int.* 25:52–55, 2002.
8. JA Stofanak, A Sharan, DA Goldstein, EA Stelts. System and method for minimizing slag carryover during the production of steel. U.S. Patent No. 6,197,086. March 6, 2001.
9. GA Meszaros, JG Estocin, FL Kemeny, DI Walker. Development and application of the Depthwave™ slag depth measuring system. *AISE Steel Technol.* 79:20–23, 2002.
10. T Iida. *The Physical Properties of Liquid Metals.* Oxford: Oxford University Press, 1988.
11. J Machini, A Asselbom. Slag detection in the electric arc furnace at the Ascometal's Hagondange Works. *SEAISI Quarterly,* 27(2):52–57, 1998.
12. C Enstrom. MPC tools for casting control. In: *Process Control in the Steel Industry.* Vol. II, Lulea, Sweden, 11–12 Sept. 1986, 635–645.
13. E Fritz, H Grabner. A system for slag separation, slag coating and automatic tapping of the BOS converter. *Steel Times* 222(4):143–146, 1994.
14. J Coveney, A Peyton, M Pham, A Kyllo, N Gray. Development of an eddy current system for monitoring taphole operations. *3rd World Congress on Industrial Process Tomography,* Banff, Canada, 2003, pp. 828–833.

15. L Heaslip, J Dorricott. Method and apparatus for detecting the condition of the flow of liquid metal in and from a teeming vessel. U.S. Patent No. 5,633,462. May 27, 1997.
16. R Ardell, A Kursfeld. Process and equipment to determine disturbance variables when pouring molten metal from a container. U.S. Patent No. 5,042,700. August 27, 1991.
17. L Heaslip, J Dorricott. Liquid metal flow condition detection. U.S. Patent No. 6,539,805. April 1, 2003.
18. S Mazumdar, N Pradhan, P Bhor, K Jagannathan. Entrainment during tapping of a model converter using two liquid phases. *ISIJ Int.* 35:92–94, 1995.
19. P Hammerschmid, K Tacke, H Popper, L Weber, M Dubke, K Schwerdtfeger. Vortex formation during drainage of metallurgical vessels. *Ironmaking Steelmaking* 11:332–339, 1984.
20. A Rex, P Zulli. Determination of the state of the hearth of No. 3 blast furnace, Rod and Bar Products Division, Newcastle, BHP Steel. *19th Australasian Chemical Engineering Conference,* Newcastle, Australia, 1991, pp. 484–491.
21. K Shibata, Y Kimura, M Shimizu, S Inaba. Dynamics of dead-man coke and hot metal flow in a blast furnace hearth. *ISIJ Int.* 30:208–215, 1990.
22. R Bird, W Stewart, E Lightfoot. *Transport Phenomena,* 2nd edition. New York: John Wiley and Sons, 1960.
23. AJ Rex, P Zulli, P Plat, F Tanzil, A Skimmings, L Jelenich. Determination of the state of the hearth of BHP Steel's blast furnaces. *Proc. 52nd ISS-AIME Ironmaking Conf.,* Warrendale, PA, 1993, pp. 603–609.
24. AS Fiorillo, N Lamberti, M Pappalardo. A piezoelectric range sensor for thickness measurements in refractory materials. *Sens. Actuators A Phys.* 37–38:381–384, 1993.
25. DA Bell, AD Deighton, FT Palin. Non-destructive testing of refractories. *International Symposium on Advances in Refractories for the Metallurgical Industries II,* Montreal, 1996, pp. 191–207.
26. SM Pilgrim. High temperature materials characterization enhanced by nondestructive imaging. *Glass Res.* 11:42–53, 2001.
27. A V Vachaev. Utilisation of the physical properties of the refractories for measuring the thickness of a single layer lining of metallurgical furnaces. *Izv. V.U.Z. Chernaya Metall.* No. 1, 93–95, 1990.
28. S Fujiyoshi, M Inoue, H Kamiyama. The development of nondestructive measuring on refractories of lining. *Taikabutsu (Refractories)* 43:590–506, 1991.
29. AS Fiorillo, N Lamberti, M Pappalardo. Time of arrival discrimination of interfering waves for thickness measurements in composite media. *1992 Ultrasonics Symposium, IEEE,* 1992, pp. 871–875.
30. AS Fiorillo, N Lamberti, M Pappalardo. Thickness characterization of refractory linings by low frequency ultrasonic waves. *Proc. SPIE* 1733:175–182, 1992.
31. W Schwenzfeier, F Kawa. Laser techniques of measurement used in metallurgical industry. *Hutnik (Katowice)* 48:151–157, 1981.
32. MA Rigaud, RA Landy, eds. *Pneumatic Steelmaking, Vol. 3: Refractories.* Warrendale, PA: Iron and Steel Society, 1996, pp. 98–100.
33. LM Sheppard. Laser systems help improve refractories performance and productivity. *Refractories Applications* 3:15–16, 2000.
34. AK Chatterjee, BN Panda, C Mishra. Condition monitoring of oxygen steel making vessel by laser technique. *Proceedings of the 14th World Conference on Non-Destructive Testing,* New Delhi, 1996, Vol. 2, pp. 1073–1078.

35. R Lamm. High speed laser-scanning station with LaCam designed for noncontact areas of steelcasting ladle refractory linings at Corus Group Aldwarke Works, U.K. *84th Steelmaking Conference Proceedings,* Baltimore, 2001, pp. 417–426.
36. V Panjkovic, J Truelove, P Zulli. Numerical modeling of iron flow and heat transfer in blast furnace hearth. *Ironmaking Steelmaking* 29(5):390–400, 2002.
37. D Xu, WK Jones Jr., JW Evans. The use of particle image velocimetry in the physical modeling of flow in electromagnetic or direct-chill casting of aluminum: Part I. Development of the physical model. *Metall. Mater. Trans. B* 29:1281–1288, 1998.
38. JL Liow, N Lawson, G Laird, NB Gray. Experimental methods in gas injection. *The Howard Worner International Symposium on Injection in Pyrometallurgy, TMS 1996,* pp. 221–236.
39. GD Rigby, CD Rielly, SD Sciffer, JA Lucas, GM Evans, L Strezov. Hydrodynamics of fluid flow approaching a moving boundary. *Metall. Mater. Trans. B* 31(5): 1117–1123, 2000.
40. H Bai, BG Thomas. Turbulent flow of liquid steel and argon bubbles in slide-gate tundish nozzles: Part I. model development and validation. *Metall. Mater. Trans. B* 32:253–267, 2001.
41. WK Jones Jr., D Xu, JW Evans. Physical modeling of the effects of thermal buoyancy driven flows in aluminum casters. *Metall. Mater. Trans. B* 33:321–324, 2002.
42. SJ Lee, S Kim, MS Koh, JH Choi. Flow field analysis inside a molten Zn pot of the continuous hot-dip galvanizing process. *ISIJ Int.* 42:407–413, 2002.
43. ID Sutalo, JA Harris, FRA Jorgensen, NB Gray. Modeling studies of fluid flow below flash-smelting burners including transient behavior. *Metall. Mater. Trans. B* 29:773–783, 1998.
44. BG Thomas, Q Yuan, S Sivaramakrishnan, SP Vanka. Transient fluid flow in a continuous steel-slab casting mold. *JOM-e,* January 2002.
45. BG Thomas. Modeling of the continuous casting of steel: past, present and future. *Metall. Mater. Trans. B* 33:795–812, 2002.
46. MB Assar, PH Dauby, GD Lawson. Opening the black box: PIV and MFC measurements in a continuous caster mold. *2000 Steelmaking Conference Proceedings,* Pittsburgh, 2000, pp. 397–411.
47. KU Köhler, P Anderzejewski, E Julius, H Haubrich. Steel flow velocity measurement and flow pattern monitoring in the mold. *78th Steelmaking Conference,* Nashville, 1995, pp. 445–450.
48. E Julius, H Haubrich. Flow meter. U.S. Patent No. 5,426,983. June 27, 1995.
49. H Bai, BG Thomas. Effects of clogging, argon injection, and continuous casting conditions on flow and air aspiration in submerged entry nozzles. *Metall. Mater. Trans. B* 32:707–722, 2001.
50. NJ Lawson, MR Davidson. Self-sustained oscillation of a submerged jet in a thin rectangular cavity. *J. Fluids Struct.* 15:59–81, 2001.
51. LJ Heaslip, J Schade. Physical modeling and visualization of liquid steel flow behavior during continuous casting. *McLean Symposium Proceedings 1998,* pp. 223–237.
52. M Burty, M Larrecq, C Pussé, Y Zbaczyniak. Experimental and theoretical analysis of gas and metal flows in submerged entry nozzles in continuous casting. *Rev. Metall.* 93:1249–1255, 1996.
53. R Binns, ARA Lyons, AJ Peyton, WDN Pritchard. Imaging molten steel flow profiles. *Meas. Sci. Technol.* 12:1132–1138, 2001.

54. X Ma, AJ Peyton, R Binns, SR Higson. Imaging the flow profile of molten steel through a submerged pouring nozzle. *3rd World Congress on Industrial Process Tomography,* Banff, Canada, 2003, pp. 736–742.
55. M Pham, Y Hua, N Gray. Imaging the solidification of molten metal by eddy currents: I. *Inverse Probl.* 16:469–481, 2000.
56. M Pham, Y Hua, N Gray. Imaging the solidification of molten metal by eddy currents: II. *Inverse Probl.* 16: 481–494, 2000.
57. MH Pham, Y Hua, SH Lee, NB Gray. Computing the solidification of molten metal by eddy currents. *TMS EPD Congress,* New Orleans, LA, February 11–15, 2001, pp. 535–546.
58. ML Trapani, AK Kyllo, NB Gray. Improvements to taphole design, sulfide smelting, *TMS Annual Meeting,* Seattle, 2002, pp. 339–348.
59. GA Brooks, Subagyo. Advances in ladle metallurgy control. *Ladle and Tundish Metallurgy, COM 2002,* pp. 41–53.
60. RL Minion, CF Leckie, KJ Legeard, BD Richardson. Improved ladle stirring using vibration technology at Stelco Hilton Works. *Iron Steelmaker* 25:25–31, 1998.
61. FL Kemeny, DI Walker, JAT Jones. Accurate argon stirring in the ladle by vibration measurement. *58th Electric Furnace Conference,* Orlando, 2000, pp. 723–733.
62. K Yonezawa, K Schwerdtfeger. Spout eyes formed by an emerging gas plume at the surface of a slag-covered metal melt. *Metall. Mater. Trans. B* 30:411–418, 1999.
63. K Yonezawa, K Schwerdtfeger. Height of the spout of a gas plume discharging from a metal melt. *Metall. Mater. Trans. B* 30:655–660, 1999.
64. Subagyo, GA Brooks, GA Irons. Spout eyes area correlation in ladle metallurgy. *ISIJ Int.* 43:262–263, 2003.
65. Subagyo, GA Brooks. Advanced image analysis of molten slag for process control. *Yazawa International Symposium, TMS 2003* 1:475–483, 2003.
66. P Geladi. Principle component analysis of multivariate images. *Chemometrics Intelligent Lab. Syst.* 5:209–220, 1989.
67. MH Bharati, JF MacGregor. Multivariate image analysis for real-time process monitoring and control. *Ind. Eng. Chem. Res.* 37:4715–4724, 1998.
68. GH Golub, CF Van Loan. *Matrix Computations.* Oxford: North Oxford Academic, 1983.
69. D Guo, GA Irons. A water model and numerical study of the spout height in a gas-stirred vessel. *Metall. Mater. Trans. B* 33:377–384, 2002.
70. *Iron and Steel Industry in 2001.* 2003 ed. Paris: OECD Publications Service, 2003, pp. 33–34.
71. E Sharif, C Bell, PF Morris, AJ Peyton. Imaging the transformation of hot strip steel using magnetic techniques. *J. Electron. Imaging* 10:669–678, 2001.
72. RJ Beichner and RA Serway. *Physics for Scientists and Engineers with Modern Physics.* 5th ed. London: Saunders College Publishing, 2000, p. 63.
73. S Wilmotte, F Degee, C Van Den Hove, M Economopoulos. Improved technology and software for the control of the coiling temperature in a hot strip mill. *Proc. Int. Conf. Steel Rolling, Sci. Technol. Flat Rolled Prod.* 2:1342–1352, 1980.
74. VW Tychowsky, JA Rigde, HJ Turner, J Gregorio, JP Chaput. Development and application of a gold cup sensor for measurement of strip temperatures on a continuous galvanizing line. *Iron Steel Eng.* 75:37–42, 1998.
75. M Dubois, A Moreau, M Militzer, JF Bussière. Laser-ultrasonic monitoring of phase transformation in steels. *Scripta Mater.* 39:735–741, 1998.

76. CL Davis, MP Papaelias, M Strangewood, AJ Peyton. Measurement of phase transformation in steels using electromagnetic sensors. *Ironmaking Steelmaking* 29:469–476, 2002.
77. K Yairo, J Yamasaki, M Furukawa, K Arai, M Morita, M Obashi. Development of coiling temperature control system on hot strip mill. *Kawasaki Steel Tech. Rep.* 24:32–40, 1991.

Index

D

E

F

G

S

T

U

V

W

Y